Renewable Energies for Sustainable Development

Renewable Energies for Sustainable Development

Editors

M. Dolores Esteban
José-Santos López-Gutiérrez
Vicente Negro

MDPI • Basel • Beijing • Wuhan • Barcelona • Belgrade • Manchester • Tokyo • Cluj • Tianjin

Editors
M. Dolores Esteban
Universidad Europea de Madrid
Spain

José-Santos López-Gutiérrez
Universidad Politécnica de Madrid (UPM)
Spain

Vicente Negro
Universidad Politécnica de Madrid (UPM)
Spain

Editorial Office
MDPI
St. Alban-Anlage 66
4052 Basel, Switzerland

This is a reprint of articles from the Special Issue published online in the open access journal *Sustainability* (ISSN 2071-1050) (available at: www.mdpi.com/journal/sustainability/special_issues/Renewable_Energies_Sustainable_Development).

For citation purposes, cite each article independently as indicated on the article page online and as indicated below:

LastName, A.A.; LastName, B.B.; LastName, C.C. Article Title. *Journal Name* **Year**, *Volume Number*, Page Range.

ISBN 978-3-0365-1382-9 (Hbk)
ISBN 978-3-0365-1381-2 (PDF)

Contents

About the Editors

María Dolores Esteban

María Dolores Esteban (Ph.D.) is an associate professor of coastal eEngineering at the Universidad Politécnica de Madrid and of coastal engineering and renewable energies at the Universidad Europea. She has more than 15 years of experience in the fields of maritime engineering and renewable energies as an engineer at a private company and as a lecturer and researcher at universities. She has published more than 50 technical papers in SCI indexed journals and has presented at numerous international conferences. Her main fields of research are coastal and maritime engineering; harbor and ports; beaches; loads actions and effects in maritime structures; and renewable energies such as offshore wind, waves, and currents. She is very active in R&D projects and is an important asset to her research group the Environmental, Coast, and Ocean Research Laboratory (ECOREL) at the Universidad Politécnica de Madrid.

José-Santos López-Gutiérrez

José-Santos López-Gutiérrez (Ph.D.) is an associate professor of coastal engineering at the Universidad Politécnica de Madrid. He has more than 25 years of experience in the fields of maritime engineering and renewable energies as a lecturer and researcher at the university. He has supervised five Ph.D. dissertations. He has more than 60 technical papers published in SCI indexed journals and has presented at more than 60 national and international conferences. He is responsible for the research line on marine energies within the group of researchers of the Environmental, Coast, and Ocean Research Laboratory (ECOREL) at Universidad Politécnica de Madrid. Apart from research fields related to marine energies, such as load and effect actions on marine structures for offshore wind installations, and wave and current exploitation, other research fields in which he is involved include coastal and maritime engineering, ports, and beaches.

Vicente Negro

Vicente Negro (Ph.D.) is a full professor of coastal engineering and the head of the Environmental, Coast, and Ocean Research Laboratory (ECOREL) at Universidad Politécnica de Madrid. He has supervised more than 10 Ph.D. dissertations. He has published more than 70 technical papers in SCI indexed journals and more than 40 scientific and dissemination articles at the national level. He has more than 40 contributions at national and international conferences with peer arbitration. He is an expert in marine sciences for the Program of the Directorate General XII of Research in Brussels, Drafting Commission of the ROM (Puertos del Estado) and is an editorial member of the ROM0.3/91, Environmental Actions I: Waves; ROM 0.0/2002, General procedure and calculation bases in the maritime and port works project; ROM 1.0/09, Recommendations for the design of Breakwaters; and ROM 2.0/11, Recommendations for the project and execution of Docking and Mooring Works.

Preface to "Renewable Energies for Sustainable Development"

In 2019, we started working together with the editorial team of the journal *Sustainability* (MDPI editorial) as guest editors of a Special Issue related to renewable energies. This opportunity was welcomed with great enthusiasm by the Guest Editorial Team. As soon as possible, we started working on the project, with the great support of the main editor, who guided us in publishing a Special Issue that exceeded our initial expectations. This Special Issue was entitled "Renewable Energies for Sustainable Development". It focused on the seventh Sustainable Development Goal (SDG), which is to ensure access to affordable, reliable, sustainable, and modern energy for all. In the current scenario in which climate change dominates our lives and in which we all need to combat and drastically reduce the emission of greenhouse gases, renewable energies play key roles as present and future energy sources. Renewable energies vary across a wide range, and therefore, there are related studies for each type of energy. The Special Issue called for studies integrating the latest research innovations and knowledge focused on all types of renewable energy: onshore and offshore wind, photovoltaic, solar, biomass, geothermal, waves, tides, hydro, etc. Authors were invited submit review and research papers focused on energy resource estimation, all types of TRL converters, civil infrastructure, electrical connection, environmental studies, licensing and development of facilities, construction, operation and maintenance, mechanical and structural analysis, new materials for these facilities, etc. Analyses of a combination of several renewable energies as well as storage systems to progress the development of these sustainable energies were welcomed. In the end, this Special Issue published 20 papers. The papers are of very good quality and are titled as follows: "An Intelligent Approach to Active and Reactive Power Control in a Grid-Connected Solar Photovoltaic System"; "Optimizing Wave Overtopping Energy Converters by ANN Modelling: Evaluating the Overtopping Rate Forecasting as the First Step"; "Emergy and Sustainability Ternary Diagrams of Energy Systems: Application to Solar Updraft Tower"; "Effect of Spatial and Temporal Resolution Data on Design and Power Capture of a Heaving Point Absorber"; "Hybrid Wind–PV Frequency Control Strategy under Variable Weather Conditions in Isolated Power Systems"; "From NEDC to WLTP: Effect on the Energy Consumption, NEV Credits, and Subsidies Policies of PHEV in the Chinese Market"; "A New Robust Energy Management and Control Strategy for a Hybrid Microgrid System Based on Green Energy"; "Revisiting the Relation between Renewable Electricity and Economic Growth: A Renewable–Growth Hypothesis"; "On the Design of an Integrated System for Wave Energy Conversion Purpose with the Reaction Mass on Board"; "Probabilistic Assessment of Hybrid Wind-PV Hosting Capacity in Distribution Systems"; "Residential Energy-Related CO_2 Emissions in China's Less Developed Regions: A Case Study of Jiangxi"; "Feasibility and Cost Analysis of Photovoltaic-Biomass Hybrid Energy System in Off-Grid Areas of Bangladesh"; "Accurate Sizing of Residential Stand-Alone Photovoltaic Systems Considering System Reliability"; "Strategic Planning of Offshore Wind Farms in Greece"; "A Compact Pigeon-Inspired Optimization for Maximum Short-Term Generation Mode in Cascade Hydroelectric Power Station"; "Evaluation of the Operating Efficiency of a Hybrid Wind–Hydro Powerplant"; "A Residential Load Scheduling with the Integration of On-Site PV and Energy Storage Systems in Micro-Grid"; "Thermo-Poroelastic Analysis of Induced Seismicity at the Basel Enhanced Geothermal System"; "Renewable Energy for Sustainable Growth and Development: An Evaluation of Law and Policy of Bangladesh"; and "Assessment on Global Urban Photovoltaic Carrying Capacity and Adjustment of Photovoltaic Spatial Planning". It was a pleasure for the Guest Editorial Team to put all of these interesting manuscripts together. We thank the authors of these articles who allowed this Special Issue to be of such a high quality.

M. Dolores Esteban, José-Santos López-Gutiérrez, Vicente Negro
Editors

sustainability

MDPI

Article

A Compact Pigeon-Inspired Optimization for Maximum Short-Term Generation Mode in Cascade Hydroelectric Power Station

Ai-Qing Tian [1], Shu-Chuan Chu [1,2], Jeng-Shyang Pan [1,*], Huanqing Cui [1] and Wei-Min Zheng [1]

1 College of Computer Science and Engineering, Shandong University of Science and Technology, Qingdao 266590, China; stones12138@163.com (A.-Q.T.); scchu0803@gmail.com (S.-C.C.); cuihq@sdust.edu.cn (H.C.); zhengweimin@sdust.edu.cn (W.-M.Z.)

2 College of Science and Engineering, Flinders University, 1284 South Road, Clovelly Park SA 5042, Australia

* Correspondence: jspan@cc.kuas.edu.tw

Received: 19 November 2019; Accepted: 4 January 2020; Published: 21 January 2020

Abstract: Pigeon-inspired optimization (PIO) is a new type of intelligent algorithm. It is proposed that the algorithm simulates the movement of pigeons going home. In this paper, a new pigeon herding algorithm called compact pigeon-inspired optimization (CPIO) is proposed. The challenging task for multiple algorithms is not only combining operations, but also constraining existing devices. The proposed algorithm aims to solve complex scientific and industrial problems with many data packets, including the use of classical optimization problems and the ability to find optimal solutions in many solution spaces with limited hardware resources. A real-valued prototype vector performs probability and statistical calculations, and then generates optimal candidate solutions for CPIO optimization algorithms. The CPIO algorithm was used to evaluate a variety of continuous multi-model functions and the largest model of hydropower short-term generation. The experimental results show that the proposed algorithm is a more effective way to produce competitive results in the case of limited memory devices.

Keywords: compact pigeon-inspired optimization; maximum short-term generation; swarm intelligence; hydroelectric power station

1. Introduction

The metaheuristic algorithm [1] has emerged as a very promising tool to solve complex optimization problems. Original pigeon-inspired optimization (OPIO) is a new type of metaheuristic search algorithm [2]. The algorithm simulates the behavior of pigeons going home. Preliminary studies indicate that it is a very promising optimization algorithm and can outperform excellent existing algorithms [3]. OPIO exploits a population of pigeons as candidate solutions by setting boundaries and optimizing the problem by moving the candidate solutions to approach the best solutions based on a given measure of quality. The general steps of the algorithm are described below.

OPIO can solve continuous solution space problems. In addition, many versions of OPIO in the literature are proposed to solve the problem of continuous and discrete solution spaces in recent years. An improved Gaussian pigeon inspired optimization algorithm preserves the diversity of early evolution to avoid premature convergence. The entire algorithm shows excellent performance in global optimization and is effective for solving multimodal and non-convex problems with higher dimensions. Multi-objective pigeon-inspired optimization (MPIO) is used for multi-objective optimization in designing the parameters of brushless direct current motors [4]. The multimodal multi-objective pigeon-inspired optimization algorithm (MMPIO) was proposed to figure out the multimodal multi-objective optimization problems [5,6].

With the continuous development of metaheuristic algorithms, intelligent group optimization has become an emerging technology to solve many engineering problems. Metaheuristic algorithms perform well on wireless sensor networks [7,8]. Since 2000, many scholars have designed many ant colony optimization algorithms, particle swarm optimization algorithms (PSO) [9], gray wolf optimization algorithms (GWO) [10,11], bat inspired algorithms (BA) [12,13], flower pollination algorithms (FPA) [14,15], cat swarm optimization (CSO) [16,17], differential evolution algorithm (DE) [18,19], quasi-affine transformation evolution algorithms (QUATRE) [20,21], genetic algorithms (GA) [22,23], etc. Based on the simulation of the above-mentioned functional mechanisms through an in-depth study, it is easy to observe that the adaptive phenomenon can widely exist in nature. Among them, OPIO was proposed by Duan and other scholars in 2014 [24]. It is a new intelligent optimization algorithm based on the homing behavior of pigeons. While it has not been long since its introduction, this algorithm has been used in model improvement and application, obtaining many research results. Because the algorithm has good adaptability and high calculation accuracy, various optimizations have been carried out in the fields of unmanned aerial vehicle (UAV) formation [25], control parameter optimization [26], and image processing [27].

A country's development and social progress are inseparable from its demand for energy [28,29]. Electric energy is a very flexible form of energy that is increasingly obtained from the sun. Electrical energy can be converted into heat, chemical energy, and mechanical energy. Its power is also convenient to use, easy to control, safe, and clean. More importantly, most of the development of today's society relies on the development of science and technology. The main ways to generate electricity are thermal, wind, hydropower, and nuclear power generation. Hydropower is a renewable energy source that can continuously generate and deliver electricity. The main advantage of hydropower generation is that it can eliminate fuel cost. The cost of operating a hydropower station is not affected by rising fossil fuel prices such as oil, natural gas, and coal. Hydroelectric power stations do not require fuel. The economic life of a hydroelectric power station is longer than those of fuel-fired power plants. In addition, hydropower stations are mostly operated automatically and normally. In addition, such power plants have low operating costs [30,31].

The short-term optimal dispatching of cascade hydropower stations refers to the maximum value of the objective function in the case of meeting the various constraints of the cascade hydropower stations on one or several days [32]. In general, this mainly refers to the following three mathematical models: The model with the shortest power generation, the short-term water consumption minimum model [33], and the short-term peak power maximum model [34]. These three mathematical models have the same properties, i.e., under certain constraints, the nonlinear multi-stage optimization problem is obtained. This paper only analyzes the model with the maximization of the short-term power generation.

In this paper, we combine the compact technique with the pigeon-inspired optimization to propose the compact pigeon-inspired optimization algorithm. The proposed CPIO not only improves the time efficiency but also reduces the hardware memory. The algorithm proposed in this paper has very good spatial complexity. The algorithm only has one particle to update, and the original algorithm uses the population to update. After expanding our work, our goal is to solve the problem of reducing memory usage and parameter selection in optimizing the short-term power generation of cascade hydropower stations. The reasons for expanding our work include adding sample probability functions that must control the perturbation vector and comparing them with other compact algorithms in this article. The probability function operates to solve the optimal value of the compact pigeon-inspired optimization (CPIO) algorithm, and uses a real-valued prototype vector to generate each candidate solution. The algorithm has been tested on multiple continuous multi-modal functions as well as the short-term power generation of cascade hydropower stations [35–37].

2. Related Work

2.1. Principle of Electricity Generation of Cascade Hydropower Station

The hydropower process is actually a process of energy conversion. By constructing a hydraulic structure on a natural river, concentrating the water head, and then guiding the high water to the low-position turbine through the water channel, the water energy is converted into rotational mechanical energy, and the generator coaxial with the turbine is used to generate electricity. It is pivotal to note the conversion from water energy to electricity. The electricity generated by the generator is sent to the user through the transmission line to form the entire process of for generating electricity, as listed in Figure 1. The water body in the high-altitude reservoir has a large potential energy. When the water body flows into the downstream of the hydropower station through the hydraulic pipe installed in the hydropower station, the water flow drives the runner of the water turbine to rotate, so that the hydrodynamic energy is converted into the rotating mechanical energy. The turbine drives the coaxial generator rotor to cut the magnetic lines of force, and generates an induced electromotive force on the stator winding of the generator. When the stator winding is connected to the external circuit, the generator supplies power to the outside. This way, the selected mechanical energy of the turbine is converted into electrical energy by the generator.

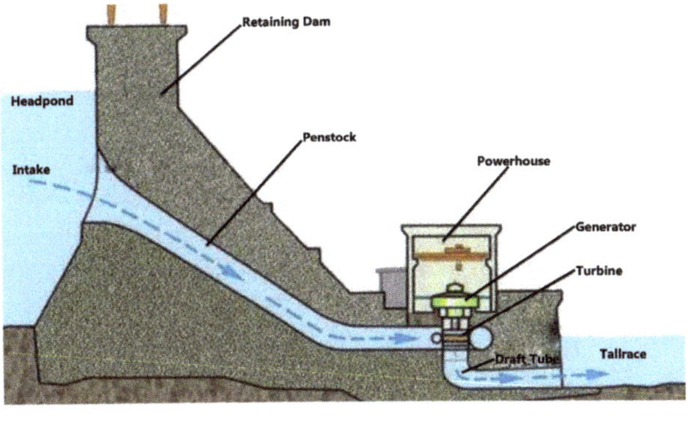

Figure 1. The principle of hydroelectricity.

Water energy resources are potential energy and kinetic energy existing in rivers and are a part of renewable resources, but the reserves of water energy are related to factors such as river flow, evaporation, precipitation, etc. Rivers vary greatly from region to region and climate varies. There is also a large difference in the amount of water energy resources in the region. Due to the current technical conditions, the water volume and the drop of the river part will not be utilized, and the mutual transition between energy also has a certain loss. Therefore, the technically developable hydropower resources are usually lower than their theoretical reserves. Taking the amount of technology developable resources as the basis, the economically available hydropower resources obtained by considering factors such as transmission distance, cost, and flooding loss are smaller than the technically exploitable amount.

Hydropower station reservoirs are generally divided into two categories, one is conventional scheduling and the other is optimized scheduling. Conventional scheduling is a commonly used scheduling method. It can also be called traditional scheduling. It is the most basic adjustment method. It only bases on historical data, no longer considers any other relevant factors, and then uses classical hydraulics and runoff regulation. The scheduling diagram and scheduling rules are used to guide the operation of the reservoir, and the water level of the reservoir is calculated and then expressed as the objective function, that is, the most basic scheduling method is used to ensure the operation of the hydropower station. These schedules are drawn based on past hydrological data and tasks when the reservoir is at different water storage levels, that is, the reservoir scheduling rules when the reservoir is in different states.

The optimal dispatching model of the hydropower station reservoir is based on the optimal theory to establish a mathematical model based on the actual conditions of the hydropower station reservoir, and then using modern computer technology to find the optimal scheduling method that meets the scheduling principle in the process of establishing the optimal scheduling model of the hydropower station. In addition, there are many factors to be considered, including the connection between water and electricity as well as numerous constraints on cascade water inventory, This will minimize pollution in terms of ecological environment, and maximize profits and social benefits in terms of economic benefits. In order to meet the above requirements to the maximum extent and to minimize the water abandonment of hydropower stations, this paper establishes a short-term power generation model suitable for the optimal operation of cascade hydropower stations.

2.2. Maximum Short-Term Generation Model

With the completion and operation of a large number of hydropower stations, the effective solution of large-scale reservoir group optimization scheduling models has become an urgent problem to be solved. The reservoir optimization problem proposes a higher solution quality and running speed for the meta heuristic algorithm. Therefore, this paper proposes a new algorithm called CPIO. This aspect of research not only improves the speed of the operation, but also ensures that the quality of the solution is not worse than the original algorithm and can suppress the premature phenomenon.

Under the given inner diameter flow of the control period, the objective function of the long-term optimal scheduling model of the cascade hydropower station group is defined as: The maximum amount of cascade power generation under the condition of ensuring the output of the cascade is considered during the control period. Under the premise of satisfying the actual situation, this paper selects the maximum benefit of cascade power generation as one of the objective functions. At the same time, it is necessary in the medium and long-term optimization scheduling, it is necessary to consider the output of the period with the least output during the year as much as possible. Medium and long-term optimized dispatching provides the largest possible uniform and reliable output for the power grid, giving full play to the capacity benefits of hydropower generation which can replace thermal power.

$$E = maxF = max\sum_{i=1}^{N}\sum_{j=1}^{T} A \times q \times \Delta h \times \Delta t \tag{1}$$

In the formula, A, q, Δh, Δt represents the output coefficient, the outflow rate, the upstream and downstream water level difference respectively. In addition, unit time E is the maximum annual power generation benefit of the cascade hydropower station, N is the total number of cascade hydropower stations, and T is within one year. Calculate the total number of time slots ($T = 12$).

In the process of optimizing the power generation of cascade hydropower stations, it is necessary to understand the water reservoir data in order to first calculate the reservoir upstream capacity and the outflow flow value, and then calculate the downstream water level value based on the outflow flow value.

$$V_{i,j} = \frac{c_1 + h^1_{i,j} - c_2}{c_3 - c_4} * (c_5 - c_6) * 10^8 \tag{2}$$

In the formula $c_1, c_2, c_3, c_4, c_5, c_6$ are a set of variable constants. This set of constants has different values according to the water level in each interval, and $h^1_{i,j}$ is the upstream water level value of the j-th time period of the i-th hydropower station.

$$q_{i,j} = h^o_j - \frac{V_{i,j+1} - V_{i,j}}{t_j * 3600} \tag{3}$$

In the Equation (3), $q_{i,j}$ is the outflow of the j-th time period of the i-th hydropower station, h_j^o is the initial flow of the j-th hydropower station, and t_j is the number of hours of the j-th time period.

$$h_{i,j}^2 = \frac{c_7 + (q_{i,j} - c_8)}{c_9 - c_{10}} * (c_{11} - c_{12}) \tag{4}$$

In this Equation (4), $c_7, c_8, c_9, c_{10}, c_{11}, c_{12}$ are a set of variable constants. This set of constants has different values according to the water level in each interval, and $h_{i,j}^2$ is the downstream water level value of the j-th time period of the i-th hydropower station.

$$\Delta h_i = \frac{h_{i,j}^1 + h_{i+1,j}^1}{2} - h_{i,j}^2 \tag{5}$$

In the Equation (5), Δh_i is the upstream and downstream water level difference of the i-th hydropower station. When the upstream and downstream water level difference is obtained, since the upstream water level is greatly affected, the average value of the upstream water level is selected for calculation in this paper.

Let us introduce the constraints of the objective function:

$$h_{i,min} \leq h_{i,j} \leq h_{i,max} \tag{6}$$

The level of the water level needs to be limited between h_{min} and h_{max}.

$$q_{i,min} \leq q_{i,j} \leq q_{i,max} \tag{7}$$

The outflow of the reservoir must fluctuate between q_{min} and q_{max}. In order to ensure the stable operation of the power generation of the entire cascade hydropower station, the output of the power has to be relatively stable, so the outflow of the reservoir cannot be lower than the minimum flow. In addition, in order to stabilize the life of the turbine and generator, the outflow of the reservoir cannot be higher than the maximum flow.

$$V_{i,min} \leq V_{i,j} \leq V_{i,max} \tag{8}$$

The capacity of the reservoir should fluctuate between the V_{min} and V_{max}. In order to ensure that the reservoir will continue to work under special circumstances, the reservoir's capacity cannot be lower than the originally set value to ensure the safe operation and that the downstream organisms are safe, so the capacity cannot be higher than the maximum reservoir capacity.

3. Pigeon-Inspired Optimization

Without prejudice, the minimization problem of the objective function $f(x)$ is discussed in this paper, where x is the vector that defines the n design variables in the domain D in the decision space.

The pigeon-inspired optimization is a meta-heuristic algorithm that is inspired by the behavior of the pigeons returning home and is widely used in most continuous or discrete optimization problems. This article mainly introduces continuity problems. Referring to extensive literature reviews, a group of pigeons move in decision space D according to the update rules to find the optimal value when looking for the solution of the problem. More formally, in order to gain the satisfactory value of the objective function $f(x)$, the population of the pigeons is randomly sprinkled in the previously set search space. The objective function judges the equivalent quality of solution based on the position information of each pigeon. At any stage t, the i-th pigeon has its own position vector x_k^t and velocity vector v_k^t. For each pigeon, the best solution is the value of the objective function. The best position of the position where the pigeon has passed will be stored. The global optimal solution is continuously updated. To transition from the t step to the $t+1$ step, a more competitive solution will be taken, and each particle is perturbed according to the following formula:

$$v_k^{t+1} = e^{-R*t} * V_k^t + \phi_1 * (x_{gbest} - x_k) \tag{9}$$

and:

$$x_k^{t+1} = \phi_2 * x_k^t + \phi_3 * v_k^{t+1} \tag{10}$$

As the formula above suggests, x_k^t refers to the current position of the k-th pigeon, and x_{gbest} is the best position ever found in the entire herd, and the vector v_k^t is a perturbation vector, namely velocity. Finally, ϕ_1 is a variable constant, is a variable amount limited to 0–1, and ϕ_2, ϕ_3 are two weight factors can be constants or variables. This stage belongs to the map and the compass operator. When the pigeon approaches the destination, the dependence on the sun and the magnetic object is reduced, and then the landmark operator is entered.

$$x_{center}^t = \frac{\sum_{k=1}^{N^t} x_k^t * F(x_k^t)}{N^t * \sum_{k=1}^{N^t} F(x_k^t)} \tag{11}$$

From here on, the landmark operator is entered. In this operator, the pigeons continue to iterate according to the pigeons or landmarks of the roads understood by the population. In the above formula, the purpose of this operation is to find out the pigeons with a high fitness value in the flock. This pigeon is then considered to be the pigeon that knows the road, and the pigeons are iterated according to the pigeon. N^t is the population number at the t-th iteration, and $F(X_k^t)$ is the fitness function value of the k-th pigeon position.

$$N^t = \frac{N^{t-1}}{2} \tag{12}$$

The significance of this operation is to halve the pigeons and discard the pigeons that do not have the way to know, to prevent such pigeons from misleading the population into local optimum.

$$x_i^{t+1} = x_i^t + \phi_4 * (x_{center}^t - x_i^t) \tag{13}$$

In the formula, ϕ_4 is a variable constant that value is a randomly generated value from $(0,1)$. In this operation, all pigeons that do not know the road will be iterated according to the pigeons that know the road.

$$F(x_i^t) = \frac{1}{Fitness(x_k^t) + \chi} \quad \text{For minimization problem} \tag{14}$$

$$F(x_i^t) = Fitness(x_k^t) \quad \text{For maximization problem} \tag{15}$$

For maximizing the problem, OPIO uses Equation (14) to calculate the value of $F(x_k^t)$ to find the pigeon with the ability to identify the function. For the minimization problem, OPIO uses Equation (15) to calculate the value of $F(x_k^t)$ to find the pigeon with the function of identifying. In Equation (14), χ is a non-zero constant whose purpose is to prevent the denominator from being zero.

4. Compact Pigeon-Inspired Optimization

The compact approach replicates the operation of the population-based algorithm by building the probability of a total solution. The optimal process encodes the probability representation of the actual population as a virtual counterpart. Compact pigeon-inspired optimization is a model built on a pigeon-inspired optimization-based framework. In the OPIO algorithm, the concept and design of the CPIO algorithm will be explored in more detail.

The purpose of CPIO is to simulate the operation of OPIO underlying overall algorithm in a smaller version of memory variable memory. By constructing a distributed data structure, the actual solution of the OPIO is transformed into a compact algorithm, the perturbation vector. The PV vector is a probabilistic model for the solution of the population.

$$PV^t = [\mu^t, \delta^t] \tag{16}$$

In the formula, μ, δ are two parameters of the standard deviation and the average of the vector PV, and t is the current number of iterations. The value of μ, δ is limited to probability density functions (PDF) [38] and is changed within [−1, 1]. The magnitude of the PDF is normalized by keeping the area to 1, because by obtaining approximately sufficient in well it is the uniform distribution with a full shape.

The initialization of the virtual population is performed as follows. For each design variable $\mu = 0$ and $\delta = \lambda$, where λ is a large constant (λ = 10). This value is initialized to initially obtain a normal distribution of truncated wide shapes.

The sampling mechanism of the design variable x_k^t associated with the generic candidate solution x in PV is not a simple process and requires extensive interpretation. For each design variable indexed by k, a truncated Gaussian PDF with the mean μ and standard deviation δ is associated, The PDF is described by the following formula:

$$PDF = \frac{e^{-\frac{(x-\mu[k])^2}{2\times\sigma[k]^2}} \times \sqrt{\frac{2}{\pi}}}{\sigma[k] \times (erf(\frac{\mu[k]+1}{\sqrt{2}\times\sigma[k]}) - erf(\frac{\mu[k]-1}{\sqrt{2}\times\sigma[k]})))} \tag{17}$$

PDF is the probability distribution function of PV, and a truncated Gaussian PDF-related μ, and δ are formulated. A new candidate solution is generated by iteratively biasing towards a promising region of the optimal solution. Every component of the probability vector may be acquired by learning the previous generations. erf is the error function established by [39]. PDF corresponds to the cumulative distribution function (CDF) by constructing a Chebyshev polynomial [39], and the upper domain of the CDF is randomly changed between 0 and 1. CDF can be described as a real-valued random variable x with a probability distribution, and the value that can be obtained can be less than or equal to x_k^t.

The relationship between CDF and PDF can be defined as $CDF = \int_0^1 \text{PDF}(x)dx$, and PV operations can sample the design variable x_k^t by randomly generating values within the range of $(0, 1)$.

In the iterative process of the compact algorithm, in order to find a better individual, a function that can be compared by two parameters is proposed in this paper. The two variable pigeon parameters are two sample individuals of the PV operation. The vector represented by the winner is the value of the fitness function. This value is higher than other virtual members, and the vector represented by the loser is that the individual fitness value is lower than the fitness evaluation standard. Two variables with return values, the winner and the loser are obtained from the calculation of the objective function, and a new candidate solution is generated to compare with the original global optimal solution to generate new winners and losers. For updating PV operations, μ and σ can be considered for updates according to the rules below. If the mean value of μ is 1, Then the update rule becomes μ^t and σ^t for each of its elements μ^{t+1} and σ^{t+1} [40] as described in the following:

$$\mu_i^{t+1} = \mu_i^t + \frac{winner_i - loser_i}{N}, \tag{18}$$

where N is the virtual population size and the value of δ is described below. The update rule for each element is given in the formula below.

$$\sigma_i^{t+1} = \sqrt{(\sigma_i^t)^2 + (\mu^{t+1})^2 + \frac{winner_i^2 - loser_i^2}{N}} \tag{19}$$

Mathematical details about construction Equations (18) and (19) have been given. The persistent and non-persistent structures of rcGA have been tested and can be seen in [41]. Seeing the virtual population size N as a parameter of a compression algorithm is not a true population-based algorithm. The virtual population size, in the real-valued compression algorithm, is an algorithm that depends on the convergence speed.

In general, a probabilistic model for compact OPIO is hired to represent all of the set of solutions for the pigeon group, neither storing location information nor storing speed information; however, storing newly generated candidate solutions. Therefore, the limited storage space is required to achieve the algorithm requirements which saves a lot of time and hardware resources for the cascade hydropower station to optimize the short-term power generation model.

CPIO uses a perturbation vector PV that has the same structure as the one shown in Equation (16), at the beginning of the optimization algorithm, just like the process described in Equation (17). The PV initialization is designed as $(\forall k, \mu[k] = 0, \sigma[k] = \lambda, \lambda = 10)$ and the variables of each design are limited to one continuous space $[-1, 1]$, and in addition, the position x and the velocity v are randomly initialized within a certain range.

Update velocity vector and position vector by slightly revised pigeon-inspired algorithm:

$$v_k^{t+1} = \omega_1 * e^{-R*t} * V_k^t + \omega_2 * (x_{gbest} - x_k) \tag{20}$$

and:

$$x_k^{t+1} = \xi_1 * x_k^t + \xi_2 * v_k^{t+1} \tag{21}$$

ω_1 is an inertia weight, ω_2 is a random variable between 0 and 1, and ξ_1 and ξ_2 are weighting factors that control the position update of the pigeon.

It can be seen that the equation updating of speed (Equation (20)) and position (Equation (21)) is similar to the OPIO algorithm. In the original version, pigeon k was closely related to pigeon group N. In the compact version, there was no real population, but the relevance of a virtual population pigeon to the virtual population was not that great. It is easy to see that compact OPIO is just a pigeon that uses the update formula to update it, so updating it once produces a solution that saves a lot of memory.

In the landmark operator entering the second stage of CPIO, the original algorithm uses the Equation (11) to determine the pigeon with the function of identifying the function based on the fitness solution of each pigeon position, so that it becomes the center point and continues to update. Since the CPIO has only one particle to update, it is not suitable when selecting a pigeon with a path function. In this paper, a center point suitable for CPIO is proposed. By setting a virtual center position point, the guiding pigeon is updated. When the virtual center position is established, it is based on the historical fitness value of the pigeon. The number selected is also based on the size of the virtual population.

$$x_{center}^{t+1} = \frac{\sum_{i=l-N+1}^{N} F(x_k^t)}{N} \tag{22}$$

l is the number of iterations until now, N is the number of virtual populations. According to the Equation (22), the historical virtual center points of the pigeons can be selected, and it is known that they continue to iterate.

$$x_k^{t+1} = x_k^t + \omega_3 * (x_{center}^t - x_k^t) \tag{23}$$

It is easy to see that CPIO saves a lot of memory space, so this approach can be applied to other variations of OPIO.

5. Numerical Results

The test results of the CPIO that have been tested by 29 test functions, and these test functions come from [42]. Each test function has a very detailed introduction in Tables 1 and 2. Among these groups of questions, they have different search range and different expressions.

In Equations (20)–(23) and Algorithm 1, the parameters of the CPIO proposed herein are: $N = 120$, $R = 0.2$. The values of these parameters are referred to [43] and have a slight change. More specifically, in order to make CPIO work better, we modeled the virtual population size proposed by OPIO. In this article, CPIO is compared to the OPIO. In all test functions, CPIO is run 30 times and averaged. Take the minimum value of CPIO in all test functions.

Algorithm 1 Compact pigeon-inspired optimization (CPIO) pseudo-code.

Map and compass factors R, The maximum number of iterations in the first stage $MaxDT_1$ and The maximum number of iterations in the second stage $MaxDT_2$, dimension is dim;
for $k = 1$ to dim **do**
    // PV operation initialization; N is the total number of pigeons
    initialize $\mu[k] = 0$
    initialize $\sigma[k] = 10$
end for
// Global Best initialization
Generate the global best solution x_{best} by means of perturbation vector PV
// Local Best Solution initialization
Generate the local best solution x_k by PV
for $t = 1$ to $MaxDt_1$ **do**
    x_k^t = Generate from PV operation
    // Update position and velocity
    $v_k^{t+1} = \omega_1 * e^{-R*t} * V_k^t + \omega_2 * (x_{gbest} - x_k)$
    $x_k^{t+1} = \xi_1 * x_k^t + \xi_2 * v_k^{t+1}$
    // Best Selection
    $[winner, loser]$=compete(x_k^{t+1}, x_{gbest})
    // Update PV operation
    for $k = 1 : dim$ **do**
        $\mu^{t+1}[k] = \mu^t[k] + \frac{winner[k] - loser[k]}{N}$
        $\sigma^{t+1}[k] = \sqrt{(\sigma^t[k])^2 + (\mu^t[k])^2 - (\mu^{t+1}[k])^2 + \frac{winner_k^2 - loser_k^2}{N}}$
    end for
end for
for $t = MaxDt_1 + 1$ to $MaxDt_1 + MaxDt_2$ **do**
    // Enter the second stage
    // Select the virtual center pigeon from the historical best points
    $x_k^{t+1} = x_k^t + \omega_3 * (x_{center}^t - x_k^t)$
    // Best Selection
    $[winner, loser]$=compete(x_k^{t+1}, x_{gbest})
    // Update PV operation
    for $i = 1 : dim$ **do**
        $\mu^{t+1}[k] = \mu^t[k] + \frac{winner[k] - loser[k]}{N}$
        $\sigma^{t+1}[k] = \sqrt{(\sigma^t[k])^2 + (\mu^t[k])^2 - (\mu^{t+1}[k])^2 + \frac{winner_k^2 - loser_k^2}{N}}$
    end for
end for

Table 1. Details of 29 test functions.

Name	Test Functions	Range	Global Minimum
Sphere	$f(x) = \sum_{i=1}^{d} x_i^2$	± 5.12	0
Rastrigin	$f(x) = 10d + \sum_{i=1}^{d}[x_i^2 - 10 * cos(2\pi x_i)]$	± 5.12	0
Rosenbrock	$f(x) = \sum_{i=1}^{d-1}[100 * (x_{i+1} - x_i^2) + (x_i - 1)^2]$	$x_i \in [-5, 10]$	0
Griewank	$f(x) = \sum_{i=1}^{d} \frac{x_i^2}{4000} - \prod_{i=1}^{d} cos(\frac{x_i}{\sqrt{i}}) + 1$	± 600	0
Ackley	$f(x) = -a * exp(-b\sqrt{\frac{\sum_{i=1}^{d} x_i^2}{d}})$ $-exp(\frac{\sum_{i=1}^{d} cos(cx_i)}{d}) + a + exp(l)$	± 32.768	0
Quadric	$f(x) = \sum_{i=1}^{n} \sum_{k=1}^{i} x_i$	± 32.768	0
Bukin6	$f(x) = 100 * \sqrt{\lvert x_2 - 0.01 * x_1^2 \rvert} + 0.01 * \lvert x_1 + 10 \rvert$	$x_1 \in [-15, 5] x_2 \in [-3, 3]$	0
Crossit	$f(x) = -0.0001*$ $(\lvert sin(x_1)sin(x_2) * exp(\lvert 100 - \frac{\sqrt{x_1^2+x_2^2}}{\pi} \rvert) + 1 \rvert)^{0.1}$	± 10	-2.06261
Drop	$f(x) = -\frac{1+cos(12\sqrt{x_1^2+x_2^2})}{0.5(x_1^2+x_2^2)+2}$	± 5.12	-1
Egg	$f(x) = -(x_2 + 47) * sin(\sqrt{\lvert x_2 + \frac{x_1}{2} + 47 \rvert})$ $-x_1 * sin(\lvert x_1 - (x_2 + 47) \rvert)$	± 512	-959.6407
Holder	$f(x) = -\lvert sin(x_1)cos(x_2)exp(\lvert 1 - \frac{\sqrt{x_1^2+x_2^2}}{\pi} \rvert) \rvert$	± 10	-19.2085
Levy	$f(x) = sin^2(\pi w_1)$ $+\sum_{i=1}^{d-1}(w_i - 1)^2 [1 + 10sin^2(\pi w_i + 1)]$ $+(w_d - 1)^2 [1 + sin^2(2\pi w_d)]$	± 10	-19.2085
Levy13	$f(x) = sin^2(3\pi w_1) + (x_1 - 1)^2[1 + sin^2(3\pi x_2)]$ $+(x_2 - 1)^2[1 + sin^2(2\pi x_2)])$	± 10	0
Schaffer2	$f(x) = 0.5 + \frac{sin^2(x_1^2 - x_2^2) - 0.5}{[1+0.001(x_1^2+x_2^2)]}$	± 100	0

When initializing the two algorithms CPIO and original pigeon-inspired optimization (OPIO) , the map and compass factor R are set to 0.2, and the result is to compare CPIO and OPIO. The quality of solution and the number of runs of CPIO and OPIO optimal solutions are compared as below described. The CPIO and OPIO data results are the average of 30 runs. All algorithms operate 500 times, including 300 in the first phase and 200 in the second phase.

In Table 3, CPIO performs better than OPIO in many test functions, and most of the values perform well. In terms of the time cost comparison, it is easy to see that CPIO time spent is much better than PIO, especially in several of them, and the time spent is more than a hundred times more.

According to the comparison of the two algorithms, it can be concluded that the running time of CPIO is much lower than that of the original algorithm. This is because the number of population used in the process of iteration is different. In the new algorithm, it uses an example to keep iterating, constantly adjusting the probability distribution according to the path that has been iterated, and the greater the possibility of generating particles where the function values are superior. However, this method also has a big problem, since in the search process of a single particle, randomness is often large, and it is thus easy to fall into the local optimal. It is also relatively simple to achieve the optimal, in the case of small dimension settings, the advantages of the algorithm are not obvious. Because of this characteristic of the new algorithm, it is easy to save time and reduce the time complexity of the algorithm.

Table 2. Details of 29 test functions.

Name	Test Functions	Range	Global Minimum
Schaffer4	$f(x) = 0.5 + \frac{cos(sin(\lvert x_1^2 - x_2^2 \rvert)) - 0.5}{[1 + 0.001(x_1^2 + x_2^2)]}$	± 100	
Schwef	$f(x) = 418.9829d - \sum_{i=1}^{d} x_i sin(\sqrt{\lvert x_i \rvert})$	± 500	0
Shubert	$f(x) = (\sum_{i=}^{5} icos((i+1)))(\sum_{i=1}^{5} icos(i+1)x_2 + i)$	± 5.12	-186.7309
Bohal	$f(x) = x_1^2 + 2x_2^2 - 0.3cos(3\pi x_1) - 0.4cos(4\pi x_2) + 0.7$	± 100	0
Perm0db	$f(x) = \sum_{i=1}^{d} (\sum_{j=1}^{d} (j + \beta)(x_j^i - \frac{1}{j^i}))$	± 30	0
Rothyp	$f(x) = \sum_{i=1}^{d} \sum_{j=1}^{i} x_j^2$	± 65.536	0
Sumpow	$f(x) = \sum_{i=1}^{d} \lvert x_i \rvert^{i+1}$	$x_i \in [-1, 1]$	0
Sumsqu	$f(x) = \sum_{i=1}^{d} ix_i^2$	± 10	0
Trid	$f(x) = \sum_{i=1}^{d} (x_i - 1)^2 - \sum_{i=2}^{d} x_i x_{i-1}$	± 30	0
Booth	$f(x) = (x_1 + 2x_2 - 7)^2 + (2x_1 + x_2 - 5)^2$	± 10	0
Matya	$f(x) = 0.26(x_1^2 + x_2^2) - 0.48x_1 x_2$	± 10	0
Mccorm	$f(x) = sin(x_1 + x_2) + (x_1 - x_2)^2 - 1.5x_1 + 2.5x_2 + 1$	$x_1 \in [-1.5, 4] x_2 \in [-3, 4]$	-1.9133
Camel3	$f(x) = 2x_1^2 - 1.05x_1^4 + \frac{x_1^6}{6} + x_1 x_2 + x_2^2$	± 5	0
Beale	$f(x) = (1.5 - x_1 + x_1 x_2)^2 + (2.25 - x_1 + x_1 x_2^2)^2 + (2.625 - x_1 + x_1 x_2^3)^2$	± 4.5	0
Stybtang	$f(x) = \frac{1}{2} * \sum_{i=1}^{d} (x_i^4 - 16x_i^2 + 5x_i)$	± 5	-1174.9797

Figure 2 shows the convergence trend of CPIO and OPIO. Best score obtained so far refers to the optimal value obtained by the algorithm during the iteration process. While the convergence speed of OPIO and the algebra needed to achieve optimal are small, the optimal value of CPIO is better or nearly equal to the value of OPIO. Here, CPIO uses one particle for updating and iteration, while OPIO uses the entire population for optimization. CPIO is far less than OPIO search capability, but CPIO can save a lot of memory and time to find excellence.

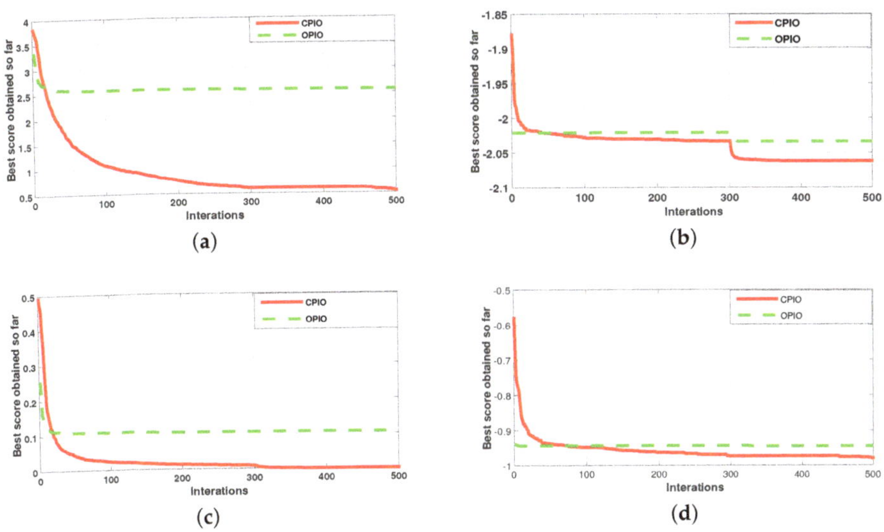

Figure 2. Compact pigeon-inspired optimization (CPIO) and original pigeon-inspired optimization (OPIO) performance in test functions. (**a**) Ackley; (**b**) Crossit; (**c**) Drop; (**d**) Griewank.

Table 3. Comparison, evaluation and speed of quality performance between CPIO and OPIO.

Test Functions	Fitness Function Value		Time	
	CPIO	OPIO	CPIO	OPIO
Sphere	1.96×10^{-1}	2.09×10^{0}	1.62×10^{-1}	4.23×10^{-1}
Rastrigin	2.13×10^{1}	4.42×10^{1}	1.37×10^{-1}	7.92×10^{-1}
Rosenbrock	4.45×10^{1}	7.92×10^{1}	1.35×10^{-1}	7.50×10^{-1}
Griewank	5.51×10^{-5}	1.04×10^{-1}	1.67×10^{-1}	1.15×10^{0}
Ackley	5.93×10^{-1}	2.43×10^{0}	1.88×10^{-1}	1.23×10^{1}
Quadric	6.50×10^{-1}	2.5×10^{1}	1.84×10^{-1}	1.03×10^{1}
Bukin6	1.96×10^{0}	6.24×10^{0}	1.40×10^{-1}	6.06×10^{-1}
Crossit	-2.06×10^{0}	-2.06×10^{0}	1.57×10^{-1}	8.48×10^{-1}
Drop	-9.86×10^{-1}	-9.36×10^{-1}	1.48×10^{-1}	7.93×10^{-1}
Egg	-5.49×10^{1}	-9.36×10^{-1}	1.49×10^{-1}	8.36×10^{-1}
Holder	-1.73×10^{0}	-1.73×10^{0}	1.46×10^{-1}	7.77×10^{-1}
Levy	3.58×10^{-1}	1.27×10^{-1}	1.58×10^{-1}	4.00×10^{0}
Levy13	1.51×10^{-1}	1.35×10^{-31}	1.54×10^{-1}	5.47×10^{-1}
Schaffer2	3.12×10^{-8}	0	1.63×10^{-1}	5.18×10^{-1}
Schaffer4	6.39×10^{-1}	5.40×10^{-1}	1.47×10^{-1}	5.26×10^{-1}
Schwef	-7.71×10^{96}	1.25×10^{4}	1.75×10^{-1}	1.13×10^{0}
Shubert	-1.57×10^{2}	-7.45×10^{0}	1.49×10^{-1}	6.31×10^{-1}
Bohal	1.92×10^{-4}	3.33×10^{-17}	1.61×10^{-1}	4.82×10^{-1}
Perm0db	2.24×10^{-4}	0	1.60×10^{-1}	4.76×10^{-1}
Rothyp	3.24×10^{3}	4.33×10^{3}	4.10×10^{-1}	7.80×10^{0}
Sumpow	2.58×10^{0}	2.84×10^{1}	1.45×10^{-1}	2.37×10^{0}
Sumsqu	2.05×10^{-6}	9.14×10^{-3}	1.70×10^{-1}	1.74×10^{0}
Trid	3.08×10^{0}	2.37×10^{1}	1.39×10^{-1}	6.03×10^{-1}
Booth	-2.21×10^{2}	-2.90×10^{1}	1.47×10^{-1}	6.92×10^{-1}
Matya	1.51×10^{0}	2.97×10^{-1}	1.53×10^{-1}	4.54×10^{-1}
Mccorm	1.80×10^{-6}	7.74×10^{-11}	1.59×10^{-1}	4.43×10^{-1}
Camel3	-1.79×10^{0}	-1.90×10^{0}	1.57×10^{-1}	4.56×10^{-1}
Beale	8.34×10^{-6}	5.35×10^{-25}	1.61×10^{-1}	5.58×10^{-1}
Stybtang	2.77×10^{0}	5.71×10^{-2}	1.53×10^{-1}	5.34×10^{-1}

Among the four selected functions, Figure 3 shows the time trend of the four functions running 30 times. In general, the time spent by the CPIO and PIO algorithms does not change much, but the two algorithms compare. It is easy to see that CPIO runs much faster than the PIO.

Table 4 shows the comparison of CPIO and PIO mentioned above in the memory variables, which makes it very convenient to implement the calculation algorithm. The number of variables of the two algorithms of CPIO and PIO proposed in this paper is calculated by the equation used in the computational optimization. In Table 4, it is easy to see that in the same computing situation, CPIO uses less memory than PIO. For example, during an iteration, CPIO uses an iteration Equations (16)–(23); the formula for PIO update iteration is Equations (9)–(13).

Table 4. The space complexity of the two algorithms.

Algorithm	Particle	Memory Size	Computing Complexity	Use Equations
CPIO	1	8	$8 \times T \times iteration$	(16), (17), (18), (19), (20), (21), (22), (23)
OPIO	N	$5 \times N$	$5 \times T \times N \times iteration$	(9), (10), (11), (12), (13)

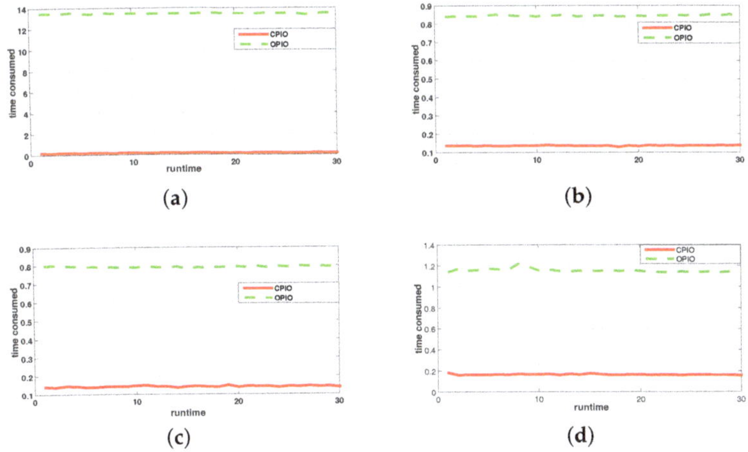

Figure 3. CPIO and OPIO performance in test functions. (**a**) Ackley; (**b**) Crossit; (**c**) Drop; (**d**) Griewank.

As can be seen from Table 4, the actual population size of the PIO is N, but the actual population size in the CPIO is 1, and the virtual population number is N. In the case where the number of iterations l and the running time t are the same, the memory usage of the variables of OPIO and CPIO is iterated by $4 \times t \times N$ and $8 \times t$, respectively. Here, it is seen that the memory occupancy of the PIO is larger than the memory usage of the CPIO.

In Figure 4, the consequence of the presented algorithm and the else three meta-heuristics are shown. According to Table 5, the trend and optimal value of CPIO are fundamentally better than the other three algorithms, and have a superior performance. Table 5 shows the comparison of CPIO, OPIO and other algorithms, such as CPSO and PSO algorithms. Among the four meta-heuristic algorithms, the performance is as follows in 29 test functions. In the process of algorithm simulation, as part of the images are not so obvious, four relatively obvious images are extracted for display.

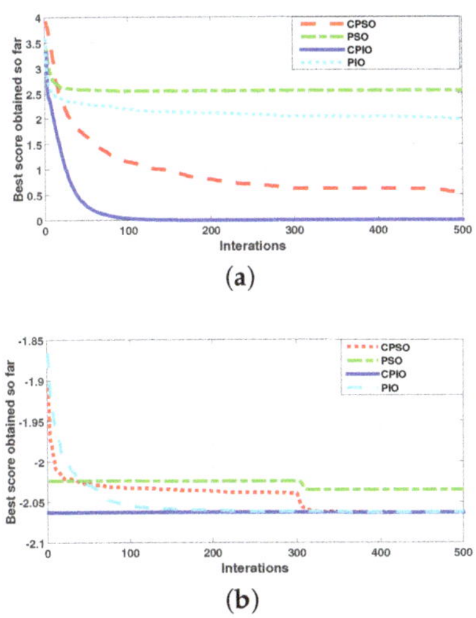

Figure 4. *Cont.*

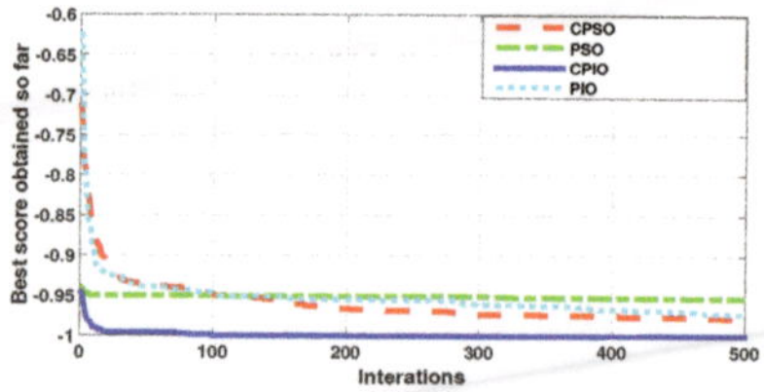

(c)

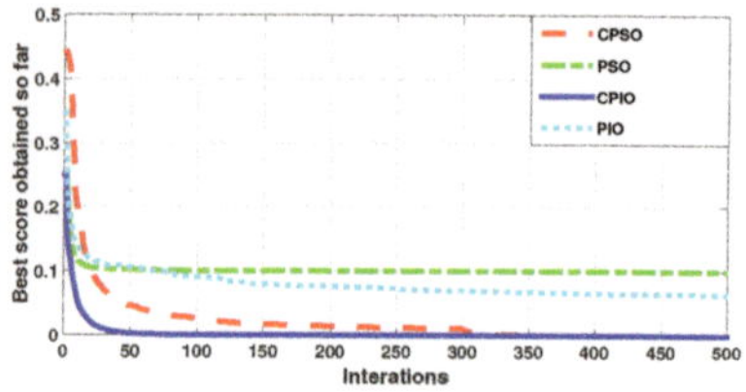

(d)

Figure 4. CPIO and three other meta-heuristic algorithms for testing function performance. (**a**) Ackley; (**b**) Crossit; (**c**) Drop; (**d**) Griewank.

Table 5. The optimal value and time cost of the four algorithms.

Test Functions	Fitness Function Value				Time			
	CPIO	OPIO	PSO	CPSO	CPIO	OPIO	PSO	CPSO
Sphere	5.16×10^{-1}	9.79×10^{-1}	3.73×10^{-6}	4.69×10^{0}	1.45×10^{-1}	4.50×10^{0}	9.94×10^{0}	1.70×10^{0}
Rastrigin	1.95×10^{2}	1.68×10^{1}	2.56×10^{1}	2.14×10^{2}	1.40×10^{-1}	4.54×10^{0}	1.63×10^{1}	1.71×10^{-1}
Rosenbrock	1.40×10^{2}	7.21×10^{2}	2.13×10^{4}	4.68×10^{2}	1.38×10^{-1}	4.52×10^{0}	1.21×10^{1}	1.71×10^{-1}
Griewank	9.53×10^{-6}	1.78×10^{1}	4.25×10^{2}	1.38×10^{-1}	1.14×10^{-1}	4.42×10^{0}	2.26×10^{1}	8.34×10^{-2}
Ackley	1.11×10^{0}	3.12×10^{0}	3.95×10^{0}	3.12×10^{0}	2.51×10^{-1}	7.22×10^{0}	1.97×10^{2}	2.30×10^{-1}
Quadric	1.49×10^{0}	3.25×10^{0}	2.89×10^{0}	2.99×10^{0}	2.30×10^{-1}	7.22×10^{0}	1.98×10^{2}	2.29×10^{-1}
Bukin6	3.66×10^{0}	4.28×10^{-2}	3.72×10^{-2}	3.36×10^{0}	1.41×10^{-1}	4.47×10^{0}	1.44×10^{1}	1.67×10^{-1}
Crossit	-2.06×10^{0}	-2.06×10^{0}	-2.06×10^{0}	-2.06×10^{0}	1.58×10^{-1}	4.48×10^{0}	1.67×10^{1}	1.74×10^{-1}
Drop	-9.76×10^{-1}	-9.58×10^{-1}	-9.93×10^{-1}	-9.44×10^{-1}	1.51×10^{-1}	4.50×10^{0}	1.60×10^{1}	1.68×10^{-1}
Egg	-3.85×10^{2}	-8.90×10^{2}	-9.15×10^{2}	-5.80×10^{7}	1.58×10^{-1}	4.31×10^{0}	1.64×10^{1}	1.14×10^{-1}
Holder	-6.24×10^{1}	-1.92×10^{1}	-1.92×10^{1}	-1.92×10^{1}	2.91×10^{-2}	4.53×10^{-1}	1.08×10^{0}	3.42×10^{-2}
Levy	1.78×10^{-3}	1.22×10^{-30}	1.50×10^{-32}	1.30×10^{-3}	4.07×10^{-2}	5.93×10^{-1}	2.12×10^{0}	3.45×10^{-2}
Levy13	7.49×10^{-2}	2.39×10^{-19}	1.35×10^{-31}	6.53×10^{-2}	2.56×10^{-2}	3.98×10^{-1}	8.40×10^{-1}	2.60×10^{-2}
Schaffer2	4.00×10^{-9}	1.42×10^{-3}	0	1.65×10^{-4}	2.60×10^{-2}	4.09×10^{-1}	7.90×10^{-1}	2.17×10^{-2}
Schaffer4	5.09×10^{-1}	5.00×10^{-1}	5.00×10^{-1}	5.00×10^{-1}	2.95×10^{-2}	4.07×10^{-1}	7.63×10^{-1}	3.25×10^{-2}
Schwef	1.44×10^{3}	9.30×10^{4}	9.26×10^{4}	1.40×10^{3}	1.69×10^{-1}	4.39×10^{0}	2.15×10^{1}	1.03×10^{-1}
Shubert	-1.47×10^{2}	-1.86×10^{2}	-1.86×10^{2}	-1.36×10^{2}	2.62×10^{-2}	4.09×10^{-1}	8.83×10^{-1}	2.64×10^{-2}
Boha1	4.14×10^{-4}	0	0	2.24×10^{-1}	2.38×10^{-2}	3.95×10^{-1}	7.87×10^{-1}	2.05×10^{-2}
Perm0db	4.69×10^{-1}	5.04×10^{2}	1.24×10^{3}	1.67×10^{2}	1.91×10^{-1}	4.07×10^{0}	7.78×10^{1}	1.59×10^{-1}
Rothyp	1.69×10^{1}	5.64×10^{4}	4.82×10^{4}	7.19×10^{1}	1.44×10^{-1}	4.78×10^{0}	4.02×10^{1}	1.34×10^{-1}
Sumpow	1.19×10^{-5}	1.87×10^{-1}	1.64×10^{-1}	4.71×10^{-1}	1.79×10^{-1}	4.49×10^{0}	3.04×10^{1}	1.79×10^{-1}
Sumsqu	1.52×10^{1}	5.80×10^{2}	6.94×10^{3}	5.3×10^{1}	1.41×10^{-1}	4.26×10^{0}	1.28×10^{1}	1.71×10^{-1}
Trid	-8.08×10^{0}	-7.35×10^{0}	-1.99×10^{2}	-2.88×10^{1}	5.20×10^{-2}	1.55×10^{0}	4.03×10^{0}	4.41×10^{-2}
Booth	9.31×10^{-1}	1.31×10^{-3}	0	1.45×10^{1}	2.44×10^{-2}	3.82×10^{-1}	6.90×10^{-1}	2.31×10^{-2}
Matya	2.97×10^{-1}	1.51×10^{0}	1.67×10^{0}	9.72×10^{-1}	1.53×10^{-1}	4.51×10^{-1}	7.56×10^{0}	1.75×10^{-1}
Mccorm	-1.82×10^{0}	-1.81×10^{0}	-1.83×10^{0}	-1.64×10^{0}	2.90×10^{-2}	3.85×10^{-1}	6.72×10^{-1}	2.58×10^{-2}
Camel3	8.12×10^{-6}	1.29×10^{-1}	2.80×10^{-1}	4.41×10^{-1}	2.79×10^{-2}	3.88×10^{-1}	7.88×10^{-1}	2.69×10^{-2}
Beale	1.89×10^{0}	6.78×10^{-1}	5.10×10^{-1}	8.17×10^{0}	2.55×10^{-2}	3.77×10^{-1}	7.66×10^{-1}	2.51×10^{-2}
Stybtang	-3.18×10^{2}	-2.37×10^{2}	-2.33×10^{2}	-2.81×10^{2}	5.99×10^{-2}	1.55×10^{0}	5.42×10^{0}	6.77×10^{2}

6. Experiments of Short-Term Power Generation Model for Cascade Hydropower Stations

Wanjiazhai Water Conservancy Project: The Wanjiazhai Water Conservancy Project is located in the canyon of the Tuoketuo to Longkou section of the Yellow River in the north of the Yellow River. It is the first of the eight cascades planned for the development of the middle reaches of the Yellow River. and also the Shanxi Yellow River Diversion Project. The starting point of the project the left bank is affiliated to the Pianguan County of Shanxi Province, and the right bank is subordinate to the Zhungeer Banner of Inner Mongolia Autonomous Region. The dam site controls a drainage area of 395,000 square kilometers, with a total storage capacity of 896 million cubic meters and a storage capacity of 445 million cubic meters. It has comprehensive benefits such as water supply, power generation, flood control and anti-icing.

Longkou Hydropower Station is located at the junction of two provinces, Hequ County, Shanxi Province and Zhungeer Banner, Inner Mongolia. It is 25.6 km from the upstream Wanjiazhai Water Control Project and 70 km from the downstream Tianqiao Hydropower Station. It is the regional center of energy and chemical bases in Shanxi Province and Inner Mongolia Autonomous Region, and controls the drainage area of 397, 406 square kilometers.

Table 6 shows the monthly inflow values of the two cascade hydropower stations in the wet years, the flat water years and the dry years. ASP is Annual scheduling period.

Table 6. Cascade hydropower station monthly water supply.

ASP	1	2	3	4	5	6	7	8	9	10	11	12
high flow	149.7	193	176.2	900.1	1077.1	1441.7	343.9	318.1	177.9	36.6	28.6	45.9
median water	82.9	243.9	598.1	554	203.5	146.2	491.7	208.3	147.8	340.9	573.3	104.5
low flow	188.2	251	255.8	550.2	406.6	849.2	132.9	81	59	121.3	11.6	7.8

The short-term power generation model of cascade hydropower stations has been introduced above. Figure 5 showcases the main flow of the algorithm. In this paper, the three periods of the two cascade hydropower stations are scheduled and modeled by Equations (1)–(8) and the sum of the power generation of the two cascade hydropower stations is the largest. As shown in Figure 6, at any stage, CPIO has the largest scheduling capacity for the two cascade hydropower stations, and the total power generation is also relatively huge. CPIO dispatched the two cascade hydropower stations. The final result has the power generation at 3.968×10^{17} KWH in the high flow year, and the total power generation at 3.108×10^{17} KWH in the year of the median water. The power generation at 2.396×10^{17} KWH in the low year.

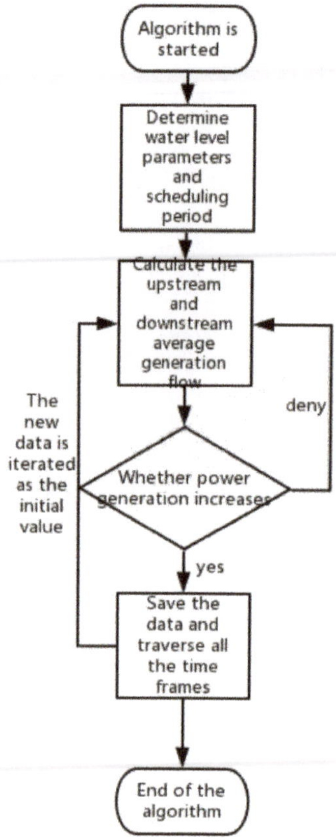

Figure 5. The main process of optimizing hydropower station.

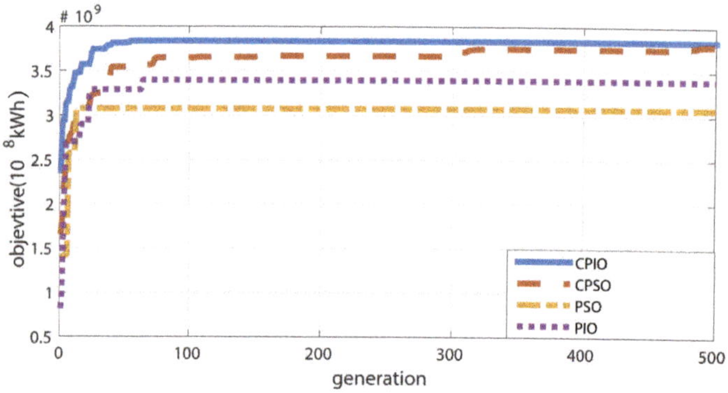

(a)

Figure 6. *Cont.*

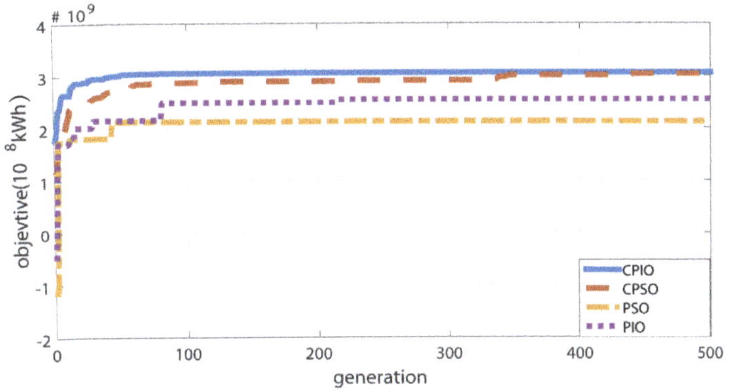

(**b**)

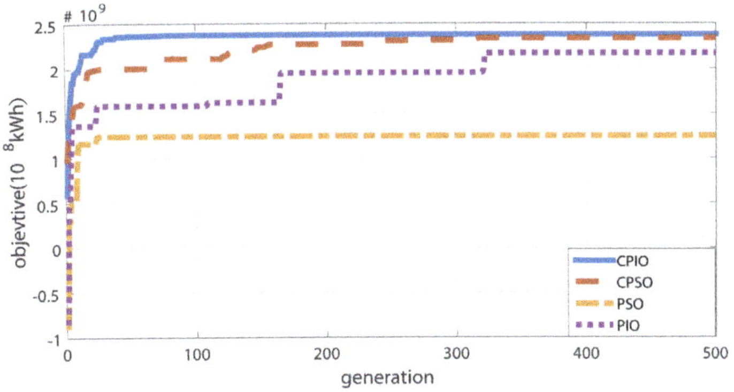

(**c**)

Figure 6. Comparison of four meta-heuristic algorithms in cascade hydropower stations: (**a**) Wet water years schedule; (**b**) flat water years schedule; and (**c**) dry water years schedule;

7. Conclusions

A novel optimization approach called compact pigeon-inspired optimization (CPIO) is proposed. The proposed CPIO was tested on 29 classical test functions to demonstrate the usefulness of the proposed optimization method. A compact method is successfully used in the pigeon-inspired optimization algorithm to reduce the usage of the memory size. The proposed CPIO was also applied to cascade hydroelectric power generation. Simulation results show the CPIO may reach better results compared with some existing algorithms for the cascade hydroelectric power station.

Author Contributions: Conceptualization, J.-S.P. and W.-M.Z.; Data curation, A.-Q.T. and H.C.; Formal analysis, A.-Q.T., S.-C.C., J.-S.P., H.C., and W.-M.Z.; Investigation, A.-Q.T.; Methodology, A.-Q.T., S.-C.C., J.-S.P., H.C., and W.-M.Z.; Software, A.-Q.T.; Validation, J.-S.P.; Visualization, A.-Q.T. and S.-C.C.; Writing—original draft, A.-Q.T.; and Writing—review and editing, S.-C.C. and J.-S.P. All authors have read and agreed to the published version of the manuscript.

Funding: This research received no external funding.

Conflicts of Interest: We wish to confirm that there are no known conflicts of interest and there has been no significant financial support for this work that could have influenced its outcome. We confirm that the manuscript has been read and approved by all named authors and that there are no other persons who satisfied the criteria for authorship but are not listed.

References

1. Yang, X.S. *Nature-Inspired Metaheuristic Algorithms*; Luniver Press: Frome, UK, 2010; pp. 15–35; ISBN 978-1-905986-28-6.
2. Jia, Z.; Sahmoudi, M. A type of collective detection scheme with improved pigeon-inspired optimization. *Int. J. Intell. Comput. Cybern.* **2016**, *9*, 105–123. [CrossRef]
3. Chen, S.; Duan, H. Fast image matching via multi-scale Gaussian mutation pigeon-inspired optimization for low cost quadrotor. *Aircr. Eng. Aerosp. Technol.* **2017**, *89*, 777–790. [CrossRef]
4. Qiu, H.; Duan, H. Multi-objective pigeon-inspired optimization for brushless direct current motor parameter design. *Sci. China Technol. Sci.* **2015**, *58*, 1915–1923. [CrossRef]
5. Deng, X.W.; Shi, Y.Q.; Li, S.L.; Li, W.; Deng, S.W. Multi-objective pigeon-inspired optimization localization algorithm for large-scale agricultural sensor network. *J. Huaihua Univ.* **2017**, *36*, 37–40.
6. Fu, X.; Chan, F.T.; Niu, B.; Chung, N.S.; Qu, T. A multi-objective pigeon inspired optimization algorithm for fuzzy production scheduling problem considering mould maintenance. *Sci. China Inf. Sci.* **2019**, *62*, 70202. [CrossRef]
7. Pan, J.S.; Kong, L.; Sung, T.W.; Tsai, P.W.; Snášel, V. α-Fraction first strategy for hierarchical model in wireless sensor networks. *J. Internet Technol.* **2018**, *19*, 1717–1726.
8. Wang, J.; Gao, Y.; Liu, W.; Sangaiah, A.K.; Kim, H.J. An intelligent data gathering schema with data fusion supported for mobile sink in wireless sensor networks. *Int. J. Distrib. Sens. Netw.* **2019**, *15*, 1550147719839581. [CrossRef]
9. Wang, L.; Singh, C. Environmental/economic power dispatch using a fuzzified multi-objective particle swarm optimization algorithm. *Electr. Power Syst. Res.* **2007**, *77*, 1654–1664. [CrossRef]
10. Hu, P.; Pan, J.S.; Chu, S.C.; Chai, Q.W.; Liu, T.; Li, Z.C. New Hybrid Algorithms for Prediction of Daily Load of Power Network. *Appl. Sci.* **2019**, *9*, 4514. [CrossRef]
11. Emary, E.; Zawbaa, H.M.; Grosan, C.; Hassenian, A.E. Feature subset selection approach by gray-wolf optimization. In *Afro-European Conference for Industrial Advancement*; Springer: Berlin/Heidelberg, Germany, 2015; pp. 1–13.
12. Yang, X.S. A new metaheuristic bat-inspired algorithm. In *Nature Inspired Cooperative Strategies for Optimization (NICSO 2010)*; Springer: Berlin/Heidelberg, Germany, 2010; pp. 65–74.
13. Dao, T.K.; Pan, T.S.; Pan, J.S. Parallel bat algorithm for optimizing makespan in job shop scheduling problems. *J. Intell. Manuf.* **2018**, *29*, 451–462. [CrossRef]
14. Yang, X.S. Flower pollination algorithm for global optimization. In Proceedings of the International Conference on Unconventional Computation and Natural Computation, Orléans, France, 3–7 September 2012; Springer: Berlin/Heidelberg, Germany, 2012; pp. 240–249.
15. Nguyen, T.T.; Pan, J.S.; Dao, T.K. An Improved Flower Pollination Algorithm for Optimizing Layouts of Nodes in Wireless Sensor Network. *IEEE Access* **2019**, *7*, 75985–75998. [CrossRef]
16. Chu, S.C.; Tsai, P.W.; Pan, J.S. Cat swarm optimization. In Proceedings of the Pacific Rim International Conference on Artificial Intelligence, Guilin, China, 7–11 August 2006; Springer: Berlin/Heidelberg, Germany, 2006; pp. 854–858.
17. Kong, L.; Pan, J.S.; Tsai, P.W.; Vaclav, S.; Ho, J.H. A balanced power consumption algorithm based on enhanced parallel cat swarm optimization for wireless sensor network. *Int. J. Distrib. Sens. Networks* **2015**, *11*, 729680. [CrossRef]
18. Qin, A.K.; Huang, V.L.; Suganthan, P.N. Differential evolution algorithm with strategy adaptation for global numerical optimization. *IEEE Trans. Evol. Comput.* **2008**, *13*, 398–417. [CrossRef]
19. Meng, Z.; Pan, J.S.; Tseng, K.K. PaDE: An enhanced Differential Evolution algorithm with novel control parameter adaptation schemes for numerical optimization. *Knowl.-Based Syst.* **2019**, *168*, 80–99. [CrossRef]
20. Meng, Z.; Pan, J.S. Quasi-affine transformation evolutionary (QUATRE) algorithm: A parameter-reduced differential evolution algorithm for optimization problems. In Proceedings of the 2016 IEEE Congress on Evolutionary Computation (CEC), Vancouver, BC, Canada, 24–29 July 2016; pp. 4082–4089.
21. Liu, N.; Pan, J.S. A bi-population QUasi-Affine TRansformation Evolution algorithm for global optimization and its application to dynamic deployment in wireless sensor networks. *EURASIP J. Wirel. Commun. Netw.* **2019**, *2019*, 175. [CrossRef]
22. Koza, J.R. Genetic Programming: Automatic Programming of Computers. *EvoNews* **1997**, *1*, 4–7. [CrossRef]

23. Hsu, H.P.; Chiang, T.L.; Wang, C.N.; Fu, H.P.; Chou, C.C. A Hybrid GA with Variable Quay Crane Assignment for Solving Berth Allocation Problem and Quay Crane Assignment Problem Simultaneously. *Sustainability* **2019**, *11*, 2018. [CrossRef]
24. Duan, H.; Qiao, P. Pigeon-inspired optimization: A new swarm intelligence optimizer for air robot path planning. *Int. J. Intell. Comput. Cybern.* **2014**, *7*, 24–37. [CrossRef]
25. Li, C.; Duan, H. Target detection approach for UAVs via improved pigeon-inspired optimization and edge potential function. *Aerosp. Sci. Technol.* **2014**, *39*, 352–360. [CrossRef]
26. Deng, Y.; Duan, H. Control parameter design for automatic carrier landing system via pigeon-inspired optimization. *Nonlinear Dyn.* **2016**, *85*, 97–106. [CrossRef]
27. Duan, H.; Wang, X. Echo state networks with orthogonal pigeon-inspired optimization for image restoration. *IEEE Trans. Neural Networks Learn. Syst.* **2015**, *27*, 2413–2425. [CrossRef] [PubMed]
28. Wang, C.N.; Le, A. Measuring the Macroeconomic Performance among Developed Countries and Asian Developing Countries: Past, Present, and Future. *Sustainability* **2018**, *10*, 3664. [CrossRef]
29. Wang, C.N.; Nguyen, H.K. Enhancing urban development quality based on the results of appraising efficient performance of investors—A case study in vietnam. *Sustainability* **2017**, *9*, 1397. [CrossRef]
30. Scieri, F.; Miller, R.L. Hydro Electric Generating System. U.S. Patent 4,443,707, 17 April 1984.
31. Davison, F.E. Electric Generating Water Power Device. U.S. Patent 4,163,905, 7 August 1979.
32. Ma, C.; Lian, J.; Wang, J. Short-term optimal operation of Three-gorge and Gezhouba cascade hydropower stations in non-flood season with operation rules from data mining. *Energy Convers. Manag.* **2013**, *65*, 616–627. [CrossRef]
33. Jain, A.; Ormsbee, L.E. Short-term water demand forecast modeling techniques—CONVENTIONAL METHODS VERSUS AI. *J. Am. Water Work. Assoc.* **2002**, *94*, 64–72. [CrossRef]
34. Fan, C.; Xiao, F.; Wang, S. Development of prediction models for next-day building energy consumption and peak power demand using data mining techniques. *Appl. Energy* **2014**, *127*, 1–10. [CrossRef]
35. Fosso, O.B.; Belsnes, M.M. Short-term hydro scheduling in a liberalized power system. In Proceedings of the 2004 International Conference on Power System Technology, PowerCon 2004, Singapore, 21–24 November 2004; Volume 2, pp. 1321–1326.
36. Nguyen, T.T.; Vo, D.N. An efficient cuckoo bird inspired meta-heuristic algorithm for short-term combined economic emission hydrothermal scheduling. *Ain Shams Eng. J.* **2016**, *9*, 483–497. [CrossRef]
37. Nazari-Heris, M.; Mohammadi-Ivatloo, B.; Gharehpetian, G. Short-term scheduling of hydro-based power plants considering application of heuristic algorithms: A comprehensive review. *Renew. Sustain. Energy Rev.* **2017**, *74*, 116–129. [CrossRef]
38. Billingsley, P. *Probability and Measure*; John Wiley & Sons: Hoboken, NJ, USA, 2008.
39. Bronshtein, I.N.; Semendyayev, K.A. *Handbook of Mathematics*; Springer Science & Business: Berlin/Heidelberg, Germany, 2013; ISBN 978-3-662-21982-9.
40. Neri, F.; Mininno, E.; Iacca, G. Compact particle swarm optimization. *Inf. Sci.* **2013**, *239*, 96–121. [CrossRef]
41. Mininno, E.; Cupertino, F.; Naso, D. Real-valued compact genetic algorithms for embedded microcontroller optimization. *IEEE Trans. Evol. Comput.* **2008**, *12*, 203–219. [CrossRef]
42. Surjanovic, S.; Bingham, D. Virtual Library of Simulation Experiments: Test Functions and Datasets. Available online: http://www.sfu.ca/~ssurjano (accessed on 26 December 2019).
43. Hao, R.; Luo, D.; Duan, H. Multiple UAVs mission assignment based on modified pigeon-inspired optimization algorithm. In Proceedings of the 2014 IEEE Chinese Guidance, Navigation and Control Conference, Yantai, China, 8–10 August 2014; pp. 2692–2697.

sustainability

MDPI

Article

A New Robust Energy Management and Control Strategy for a Hybrid Microgrid System Based on Green Energy

Bilal Naji Alhasnawi [1,*], Basil H. Jasim [1] and M. Dolores Esteban [2]

1 Electrical Engineering Department, University of Basrah, Basrah 61001, Iraq; hanbas632@gmail.com
2 Civil Engineering Department, Hydraulics, Energy and Environment, Universidad Politécnica de Madrid (UPM), CP 28040 Madrid, Spain; mariadolores.esteban@upm.es
* Correspondence: bilalnaji11@yahoo.com; Tel.: +964-7809-098-542

Received: 18 June 2020; Accepted: 14 July 2020; Published: 16 July 2020

Abstract: The recent few years have seen renewable energy becoming immensely popular. Renewable energy generation capacity has risen in both standalone and grid-connected systems. The chief reason is the ability to produce clean energy, which is both environmentally friendly and cost effective. This paper presents a new control algorithm along with a flexible energy management system to minimize the cost of operating a hybrid microgrid. The microgrid comprises fuel cells, photovoltaic cells, super capacitors, and other energy storage systems. There are three stages in the control system: an energy management system, supervisory control, and local control. The energy management system allows the control system to create an optimal day-ahead power flow schedule between the hybrid microgrid components, loads, batteries, and the electrical grid by using inputs from economic analysis. The discrepancy between the scheduled power and the real power delivered by the hybrid microgrid is adjusted for by the supervisory control stage. Additionally, this paper provides a design for the local control system to manage local power, DC voltage, and current in the hybrid microgrid. The operation strategy of energy storage systems is proposed to solve the power changes from photovoltaics and houses load fluctuations locally, instead of reflecting those disturbances to the utility grid. Furthermore, the energy storage systems energy management scheme will help to achieve the peak reduction of the houses' daily electrical load demand. Also, the control of the studied hybrid microgrid is designed as a method to improve hybrid microgrid resilience and incorporate renewable power generation and storage into the grid. The simulation results verified the effectiveness and feasibility of the introduced strategy and the capability of proposed controller for a hybrid microgrid operating in different modes. The results showed that (1) energy management and energy interchange were effective and contributed to cost reductions, CO_2 mitigation, and reduction of primary energy consumption, and (2) the newly developed energy management system proved to provide more robust and high performance control than conventional energy management systems. Also, the results demonstrate the effectiveness of the proposed robust model for microgrid energy management.

Keywords: inverters; converters; distributed generators; utility grid; hierarchical control

---

1. Introduction

During the previous decade, distributed energy systems have assumed supreme importance in the electrical power infrastructure because they lead to a reduction in greenhouse gas emissions and produce quality power with increased efficiency and service reliability. Deploying distributed energy resources facilitates a modification in the way energy is transmitted through the electrical power grid and provides flexibility to the consumer concerning energy use.

To accomplish an adequately set up distributed energy ecosystem, there is a necessity to convert the power system into smaller distributed-energy systems [1]. These distributed systems may incorporate several power sources like photovoltaic systems, wind turbines, or fuel cells. These fuel-cell-based energy sources are essential since power generation based on such a system may be set up near or at the consumer premises. Recent research [2] indicates that such resources help decrease household emissions, operating costs, and primary energy required to produce electricity [3].

The integration of various distributed energy resources, distributed loads, and energy storage systems with a renewable energy ecosystem is called a microgrid [4]. In the recent past, microgrids have been explored with immense interest as power systems that offer resilience and help set up a smart active electrical grid. Additionally, it has the potential to enhance system safety, reliability, efficiency, and boost connectivity with renewable energy sources [5]. The microgrid may not be connected to the utility grid since it can be used as a standalone system supplying energy to controllable loads using several distributed energy sources. The introduction of controllable loads and distributed energy systems causes several challenges for energy management systems. The primary role of an energy management system is to calculate the adequate energy transmission from the microgrid and the primary energy network independently and meet load requirements on an hourly basis. The literature contains several proposals regarding microgrid energy management systems using varying algorithms and different microgrids [6]. The rest of this paper is organized as follows. Section 2 presents a description of the related works. Section 3 presents a description of the proposed system described in this paper. Section 4 presents the mathematical model of the distributed hybrid generation system, Section 5 presents the control methods for photovoltaics and fuel cell system, Section 6 presents the supervisory control method of the proposed system. Section 7 presents the energy management systems for houses and Section 8 presents the results of the proposed system. Finally, Section 9 concludes the paper.

2. Related Work

This section contains a review of the pertinent research regarding energy management strategies related to the topics connected to this work. In Lujano-Rojas et al. [7], the photovoltaic system with battery backup is operated in the grid-connected and islanded modes of operation. In this paper, separate control algorithms are implemented for the inverter, battery, and photovoltaic array for maximum utilization of available sources for meeting the energy requirement of the consumer. In Delgado and Navarro [8], the authors propose a linear programming-based algorithm to manage microgrid power. This algorithm facilitates adequate generator use for controllable and uncontrollable electrical loads. One example of an optimization challenge is the adequate use of diesel generators while maintaining operational and economic limitations due to energy demand and supply for each microgrid component (storage system, load, and generator). In the study conducted by Helal et al. [9], the power management system for an AC/DC hybrid microgrid was explored in an isolated area that uses a photovoltaic setup for desalination. The optimization algorithm suggested uses mixed-integer nonlinear programming, and the objective function works towards reducing the daily operational cost. In Correa et al. [10], the authors formulated a virtual power plant (VPP)-based power management system. The microgrid investigated consists of a solar panel and an energy storage system working in tandem. Linear programming methods are used to design the techniques required to decrease operating expenditure. Renewable energy is accounted for in the energy models like the one in Colombia and is typically based on hydropower resources. In Dufo and Agustín [11], the authors present a generic algorithm-based control system to optimally manage energy in a hybrid system. The system comprises renewable energy sources such as hydropower, wind turbines, and photovoltaics in conjunction with AC generators, fuel cells, and an electrolyzer. Energy management optimization leads to decreased operating costs. This facilitates the surplus produced by the renewable source to be stored in batteries and use of the electrolyzer to produce hydrogen gas. Loads that cannot be powered using a renewable energy source may be operated using a fuel cell, or by using energy from the batteries. In Das et al. [12], the authors investigate the impact of supplementing a standalone hybrid microgrid based on photoelectric sources with gas turbines and an internal combustion engine. This multi-objective generic technique was implemented to achieve a system optimized considering energy cost and efficiency. Load evaluation was conducted using thermal and electrical measurements. Combined with heating or cooling, all the systems studied were able to meet the electrical demand. In Luna et al. [13], the author introduced a real-time power management system. The three cases were investigated with full, incomplete, and accurate forecasts. The optimization model was tested on connected and isolated microgrids, and there was a significant imbalance between loads and generations. In Abedini et al. [14], the author used the particle swarm algorithm with Gaussian boom and applied the energy management algorithm to a microgrid consisting of wind turbines, diesel generators, or photovoltaics. This study suggested a reduction in both capital and operational (fuel) costs. In Marzband et al. [15], energy management in the microgrid was tested using the artificial bee colony (ABC) algorithm. Provided the transient nature of wind and solar energy resources, an approach is required to evaluate the economic distribution of energy generating units in the microgrid. The results indicated a 30% reduction in cost. The non-dispatchable generation and load uncertainty are managed using Markov chain and neural networks. In Rouholamini and Mohammadian [16], the study proposes appropriate power management in the case of grid-tied hybrid energy generation systems, which include wind turbines, photovoltaics, electrolysis, and fuel cells. Electricity is traded with the local network using the real-time pricing determined over a 24 h time slot using simulation results. This energy management optimization case used the interior search technique. In Almada et al. [17], the research proposed a centralized microgrid energy management system in both grid-connected and island mode. In the island mode, the fuel cells start supplying energy only if the battery state of charge reaches less than 80%, whereas in the grid-connected mode, a 60% state of charge is required to ensure reliability. In Merabet et al. [18], the authors recommended an energy management system to handle a hybrid microgrid system having battery power, photovoltaics, and wind turbines. The data acquisition and control systems operate in real-time. A set of rules governs the power management platform, and those enhance microgrid performance by monitoring and manage load, power generation, and storage devices. Farzin et al. [19] proposed an energy management system to handle an isolated microgrid. An isolated failure event was considered as being a natural probability distribution failure scenario in the utility grid. The objective was to decrease operating expenditure, which considers the expenditure for running wind turbines, small turbines, batteries, and electrical load. In Battistelli et al. [20], the authors suggested a management system for a hybrid DC to AC microgrid that promises economic electrical transmission despite the skepticism about the economics of renewable energy. The system has on-demand load control, and factors in generators, battery charge/discharge thresholds (electric or thermal automobiles), and controllable loads.

The method presented has the following demerits:

- Several energy management techniques were designed using the small-signal model. In spite of the straightforward design, small-signal-based control systems do not ensure stability, which is a crucial need for complex networks.
- Several present models face the challenges of partial plant dynamics because they disregard the impact of the inner controllers on the control system, thus altering the stability and performance of the converters.

To address these challenges, this paper proposes a newly designed and flexible energy management and control method applicable to a hybrid microgrid that uses green energy. The proposed technique has specific advantages which are listed below:

1. To the best of the authors knowledge, this is the first research paper that proposes an energy control system for homes, which regulates the frequency and restoration of the voltage for an islanded or a grid-connected microgrid, which is based on the use of a completely nonlinear framework, without considering uncertainties and parametric disturbances.

2. This work proposes a new energy management framework specific to a hybrid microgrid. A localized photovoltaic and fuel cell generator control technique is proposed to manage the load connected to the hybrid system to AC loads as a feasible option for several commercial, domestic, and industrial load scenarios. The performance of the microgrid, both at steady state and in a dynamic scenario, is evaluated to validate the effectiveness of the proposed energy management technique.
3. This distributed consensus-based system exhibits the preciseness of power sharing in case the electrical frequency is restored.
4. This paper proposes a sophisticated EMS (Energy Management System) capable of providing an optimal operating technique for a typical Microgrid (MG) system where cost is the competitive objective function and emissions are the constraints.
5. This work presents a hybrid microgrid architecture based on AC for a smart home, where better adoption of distributed electrical generation may isolate the grid from interference. Therefore, the system can have increased stability and safety despite the dynamic and complex grid environment.
6. The MPPT (Maximum Power Point Tracking) method is used to photovoltaic installations to extract maximum energy from hybrid systems during varying environmental aspects.

3. Proposed System Description

The hybrid microgrid evaluated comprises photovoltaic cells and fuel cells, along with super capacitors and batteries that are used for energy storage, an electrolyzer cell, and AC loads. These systems, when connected, create a small hybrid microgrid, which is connected to a low-voltage AC microgrid for distribution or islanded mode. Figure 1 depicts the configuration of such a hybrid system for the distribution microgrid case. The DC bus is used to interconnect the fuel cells to the boost converter. The battery is connected to the charging regulator using a two-way DC to DC converter, while the super capacitor has a direct connection to the DC bus. The one-way DC to DC converter is used to interconnect the FC and the DC buses, which maintains system stability in spite of varying electrical loads. Batteries are connected to the DC bus using a two-way DC to DC converter.

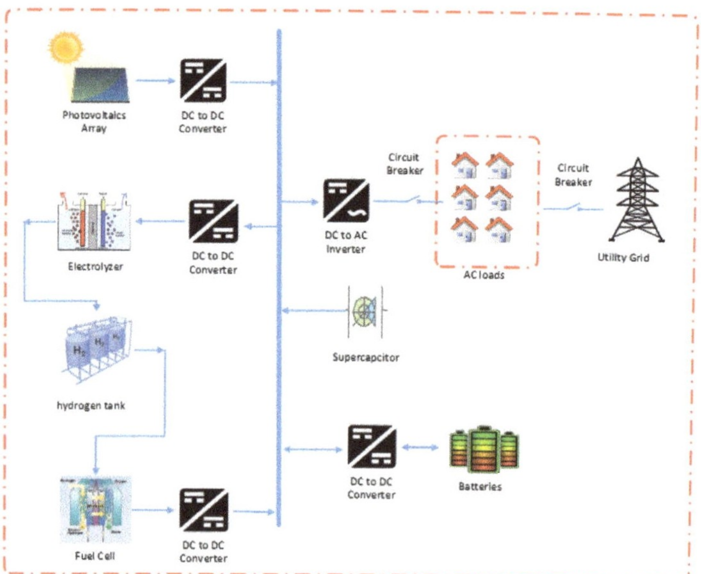

Figure 1. The proposed hybrid microgrid system structure.

4. Distributed Hybrid Energy Generation System

4.1. Photovoltaic Cell Modeling

Figure 2 shows an equivalent circuit based on the diode of the solar cell, which can be represented as a parallel resistor, current source, diode, and serial resistance. The current and voltage properties of solar cells are described by the standard mathematical equation (Equation (1)) [21]:

$$I = I_{ph,cell} - \underbrace{I_{o,cell}[\exp((q(V + IR\_(s,cell))/akT) - 1]}_{I_{d,cell}} - \frac{V + IR_{s,cell}}{R_{p,cell}} \quad (1)$$

where $I_{o,cell}$ is reversed leakage current, $I_{ph,cell}$ is photocurrent (A) of photovoltaic, k is Boltzmann's constant (1.38×10^{-23} J/K), q is the electron charge (1.602×10^{-19} C), and $R_{p,cell}$ is parallel resistance (Ω). $R_{s,cell}$ is the series resistance (Ω) and T is the diode temperature.

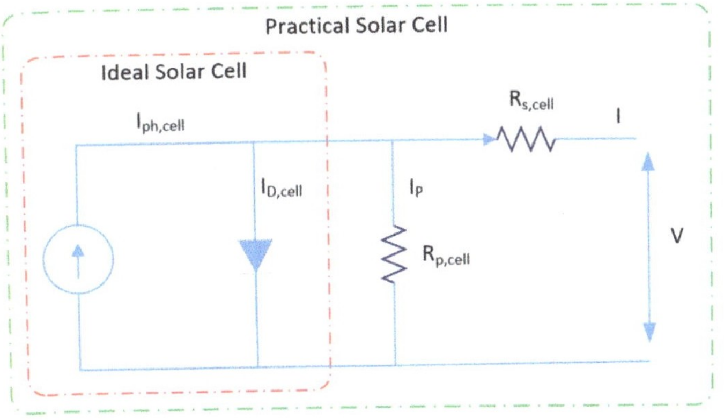

Figure 2. The circuit of photovoltaic cell based on single diode mode.

The solar cell model is ideal if the parallel and series resistance of the solar cell are not taken into account. Figure 3 illustrates the ideal electrical current and voltage curves from Equation (1) [22,23].

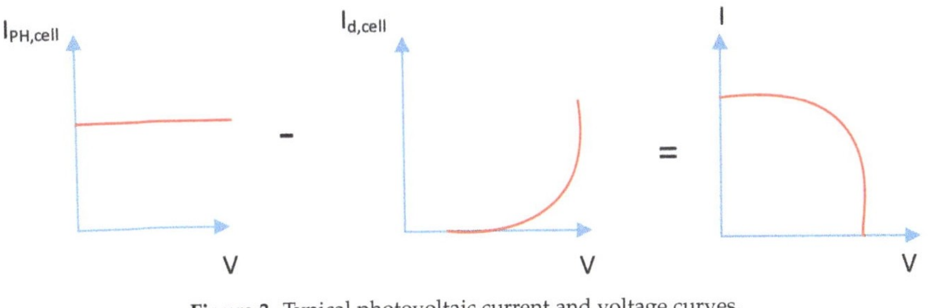

Figure 3. Typical photovoltaic current and voltage curves.

4.2. Modeling of Photovoltaic Module

As mentioned earlier, the photovoltaic modules are made up of connected photovoltaic cells in parallel form. Therefore, the standard mathematical equation is derived from Equation (2) and the characteristic description of the I–V photovoltaic unit [23] is:

$$I = I_{ph,cell} - I_O\left[exp\left(\frac{V + IR_S}{a\ Vt}\right) - 1\right] - \frac{V + IR_S}{R_p} \quad (2)$$

where $I_{ph,cell}$ is the photocurrent (A) of the photovoltaic, I_o is photovoltaic reverse leakage current, V_t is photovoltaic thermal voltage, R_p is parallel resistance, R_s is photovoltaic series resistance, Equation (2) generates the current and voltage curve as illustrated in Figure 4.

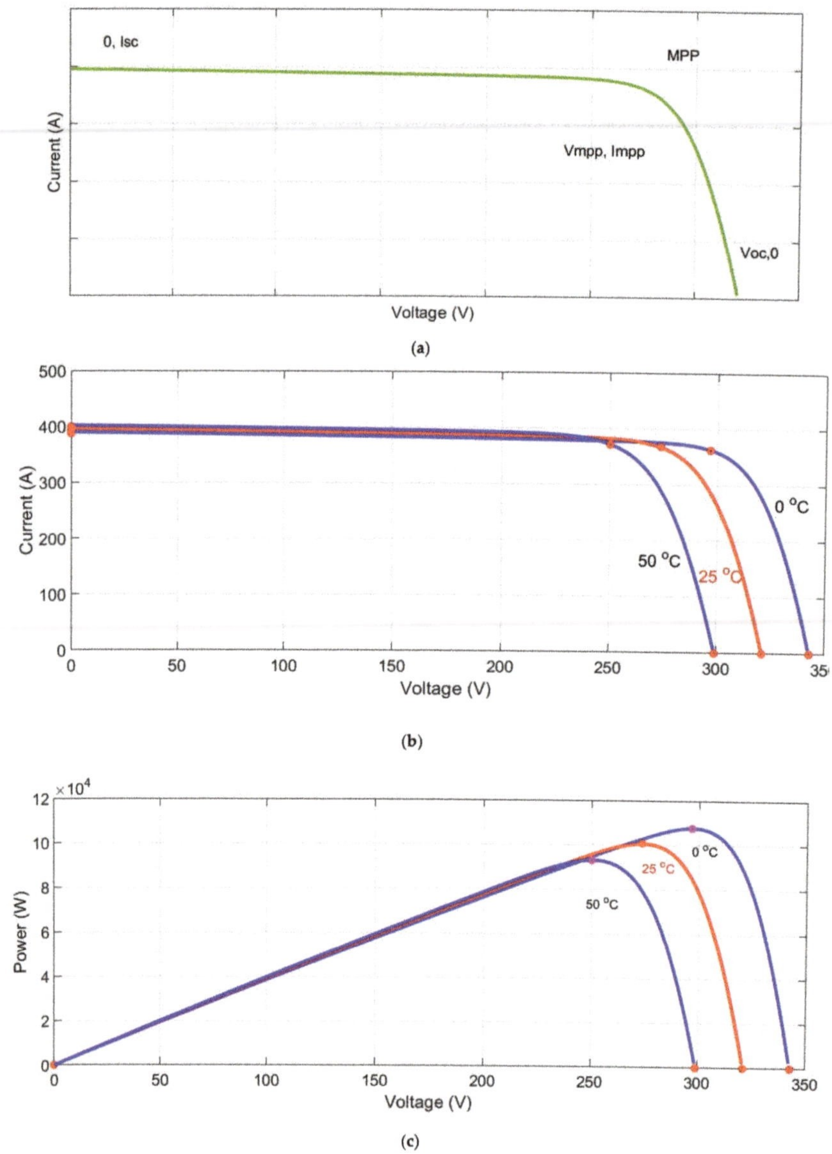

Figure 4. (**a**) A typical current–voltage curve of the photovoltaic (PV), (**b**) current–voltage curves of a photovoltaic at different temperature levels and constant irradiation, (**c**) Power–voltage curves of a photovoltaic at constant irradiation and different temperature levels.

The current of a photovoltaic (I_{PH}) based on the amount of the photovoltaic irradiance falling on the photovoltaic and the photovoltaic cell temperature corresponds to Equation (3) [24]:

$$Iph = \frac{G}{Gn}\left(I_{ph;n} + Ki\Delta T\right) \quad (3)$$

where $I_{ph;n}$ is the photocurrent, G is the photovoltaic irradiance measured in W/m^2, G_n is the nominal irradiance (1000 W/m^2), and Ki is the temperature coefficient.

V_{oc} is determined by the Equations (4) and (5):

$$V_{oc} = V_{oc;n} + K_v \Delta T \tag{4}$$

where K_v is coefficient of temperature and $V_{oc;n}$ is the open circuit voltage under the nominal conditions.

$$I_o = \frac{I_{sc;n} + K_i \Delta T}{exp\left(\frac{V_{oc;n} + K_v \Delta T}{a\ V_t}\right) - 1} \tag{5}$$

where $I_{sc;n}$ is the short-circuit current under the nominal conditions [25].

In this paper, the electrical parameters of the SPR-305E-WHT-D solar photovoltaic are extracted and used to simulate this modular model. These parameters are listed in Table 1.

Table 1. Electrical parameters of the SPR-305E-WHT-D photovoltaic.

Parameters	Value
Series connected modules	5
Parallel string	66
Voltage of open circuit (V_{oc})	64.2 (V)
Maximum voltage $\left(V_{mp}\right)$	54.7 (V)
Temperature coefficient of (V_{oc})	−0.27269 (%/°C)
Short-circuit current (I_{sc})	5.96 (A)
Maximum current I_{mp}	5.58 (A)
Temperature coefficient of (I_{sc})	0.061745 (%/°C)
Shunt resistance (R_{sh})	269.5934 Ω
Series resistance (R_s)	0.37152 Ω
Diode ideality factor	0.945
Diode saturation current I_o	6.3×10^{-1} (A)
PV type	SPR-305E-WHT-D
Number of cells	96

4.3. Battery Storage System

Battery storage systems store additional energy produced by renewable energy systems. However, if there is insufficient energy from the renewable energy generation system, the battery will be discharged to meet the load demand. It consists of a 4800 (Ah) Li-ion battery unit with a rated voltage of 96 VDC. Equations (6)–(9) govern the battery charge/discharge process [26]:

$$E_{disch}^{\text{Ni-MH}} = Eo - k\frac{Q}{Q - it}i^{*} - k\frac{Q}{Q - it}it + e^{t} \tag{6}$$

$$E_{ch}^{\text{Ni-MH}} = Eo - k\frac{Q}{|it| - 0.1Q}i^{*} - k\frac{Q}{Q - it}it + e^{t} \tag{7}$$

The equations that govern the discharging process of Li-ion batteries are:

$$E_{disch}^{\text{Li-Ion}} = Eo - k\frac{Q}{Q - it}i^{*} - k\frac{Q}{Q - it}it + Ae^{-Bit} \tag{8}$$

$$E_{ch}^{\text{Li-Ion}} = Eo - k\frac{Q}{|it| - 0.1Q}i^{*} - k\frac{Q}{Q - it}it + Ae^{-Bit} \tag{9}$$

The discharge–charge conditions are the same as those of the Ni–MH battery type. In the equations above, the variables and parameters are: k, polarization constant (Ah^{-1}); E_o, battery voltage; Q, maximum battery capacity, in Ah; i^{*}, filtered low-frequency current dynamics, in A. In this paper, the electrical parameters of the lithium-ion battery module are extracted and used to simulate this modular model. These parameters are listed in Table 2. Figure 5 illustrates the battery current curves at 0.083333C.

Table 2. Electrical parameter of lithium-ion battery module.

Parameters	Values
Capacity	4800 (Ah)
Voltage	96 (V)
Initial State of Charge (SoC)	65 (%)
Battery response time	20 (s)
Voltage of fully charged	110 (V)
Current of discharge	400 (A)
Voltage of cutoff	72 (V)
Resistance	0.012 Ω

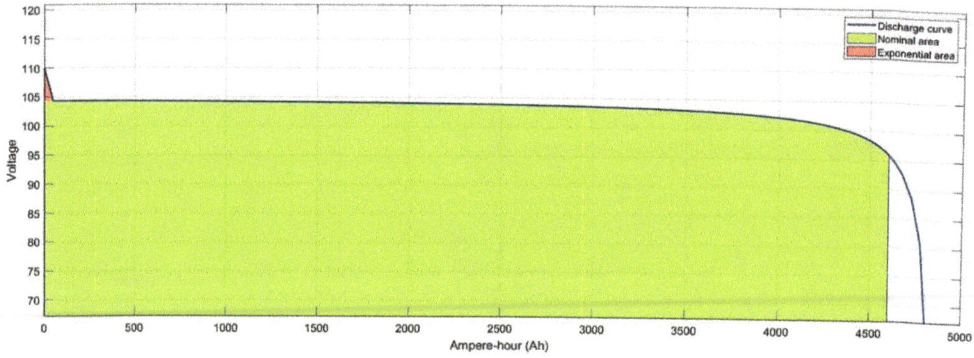

Figure 5. The battery current curves at 0.083333C (400A).

4.4. Loads

The loads consist of residential and commercial loads. Commercial loads appear on asynchronous devices to show the effect of commercial inductive loads, such as air conditioning systems, on the microgrids. Residential loads are designed according to the daily nonseasonal consumption profile of the resort island. Residential loads are simulated according to the actual difference in the specific load profile for the specified resort island.

4.5. Fuel Cell Model

Chemical energy is transformed into electrical energy using fuel cells based on a proton exchange membrane. This hybrid system, consisting of fuel cells and photovoltaics, offers better efficiency along with several other advantages of fuel cells and photovoltaics, which are reduction in carbon dioxide emission, water management, cell stack heating, and the development of cost-effective material. One cell consists of two variants of porous gas diffusion electrodes that separate the electrolyte, thus defining the cell type. The electrolyte consists of a thin conductive film, which conducts cations but restricts the flow of electrons and gases. The fuel cell is fed hydrogen gas from the electrolyzer. At the same time, oxygen enters the cell from the other side and causes a chemical reaction, where hydrogen dissociates into a proton and an electron. The chemical reaction is specified as (Equations (10) and (11)):

$$H_2 \leftrightarrow 2H^+ + 2e^- \tag{10}$$

$$\frac{1}{2}O_2 + 2H^+ + 2e^- \rightarrow H_2O \tag{11}$$

Hydrogen dissociation occurs at the left anode, causing a concentration gradient between various electrodes across the membrane. The gradient leads to the protons spreading across the membrane, while the electrons are left behind. The protons drifting towards the cathode provide it a positive charge compared to the anode. Electrons are attracted to the cathode; however, their movement is restricted by the membrane, and the only way for the electrons to reach the cathode is through an external circuit, which causes a current flow. The output voltage $V_{FC}(t)$ for the fuel cell is specified as (Equation (12)):

$$V_{FC}(t) = E(t) - V_{act} - V_{oh} - V_{con} - V_{trans} \quad (12)$$

where V_{oh}, V_{act}, V_{trans}, and V_{con} are ohmic voltage, activation voltage, mass transport loss, and concentration voltage, respectively. The voltage created by a fuel cell is specified as (Equation (12)):

$$E(t) = -\frac{\Delta G^o}{nF} + \frac{\Delta S}{nF}\left(T - T_{ref}\right) + \frac{R \times T}{nF}\ln\left(\frac{P^{0.5}_{O_2} \times P^1_{H_2}}{P^1_{H_2O}}\right) \quad (13)$$

where ΔG^o refers to electric work, also called Gibbs free energy; F is the Faraday constant; n refers to number of moles; S is the specific entropy; T refers to the operational absolute temperature; T_{ref} is set at 25 °C; $P^{0.5}_{O_2}$, $P^1_{H_2}$, and $P^1_{H_2O}$ are the respective pressures of pure oxygen, hydrogen, and water as fuel; R refers to the gas constant. For the fuel cell stack, the total power output $P_{FC}(t)$ is specified as (Equation (14)):

$$P_{FC}(t) = P_{FC} \times V_{FC}(t) \times I_{FC} \quad (14)$$

where P_{FC} and I_{FC} refer to the stacked fuel cell rating and the fuel cell current, respectively. For modeling fuel cell power generation, the primary aspect is the hydrogen mass flow rate. The flow rate, $\left(Q_{H_2}\right)^c$, required by a 1 kW fuel cell, expressed in kg/hour, is specified as (Equation (15)):

$$\left(Q_{H_2}\right)^c = \propto_1 \times P^r_{FC} + \propto_2 \times P^1_{FC}(t) \quad (15)$$

where $\propto_1$, $\propto_2$, and P^r_{FC} refer to the fuel cell intercept coefficent in kh = g = kW rated, fuel cell curve slope in kg/h/W, and the rated fuel cell capacity in kW. For a fuel cell rated at 1 kW, assuming $\propto 1$ and $\propto 2$ at 0.00031 and 0.0581 kg/h/kW, respectively, P^1_{FC} is determined to be 0.059 kg/h, which indicates that the fuel cell requires this quantity of hydrogen per hour to produce its rated power [27].

4.6. Electrolyzer Model

The electrolyzer uses surplus electricity from the photovoltaic system to produce hydrogen for use in fuel cells, which is stored in a hydrogen tank until it is required. The aspect that requires attention is the quantity of hydrogen produced by the electrolyzer. For a 1 kW electrolyzer, the hydrogen mass-flow rate $\left(m_{H_2}\right)^p_1$ in kg/hour is specified as (Equation (16)):

$$\left(m_{H_2}\right)^p_1 = \frac{3600 P^r_{el} \times n_{el}}{HV_{H_2}} \quad (16)$$

where P^r_{el} is 1 kW, HV_{H_2} refers to the heating value of hydrogen (in MJ/kg), and n_{el} represents the efficency of the electrolyte. Electrolyzer efficiency is assumed to be 90% [28]. Given the heating value of hydrogen at 142 MJ/kg, the hydrogen mass flow would be 0.02268 kg/h/kW. Therefore, the 1 kW electrolyzer will output 0.02268 kg of hydrogen per hour [29].

5. Control Method for Photovoltaics and Full Cell System

5.1. Control of the Converter Interfaced Fuel Cell

For the boost converter, general control comprising voltage stabilization and output current limitation is used (Figure 6). The current limit is fixed; however, it may be modified for varying sources (for instance, the current should be proportional to the amount of hydrogen injected in a fuel cell, depending on the attributes of the fuel cell). For the test considered in the study, the current is capped at 30 A. The DC reference voltage is aligned to the fixed value needed by the inverter connecting the energy source with the load or the grid. A deviation from the set values is assumed to be an energy deficit or surplus, and is indicated to the power management system. The reference voltage for this control system defines the maximum. Any DC voltage exceeding the maximum causes a decrease in inductor current. When State of Charge (SoC) falls below the reference, the load is powered using the fuel cells. Compared to previous schemes, the current scheme may be implemented with relative ease, while PI (Proportional Integral) gains are regulated online to provide a better response, PWM is the pulse width modulation signal, C is the capacitor, V_{dc} is the voltage of DC bus, L is the inductance.

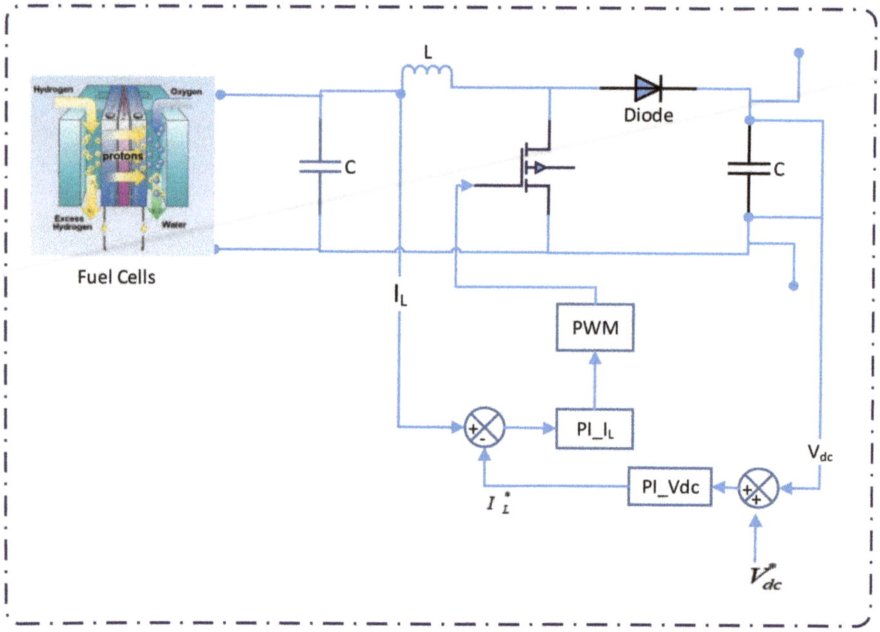

Figure 6. The control of the fuel cell converter.

5.2. Control of the Converter Interfaced Photovoltaics

A single-phase boost is used to up the voltage from the panel and control maximum power point tracking. Input current (I_{PV}) sensing happens before the measurement of input capacitance (C_i) in addition to the panel voltage (V_{PV}). The maximum power point tracking algorithm uses these two values. The MPPT algorithm determines a reference point, which, when maintained at the panel, provides maximum power to be extracted from the PV system. As depicted in Figure 7, MPPT is achieved using the voltage of the outer loop and the current flowing in the inner loop. Hence, the signs for the reference outer voltage compensator and feedback are reversed. The converter output is not adjusted. A voltage feedback mechanism provides input to the internal voltage comparators, which use Pulse-Width Modulation (PWM) in an overvoltage situation to control the voltage from rising beyond the components' rated voltage.

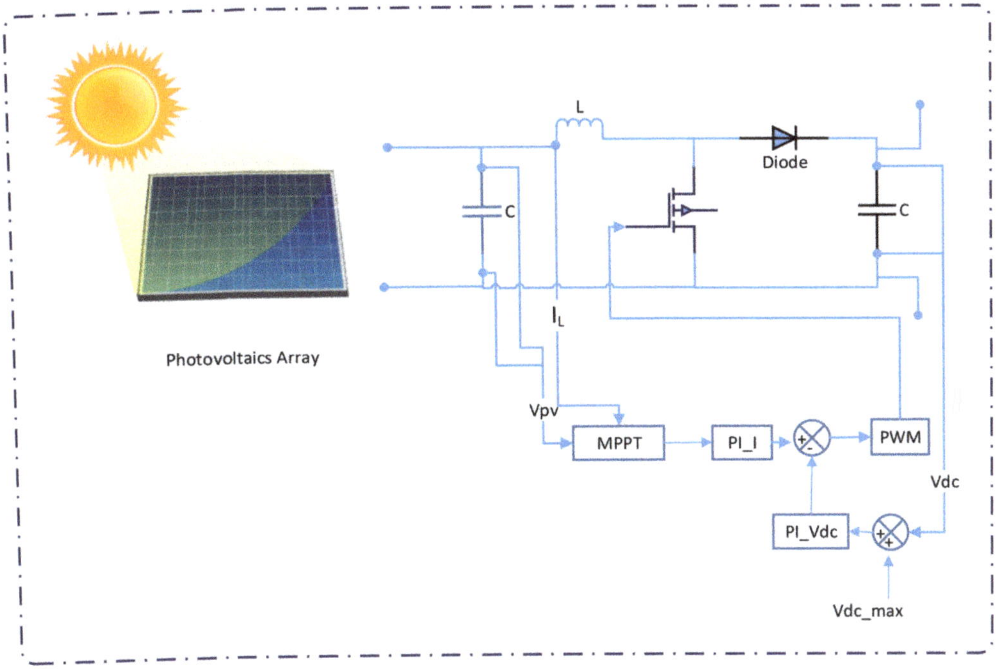

Figure 7. Control of the converter with maximum power point tracking.

5.3. *Control of Bidirectional-Converter Interfacing Battery*

In the proposed setup, as depicted in Figure 8, the bidirectional converter has an output filter capacitor (C_{dc}), a high-frequency inductor (L), and two switches (S1 and S2) that permit bidirectional flow of current. The power management system consists of a two-voltage controller with appropriate limitation blocks to provide the necessary power flow under varying conditions. The controllers output a reference current for the storage of energy. The first aspect of control is DC bus voltage regulation, while the other controllers regulate battery voltage. To facilitate better power management in the microgrid, backup energy storage is a part of the system, which comprises a battery connected to the DC bus using a bidirectional converter. The converter serves multiple purposes: it serves as a battery charge regulator in grid-connected operation, and a boost converter to deliver energy from the batteries to the microgrid when the PV and fuel cell sources have insufficient power to feed the local loads in islanded operation. During island mode operation, the optimal operating scenario is the equivalence between load energy requirement and photoelectric power generation, where there is no need for the converter to process energy. Figure 8 depicts the bidirectional control structure and a simplified phase of the power from the converter.

5.4. *Control of Converter Interfaced Electrolyzer*

The buck converter is a DC to DC power converter which steps down voltage from its input (source) to its output (load). For the buck converter, general control comprising voltage stabilization and output current limitation is used (Figure 9). The input of the buck converter in this system interfaces with the DC bus in the hybrid microgrid, and the output coupling with electrolyzer.

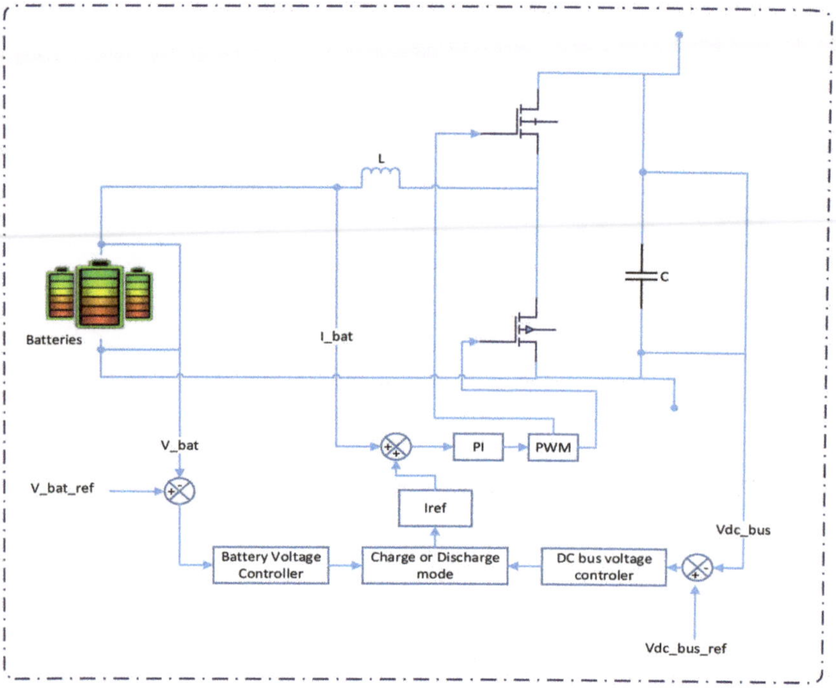

Figure 8. The control of battery storage.

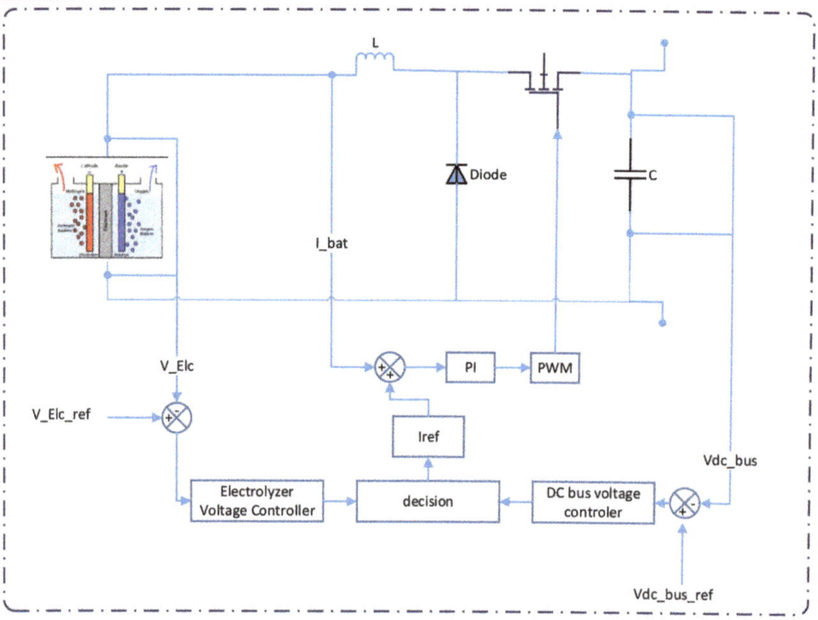

Figure 9. The control of the electrolyzer buck converter.

The control system consists of one PI controller, electrolyzer voltage controller, current controller, and DC bus voltage controller. The current control loop should be faster than the voltage control loop to minimize interaction between the two loops and therefore prevent instability. The pulse width modulation stage produces the duty cycle d that is proportional to the control voltage. The pulse width modulator makes a comparison between the control voltage and a reference waveform with peak to peak amplitude Vm. The value for Vm is selected by the designer. The frequency of the waveform corresponds to the desired converter switching frequency fs. This comparison is used to determine the switching on/off of the converter switch. The buck converter purpose, among others, is to act as an electrolyzer charge regulator during the islanded mode and grid-connected mode. Figure 9 depicts the unidirectional control structure and a simplified phase of the power from the buck converter.

5.5. Method of Hybrid Microgrid Control

The proposed microgrid has a control system implemented to stabilize the bus voltage and to regulate the power supply to serve the electrical demand during grid-connected as well as islanded mode. Table 3 lists the specifications for the design of the power management system. In such a control scheme, one unit is the master controller that controls the entire system, whereas the other units act as sources of current (slave). This case would not have a voltage differential between the outputs from the sources since the master unit is responsible for voltage regulation of all outputs. Hence, no current may circulate between the sources.

Table 3. The requirements of energy-management design.

Item	Symbols	Values
Power of fuel cell	PFC_{min}–PFC_{max}	40–50 (kW)
Depth of battery discharge	DoD_{min}–DoD_{max}	60–90 (%)
Voltage of DC bus	VDC_{min}–VDC_{max}	280–330 (V)
Nominal voltage of fuel cell	V_{nom}	250 (V)

In the microgrid, the DC bus voltage is quantified and matched against the reference voltage (300 V). Error processing is done using a compensator (PI block) to get the required reference impedance current for the loop. This compensation may be exhibited as the following Equation (17):

$$I_L = K_P\left(V_{ref} - V_{MG}\right) + K_I \int \left(V_{ref} - V_{MG}\right)dt \tag{17}$$

The current controller regulates power flow using a comparison of the impedance current flowing through the master unit to the reference value required for system stability. Error processing is conducted using another PI block to get the appropriate duty cycle for the master converter. The PI block can be depicted as Equation (18):

$$d = K_P\left(I_L^* - I_L\right) + K_I \int \left(I_L^* - I_L\right)dt \tag{18}$$

This topology presents the challenge of the master unit requirement. If a fault occurs, the control will not function correctly [30]. To enhance the reliability of the system, three individual sources may act as a master unit, thereby decreasing the chances of a fault in the microgrid control. The model studied has a voltage source converter (VSC) in the master role while the microgrid has a connection with the grid, where voltage and current loops are set up to regulate the voltage levels, so the VSC cannot change the power flow. The energy storage system uses a bidirectional converter to regulate voltage and power levels. During an islanded mode operation of the microgrid, this would be the master control where the voltage shall be maintained at 300 V while meeting the required load. Challenges include faults in the storage system or inappropriate SoC level for the system to operate in the island mode. There is a voltage control system having a voltage and current loop, as depicted in Figure 10.

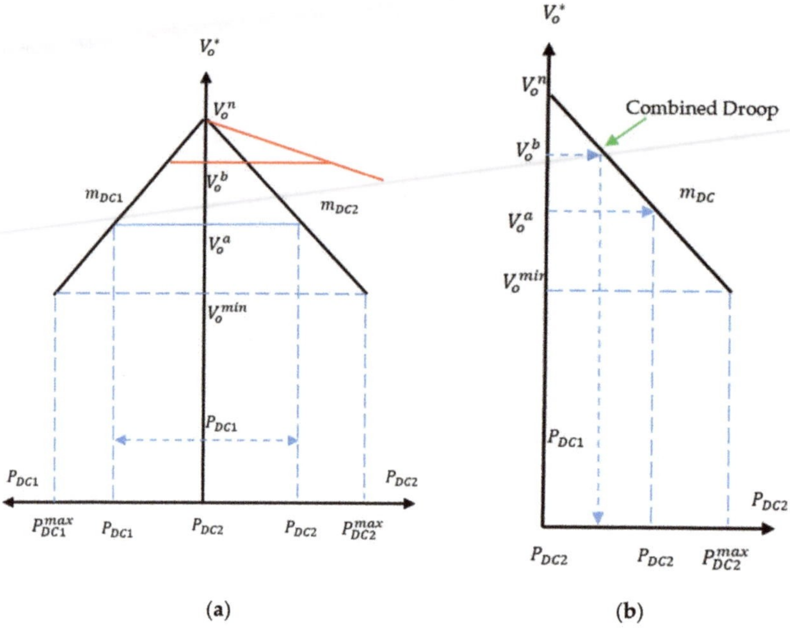

Figure 10. The droop curve: (**a**) droop curve of two distributed generators and (**b**) the droop for the entire DC subgrid.

Drooping-based power sharing is displayed in Figure 10a. Terminal voltage of the DC link for every DC–DC converter is drooped with the DC power generated (P_{dcj}), using the droop coefficient m_{dcj}, as specified below (Equation (19)):

$$V_{oj}^{*} = V_{oj}^{n} - m_{dcj}P_{dcj} \tag{19}$$

where V_{oj}^{*}, V_{oj}^{n} are the reference and no-load DC link voltage of the converter, whereas the j represents a distributed generator unit. The injected DC power from each distributed generator unit (P_{dcj}) is determined to supply the common DC load (P_{dc1}) (Equation (20)):

$$m_{dc1}P_{dc1} = m_{dc2}P_{dc2} \tag{20}$$

6. Supervisory Control Method of the Proposed System

Calculations for the energy produced using the photovoltaic generator are done using temperature and solar radiation as input data points. The photovoltaic output (P_{PV}) and the power demanded (P_L) are compared to evaluate the energy flow to the storage system and the load. The electrolyzer (P_{el}) produces hydrogen using surplus power generated by the photovoltaic system. During an energy deficit scenario, the backup fuel cell generator (P_{FC}) may supply energy using hydrogen. Figure 11 depicts the flow of in a photovoltaic-fuel-cell hybrid system control method. According to Figure 11, the control strategy is based on the following three different cases:

1. If $P_{PV} > P_L$, then $P_{el} = P_{PV} - P_L$. That is, if the radiation level is high enough, the photovoltaic powers the loads and the excess power is stored in hydrogen by the electrolyzer.
2. If $P_{PV} < P_L$ and $P_L - P_{PV} \leq P_{FC}$, then $P_L - P_{PV} = P_{FC}$. That is, if the photovoltaic cannot power the load, then the loads are connected directly to the photovoltaic and the fuel cell is switched on.
3. If $P_{PV} < P_L$ and $P_L - P_{PV} > P_{FC}$, then $P_L = P_{FC} = 0$ and $P_{el} = P_{PV}$.

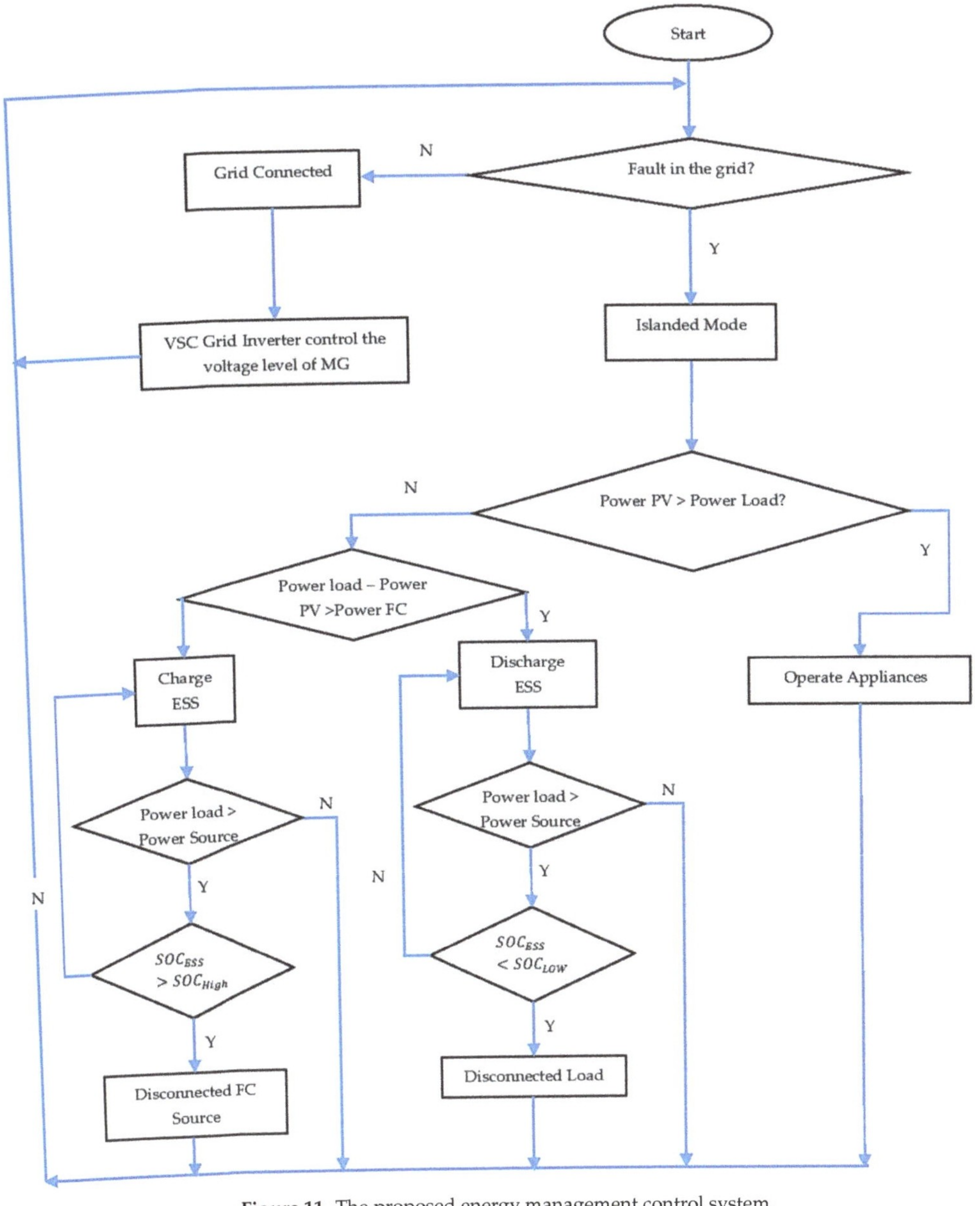

Figure 11. The proposed energy management control system.

Meaning, if the PV does not provide power for the load and the fuel cell does not start, the fuel cell and the load are separated, the electrolyzer is connected directly to the PV cells, then the load is connected according to the conditions in 1 or 2. Therefore, super capacitor units, which represent short-term volumes, provide a more stable energy response to transient load changes, or to stabilize fuel cell operation, are not considered here. Also, the electrolyzer of the proposed system is used to generate hydrogen by photovoltaic, so fuel cells can be turned on later when sufficient hydrogen is produced.

There are two levels in the higher-level controller. The first level uses individual component states to establish the operating modes. The second level is integrated with smart systems, which ascertain the way individual components behave in that operational mode. Operating modes consist of the normal mode, super capacitor charging mode, standby mode, battery charging mode, and transient mode. The standby mode has the DC–DC converter operational, while the others are disabled. When the power demand crosses a specific threshold, the controllers switch the mode from standby to transient, which sets up the console to include all controllers. A deficiency in the energy derived from the fuel cells and the battery is corrected using the super capacitor; when fuel is available, the controls switch the super capacitors to the charging mode. The super capacitor may have any magnitude of charge current until it reaches the fully charged state. Both the load and the super capacitor are powered using the fuel cells, while any shortage is supplemented using the batteries. After the super capacitor reaches 100% SoC, the controller transitions to normal load, where only the fuel cell provides power to the loads. This mode consists of the battery being disabled, while the super capacitor is still on. With a spike in load, the controller changes operation to the transient mode.

Battery charging mode may be set in two ways. The first situation, causing a switchover, is the battery state of charge falling below the minimum value. In contrast, the second situation comprises a sudden drop in load, and the batteries SoC is lesser than the maximum. Hence, with decreasing load power, the surplus flows to the battery if it is not fully charged. Splitting the complex control mechanism into several levels provides ease in modeling and controlling the individual components. Those are lower-level controllers, which include a converter controller along with the plant controller balance. The management system is examined for the proper operation of a microgrid in both grid-connected and island modes. Figure 10 depicts the general control.

In the case of grid connections, the master control is handled by the VSC inverter while the sources in the microgrid operate as slaves or sources of current. Hence, the VSC can regulate the DC voltage bus. When an unplanned event occurs, like a fault, a microgrid is required to work in the island mode and should be isolated from the grid. During such a situation, energy storage systems take master control and regulate voltage and power levels in the microgrid. Under all circumstances, the fuel cells and the photovoltaics behave as slave units. Any imbalance between the load and the power produced from the distributed generation sources is corrected by the energy management controls. Such imbalance is addressed by using battery power, given that those have an appropriate state of charge; otherwise, fuel cells are used. It must be noted that fuel cells, because of chemical reactions, respond slowly compared to an energy storage system. If the storage system is malfunctioning or has an unacceptable charge level, the fuel cell becomes the master. In case the microgrid controls are unable to balance the energy flow, and the load exceeds the power generation, the measure of last resort is load rejection. In case there is a power surplus in the grid, the DG (Distributed Generator) systems are disconnected from the microgrid, as required.

7. Energy Management Systems in Houses

In the case of houses, energy management systems incorporate intelligence to decrease energy consumption, enhance uptime, and provide added safety. Control and modeling framework for fuel cells and photovoltaics is designed. This technique uses primary performance attributes like super capacitor voltage, SoC, and DC bus voltage and subject to a proportional integral (PI). Proportional-integral control is straightforward to tune online for improved tracking. The load power is divided so as to permit the fuel cells to provide the steady load. An integrated system comprising these power sources may be considered by adding the current values where the system connects with the microgrid. The real power traded between the hybrid microgrid and the utility network $P_{grid}(t)$ is addition of power generated by the fuel cells ($P_{FC}(t)$), batteries ($P_{Bat}(t)$), and the photovoltaic system (P_{PV}), and the electrical load (Figure 1) (Equation (21).

$$P_{grid}(t) = P_{FC}(t) + P_{PV}(t) + P_{Bat}(t) + P_{SCB}(t) - P_{Load}(t) \tag{21}$$

During extended periods, rapid power variations in regard to the super capacitor $P_{SCB}(t)$ may be disregarded.

Equation (22) is specified for an extended time frame:

$$\{P_{grid}\}T = \{P_{FC} + P_{PV} + P_{Bat}(t) - P_{Load}\}T \quad (22)$$

For such a short duration, the super capacitor, due to its rapid response time, takes control of power flow. Using the inverse of Equation (21), reference power for the super capacitor $P_{SCB}(t)$ may be expressed as (Equation (23)):

$$P_{SCB}(t) = P_{grid}(t) - P_{FC}(t) - P_{PV}(t) - P_{Bat}(t) + P_{Load}(t) \quad (23)$$

Super capacitors have a very swift charge/discharge response, and, therefore, they are employed in this study for voltage control on the DC bus for the hybrid microgrid. The following constraints should be factored in when determining the energy storage capacity of super capacitors (Equation (24)):

$$E_{SC\_min} \leq E_{SC}(t) \leq E_{SC\_max} \quad (24)$$

where E_{SC_max} and E_{SC_min} are, respectively, the maximum and minimum permissible storage capacity for the supercapacitor. E_{SC_min} may be calculated as per Equation (25):

$$E_{SC\_min} = SOC_{SC} \times E_{SC\_max} \quad (25)$$

Therefore, the larger the power drawn from the super capacitor, the stored energy reduces likewise (Equations (26) and (27)).

$$E_{SC} = \int \Delta P_{FC}(t)dt \quad (26)$$

$$\Delta P_{FC}(t) = K_{pe}(e_{sc_{ref}}(t) - \hat{e}_{sc}(t)) \quad (27)$$

Battery storage capacity is subject to the following constraints (Equations (28)–(35)):

$$E_{Batt\_max} \geq E_{Batt}(t) \geq E_{Batt\_min} \quad (28)$$

$$0 \leq P_{bat,ch}(t) \leq Y_{ch}(t) \cdot P_{bat,cap} \cdot (1 - SoC(t-1)) \quad (29)$$

$$0 \leq P_{bat,disch}(t) \leq Y_{bat,disch}(t) \cdot P_{bat,cap} \cdot SoC(t-1) \quad (30)$$

$$Y_{bat,ch}(t) + Y_{bat,disch}(t) \leq 1, Y_{bat,ch}(t), Y_{bat,disch}(t) \ \{0,1\} \quad (31)$$

$$SoC(t) = SoC(t-1) - \frac{1}{P_{bat,cap}} \times \left(P_{bat,disch}(t) - P_{bat,ch}(t)\right) \quad (32)$$

$$0 \leq SoC(t) \leq 1 \quad (33)$$

$$SoC(t_o) = SoC_{initial} \quad (34)$$

$$SoC(t_{end}) = SoC_{final} \quad (35)$$

Equations (29) and (30) stand for the power capacity constraints of the battery for charging and discharging status. Equation (31) keeps the simultaneous occurrence of discharging and charging of the battery. Finally, Equations (32)–(35) are related to *SoC* or the level of the stored energy in the battery. In the above equations, $P_{bat,ch}$, $P_{bat,disch}$, and $P_{bat,cap}$ stand in the charging power, discharging power, and the maximum available power from the battery, respectively. Also, $Y_{bat,ch}$ and $Y_{bat,disch}$ are ancillary binary variables indicating the charging or discharging status of the battery.

House-based microgrids may trade power with the utility grid. Any energy surplus produced by the microgrid after the batteries are fully charged could be sold to the grid. If the total production of the microgrid is insufficient to meet the load, power should be bought from the utility grid (Equation (36)).

$$P_{FC}(t) = P_{grid}(t) - \hat{P}_{PV}(t) + \hat{P}_{Load}(t) + \Delta P_{FC}(t) \tag{36}$$

Energy management and power control systems are very important for hybrid networks based on intermittent sources. The photoelectric control provides reference voltage (V_MPP) for the unidirectional converter. With the converter controlling the PV output, PV works at the voltage that uses the maximum power. Super capacitor discharge and charge circuit control maintains a DC voltage with a constant value of 270 V. Even with DC voltage ripples, controlling the transformer at the end of the load makes the load voltage as smooth as possible. Fuel cells can instantly provide maximum energy. The boost transformer sets the performance limit. Low capacitance and induction values accelerate transients and accelerate overall response, but require faster switches.

8. Results of Proposed System

Hybrid systems are corroborated by applying a thorough model in the environment of Simulink. This model represents a different emergency power system which is based on super capacitors, fuel cells, and Li-ion batteries. The proposed model also includes different power handling systems for fuel cell fusion power supplies. The findings are attained based on the standard profile of daily load for scrutinized homes, as displayed in Figure 12. Table 4 presents the details of the power system under study.

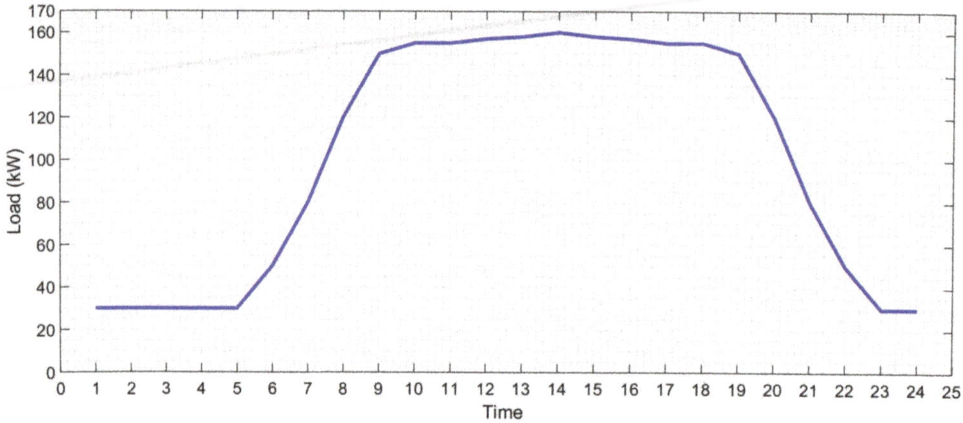

Figure 12. The curve of demand.

Table 4. The parameters of the proposed system.

Item	Description
Photovoltaics	100 kW
Proton exchange membrane (PEM) fuel cell	50 kW (peak)
Capacity of battery	4800 (Ah)
Voltage of battery	96 (V)
Initial SoC of battery	65 (%)
Voltage of the fully charged of battery	110 (V)
Current of battery discharge	400 (A)
Super capacitor	291 V, 15.6 F, (six 48.6 V cells in series)

The 25 kV utility grid is connected to houses via a "25 KV/270 V line to line (V_rms)" transformer to step down voltage to the required grid voltage, which means the output of the transformer equal to 380 V line to line peak voltage ($V_{Peak\ (L-L)}$), meaning the transformer output voltage is equal 220 V line to ground.

$$V_{Peak\ (L-L)} = \sqrt{2} \times V_{rms\ (L-L)} = \sqrt{2} \times 270 = 381.83766\ \text{V} \tag{37}$$

$$V_{Peak\ (L-N)} = \frac{V_{Peak\ (L-L)}}{\sqrt{3}} = \frac{381.837}{\sqrt{3}} = 220\ \text{V} \tag{38}$$

where $V_{Peak\ (L-L)}$ is the peak line to line voltage, $V_{rms\ (L-L)}$ is the RMS (Root Mean Square) line to line voltage, $V_{Peak\ (L-N)}$ is the peak line to neutral voltage.

The three-phase AC loads with variable power factor and apparent power are used to simulate a residential load profile. The energy management system distributes energy between sources according to the given energy management strategies. The peak electrical load is 160 kW at 3 p.m., while the lowest load occurs at 29 kW from 10:30 p.m. to 8:30 a.m. Figure 13 shows the photoelectric voltage and photoelectric current. Photovoltaic power generation is set to track the maximum power point proportional to radiation. The meteorological data for the hybrid power generation system of photovoltaic/fuel cells are the amount of solar radiation (W/m^2) and ambient temperature (°C). These typical weather data are collected every hour. The microgrid reference generated is set to one real energy step (25 s, Figure 13) and the energy detected between the microgrid and the main line is very close to the power reference. Super capacitors and batteries compensate for the differences in the reference power of the microgrid and all the differences of the negative energy of the microgrid (total load and photoelectric cells).

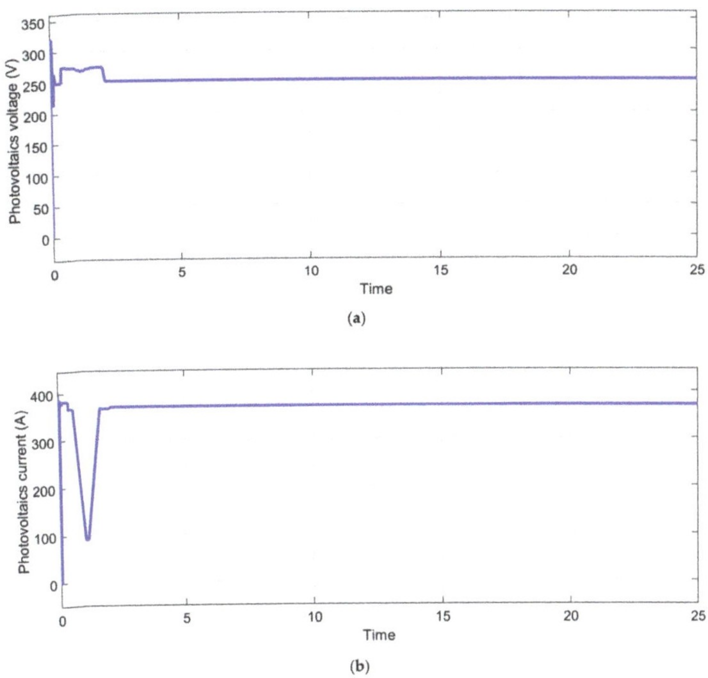

Figure 13. (**a**) Photovoltaics voltage and (**b**) photovoltaics current.

It can be seen that the fast power fluctuation due to the slow response time of the fuel cell is reduced in the short-term energy storage system, as shown in Figure 14. The fuel cell will try to charge the super capacitor to increase the SoC if the load is not too high. Figure 15 shows the fuel cell voltage and fuel cell current. Note that a positive super capacitor current means that the super capacitor will supply the load, and a negative current means that the super capacitor will be charged. In this case, the carrier voltage fluctuates in the permissible range from 226 to 270 V. The super capacitor voltage is properly controlled between 226 and 270 V, as shown in Figure 16c. SoC range for super capacitors is 85–100%. Figure 17 shows the load voltage and load current. The load current increases in 8 s and decreases in 20 s. As shown in Figure 18, the voltage decreases due to the sudden increase in the load current.

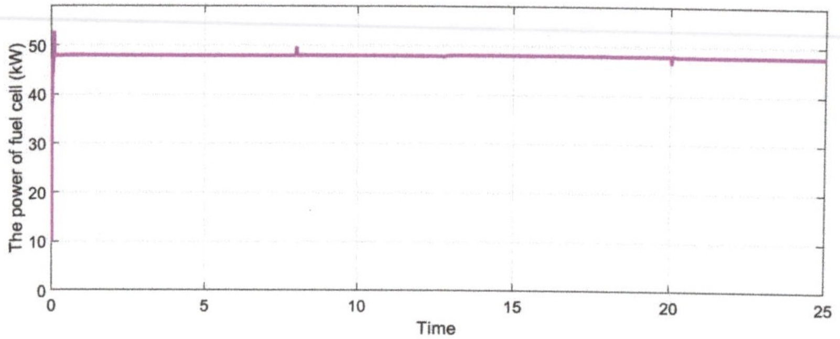

(a)

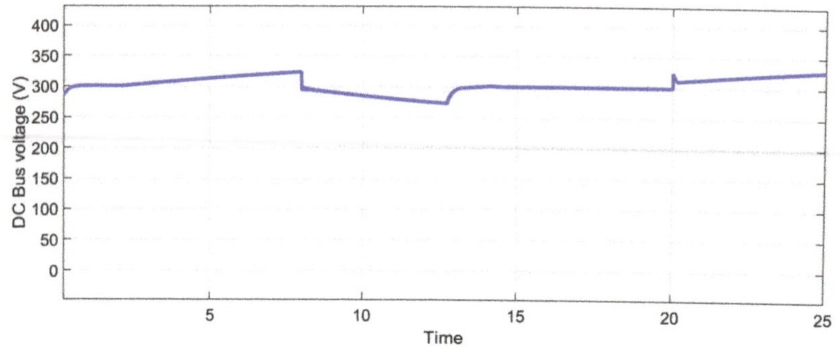

(b)

Figure 14. (**a**) Fuel cell power and (**b**) DC bus voltage.

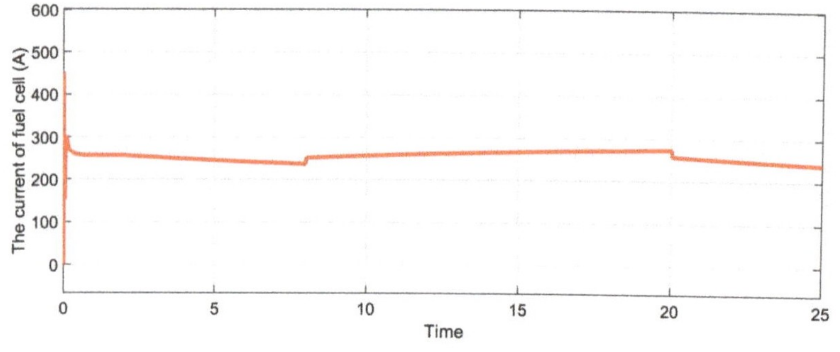

(a)

Figure 15. *Cont.*

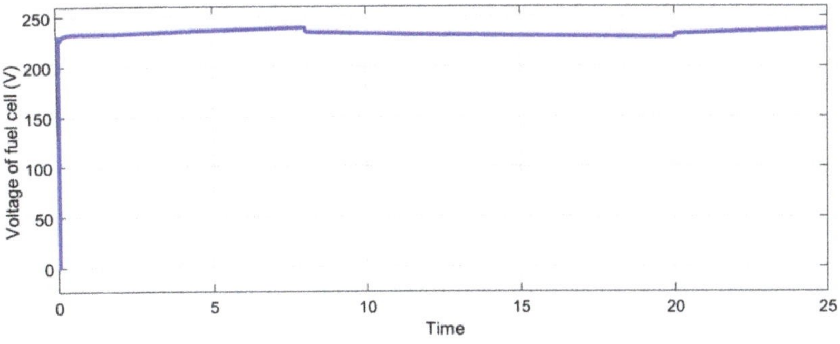

(b)

Figure 15. (**a**) Fuel cells current and (**b**) fuel cells voltage.

In this study, the fuel cell and photovoltaic hybrid power generation structure is either the island or grid-connected mode. The load demand regions are determined by a preset power. Nonetheless, most resident energy consumption of power, like lighting and apparatus, is always varying. Based on the scheme of energy management, the energy regulation system manages the power of every power supply apparatus through the reference signals of the photovoltaic and fuel cell converter, together with super capacitors, batteries, and fuel cells. A description of what happens in a virtual emergency situation follows. At t = 0 s, the primary load is provided by the chief generator, which switches on the fuel cell hybrid system, to plan for an emergency scenario. Now, the fuel cell starts recharging the battery as well as the super capacitor using its optimal power. At time t = 8 s, every load is connected and the loads soar to 160 kW. At this moment, the extra power needed is instantly provided by the super capacitor because of its fast dynamics, although the power in the fuel cell increases gradually. At time t = 13 s, the super capacitor gets discharged and its DC bus voltage goes below the required amount (270 V) and the battery starts supplying power to control the DC bus voltage to 270 V. At time t = 20 s, the DC bus or super capacitor voltage gets to 270 V and gradually the power of the battery drops down to zero. The fuel cell supplies complete load power and keep on recharging the super capacitor. At t = 13 s time, the battery also gets to its highest power and the super capacitor supplies additional load power. At time t = 20 s, the load power reduces below the fuel cell maximum power. Due to the slow fuel cell dynamics, the extra fuel cell power during transients is transferred to the supercapacitor. Owing to the fuel cells' slow dynamics, power during transients is shifted towards the super capacitor. At time t = 20 s, the power of the load is reduced and falls below the fuel cell highest power, and the additional fuel cell power is shifted towards both the super capacitor and battery. At time t = 23 s, the load power gets reduced rapidly. The additional fuel cell energy gets stored in the batteries and super capacitor. The fuel cell provides nearly all the total load power needed and decreases its power gradually to its optimum and the battery gets recharged.

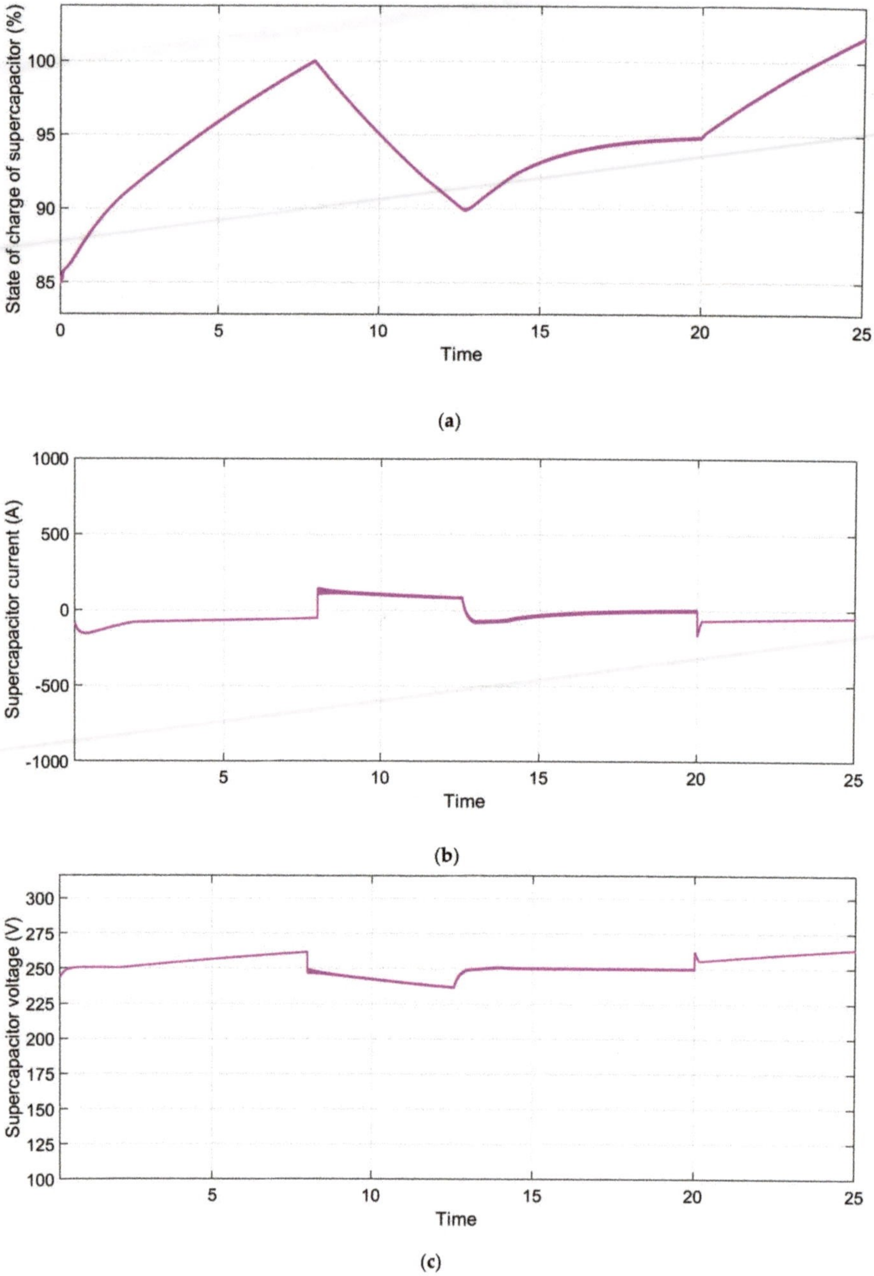

Figure 16. (**a**) State of charge of the super capacitor, (**b**) super capacitor current, (**c**) super capacitor voltage.

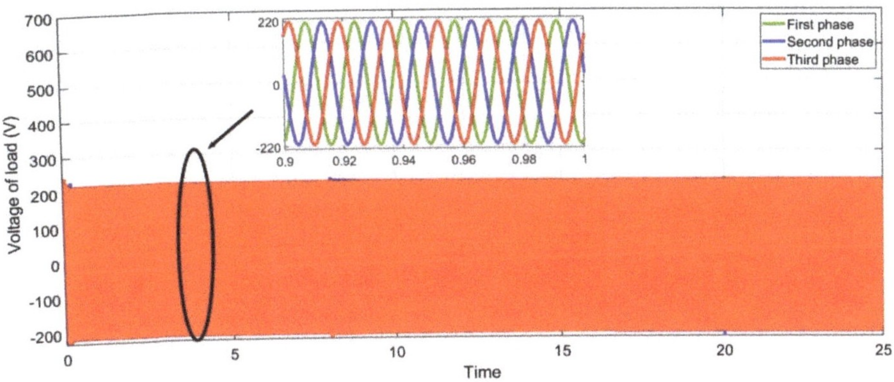

(**a**)

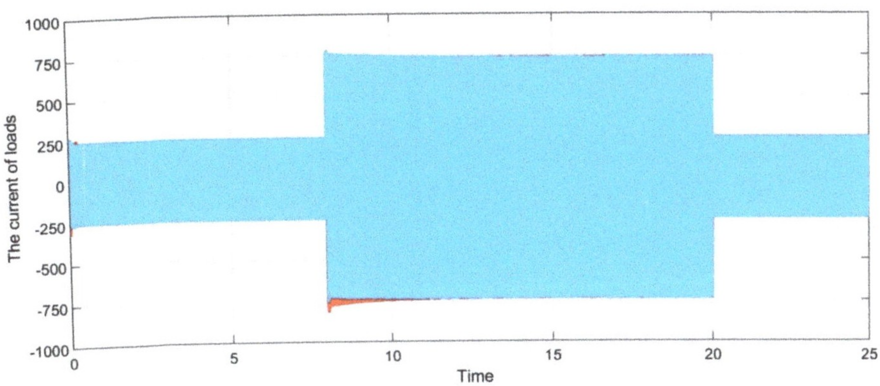

(**b**)

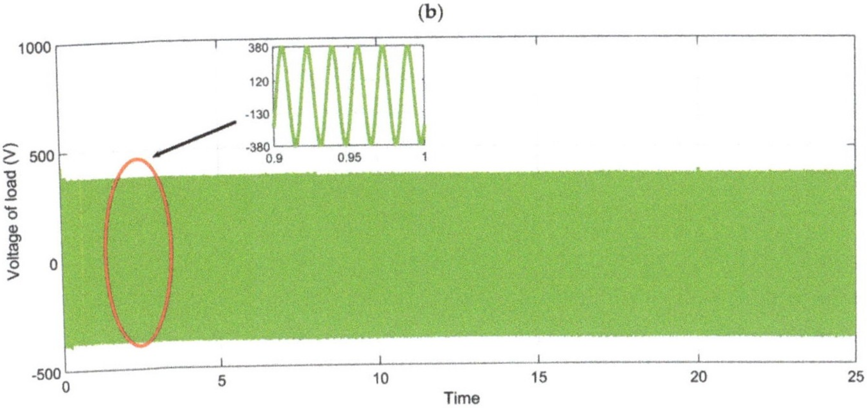

(**c**)

Figure 17. (**a**) Line to neutral loads voltage, (**b**) loads current, (**c**) line to line load voltage.

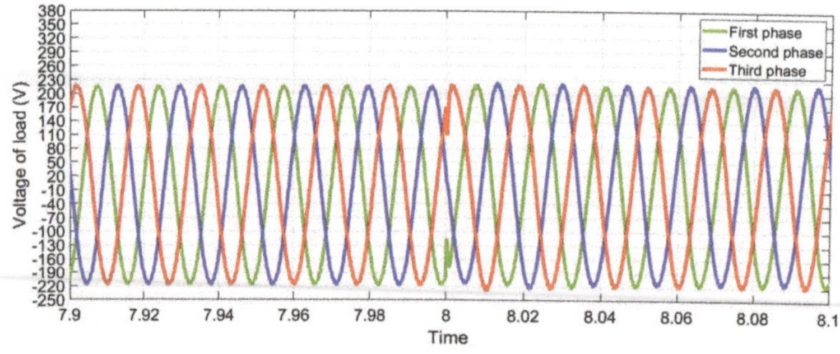

(a)

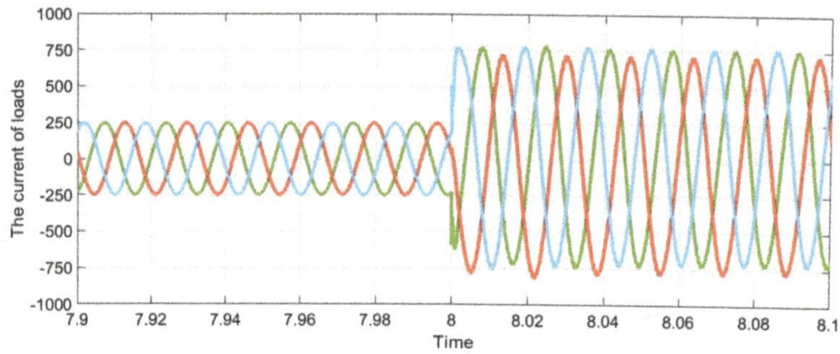

(b)

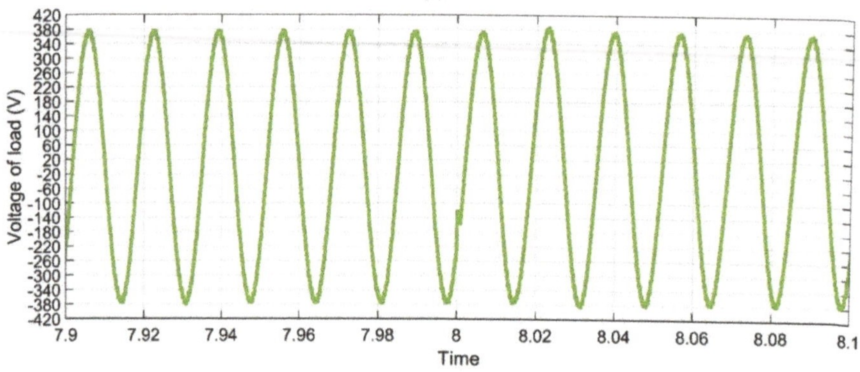

(c)

Figure 18. (**a**) Step change in load line to neutral voltage at 8 a.m., (**b**) step change in load current at 8 a.m., (**c**) load line to line voltage at 8 a.m.

9. Conclusions

In this study, the new home management system demonstrates the effectiveness of this method to deliver constant power to the customer through different controller designs. Moreover, this paper recommends an effective control scheme for the effortless shift from grid-connected mode to island mode because of unintentional islanding. Furthermore, a technique for smart management and regulation was employed to ensure optimum functioning of a hybrid microgrid formed using different renewable energy sources, as well as different energy storage devices. Using this energy management strategy as component of the control design of a grid-linked microgrid can reduce the total operational cost. The regulatory control was included in the control model to correct any variation among the primary grid power and the intended reference power by altering the reference fixed power of the battery provided by this energy management system. This study put forward a full local regulatory design for the recommended system. The local controller regulates the local power, DC bus voltage, and the current of the microgrid where the local controller's aim is to follow the fixed point of the monitoring controller. The outcomes show that the suggested robust energy management and regulation scheme in the hybrid microgrid system, which is based on ecofriendly energy, supports nominal use of the utility grid power. The proposed scheme is able to distribute power amongst the distributed generator components even under uneven conditions. Moreover, from the outcomes, it is apparent that the output voltage of the unit terminals of the distributed generator gets higher when the unit of fuel cell is attached along with the super capacitor and the photovoltaic. The loads of the houses for the hybrid microgrid were taken into account and the hybrid microgrid can swap power with the transmission grid. The hybrid microgrid used in this research is highly reliable.

The outcomes suggested that the alternating integration of solar energy sources and fuel cell in the microgrid should be devised carefully in individual operation. The recommended control schemes provide exceptional performance under various operating conditions. Batteries improve the system reliability since they store additional renewable energy when the demand is low and supply energy when the demand is high.

For future work, these results suggest (1) implementing the strategy presented on a real network and comparing the results, and (2) investigating the effects of other uncertain parameters, such as fuel cost, on microgrid planning.

Author Contributions: B.N.A.: writing—original draft, software, methodology, and validation; M.D.E.: formal analysis, investigation, resources, writing—review and editing; B.H.J.: supervision, funding, writing—review and editing. All authors have read and agreed to the published version of the manuscript.

Funding: This research received no external funding.

Conflicts of Interest: The authors declare no conflict of interest.

References

1. Justo, J.J.; Mwasilu, F.; Lee, J.; Jung, J.W. AC-microgrids versus DC-microgrids with distributed energy resources: A review. *Renew. Sustain. Energy Rev.* **2013**, *24*, 387–405. [CrossRef]
2. Basak, P.; Chowdhury, S.; Nee Dey, S.H.; Chowdhury, S.P. A literature review on integration of distributed energy resources in the perspective of control, protection and stability of microgrid. *Renew. Sustain. Energy Rev.* **2012**, *16*, 554–556. [CrossRef]
3. Moradi, M.H.; Hajinazari, M.; Jamasb, S.; Paripour, M. An energy management system (EMS) strategy for combined heat and power (CHP) systems based on a hybrid optimization method employing fuzzy programming. *Energy* **2014**, *86*, 86–101. [CrossRef]
4. Ma, X.; Wang, Y.; Qin, J. Generic Model of a Community-Based Microgrid Integrating Wind Turbines, Photovoltaic, and CHP Generations. *Appl. Energy* **2013**, *112*, 147–582. [CrossRef]
5. Alonso, M.; Amaris, H.; Alvarez-Ortega, C. Integration of Renewable Energy Sources in Smart Grids by Means of Evolutionary Optimization Algorithms. *J. Expert Syst. Appl.* **2015**, *39*, 22–55. [CrossRef]
6. Zhang, Y.; Gatsis, N.; Giannakis, B. Robust Energy Management for Microgrid with High-Penetration Renewables. *IEEE Trans. Sustain. Energy* **2015**, *4*, 45–53. [CrossRef]
7. Lujano-Rojas, J.M.; Monteiro, C.; Dufo-López, R.; Bernal-Agustín, J.L. Optimum Load Management Strategy for Wind/Diesel/Battery Hybrid Power Systems. *J. Renew. Energy* **2015**, *44*, 288–295. [CrossRef]
8. Delgado, C.; Navarro, D. Optimal Design of a Hybrid Renewable Energy System. In Proceedings of the 2014 Ninth International Conference on Ecological Vehicles and Renewable Energies (EVER), Monte-Carlo, Monaco, 25–27 March 2014. [CrossRef]
9. Helal, S.; Najee, J.; Hanna, M.; Shaaban, F.; Osman, A.; Hassan, S. An Energy Management System for Hybrid Microgrids in Remote Communities. *Can. Conf. Electr. Comput. Eng.* **2019**, 1275–1286. [CrossRef]

10. Correa, C.; Marulanda, G.; Garces, A. Optimal microgrid management in the colombian energy market with demand response and energy storage. In Proceedings of the 2016 IEEE Power and Energy Society General Meeting (PESGM), Boston, MA, USA, 17–21 July 2016; pp. 145–156. [CrossRef]
11. Dufo, R.; Agustín, B. Optimization of Control Strategies for Stand-Alone Renewable Energy Systems with Hydrogen Storage. *Renew. Energy* **2007**, *34*, 124–134. [CrossRef]
12. Das, B.K.; Al-Abdeli, Y.M.; Kothapalli, G. Effect of Load Following Strategies, Hardware, and Thermal Load Distribution on Stand-Alone Hybrid CCHP Systems. *Appl. Energy* **2018**, *4*, 34–45. [CrossRef]
13. Luna, A.; Meng, L.; Diaz, N.; Graells, M.; Vasquez, J.; Guerrero, J. Online Energy Management Systems for Microgrids: Experimental Validation and Assessment Framework. *IEEE Trans. Power Electron.* **2018**, *33*, 134–146. [CrossRef]
14. Abedini, M.; Moradi, H.; Hosseinian, S. Optimal management of microgrids including renewable energy sources using GPSO-GM algorithm. *Renew. Energy* **2016**, *56*, 45–55. [CrossRef]
15. Marzband, M.; Azarinejadian, F.; Savaghebi, M.; Guerrero, J. An Optimal Energy Management System for Islanded Microgrids Based on Multipored Artificial Bee Colony Combined with Markov Chain. *IEEE Trans. Sustain. Energy* **2017**, *45*, 67–87. [CrossRef]
16. Rouholamini, M.; Mohammadian, M. Heuristic-Based Power Management of a Grid-Connected Hybrid Energy System Combined with Hydrogen Storage. *Renew. Energy* **2016**, *89*, 12–24. [CrossRef]
17. Almada, B.; Leão, R.; Sampaio, F.; Barroso, G. A Centralized and Heuristic Approach for Energy Management of an AC Microgrid. *Renew. Sustain. Energy Rev.* **2016**, *45*, 67–87. [CrossRef]
18. Merabet, A.; Tawfique, A.; Ibrahim, K.; Beguenane, R.; Ghias, A. Energy Management and Control System for Laboratory Scale Microgrid Based Wind-PV-Battery. *IEEE Trans. Sustain. Energy* **2017**, *8*, 217–224. [CrossRef]
19. Farzin, H.; Firuzabad, F.; Moeini-Aghtaie, M. Stochastic Energy Management of Microgrids during Unscheduled Islanding Period. *IEEE Trans. Ind. Inform.* **2017**, *13*, 1079–1087. [CrossRef]
20. Battistelli, C.; Agalgaonkar, Y.; Pal, B. Probabilistic dispatch of remote hybrid microgrids including battery storage and load management. *IEEE Trans. Smart Grid* **2017**, *8*, 1305–1317. [CrossRef]
21. Hussaian Basha, C.; Rani, C.; Brisilla, R.M.; Odofin, S. Mathematical Design and Analysis of Photovoltaic Cell Using MATLAB/Simulink. Soft Computing for Problem Solving. *Adv. Intel. Syst.* **2019**, *1048*, 711–726. [CrossRef]
22. Sangwongwanich, A.; Blaabjerg, F. Mitigation of Interharmonics in PV Systems with Maximum Power Point Tracking Modification. *IEEE Trans. Power Elect.* **2019**, *34*, 8279–8282. [CrossRef]
23. Rakhshan, M.; Vafamand, N.; Khooban, M.H.; Blaabjerg, F. Maximum Power Point Tracking Control of Photovoltaic Systems: A Polynomial Fuzzy Model-Based Approach. *IEEE J. Emerg. Sel. Top. Power Electron.* **2018**, *6*, 292–299. [CrossRef]
24. Subramaniyan, A.B.; Pan, R.; Kuitche, J.; TamizhMani, G. Quantification of En-vironmental Effects on PV Module Degradation: A Physics-Based Data-Driven Modeling Method. *IEEE J. Photovolt.* **2018**, *8*, 1289–1296. [CrossRef]
25. Bilal Naji, A.; Basil, H.J.; Anvari-Moghaddam, A.; Blaabjerg, F. A New Robust Control Strategy for Parallel Operated Inverters in Green Energy Applications. *Energies* **2020**, *13*, 3480. [CrossRef]
26. Fathima, A.H.; Palanisamy, K. Optimization in Microgrids with Hybrid Energy Systems—A Review. *Renew. Sustain. Energy Rev.* **2015**, *45*, 431–446. [CrossRef]
27. Mao, L.; Jackson, L.; Davies, B. Effectiveness of a novel sensor selection algorithm in PEM fuel cell on-line diagnosis. *IEEE Trans. Ind. Electron* **2018**, *65*, 7301–7310. [CrossRef]
28. Ghenai, C.; Salameh, T.; Merabet, A. Technical economic analysis of off-grid solar PV/fuel cell energy system for residential community in desert region. *Int. J. Hydrog Energy* **2018**, *6*, 97–105. [CrossRef]
29. Ghenai, C.; Bettayeb, M. Optimized design and control of an off-grid solar PV/hydrogen fuel cell power system for green buildings. *IOP Conf. Ser. Earth Environ.* **2017**, *93*, 1–11. [CrossRef]
30. García, P.; Torreglosa, J.; Fernández, L.; Jurado, F. Control Strategies for High-Power Electric Vehicles Powered by Hydrogen Fuel Cell, Battery and Supercapacitor. *Expert Syst. Appl.* **2016**, *40*, 4791–4804. [CrossRef]

sustainability

MDPI

Article

A Residential Load Scheduling with the Integration of On-Site PV and Energy Storage Systems in Micro-Grid

Ihsan Ullah [1], Muhammad Babar Rasheed [2,*], Thamer Alquthami [3,*] and Shahzadi Tayyaba [1]

1 Department of Computer Engineering, The University of Lahore, Lahore 54000, Pakistan; ciit.ihsan@gmail.com (I.U.); shahzadi.tayyaba@hotmail.com (S.T.)
2 Department of Electronics and Electrical Systems, The University of Lahore, Lahore 54000, Pakistan
3 Electrical and Computer Engineering Department, King Abdulaziz University, Jeddah 21589, Saudi Arabia
* Correspondence: babarmeher@gmail.com (M.B.R.); tquthami@kau.edu.sa (T.A.)

Received: 24 October 2019; Accepted: 19 December 2019; Published: 25 December 2019

Abstract: The smart grid (SG) has emerged as a key enabling technology facilitating the integration of variable energy resources with the objective of load management and reduced carbon-dioxide (CO_2) emissions. However, dynamic load consumption trends and inherent intermittent nature of renewable generations may cause uncertainty in active resource management. Eventually, these uncertainties pose serious challenges to the energy management system. To address these challenges, this work establishes an efficient load scheduling scheme by jointly considering an on-site photo-voltaic (PV) system and an energy storage system (ESS). An optimum PV-site matching technique was used to optimally select the highest capacity and lowest cost PV module. Furthermore, the best-fit of PV array in regard with load is anticipated using least square method (LSM). Initially, the mathematical models of PV energy generation, consumption and ESS are presented along with load categorization through Zero and Finite shift methods. Then, the final problem is formulated as a multiobjective optimization problem which is solved by using the proposed Dijkstra algorithm (DA). The proposed algorithm quantifies day-ahead electricity market consumption cost, used energy mixes, curtailed load, and grid imbalances. However, to further analyse and compare the performance of proposed model, the results of the proposed algorithm are compared with the genetic algorithm (GA), binary particle swarm optimization (BPSO), and optimal pattern recognition algorithm (OPRA), respectively. Simulation results show that DA achieved 51.72% cost reduction when grid and renewable sources are used. Similarly, DA outperforms other algorithms in terms of maximum peak to average ratio (PAR) reduction, which is 10.22%.

Keywords: HEM; PV sizing; Load scheduling; Dijkstra Algorithm; BPSO; GA; optimization

1. Introduction and Motivation

With the rapid increase in population and economic growth, global energy consumption is also increasing drastically. High reliance on fossil fuel and increased use of thermal power resources contributed to an increase in CO_2 emissions that causes an increase in global warming [1]. While adopting the ways to tackle the increased energy demand, a reduction in greenhouse gas emissions is a major challenge that must be met on a global scale in order to promote energy sustainability. In recent years, while reducing the gap between energy generation and demand through a combination of thermal and nuclear energy resources, the pursuit of power generation through renewable energy sources, such as solar, wind, biomass and lowering environmental load, has been promoted with the aim of reducing greenhouse gas emissions, significantly. Renewable energy is becoming widely accepted as a new source of energy to provide power to residential and commercial buildings around the world. On the other hand, energy is one of the basic necessities for sustainable development of society that must surely be delivered in an efficient way in order to promote a green environment with reduced CO_2 emissions [1]. However, the great paradigm shift due to the involvement of information and communication technologies (ICTs) the traditional ways of energy generation, transmission and distribution need to be upgraded [2–8]. The author in [9] presented a hybrid opinion network containing of continuous and discrete valued agents. This model discussed a communication behavior for social dynamical systems. The scaled consensus of switching topologies with continuous and discrete time subsystems is investigated. The traditional electricity infrastructure is not fully capable of handling and managing the distributed energy resources with high quality of communication service and grid stability. This is due to the fact that increasing energy demand needs extra costlier generation or back-up reserve capacity units that lead to high tariff rates and CO_2 emissions.

Therefore, to cope with such types of situations, a smart grid (SG) concept has recently been introduced which has the capability to fulfil the load demand that benefices both the end users and utility market [10]. The users can enjoy an uninterruptible power supply with economical tariff rate and utility can avail the opportunity to improve power system stability through supply–demand balance. The former can be achieved by exploiting efficient load management through distributed as well centralized control strategies [11]. For this purpose, different researchers introduced residential and commercial load scheduling techniques considering both end user and utility objectives [12,13]. Among these solutions, the major emphasis is given to balance the load demand through market supply without heavily relying on peak power plants and reserved capacity units [14]. However, some researchers focussed on balancing the load demand through load scheduling and energy management using optimization technique. Although these kinds of techniques are efficient in managing the load demand with partial or full control on residential load [15]. The peaks and troughs in the aggregated user energy demand profile are caused by the temporal variations in energy demand. The user's participation is required in balancing the load by allowing the grid to reschedule the load. However, the end users are affected due to unplanned load scheduling by the grid [16]. For example, it is assumed that the user suffers the same degree of inconvenience if base and/or non-flexible load is shifted in either direction (before or after) of its most preferred time slot. In other words, delaying a base and/or non-flexible load by 2 time slots or advance scheduling the same load by 2 time slots will result in same amount of user inconvenience. To handle this situation, some researchers put their efforts into devising the load control strategies with major focus on load scheduling without disturbing end users' objectives [17]. Furthermore, it is also equally important to understand that world is moving from a centralized energy control system to a distributed control and management system with the objective of promoting the environment [18]. In addition, this is also due to the European vision of the year 2030 and 2050 for achieving the objective of green energy and green economy, incorporation of renewable energy resources such as PV, wind, geothermal, biomass and electric vehicles into existing electrical systems is required [19]. In achieving these objectives, numerous researchers have done some work in managing the integration of distributed energy resources through autonomous control strategies [20]. Some researchers focussed on developing the autonomous control algorithms that provide benefits to only end users [21], while others focussed on maximizing utility and energy retailers benefits [22]. Some work also tackled the utility and end user objectives as a joint objective and tried to find the solutions to provide the benefits to both parties [23].

In recent years, the resources of water for producing energy are depleting very quickly due to the effects of climate change on hydrology patterns [24,25]. It is estimated that only 200,000 km^3—1% of water is available for agriculture, residential, industrial and power sectors in order to meet everyday needs [26]. Furthermore, the per capita energy demand is expected to rise significantly due to the rapid growth of population and industries in developed and developing countries [27]. As a result of these great challenges, the efficient management of the available capacity is imperative to prevent the over-exploitation [28]. On one hand, these important challenges are considered for managing the energy resources efficiently, whether these are used used for the end user comfort or upgrading the existing standards or infrastructure [29]. On the other hand, better management of fuel intensive resources and end user consumption behaviours can alleviate the strain on these commodities [30]. Therefore, flexible control of the energy supply and use system [19,31] is crucial within the context of renewable energy and storage integration [32]. In addition, a recent study shows the water usage for carbon-based and nuclear fuel-based energy as follows: (i) oil from oil reservoirs (70–1800 L/GJ), (ii) shale gas (36–54 L/GJ), coal (5–70 L/GJ), (iii) uranium (4–22 L/GJ), and (iv) traditional oil (3–7 L/GJ) [31]. Another study reveals that 76.9% of electricity is produced from oil, natural gas, coal, and nuclear fuels in the year 2015, while only 16% is generated from hydrothermal power plants. The energy generated from hydrothermal power plants led to high emission of carbon and causes serious environment concerns [33]. Similarly, a significant amount of power is required to support industrial and residential sectors and with this level of coupling, significant synergy could be observed while analysing these types of coupled system holistically. In the meantime, the SG concept was introduced and gained significant attention is given to the integration of renewable energy resources, especially into the demand side load management as a mean of de-carbonizing the environment. The importance of producing energy from coal, natural gas, oil and nuclear commodities is decreasing day by day, thus results in moving towards integration of green energy concept. The integration of green energy (solar and wind) and storage systems are at present gaining popularity [34]. Recent studies regarding variable energy resources and energy management through electrical load scheduling show that a dynamic control mechanism is required that integrates the intermittent nature of variable energy resources and dynamic energy consumption trends. It is a complex task, however, studies showed that due to high penetration and intermittent nature of variable resources, operator and energy retailers are restricted to rely on the combined dispatch of all energy resources. Operators need to encourage the end users to install stand-alone renewable energy systems to improve the system stability through load management. Otherwise, the mismatch between demand and supply may lead to infeasible dispatch of energy and increase the marginal cost production [35–37], while the challenges of utility and residential premises in managing the residential load demand with the integration of renewable energy resources may seem unrelated. However, these are actually interlinked entities and their resolution is potentially synergistic. Renewable energy technologies are a cheap source of energy and provide with low CO_2 emissions. Since, renewable energy can easily be stored in back-up storage systems and can act as a flexible source of energy regarding supply and demand side of electricity systems [38]. Thus, demand response (DR) programs have an eminent effect on SG infrastructure in managing the end user load via market clearing prices and incentives. Because, residential sector is consuming 40% of energy and diversity in energy consumption plans poses major challenge to energy retailers to encourage them in participating energy management programs. The effective outcomes of SG concept are difficult to achieve without active participation of end users in DR-based energy management programs.

The review of relevant literature shows that homes equipped with Home energy management (HEM) architectures significantly reduce the electricity cost. HEMs equipped with renewable energy generation and storage architecture are given importance but the load matching with the renewable energy generation is ignored. Several HEM architectures efficiently shift the load from grid to an on-site energy generation and storage system, but the design of PV-based energy generation and storage setup has not yet been reported. An efficient and autonomous HEM architecture is required with twofold objectives, i.e., benefices the end user and utility. With the end user perspective, HEM needs to incorporate the on-site PV generation and storage system that optimally uses the grid and stored energy to maximize the comfort level. With the utility perspective, HEM should limit the energy consumption to a certain threshold level to avoid the load shedding, blackouts and use of peak power plants. Thus, our objectives are to design a demand side energy management (DSEM) model that (i) Categorizes, monitors and controls the household different loads according to end user preference and priority, (ii) Maximizes the end user comfort level by scheduling the load with minimum delay, (iii) Anticipates the optimum number of PV modules for an on-site load, i.e., smart home in our case, (iv) Efficiently integrates the PV generation and storage system with energy management controller

(EMC), (v) Optimally manages the grid energy and on-site stored energy, (vi) Minimizes the end user electricity cost, (vii) Minimizes the PAR to increase the grid lifetime, avoid the blackout, load shedding and use of Peak Power Plants (PPPs), and (viii) Efficiently formulates and solves the optimization problem considering the constraints.

2. Literature Review

With rapid increase in energy demand and CO_2 emissions due to massive use of fossil fuels, a significance attention needs to be given to consider renewable energy resources. Among all renewable energy resources, the most promising resources with nearly zero carbon emissions are the windmills and PV systems [39]. In comparison to both of the resources, energy obtained from PV system is highly attractive due to its abundant nature. Therefore, several incentive policies are designed especially for the end users to encourage them installing PV systems to alleviate uncertain energy demand during off peak hours. The integration of PV systems at end user side will reduce the energy cost, PAR, blackouts and carbon emissions, particularly. Consequently, PV systems tied with grid are expected to sustain popularity for the next generations [40]. One of the major challenges during the design of energy management systems (EMS) with the PV systems is its intermittent nature and thus leads to an increased design complexity. Other major challenges of PV system integration with existing electricity supply system are the variable PV generation and dynamic energy demand trends. The work reported in [41] used the on-site battery energy storage and PV system to solve the supply–demand mismatch problem. The work discussed in [42] focussed on efficient energy use for cost saving through an optimization algorithm with the integration of battery storage systems. The end user achieved the reduced energy cost by compromising on their comfort level. Thus, cost reduction and comfort maximization are contradictory objectives, hence difficult to achieve simultaneously.

Due to the battery-based ESS, autonomous control algorithms considering variable energy resources and dynamic energy demand are sought to optimally manage the available resources [43]. Therefore, combined dispatch and use of power resources considering a utility tariff scheme leads towards a complex optimization problem. Thus, finding the optimal charging and discharging patterns of battery storage system would ultimately minimize the end user electricity cost and maximize the power system stability. Numerous studies are available dealing with the complex nature of such types of problems and their possible solutions [44]. For nonlinear problems, it is less feasible to use mixed integer linear programming (MILP) and linear programming (LP) approaches without transforming the problem first [45]. The authors in [46] formulate the objective function using a binary and discrete decision variables with a low computing resources [47,48]. The authors in [49] used a hybrid computational intelligence approach for energy optimization using GA and Evolutionary algorithms. In this study, load is scheduled in such a way to minimize the electricity cost rebound peaks. Another study used ant colony optimization (ACO) algorithm to solve predefined load scheduling problem [50]. It is also concluded from the study that heuristic and metaheuristic algorithms such as; PSO, GA, ACO, are comparatively efficient in achieving the optimized solutions when uncertain decision variables are involved [51,52].

The extensive literature review regarding load scheduling has come up with individual technologies, policy recommendations, system analysis and control techniques, energy management, PV-based generation and storage integration issues [53]. Policy-based system analysis techniques sometimes tend to take statistical, qualitative and qualitative approaches while focusing on efficient resource management especially at residential premises [54]. Similarly, quantitative approaches have been case study driven and thus focusing on realistic solutions by taking into consideration utility and end user objectives. Some studies have a particular focus on devising customer centric solutions, while others tend to consider utilities concerns as well [55]. In the meantime, some researchers being able to devise autonomous energy management solutions considering utility and variable energy resources at the same time to facilitate both parties [56]. However, it is found that due to contradictory objectives, the solutions still have some gaps to be handled efficiently [57]. Another problem associated with some load management techniques is the single layer optimizations, such as; residential load scheduling [58], generation scheduling with distributed resources [59], optimal storage integrations [60], end user comfort management with particular focus on cost reduction [61]. Thus, due to lack of generic techniques considering respective objectives, most of these techniques neither feasible nor generally extensible to other case study geographies [62].

To complete the discussion, several researchers used different algorithms and energy management techniques to achieve energy saving and optimization objectives. Statistical results show that the building equipped with EMSs have energy saving factor from 11.39% to 16.22% each year. The EMS used for artificial lighting system reported the highest energy saving factor, of up to 39.5%. Similarly, the energy saving factor for HVAC and other equipments are reported as 14.07% and 16.66%, respectively. Furthermore, the published work also reported that EMS based on different optimization algorithms, such as the harmony search algorithm, enhanced differential evolution, and harmony search differential evolution, efficiently reduced the end user electricity cost by a factor of 17.84%, 11.12%, and 13.2%, respectively. Homes and buildings equipped with EMS architectures significantly reduce the electricity cost. However, EMS equipped with renewable energy generation and storage system and load matching with the PV generation system is not yet reported in the residential sector.

2.1. Contributions

The review of relevant literature shows that homes equipped with HEM architectures significantly reduce the electricity cost, although HEM equipped with renewable energy generation and storage units are given importance. However, the estimation of installed capacity of renewable and storage units as per load requirements is yet to be explored and significant work is required in this area. The literature reveals that HEM architectures efficiently manage the load considering grid and on-site energy generation and storage systems. However, the design of PV-based energy generation and storage systems need to be explored in order to get the full benefits of SG technology. Thus, an efficient and autonomous HEM architecture is required to provide benefits to both end user and utility. With the end user perspective, HEM needs to incorporate the on-site PV generation and storage system that optimally uses the grid and stored capacity to maximize the comfort level. With the utility perspective, HEM should be able to minimize the energy consumption to a certain level to avoid the load shedding, blackouts and use of expensive peak power plants. Thus, on the basis of aforementioned discussion, this work proposes a new home energy management (HEM) mechanism to address the limitations and drawbacks. The proposed work has the following capabilities; it facilitates the end users to decide whether they want to maximize comfort or cost (through a waiting time parameter), efficient load management with the integration of RES and user involvement significantly improved the power system stability through balanced supply–demand profile and it helps in reducing CO_2 emissions as renewable energy and storage units act as "first choice" in the proposed work. The key contributions of this paper are discussed as follows:

- We first categorize the loads based on user preference requirement and priority of operation. Then based on this, mathematical models of each individualized load category are presented to help optimized working patterns.
- To facilitate end users in terms of less cost and high comfort, operational time of each load is modelled as a delay parameter, where high priority is given to critical load, while low priority loads are given flexibility in their operating horizon with the objective of minimized cost and delay.
- To avoid users to buy costlier electricity tariff, on-site PV generation and back-up storage systems are integrated in our model. For this purpose, PV generation and storage systems are optimally designed in such a way to provide baseline load capacity to all respective loads without heavily relying on grid energy source. This helps in minimizing the cost and discomfort of end users along with reduced CO_2 emissions.

- Then based on mathematical models of load consumption, delay and ESS, the optimization problem is formulated as a multi-objective optimization problem which is solved by using DA. However, for optimality analysis, we compare the results of proposed DA with GA, BPSO and OPRA. Furthermore, the performance and complexity of the algorithms are also analysed in terms of large number of loads and time slot variations.
- The simulations are performed against different case studies and results are obtained for cost, PAR, CO_2 and discomfort reduction.

2.2. Paper Outline

The rest of the paper is organized as: Section 3 discusses the DSEM model describing the mathematical description of different loads, on-site PV generation and storage systems, hybrid energy system and cost calculation mechanism. Section 4 presents the formulation of multi-objective optimization problem. Section 5 provides the description of different algorithms which are used to solve the proposed problem. Section 6 presents the results obtained from different algorithms focusing on load scheduling, cost and PAR reduction, optimal PV energy and storage integration and use. Finally, the paper is concluded in Section 7.

3. DSEM Model

The energy of a residential grid network is supplied by the utility and shared by several users through the power line. DSEM is equipped with EMC which is connected to household load on one side and smart meter on other side. The working of EMC is to: (i) collect the user preferences and demand information, (ii) send the control signal to each load via wireless network controller (WNC) for optimal consumption of energy, (iii) gather the energy generation and storage information from the on-site energy generation and ESS, (iv) optimally use the grid energy and ESS, and (v) receive the time variant electricity price signal from the utility. Based on the collected information, EMC sends an on-site energy demand to the smart meter. The smart meter bi-directionally communicates through communication line with the utility. The DSEM architecture is shown in Figure 1.

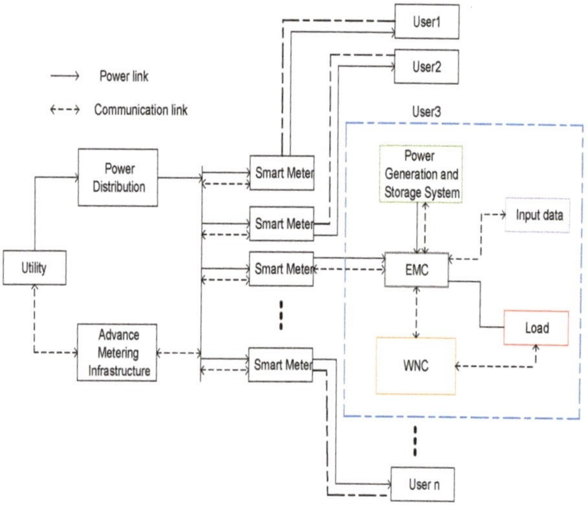

Figure 1. DSEM Architecture.

3.1. Modelling Consumption Behaviour

This section discusses the load consumption behaviour of the considered load. We also discuss the mathematical models of all load types and their respective constraints and limits. Let **A** denotes a set of automatic and manual operated residential loads such that:

$$\mathbf{A} = \mathbf{B} \cup \mathbf{C} \cup \mathbf{D} \cup \mathbf{E}, \tag{1}$$

where, **B**, **C**, **D** and **E** represent sets of base, delay tolerant, semi delay tolerant and critical load, respectively. It is considered that **B** and **E** are zero-delay tolerant load, since the end user is not willing to give-up on the comfort level. The category **C** includes elastic loads in such that their operation time can be interrupted or shifted to any other time slot over the given time horizon $\mathcal{H}$. The category **D** comprises semi-delay tolerant loads that are assumed to be operated within a specific time window. The load in category **E** can be added any time as per user needs. The loads in **B** and **E** have the highest priority and zero tolerance towards delay. Moreover, the loads in **D** has the second highest priority and can bear some delay on the basis of predefined intervals $[\alpha_a, \beta_a]$. The category **C** is maximum delay tolerant and has the minimum priority. For each load $a \in \mathbf{A}$, the power consumption vector over the given time horizon $t \in \mathcal{H}$ is written as;

$$P_{a,t} \triangleq [p_{a_1,t1}, p_{a_2,t_2}, ..., p_{a_n,\mathcal{H}}]. \tag{2}$$

The loads in **B**, **C**, **D** and **E** consume certain amount of power during time t, can be written as:

$$\sum_{t=1}^{\mathcal{H}} \sum_{a \in \mathbf{A}} P_{a,t} \times \sigma_{a,t}, \tag{3}$$

subject to:

$$\underline{P_a} \leq P_{a,t} \leq \overline{P_a}, \forall a \in \mathbf{A}, \tag{3a}$$
$$\sigma_{a,t} = 1, \forall t \in \{t_1, t_2, t_3..., t_n\}, \forall a \in \mathbf{B}, \tag{3b}$$
$$\sigma_{a,t} = 0, \forall t \in \mathcal{H} \backslash t, \forall a \in \mathbf{B}, \tag{3c}$$
$$\sigma_{a,t} = 1, \forall t \in \mathcal{H}', \forall a \in \mathbf{C}, \tag{3d}$$
$$\sigma_{a,t} = 0, \forall t \in \mathcal{H} \backslash \mathcal{H}', \forall a \in \mathbf{C}, \tag{3e}$$
$$\sigma_{a,t} = 1, \forall \gamma_a \leq t \leq \delta_a, \forall a \in \mathbf{E}, \tag{3f}$$
$$\sigma_{a,t} = 0, \forall t = \mathcal{H} - (\gamma_a + l_a), \forall a \in \mathbf{E}, \tag{3g}$$

where, $\sigma_{a,t}$ is the status of load a during time slot t, $\underline{P_a}$ and $\overline{P_a}$ is the minimum and maximum power consumption, γ_a is the operation start time, δ_a is the operation finish time, and l_a is the length of operation time of load a. Equation (3a) gives upper and lower limits on power, while Equations (3b)–(3f) denote power ON/OFF status of various loads according to the respective limits and conditions. The category **E** includes critical load, that can be plugged-in to the HEM architecture any time $\mathcal{H}$. The semi-delay tolerant load has a specific operation window e.g., for load z_1, the lower and upper bound are $\underline{z}_1$ and $\overline{z}_1$, respectively, in which load has to complete its task. The duration between $\underline{z}_1$ and $\overline{z}_1$ is divided as $[\underline{z}_1, \underline{z}_1 + t]$, $[\underline{z}_1 + t, \underline{z}_2]$, $[\underline{z}_2, \underline{z}_2 + t]$, and $[\underline{z}_2 + t, \overline{z}_1 - t]$, and $[\overline{z}_1 - t, \overline{z}_1]$. During the interval $[\underline{z}_1, \underline{z}_1 + t]$, $[\underline{z}_1 + t, \underline{z}_2]$, $[\underline{z}_2, \underline{z}_2 + t]$, and $[\underline{z}_2 + t, \overline{z}_1 - t]$, and $[\overline{z}_1 - t, \overline{z}_1]$, the load may be either in ON or OFF state. After the time $\overline{z}_1$, the load must be in OFF state. The power scheduling of load category **D** over the specified time range $[\alpha_a, \beta_a]$ is given as:

$$\alpha_a \leq t \leq \alpha_a - \beta_a - \rho_a, \quad \forall\, a \in \mathbf{D}, t \in \mathcal{H}. \tag{4}$$

where, ρ_a gives the amount of time delay in regard when load is rescheduled to other feasible time slot during peak hours or overload conditions. The Equation (4) denotes that the load $a \in \mathbf{D}$ is restricted in between α and β time intervals to complete its task. The allowed time horizon for any load $a \in \mathbf{A}$ should not be less than the length of its operation time and is expressed as:

$$l_a \leq \beta_a - \alpha_a, \forall \mathcal{H}. \tag{5}$$

Then based on the aforementioned limits and constraints associated with respective load types, the information is transmitted to EMC via home area network as shown in Figure 2.

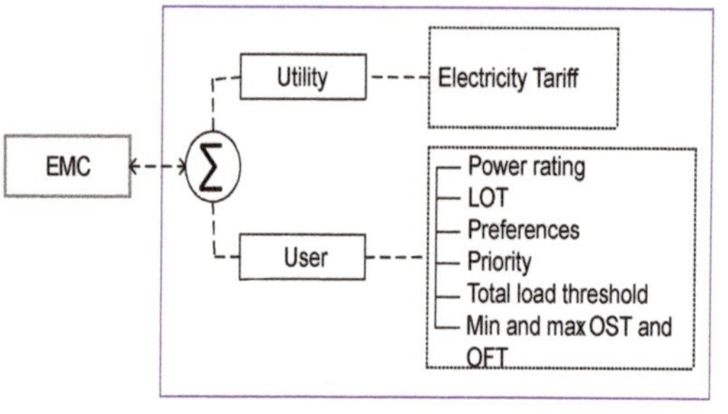

Figure 2. Input set by the user.

The operation start and finish time of load a ranges between γ and δ and is calculated as:

$$\gamma_a \leq r_a \leq \delta_a, \forall \ a \in \mathbf{A} \tag{6}$$

subject to:

$$\alpha_a \leq \gamma_a \leq \beta_a - l_a, \tag{6a}$$

$$\delta_a \leq \beta_a, \tag{6b}$$

where, Equation (6a) gives the operation start time of γ_a and Equation (6b) shows that each load must complete its task during allowed time horizon. Our objective is to optimize the power consumption vector of each load type over the given time interval. Therefore, in order to achieve this objective, the major emphasis is given to obtain optimized scheduling patterns of each load, especially elastic and semi elastic loads.

The category **E** has the highest priority and EMC needs to operate immediately whenever any service is required. EMC manages, monitors and controls the household loads and its working operation is shown in Figure 3.

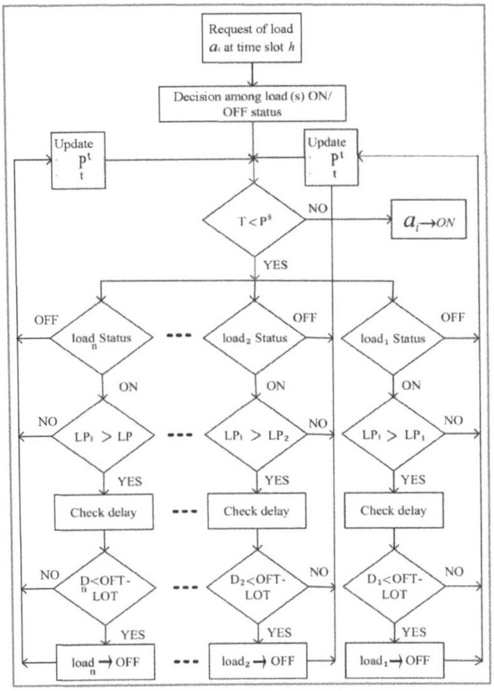

Figure 3. EMC Load Management and Control Architecture.

Load Delay

It is usually expected that residential load can finish their working within the permissible time limits in order to avoid extra delay, which may disturb end user comfort level. Keeping this objective in mind, the maximum acceptable delay ($\overline{\rho}$) that any load can bear is calculated as:

$$\overline{\rho_{a,t}} = 0, \forall\, a \in \{\mathbf{B},\, \mathbf{E}\}, t \in \mathcal{H} \tag{7}$$

$$\overline{\rho_{a,t}} = \mathcal{H} - l_a, \forall\, a \in \mathbf{C}, t \in \mathcal{H} \tag{8}$$

$$\overline{\rho_{a,t}} = (\beta_a - \alpha_a) - l_a, \forall\, a \in \mathbf{D}, t \in \mathcal{H} \tag{9}$$

whereas, the minimum delay ($\underline{\rho}$) for each load over the given time interval is:

$$\underline{\rho_{a,t}} = 0, \forall\ a \in \mathbf{A}, t \in \mathcal{H}, \tag{10}$$

where, Equation (7) shows that load a belongs to the load category **B** and **E** and thus have zero tolerance towards delay, Equation (8) represents the load category **C** that has the minimum priority and can be scheduled anytime in the entire time horizon and Equation (9) shows that load a belongs to the semi-delay tolerant category and can only be scheduled in the pre-set allowed time slots. Where, $\Gamma_{req,\,a}$ denotes the required operation time of load a, and Γ_{EMC_a} provides the scheduled time interval. Hence, Equation (10) elucidates that scheduled time interval is equal to the required amount of time needed to perform a specific task.

$$\Gamma_{req,\,a} = \Gamma_{EMC_a} \tag{11}$$

Therefore, in order to achieve this objective, EMC seeks to minimize the mismatch between required and optimized time intervals in association with limits and constraints.

3.2. On-Site Energy Generation and Storage System

In communication and networks, the coexistence of both discrete and continuous valued agents is interpreted as hybrid system [9]. Whereas in SG, a hybrid energy system is defined as the existence of more than one energy system to accommodate the end users energy demand. As was discussed, total load demand is fulfilled by the grid, renewable energy and storage systems and this complete architecture is considered to be a hybrid energy system, wherein renewable energy and battery storage systems act as "primary choice", while the grid source acts as secondary choice in fulfilling the energy demand. In doing so, the end user can enjoy the reduced electricity cost, while utility gains the benefits in terms of power system stability as high peaks are eliminated during demanding hours. This proposed system architecture is shown in Figure 4.

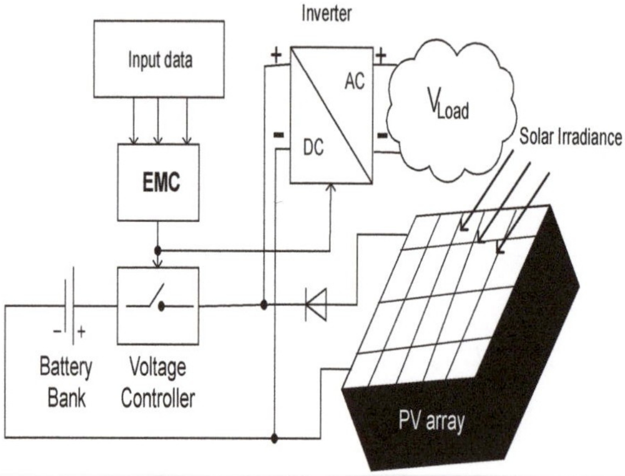

Figure 4. A conceptual diagram of proposed hybrid energy system.

3.2.1. PV Generation

The average power obtained from installed PV modules is calculated as:

$$\kappa_{avg} = P_i \times \varsigma_i, \tag{12}$$

where, P_i denotes the power produced by PV cell at irradiance level i and ς_i gives the probability of irradiance. The Equation (12) for irradiance spectrum can be written as [46]:

$$\kappa_{avg,t} = \int P_{i,t} \times \varsigma_{i,t}\, dt, \forall t \in \mathcal{H}, \tag{13}$$

The Equation (13) denotes the average power output of a PV module for all the possible irradiance spectrums in the integral form. Where, $P_{i,t}$ can be written in terms of voltage V and current I as under:

$$P_{i,t} = V_{i,t} \times I_{i,t}, \forall t \in \mathcal{H}. \tag{14}$$

The Equation (14) gives average power obtained form PV array against specific temperature and irradiance over the given time t. Thus, the total obtained power as per dynamic temperature and irradiance can be calculated using [63–65]. Whereas, the rate of change in temperature with respect to the reference temperature is calculated as:

$$\delta T = T - T_r \tag{15}$$

Equation (16) gives the rate of change in current due to variation in PV module tilt and temperature:

$$\delta I = a \times \left(\frac{S}{S_r}\right) \times \delta T + \left(\frac{S - S_r}{S_r}\right) \times I_{sc} \tag{16}$$

Equation (17) is the change in voltage due to change in current during different time horizon:

$$\delta V = -b \times \delta T - R_s \times \delta I \tag{17}$$

This rate of change results in changing the value of V and I, which are given as:

$$V_{new} = V_r + \delta V \tag{18}$$

$$I_{new} = I_r + \delta I \tag{19}$$

The insolation and cell temperature level relation is defined as the sum of ambient temperature and two percent of the irradiance level and is given as under:

$$T = \frac{s}{50} + T_A \tag{20}$$

The optimum number of PV modules required for the household maximum load shifted from grid to ESS at any time slot t is anticipated as: if $n\kappa_t < P_{D,t}$, increase the value of n. The quality of load matching depends on PV arrays, load characteristics, and isolation profile. Based on PV supply/demand match evaluation criteria, an optimum PV-site matching technique is used to select the module type with the highest average capacity and lowest cost. The current I of the PV module as a function of V can be expressed as [66]:

$$I = I_{sc}\left\{1 - g_1\left[e^{(g_2 V^n)} - 1\right]\right\} \tag{21}$$

The values of constants, g_1, g_3 and n are obtained from [63]. Finally, the total renewable energy $\kappa_{tot,t}$ obtained from PV system is used in two ways; some amount $\kappa_{a,t}$ is directly provided to fulfil the residential load demand, while, the remaining energy ($\kappa_{st_{s,t}} = \kappa_{tot,t} - \kappa_{a,t}$) is stored in backup storage system. The relationship between these two amounts is given as:

$$\sum_{t=1}^{\mathcal{H}}\left(\sum_{a\in\mathbf{A}} \kappa_{a,t} + \sum_{s=1}^{N} \kappa_{st_{s,t}}\right) \leq \kappa_{tot,t} \tag{22}$$

In contrast to renewable energy, the net amount of energy obtained from grid at any time slot $t \in \mathcal{H}$ cannot exceed a certain level P^{grid}_{max}, which is given as:

$$\left(\sum_{t=1}^{\mathcal{H}}\sum_{a\in\mathbf{A}} P_{a,t} + P_{c_{s,t}}\right) - \kappa_{st_{s,t}} \leq P^{grid}_{max}, \quad \forall\, \sigma_{a,t} = 1 \tag{23}$$

where, $P_{c_s,t}$ denotes the electrical charge in the battery power. Furthermore, the ESS must satisfy Equation (24) constraint for any type of load that is to be shifted from the grid to ESS:

$$\kappa_{st_{s,t}} - (P_{a,t} \times \sigma_{a,t}) \geq 0, \quad \forall\, a \in \mathbf{A}, \quad \forall\, 1 \leq t \leq \mathcal{H}. \tag{24}$$

where, the remaining power obtained from PV generation must not exceed the ESS capacity:

$$P_{c_{s,t}} - \kappa_{st_{s,t}} \geq 0, \quad \forall\, s, t \in \mathcal{H}. \tag{25}$$

3.2.2. ESS

As it is understood, ESS has the upper and lower levels to store the surplus energy, which is expressed as:

$$\underline{\kappa L_{s,t}} \leq \kappa L_{s,t} \leq \overline{\kappa L_{s,t}} \quad \forall 1 \leq t \leq \mathcal{H}. \tag{26}$$

The total amount of surplus energy which can be stored in ESS at time t, which is greater than $\kappa L_{s,I}$ is given as:

$$\left(\kappa L_{s,t} = \kappa L_{s,t-1} + (P_{c_{s,t}} \times \eta_s) - \frac{P_{d_{s,t}}}{\eta_s} \right) \geq \kappa L_{s,I} \quad \forall t \in \mathcal{H}. \tag{27}$$

where, η_s, $P_{d_{s,t}}$ and I denote efficiency of battery s, discharge in power of the battery and pre-defined storage quantity, respectively. To prolong the battery life time, it is recommended to avoid deep discharge of a battery. Therefore, we defined upper and lower limits on battery charge and discharge rate, which are given as:

$$\underline{P_{c_{s,t}}} \leq P_{c_{s,t}} \leq \overline{P_{c_{s,t}}}, \forall t \in \mathcal{H}. \tag{28}$$

$$\underline{P_{d_{s,t}}} \leq P_{d_{s,t}} \leq \overline{P_{d_{s,t}}}, \forall t \in \mathcal{H}. \tag{29}$$

During the time when load demand exceeds a certain level and electricity tariff is also high, then ESS acts as "first choice" to fulfil the load demand. Therefore, ESS starts providing the energy as per schedules given by EMC and is given as:

$$\kappa_{st_{s,t+1}} = \kappa_{st_{s,t}} - P_{d_{s,t}}, \quad \forall \quad t \in \mathcal{H} \tag{30}$$

whereas,

$$P_{d_{s,t}} = \sum_{a \in \mathbf{A}} (P_{a,t} \times \sigma_{a,t}), \quad \forall \quad t \in \mathcal{H} \tag{31}$$

The Equation (30) denotes the remaining stored energy in the ESS system. Whereas, the Equation (31) represents the amount of energy dissipated during time slot t for the load $a \in \mathbf{A}$ shifted from grid to ESS system.

3.2.3. Evaluation of Hybrid Energy System

The daily load demand of the end user as a function of t and user preferences are shown in Table 2. End users are hoping to fulfil their demand using PV system, ESS and grid resources. Whereas, the choices to use these resources are mentioned above. Therefore, in order to manage these energy resources efficiently, with the objective of overall cost reduction and power system stability, autonomous energy management algorithm is needed. In this regards, this work considers all the objectives and limitations and designs an autonomous load scheduling algorithm(s). As the primary source of energy is PV and ESS system, therefore, the total PV modules required to fulfil the load demand and the energy obtained from the grid are calculated as:

$$j_{pv} \times P_{o,t} \times \phi_{pv} = P_{D,t} \tag{32}$$

$$P_{g,t} \times \phi_g = P_{D,t} \tag{33}$$

where, j_{pv} denotes the total number of PV modules required for the energy demand D, ϕ_{pv} and ϕ_g define efficiency of PV modules and grid system, and $P_{o,t}$ and $P_{g,t}$ are power imported from single PV module and grid at time slot t, respectively. The equation formulated for a hybrid energy system is given as:

$$z \times P_{o,t} \times \phi_{pv} + (1 - z) P_{g,t} \times \phi_g = P_{D,t}, \tag{34}$$

where, z denotes the ratio which is used to define the amount of PV and grid energy. The best fit ratio z is anticipated by using LSM and is given as [64]:

$$z_{bf} = \left[(\phi_{pv} \times P_{o,t} - \phi_g \times P_{g,t})^t (\phi_{pv} \times P_{o,t} - \phi_g P_{g,t})\right]^{-1} \times (\phi_{pv} \times P_{o,t} - \phi_g P_{g,t})^t (P_{D,t} - (\phi_g \times P_{g,t})) \quad (35)$$

It reduces the sum of squares of deviations of generated power and household load, which is expressed as [64–66]:

$$(\phi_{pv} \times P_{o,t} - \phi_g \times P_{g,t}) \times z = P_{D,t} - (\phi_g \times P_{g,t}) \quad (36)$$

Based on Equations (35) and (36), EMC takes decision about when to use ESS along with grid energy. For example, if $0 < z < 1$, then it would be beneficial to usde ESS along with grid energy.

3.3. Energy Consumption and PAR Model

For each load $a \in \mathbf{A}$, the energy consumed by load is given as:

$$E_{a,t} = \sum_{t=\gamma_a}^{\delta_a - \rho} \sum_{a \in \mathbf{A}} P_{a,t} \times \sigma_{a,t} \times l_a, \ \forall \sigma_{a,t} = 1. \quad (37)$$

Let the house have N number of loads, then the total daily energy demand $E_{tot,t}$ of a single household is written as:

$$E_{tot,t} = \sum_{t=1}^{\mathcal{H}} \sum_{a \in \mathbf{A}} E_{a,t} \times \vartheta_{a,t}. \quad (38)$$

where, $\vartheta_{a,t}$ is the operation time of load a during time slot t. Generally, if we consider only the users' benefits, then the load scheduling has to be performed in a way that most of the load during high peak hours will be shifted to off-peak hours leading towards reduced billing cost. However, it may create rebound peaks which ultimately disturb power system stability regarding high PAR. In contrast, the effective load management mechanism has to reschedule the load to avoid the rebound peaks. Otherwise, the energy retailers and service providers keep backup generation capacity to protect electrical network. Eventually, this may lead costlier generation and hence increase the tariff price which would ultimately disturb end user benefits. Mathematically, the PAR is calculated as follows:

$$PAR = \frac{\overline{P}}{P_{avg}} \quad (39)$$

where,

$$\overline{P} = max_{t \in \mathcal{H}} \left(\sum_{t=1}^{\mathcal{H}} \sum_{a \in \mathbf{A}} P_{a,t} \right) \quad (40)$$

and

$$P_{avg} = \frac{\sum_{t=1}^{\mathcal{H}} \sum_{a \in \mathbf{A}} P_{a,t}}{\mathcal{H}}. \quad (41)$$

Equation (39) shows the ratio between the maximum power and average power consumed by household load a at a given time horizon $\mathcal{H}$. An efficient energy consumption leads to a minimized PAR and thus our third objective is to minimize the maximum PAR over the given time slot.

3.4. Cost Model

The daily energy consumption cost of N household loads during time slot $t \in \mathcal{H}$ is calculated by EMC as:

$$C_{1,t} \triangleq \sum_{t=1}^{\mathcal{H}} \sum_{a \in \mathbf{A}} ([P_{a,t} \times l_{a,t}] \times \sigma_{a,t}) \times \psi_t, \quad \forall\ \sigma_{a,t} = 1. \tag{42}$$

where, ψ_t is the electricity cost charged by the utility for the amount of energy consumed at time slot t. It is, therefore, reasonable to assume that the total cost anticipated by utility for load $a \in \mathbf{A}$ would become:

$$C_{2,t} = \sum_{t=1}^{\mathcal{H}} \sum_{a \in \mathbf{A}} E_{tot,t} \times \psi_t. \tag{43}$$

Now, the fair billing $\mathcal{B}$ is defined as:

$$\mathcal{B} \triangleq \frac{\sum_{t=1}^{\mathcal{H}} C_{2,t}}{\sum_{t=1}^{\mathcal{H}} C_{1,t}} \geq 1. \tag{44}$$

The fair billing is validated by the end user if $\mathcal{B} = 1$, and if $\mathcal{B} > 1$, then user's bill includes the cost $\mathcal{E}_{cost}$ due the losses that occurred in between power distribution and smart meter at any time t and is calculated by:

$$\mathcal{E}_{cost,t} = C_{2,t} - C_{1,t}, \forall t \in \mathcal{H} \tag{45}$$

4. Formulation of Multi-Objective Optimization Problem

The objective of proposed model is to optimally schedule the household load over the given time period by taking into consideration respective constraints and limits. In this regard, the proposed algorithm is designed to optimally integrate PV system, ESS, and grid energy resources with the objective of cost and PAR reduction with increased comfort level. Hence, the primary objective is to fulfil the load demand through PV and ESS system without heavily relying on grid energy source.

$$P1 = minimize\{\sum_{t=1}^{\mathcal{H}} \sum_{a \in \mathbf{A}} ([P_{a,t} \times \vartheta_{a,t}] - \kappa_{tot,t})\}. \tag{46}$$

subject to:

$$0 \leq \overline{\rho_a} \leq \delta_a - \gamma_a - l_a, \forall\ a \in \mathbf{A} \tag{47}$$

$$\alpha_a \leq \vartheta_a \leq \beta_a, \forall\ a\ \in\ \mathbf{A} \tag{48}$$

$$E_{a,t} = \begin{cases} P_{a,t},\ t_a \in \delta_a - \gamma_a - \rho_a, \forall\ a \in \mathbf{A},\ t \in \mathcal{H}, \\ 0, \quad \mathcal{H} \setminus t_a \end{cases} \tag{49}$$

$$\sigma_{t,a}\ \in\ \{0, 1\} \tag{50}$$

The constraints shown in Equations (47)–(50) define: (i) maximum load delay between starting and finish time while completing its task, (ii) load a must complete its assigned task within the allowed time (α and β), (iii) load a consumes certain amount of power during its operation time, and (iv) status of the load a (0 for idle mode, 1 for operation mode).

Therefore, the optimal management and use of the energy resources lead to effective cost minimization, because the electricity prices are dynamic in nature and usually known in advance to customers in a day-ahead market. However, prices are unknown to users in a real time pricing (RTP) environment, where it is more difficult to optimally manage the loads in conjunction with the integration of distributed energy resources (PV and ESS). Therefore, our next objective is to reduce the electricity cost through scheduling the load in optimal time slots without curtailing the load demand and is given as:

$$P2 = minimize\{\sum_{t=1}^{\mathcal{H}} \sum_{a\in\mathbf{A}} \left([P_{a,t} \times \vartheta_{a,t}] - \kappa_{tot,t}\right) \times \psi_t\}. \tag{51}$$

where, Equation (51) denotes the electricity cost charged by the utility against the amount of energy consumed by load a during time slot t. However, it is worth mentioning here that in order to avoid the situations when high peak can be generated due to turning-on high load during low pricing hours, a threshold is imposed to limit the maximum upper limit of energy consumption. Thus, an efficient energy consumption leads to a minimized PAR and thus our next objective is to minimize the maximum PAR over the given time slot t and is formulated as:

$$P3 = minimize(PAR) \tag{52}$$

subject to:

$$\sum_{t=1}^{\mathcal{H}} \sum_{a\in\mathbf{A}} P_{a,t} \leq \tau_{max}. \tag{53}$$

where, τ_{max} is the threshold value used to avoid different risks, such as blackout, use of expensive backup generation plants, and load shedding. Generally, each user has a desire to complete the tasks of each load with minimum delay subject to minimized electricity bill. Thus, our next objective is to minimize the load operation delay in order to achieve maximum end user comfort and is described as:

$$P4 = minimize\{\sum_{t=1}^{\mathcal{H}} \sum_{a\in\mathbf{A}} \rho_{a,t}\}. \tag{54}$$

subject to:

$$\rho_{a,t} = \mathcal{H} - l_a,\ \forall\, a \in \mathbf{C}, t \in \mathcal{H} \tag{55}$$

$$\rho_{a,t} = (\beta_a - \alpha_a) - l_a,\ \forall\, a \in \mathbf{D}, t \in \mathcal{H} \tag{56}$$

$$\rho_{a,t} = 0,\ \forall\, a \in \{\mathbf{B},\mathbf{E}\}, t \in \mathcal{H} \tag{57}$$

The constraints in Equations (55)–(57) define the allowed scheduling horizon for each load $a \in \mathbf{A}$. Mathematically, the cost minimization objective function in conjunction with PV and ESS integration is written as:

$$\begin{gathered} P5 = minimize(P1 + P2 + P3 + P4). \\ subject\ to: Equations\ (7)\text{–}(11), (28), (29), (47)\text{–}(50) \end{gathered} \tag{58}$$

5. Proposed Solutions

In the following sections, we describe the working and coping of the algorithms used to solve the multi-objective optimization problem. For detailed analysis, we used heuristic and robust optimization algorithms and provided the merits and demerits on the basis of input and outputs.

5.1. Optimal Load Scheduling Using DA

DA is a greedy graph search algorithm that is used to find the shortest path for load scheduling problems in micro grid [56,67,68]. It calculates the shortest path (with lowest cost in our case) for a start point (source node; operation start time of load) to the end point (destination node; finish time of load) in the graph. The DA anticipates the shortest path, i.e., minimum cost from any specific source node s_1 to any node in a parent weighted graph such as; $G_w = (\mathbf{F}, \mathbf{e})$, where $\mathbf{F}$ and $\mathbf{e}$ are the sets of vertices and edges, respectively. Initially, the algorithm calculates weights as it traverses from the source node to all the connected nodes and saves the best values. Then it moves through all other connected nodes to find the next best value. Meanwhile, it also avoids saving all the previously saved best value and calculates the cost by comparing with previously saved values. The weighting factor x is the sum of edges weights from the node s_1 to d and is given as:

$$\begin{gathered} if(u,f) \subset \mathbf{e} \rightarrow x(u,f) > 0 \forall f, u \in \mathbf{F} \\ x(f_1, f_2, \ldots, f_n) = \sum_{i=0}^{i=1} x(f_i, f_{i+1}) \end{gathered} \tag{59}$$

where, $x(f_i, f_{i+1})$ is the weight (cost) between f_i and f_{i+1}. The shortest path from u to f, $d(u,f)$, in G has the minimum weight among all nodes. If there is no path between u and f, then $d(u,f) = \infty$.

$$\begin{gathered} \mathbf{S} = \{u | u \in \mathbf{F}\} \\ f \in \mathbf{F} \rightarrow f(u,f)\, is\, defined, \forall u \in \mathbf{S} \end{gathered} \tag{60}$$

where, $\mathbf{S}$ is a subset of $\mathbf{F}$, made-up gradually in an iterative process as follows: (*a*) start node y is selected from set $\mathbf{F}$ and inserted in $\mathbf{S}$, (*b*) the shortest path from y to all available nodes in set $\mathbf{F} - \mathbf{S}$ are determined as:

$$d(y,f) = d(y,u) + min[x(u,f)], \forall f \in \mathbf{F} \setminus \mathbf{S} \tag{61}$$

where, u is the node between f and y. In the next step, the marked best fit solution u and the related edges to this node are added to set $\mathbf{S}$ and $\mathbf{e}$, respectively. The algorithm continues until it updates all the nodes in set $\mathbf{S}$, which means the final result is achieved. The time complexity of DA is $(O(N + N(N-1) + N(N-1)(N-2) + \ldots + N(N-1)(N-2)\ldots(N-N-1)))$. The pseudo-code of DA is shown in Algorithm 1.

Algorithm 1 DSEM model based on DA.

Require: a, b, s, d, g
$d_a = 0, g \geq 0, P_a = \infty, \forall a \in \{1,2,3,\ldots,N\}$
$T_c = mini(g,1), \% T_c \rightarrow currentTier$
$T_c{+}{+}$
$T_n = mini(g, T_c), TN_c = d_1, d_2, d_3, \ldots, d_N; \% \ TN_c \rightarrow$ current TierNode
$TN_n = \infty$
Find $e = arg\, mini_a\, (d_e^{d_a})$
Update P = P $\cup\, d_e; \% P \rightarrow Path$
$TN_n = TN_c - \{d_e\};$
while $TN_c \neq \infty$ **do**
    $TN_c = TN_n;$
    $s = d_e;$ %$d \rightarrow$ Destination
    $e = arg\, mini\, (d_s^{d_a}); \forall d_a \in TN_n$
    Update P = P $\cup\, d_e;$
    $TN_n = TN_c - \{d_e\}$
end while
P_r is obtained; % $P_r \rightarrow$ resultant Path

5.2. Optimal Load Scheduling Using BPSO

BPSO is a population-based heuristic optimization technique used in micro grid for load scheduling [69–71] in which each possible solution in entire search space is represented by a particle. Initially, each particle is assigned a random position and velocity. Each particle represents the status of the electrical load. In this work, a pattern of n number of particles is selected that represents the n number of household electrical loads. For example, if the household has 10 electrical loads then a pattern of 10 number of particles is selected to solve the problem. These particles then fly and move towards the optimal solution in the entire search space. Their flight of movement can be affected by local and global best positions.

$$l_{best}(a,t) = arg_{t\in\mathcal{H}}\, min\,[f_a(t)],\ \forall a \in \{1,2,3,...,M\}, \tag{62}$$

$$g_{best}(t) = arg_{\substack{a\in M\\ t\in\mathcal{H}}}\quad min \quad l_{best}(a,t). \tag{63}$$

The velocity g of each particle is updated as [32]:

$$g_{i,t+1}(j) = \sum_{t=1}^{\mathcal{H}}\sum_{i=1}^{M}\Big(wg_{i,t}(j) + c_1r_1(l_{best,i,t}(j) - \sigma_{i,t}(j)) + c_2r_2(g_{best,i,t}(j) - \sigma_{i,t}(j))\Big), \forall\ r_1, r_2 \in [0,1] \tag{64}$$

where, $\sigma_{i,t}(j)$ is the jth element of ith particle in tth iteration of the algorithm and r_1 and r_2 are random variables. The pulls of the local and global best position are represented by constants c_1 and c_2. The weight w of the particle momentum is mathematically defined as:

$$w = w_i + \frac{\xi_t}{\overline{\xi}} \times (w_f - w_i). \tag{65}$$

where,

$$g_{i,t+1} = \begin{cases} \overline{g}, & if\ \overline{g} < g_{i,t+1} \\ \underline{g}, & if\ g_{i,t+1} \leq \underline{g}. \end{cases} \tag{66}$$

The position λ of each particle is updated as:

$$\lambda_{i,t+1}(j) = \begin{cases} 1, & if\ sig\ (g_{i,t+1}(j)) > r_{ij} \\ 0, & otherwise. \end{cases} \tag{67}$$

where, sigmoid sig function is defined as:

$$sig\ (g_{i,t+1}(j)) = \frac{1}{(1+\exp(-g_{i,t+1}(j)))}. \tag{68}$$

Equation (68) shows the sigmoid function and is used to map velocity of a particle coordinate with probability. The resulting probability determines whether the coordinates take values of 1's or 0's. Each particle is evaluated and ranked based on their fitness value. After n-iterations, the g_{best} is selected, which is known as best optimal solution. The g_{best} is the pattern of N bits representing the status of N number of household loads. The pseudo-code of BPSO is presented in Algorithm 2.

Algorithm 2 Load management mechanism based on BPSO.

Require: p, i, s, l, ep, pos, g
$g_{best} = \infty$ && $l_{best} = zero$
for $i = 1$ **to** s **do**
    Generate population
    $g \leftarrow rand(\overline{g}, \underline{v})$
    $pos \leftarrow rand(s)$
    $l_{best} \leftarrow pos$
end for
update g using Equation (64)
Update pos using Equation (67)
for $k = 1$ **to** i **do**
    $g_{best,i} \leftarrow argmini\ (l_{best,i})$
    $l - -$
end for
g_{best} is obtained; % $g_{best} \rightarrow$ resultant best solution

5.3. *Optimal Load Scheduling Using OPRA*

To develop an understanding of the performance of different load scheduling algorithms, we implement and solve load scheduling problem using OPRA. OPRA consists of N number of binary bits and each bit represents the status of the load $a \in \mathbf{A}$, which are evaluated in each iteration using fitness function. Where, each binary bit represents the status of household load (ON or OFF) and 2^N number of possible combinations of N bits are generated to find best optimal solution. To find and verify best optimal solution, which is also known a global optimum, the fitness function against all the constraints is evaluated. The very initial pattern with minimum electricity cost is selected as the best pattern. This selection of patterns is shown in Figure 5. The algorithm stops working when the required optimal number of patterns from entire search space are obtained. Alternatively, we can also define total number of iterations as a stopping criteria. The pseudo-code of OPRA is shown in Algorithm 3.

Algorithm 3 Load management mechanism based on OPRA.

Require: l_n, i_max, e_p, l and p_a
Generate the N bit pattern
for $h = 1$ **to** H **do**
    $G_{best} \leftarrow \infty$
    for $k = 1$ **to** 2^N **do**
        $\sigma \leftarrow \sigma^{2^N}$
    end for
    for t = 1 **to** T **do**
        Validate all constraints
    end for
    for $i = 1$ **to** 2^M **do**
        if $f(\sigma_i) < f(G_{best})$ **then**
            $G_{best} \leftarrow \sigma_i$
        else
            $G_{best,i} \leftarrow G_{best,i}$
        end if
        $l_a \leftarrow l_a - 1$
    end for
end for

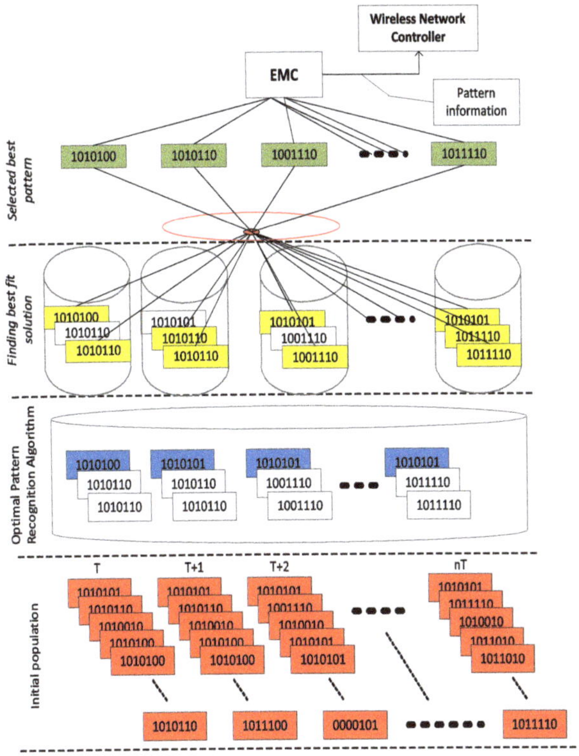

Figure 5. Schematic diagram showing the selecting of optimal load patterns using OPRA algorithm.

5.4. *Optimal Load Scheduling Using GA*

GA algorithm is used to optimally schedule the household load [72–74] and generates the optimal patterns using objective function without violating the constraints. The algorithm usually adapts heuristics in the problem to provide the cost effective solutions. Initially, a population is randomly generated that consists of $\mathcal{M}$ number of chromosomes. Each chromosome represents a solution to the problem and is constructed as an array of bits (combination of genes). Each gene represents the ON/OFF status of the household electrical load. The length of chromosome is equal to the required number of bits that represent the household total electrical loads. These chromosomes are evaluated using fitness evaluation function and fitness of each chromosome is ranked, accordingly. After calculating the fitness value of each chromosome, the reproduction, crossover, and mutation operators are applied to form new off-springs from the previous population. The population of new off-springs from the existing ones have possibly higher fitness and this operation is known as crossover. The generation of new off-springs from the parent chromosomes is shown in Figure 6.

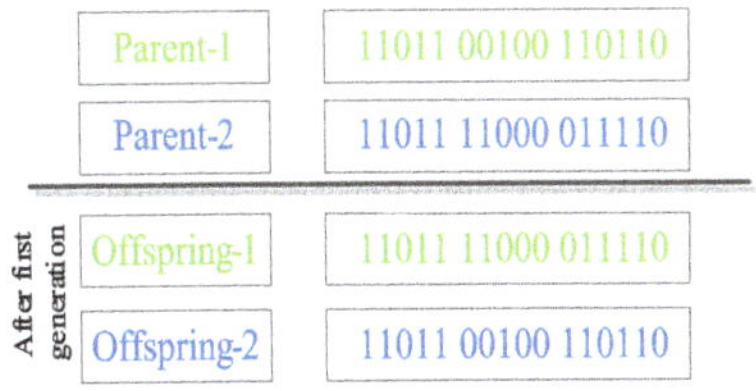

Figure 6. Generation of new off-springs from parent chromosomes through crossover operation.

Selection allocates more copies of those solutions with higher fitness values and thus imposes the survival of the fittest mechanism on the candidate solutions. To avoid same off-springs in the next population and pre-mature convergence, a little bit randomness needs to be added to avoid repetition and is demonstrated in Figure 7.

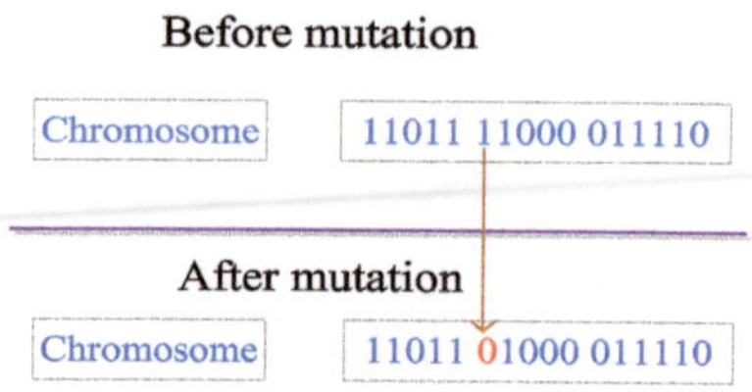

Figure 7. Single point mutation.

In this work, a single point is considered for crossover, the Roulette Wheel selection method for reproduction, and the single bit is flipped during mutation operation. After crossover and mutation, all the chromosomes are screened with the pre-defined constraints to filter out the best chromosomes among the population. The chromosomes with best fitness during n number of generations are selected as an optimal solution. The GA process is stopped after reaching $n \times n$ number of iterations. The pseudo-code of GA-based DSEM is shown in Algorithm 4.

Algorithm 4 Load management mechanism based on GA.

Require: p, g, m, n, t, c, l, e, e_m, e_c
Generate initial population
for $i = 1$ **to** t **do**
    $best = arg\,mini\,(e)$
    if $best == 1$ **then**
        $e--$
    end if
    for $j = 1$ **to** l **do**
        $m = arg\,mini\,(e, t)$
    end for
    if $c(i) < c_{max}$ **then**
        $RE = RE - c(i)$
    else
        $e(i) = e_g$
    end if
    do selection process
    if $rand < e_i$ **then**
        do crossover
    end if
    if $rand < e_m$ **then**
        do mutation
    end if
end for

5.5. Feasibility Region

A set of optimal points that contains all the possible solutions satisfying all the constraints, equalities and in-equalities in a given scenario is known as feasible region. The main objective of the DSEM is to maximize the end user comfort with minimized electricity cost and PAR. Thus, electricity cost is comprised of two components, i.e., amount of energy consumed during certain time slot t and electricity price at that time. Four scenarios are considered to find the feasible region for energy consumption and electricity cost and are given as: (i) S1: Minimum load, minimum electricity price, (ii) S2: Minimum load, maximum electricity price, (iii) S3: Maximum load, minimum electricity price and (iv) S4: Maximum load, maximum electricity price. The per hour electricity cost charged by the utility against any time slot is anticipated as:

$$C_t \triangleq \sum_{a \in \mathbf{A}} P_{a,t} \times \psi_t, \quad \forall\ \sigma_{a,t} = 1,\ h\ \in\ \mathcal{H}. \tag{69}$$

subject to:

$$C1 : 0\ \leq\ C_t\ \leq\ 64 \quad \forall\, t \in \{1, 2, 3, ..., \mathcal{H}\} \tag{70}$$

The daily electricity cost is calculated using Equation (42) and minimized with subject to the following constraints:

$$C2 : C_{1,t}\ \leq\ 232 \tag{71}$$

$$C3 : 0\ \leq\ E_{a,t}\ \leq\ 9 \tag{72}$$

The feasible region of the electricity cost against the consumed energy subject to the constraints (C1, C2 and C3) is demonstrated in Figure 8. In Figure 8, the overall region for the load operation is an area covered by points $p_1(2,4), p_2(2,16), p_3(9,18), p_4(9,72), p_5(9,64)$ and $p_6(8,64)$. The points p_1 and p_2 represent the scenario *S1* and *S2* and p_3 and p_4 represent the scenario *S3* and *S4*. This leads to the result that the scheduling horizon must not: (i) exceed the maximum unscheduled cost (232 cents) and (ii) violate the constraints given in Equations (70)–(72). After implementing this threshold level, feasible region of the load scheduling is shown by the pentagon of points p_1, p_2, p_3, p_5, and p_6.

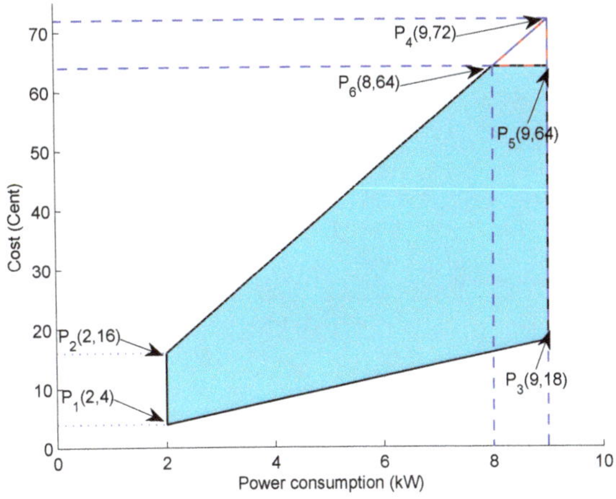

Figure 8. Feasible region of Energy consumption.

6. Simulation Results and Discussion

In this section, we analyse the performance of proposed model using different algorithms in terms of managing grid energy, electricity cost, comfort and PAR minimization. The performance analysis is done in software tool MATLAB version 2014b on Sony VAIO VPC-EB47GM machine with Intel(R) Core(TM) i5 CPU M 480 2.67 GHz and 6 GB of memory on Windows platform. Furthermore, the scalability of the algorithms in terms of total number of time slots and computation time is also measured, so as to analyze the optimality. We verify the applicability of proposed algorithms through optimally managing the energy consumption by finding the tuple of time shifts of various loads. In a result, minimum cost and PAR objectives are achieved along with high user comfort. For experiments, the electricity price data is obtained from a day-ahead market showing real time consumption trends, which is shown in Figure 9.

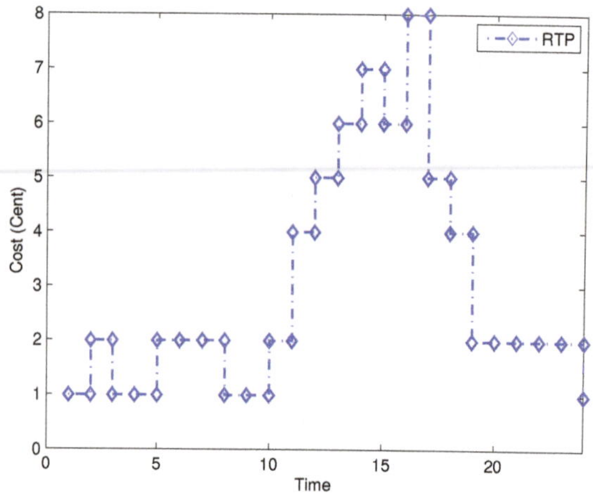

Figure 9. RTP scheme obtained from a day-ahead electricity market.

6.1. Simulation Setup

Table 1 provides the parameters of a single household, which are used for experimentation purpose. Generally, each user may have a variable load pattern, on the basis of their comfort level.

Table 1. Simulation parameters.

Parameters	Values
Number of devices in single home	10
Mini and Max power rating	0.5–2 kW
Mini and Max operation time	1, 5 time slots
Delay	As per user's priority
Pricing scheme	RTP
Parameters of GA	
population size	10,000
Swarm size	1000
Velocity ranges	[−4 4]
Crossover probability	0.9
Mutation probability	0.01

For simulation, we considered four different type of loads that are used in different seasons. In the proposed work, each user has 11 fixed loads with different power rating, operation time and comfort requirements. The total household load is managed and operated to complete the assigned task as under:

- Out of 11, the 4 loads, i.e., l_1, l_2, l_3 and l_{11} have the highest priority and will operate when required.
- l_4, l_5 and l_6 have the second highest priority and will operate in the predefined time slots.
- l_7, l_8, l_9 and l_{10} have zero priority and can be operated at any time within the given scheduling horizon.

The power rating and load demand are summarized in Table 2.

Table 2. Load categorization and user comfort level.

Load	Power Rating (kW)	Priority (3 = high, 1 = Low)	Delay Tolerant	Time Restriction
l_1	0.5	3	No	As required
l_2	0.5	3	No	As required
l_3	1	3	No	As required
l_4	0.5	2	Yes	t_1–t_7
l_5	2	2	Yes	t_3–t_8
l_6	0.5	2	Yes	t_8–t_{10}
l_7	1	1	Yes	t_1–t_n
l_8	2	1	Yes	t_1–t_n
l_9	1.5	1	Yes	t_1–t_n
l_{10}	1.5	1	Yes	t_1–t_n
l_{11}	1	1	Yes	As required

6.2. User Comfort

The comfort level is defined as the services required by the end user to fulfil their load demand with reduced electricity cost and operating delay time, and is demonstrated in Table 2. Based on end user comfort requirement, each load has a predefined start and finish operation time horizon. In this work, the end user defines its comfort level by defining certain delay level in the operation of household load. EMC needs to schedule the load in the pre-defined window to complete the desired task while achieving the aforementioned objectives.

From Table 2, the *l* of load 4, 5 and 6 are 3 time slots each, which lies in semi-delay tolerant category. In the absence of DSEM architecture, each load is assigned t_5, t_6 and t_7 time slots for operation and is shown in Figure 10a. The user comfort is also a major objective of our proposed work and is demonstrated in Figure 10b–d. It is evident from Figure 10b–d that our proposed model efficiently scheduled the end user loads in the allowed time horizon and achieved maximum user comfort. The operation time assigned by each optimization algorithm is tabulated in Table 3.

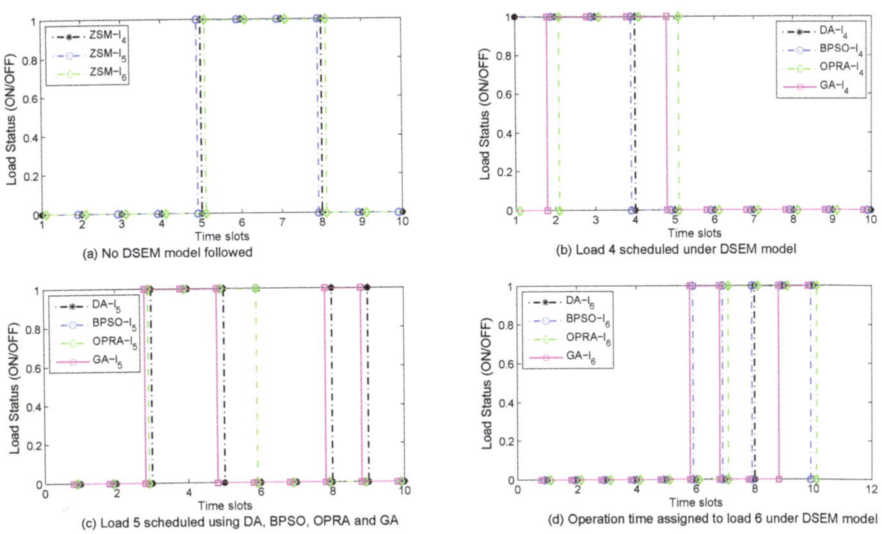

Figure 10. User comfort profiles of different loads over the given time period.

Table 3. DSEM based semi-delay tolerant load scheduling.

Load	Optimization Technique	Operation Time Slots
load 4	DA	t_1, t_2, t_3
	BPSO	t_1, t_2, t_3
	OPRA	t_2, t_3, t_4
	GA	t_2, t_3, t_4
load 5	DA	t_3, t_4, t_8
	BPSO	t_3, t_4, t_5
	OPRA	t_3, t_4, t_5
	GA	t_3, t_4, t_8
load 6	DA	t_8, t_9, t_{10}
	BPSO	t_6, t_8, t_9
	OPRA	t_7, t_8, t_9
	GA	t_6, t_9, t_{10}

6.3. *Energy Consumption Profile*

Regarding energy consumption, three sets of evaluations are carried out to examine the performance of the proposed DSEM and optimization algorithm.

1. **Zero shift method (ZSM)**: Household loads operate under no DSEM and optimization algorithms. Users operate household load irrespective of the electricity cost to achieve the maximum comfort.
2. **Fine shift method (FSM)**: Household loads are being used under DSEM and optimization algorithms. The residents are concerned with electricity cost along with the comfort level.
3. **FSM equipped with PV generation and storage system**: Household loads run under DSEM and optimization algorithm along with the integration of PV generation and storage system.

In the first case, the household loads are operated using grid energy only while ignoring the electricity tariff. It is clear from Figure 9 that the electricity tariff is at peak from 10:00 to 20:00. For simulation, we considered 10 time slots, i.e., $H = t_1, t_2, t_3, ..., t_8, t_9, t_{10}$ to schedule the loads. The load demand in different time slots under the ZSM is tabulated in Table 4. It is observed in Table 4 that the household selected the loads to operate during some time slots (t_3–t_8) rather than scheduling in the entire allowed time horizon (t_1–t_{10}). From Table 4, it is also clear that most of the high loads are operated during high peak hours (t_5–t_7) which will ultimately result in high electricity cost.

Table 4. Load demand of different loads for ZSM.

Power Rating (kW)/LoT	t_1	t_2	t_3	t_4	t_5	t_6	t_7	t_8	t_9	t_{10}
0.5/3	-	-	-	-	l_4	l_4	l_4	-	-	-
2/3	-	-	-	-	l_5	l_5	l_5	-	-	-
0.5/3	-	-	-	-	l_6	l_6	l_6	-	-	-
1/4	-	-	l_7	l_7	l_7	l_7	-	-	-	-
2/4	-	-	-	-	l_8	l_8	l_8	l_8	-	-
1.5/5	-	-	l_9	l_9	l_9	l_9	l_9	-	-	-
1.5/4	-	-	l_{10}	l_{10}	l_{10}	-	l_{10}	-	-	-
1/1	-	-	-	-	-	l_{11}	-	-	-	-
Total Demand (kW)	**0**	**0**	**4**	**4**	**9**	**8.5**	**8**	**2**	**0**	**0**

However, in the second case, the loads are operated under DSEM and optimization algorithm. The energy consumed by ZSM and FSM is shown in Figure 11. BPSO meets the DA at some most costly time slots while the other two schemes exhibit different scheduling plans.

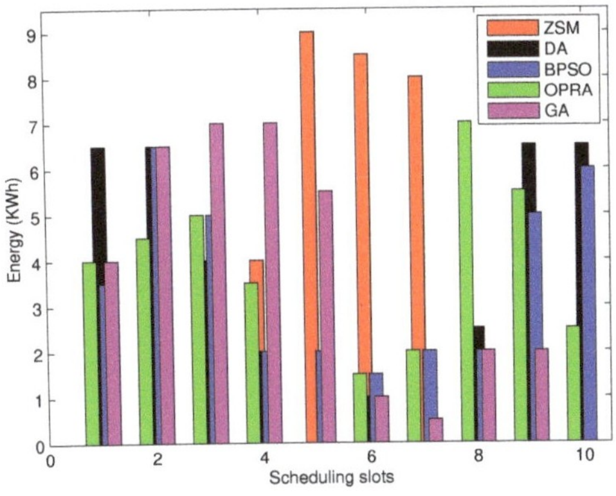

Figure 11. Energy Consumption profiles using different optimization techniques over the given scheduling time period.

For the highest tariff slots, i.e., 4, 5, 6 and 7, DA performed significantly better in comparison with other optimization algorithms. It can be seen in Figure 11 that DA managed the entire load in the low and off peak hours. From Figure 9, there are 10 peak hours starting from 10:00 to 20:00, i.e., electricity price is high during these time slots. It is clear from Figure 11 that FSM has optimally shifted the load from high peak hours to low peak hours. In case of FSM equipped with PV generation and storage system, the load during high peak hours are shifted from grid to ESS and draws energy from the ESS.

In this work, the storage capacity ranges from 3 KWh to 4 KWh depending upon the solar irradiance level and status of the battery cells. The health of the battery is good when it is kept at near 100% of its capacity or to be charged immediately as it discharges completely or partially. The output power from a PV array is directly proportional to the solar irradiance received from the sun. The output of the PV module starts from zero during sunrise, increases till noon and declines to zero till evening. The rated maximum output of the PV module depends on the actual site condition and is achieved only occasionally. In our scenario, smart home is connected to gird and PV, daily sun peak hours can be used for calculation and the number of sun peak hours is thus numerically equal to the daily solar irradiance measured in KWh/m^2. The amount of usable energy harvested from PV modules is lower than the output of the PV modules due to the energy losses in the system components. The solar radiation observed greater than the 250 W/m^2 is taken in simulation. The solar irradiance and the ambient temperature for a typical day of June is shown in Figure 12. In our case, PV modules are selected using Module-site matching technique with an assumption of 80% efficiency, and parameters of the PV modules are tabulated in Table 5.

The ESS is exploited by the proposed DA and the effectiveness of the model is shown in Figure 13. It is evident from Figure 13 that the proposed model optimally used the grid and stored energy. The load that needs to be operated during high peak hours on grid energy is shifted to ESS by EMC thus results in optimal use of both energy resources.

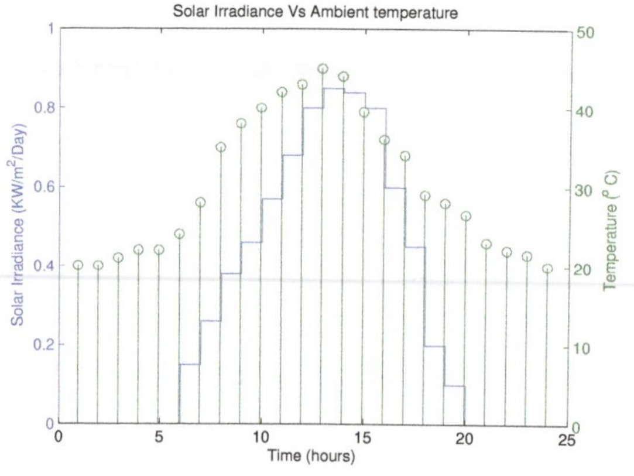

Figure 12. Solar irradiance and ambient temperature profiles over the period of 24 h.

Table 5. Parameters of PV module at standard test conditions.

Parameters	Values
Manufacturer	Trinasolar
Model	TSM-245 PC/PA05
Number of PV modules	18
(P_{max})	245
(V_{max}) (V)	30.2
(I_{max}) (A)	8.13
V_{oc} (V)	37.5
I_{sc} (A)	8.68
Efficiency of the module (%)	15
Cell orientation	60 cells
Temperature co-efficient of P_{max}	−0.43%/°C
Temperature co-efficient of V_{oc}	−0.32%/°C
Temperature co-efficient of I_{sc}	0.047%/°C

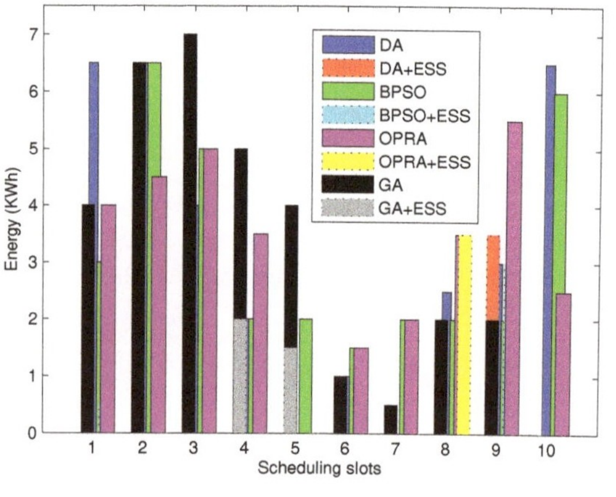

Figure 13. Energy consumption profile against different optimization techniques with the integration of ESS.

6.4. Cost Profile

The cost against the amount of energy consumed by the household loads and the performance of each optimization technique is shown in Figure 14. According to the electricity tariff, the highest cost needs to be paid against the time slots 5 and 7. It is evident from the results that DA outperformed all the traditional techniques by shifting all the loads to other time slots except the peak cost slots. Similarly, BPSO exhibits better results in comparison with OPRA and GA and is demonstrated in Figure 14.

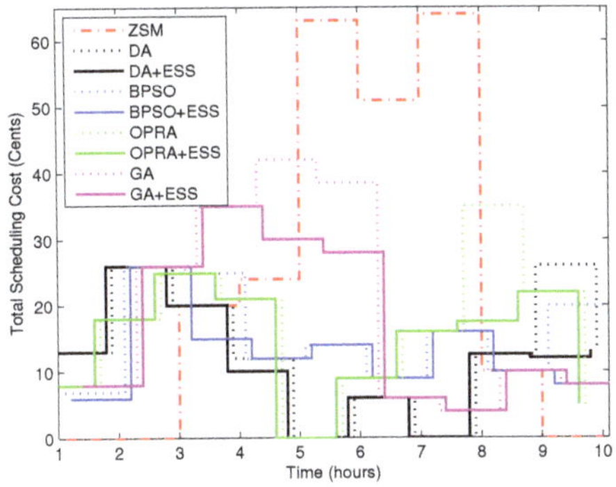

Figure 14. Total cost incurred over the given time period using different optimization algorithms.

6.5. Scalability Factor and Computation Time

The scalability factor of the proposed schemes are evaluated in terms of:

1. Number of time slots,
2. Number of loads.

In the first case, the performance of each optimization algorithm is evaluated based on increasing the number of time slots for the load given in Table 2. By increasing the number of time slots and following the RTP pricing scheme, more scheduling opportunities are available for each algorithm. This higher degree of freedom for each optimization algorithm results in better scheduling the loads as shown in Figure 15. It is clear from Figure 15 that DA-based DSEM performed better under wide range of scheduling horizons.

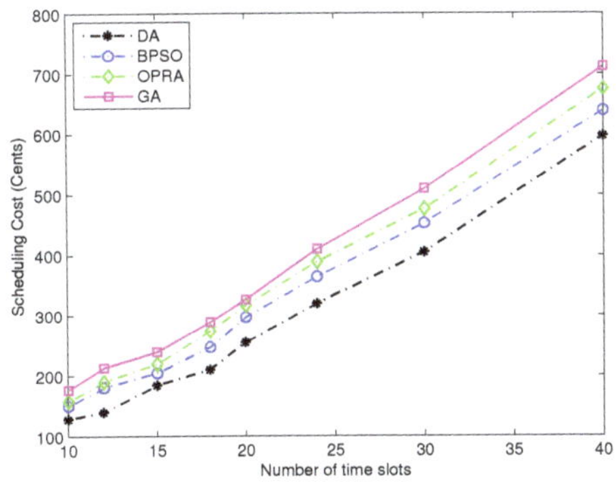

Figure 15. A comparison between energy consumption cost and total number of time slots against optimization algorithms (i.e., scalability w.r.t number of time slots).

In the second case, simulation is done for a fixed number of scheduling slots, i.e., 10 while the number of loads vary from 27 to 100. The total cost increases as the number of loads increases (greater the load, greater the energy consumption) and the result is shown in Figure 16.

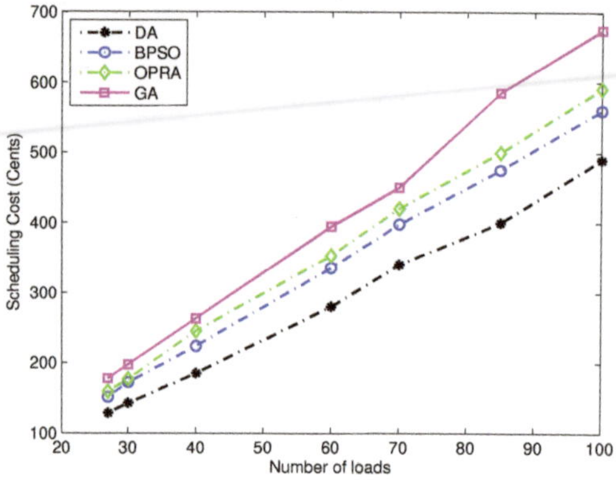

Figure 16. A comparison between energy consumption cost and total number of time slots against optimization algorithms (i.e., scalability w.r.t number of time loads).

The results demonstrated in Figure 16 shows that the DA outperformed all the candidate solutions followed by BPSO. Thus, we summarize that our optimization schemes provide the low cost solutions irrespective of the number of scheduling time slots and loads.

Simulation is run for two different scenarios in order to calculate the computational time for each algorithm. In the first scenario, a fixed number of loads (30) with varying time slots (50 to 300) is considered and the computation time taken by each optimization technique is shown in Figure 17.

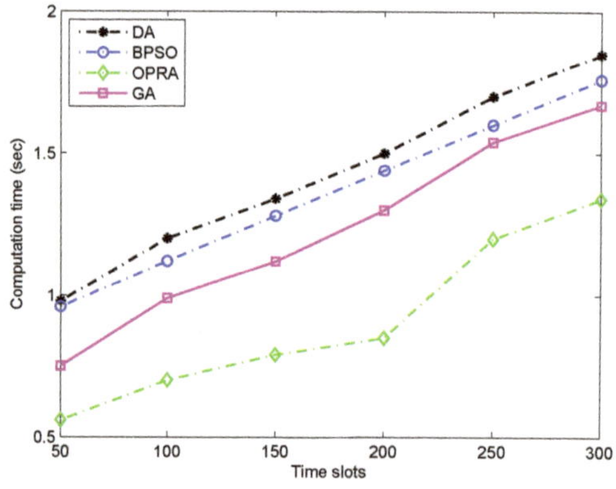

Figure 17. A comparison between computation time and total time slots.

In the second case, iteration number is fixed to 100 and number of load varies from 30 to 65. The computation time taken by each algorithm is depicted in Figure 18.

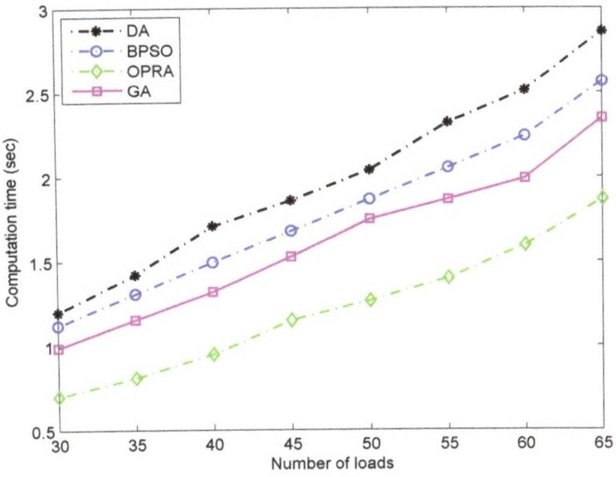

Figure 18. A comparison between computation time and total number of loads.

It is clear from Figure 17 that the OPRA outperformed all the candidate solution techniques and scheduled the load with minimum execution time. From Figures 17 and 18, BPSO took less computation time in comparison with DA due to its low complexity factor, i.e., $O(I \times N)$. From Figure 18, the complexity of DA increases as the number of nodes (loads) increases. The entire network has N different costs, i.e., each node in different tier has a different cost. The distance to reach a node in any tier does not depend on the previous starting node. One node will be selected in one tier that finds the next and current distance from the starting node. Thus, DA takes a bit high execution time due to its highest complexity nature $(O(N + N(N-1) + N(N-1)(N-2) + ... + N(N-1)(N-2)...(N-N-1)))$.

6.6. PAR Discussion

The relationship between the ZSM and FSM with respect to the PAR is shown in Figure 19. Simulation is run for a fixed time slots of 24 hours with total number of loads varying from 11 to 20 each day.

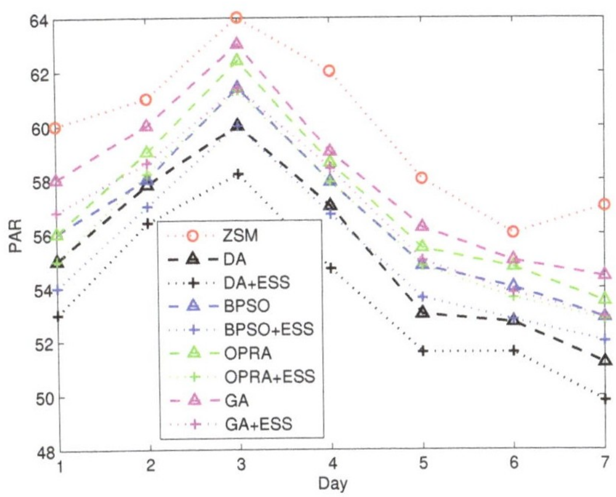

Figure 19. PAR profiles of different algorithms over the period of one week.

The purpose of minimizing PAR is a two-way concept, i.e., for utility and end user. With utility prospects, the lifetime of the grid is increased as the demand curve always remains under the supply curve. The amount of energy produced by utility during any time slot is sufficient for the users demand thus resulting in no activation of peak power plants. On the other hand, the end user consumption is limited to a certain threshold level. This threshold level helps the users in getting the desired energy level from the utility with no load shedding and blackouts issues. With the addition of ESS, it is clear from Figure 19 that PAR is minimized as compared to the case where there is no ESS.

6.7. Overall Analysis of the Proposed Schemes

To complete the discussion, the proposed schemes are analysed based on: (i) maximizing user comfort, (ii) optimally consuming grid energy, (iii) minimizing energy cost, (iv) minimizing PAR and (v) optimally usage of ESS. In the first scenario, cost comparison is done by scheduling the load ($l_1 - l_{11}$) during high peak hours, i.e., 10:00–20:00. The results obtained from the simulation are tabulated in Table 6.

Table 6. A comparison of the proposed algorithm with conventional algorithms in terms of performance parameters.

Optimization Technique	Scheme	DSEM	ESS	$Cost_{max}$	$Cost_{avg}$	$Cost_{total}$	$Cost_{red(\%)}$	PAR_{max}	PAR_{avg}	PAR_{total}	$PAR_{red(\%)}$
- -	ZSM	×	×	64	23.2	232	- -	64	59.7	418	- -
DA	FSM	√	×	26	12.85	128.5	44.61	60	55.24	386.7	7.49
		√	√	26	11.25	112.5	51.72	58.2	53.6	375.3	10.22
BPSO	FSM	√	×	26	15.1	151	34.91	61.4	56.4	394.97	5.51
		√	√	26	12.8	128	44.83	60	55.16	386.1	7.63
OPRA	FSM	√	×	35	15.9	159	31.47	62.4	57.09	399.67	4.39
		√	√	25	14.15	141.5	39.01	61.3	56.22	393.6	5.84
GA	FSM	√	×	42	17.75	177.5	23.49	63	57.94	405.6	2.97
		√	√	35	15.5	155	33.19	61.4	56.7	389.9	6.72

The results show that DA has considerably better results for all the aforementioned objectives. DA optimally managed the grid and ESS energy in comparison with ZSM and thus leads to low electricity cost and maximum user comfort level. Keeping the trend, BPSO meets the DA in some case (FSM with DSEM) while OPRA and GA provide different patterns each time. It is evident from Table 6 that DSEM model based on DA has efficiently scheduled the load and reduced the end user electricity cost by 44.61% and 51.72% in the presence of only DSEM architecture and DSEM along with ESS, respectively. The cost reduction between the DA and the next best optimization technique, i.e., BPSO, is 9.7% and 6.89% in case of only DSEM architecture and DSEM along with ESS, respectively.

In the second scenario, simulation is run for 24 h, 7 days per week to analyse the PAR response and results obtained are tabulated in Table 6. It is evident from the Table 6 that DA efficiently reduced the PAR by a factor of 7.49% and 10.22% in case of DSEM and DSEM with ESS, respectively. The overall trade-off among all optimization schemes are given in Table 6. Thus, we summarize that our proposed DSEM based on DA provides low cost and PAR solutions independent of the number of scheduling time slots and loads.

7. Conclusions and Future Work

A novel DSEM model and EMC based on DA were proposed and implemented for the optimal energy management of residential user in this paper. The novel aspect of this work is the optimal selection of PV arrays and ESS, based on which the optimal load scheduling algorithms have been designed. Prior to this, we categorized the load into base, schedulable, semi-schedulable and critical according to user preferences and assigned predefined priorities in order to provide operation flexibility. Then, these loads were mathematically modelled in such a way that optimization algorithm can have better control. EMC is supported by optimum PV module selection and use scheme that efficiently shifts the high cost load from grid to an on-site PV and ESS. The cost reduction objective is achieved through optimal use of grid, PV and ESS sources, where PV units and ESS act as a "primary source" of energy during peak hours. The intuition behind our model is to address the optimal management of base, semi-schedulable, schedulable, and critical energy demands, time varying price signal, optimal selection and integration of PV modules, energy harvesting from PV modules and use of stored energy to maximize comfort level and minimize the end user electricity cost and PAR. Simulation results show that EMC based on DA achieved the maximum comfort level and also optimally scheduled the load, minimized the electricity cost and PAR in comparison with BPSO, OPRA and GA. An evaluation of the proposed algorithm in comparison with other algorithms is done based on an increased number of time slots and household loads and the presented results validates that the proposed algorithm leads to better solution. In addition, incorporating PV-based energy generation and storage system in the HEM architecture leads to better synergy between electricity consumption and PV production. The result is that the end user electricity cost is reduced by 52% and grid stability is increased by minimizing the PAR to 10.22% in comparison with ZSM. In future, we will update this model with other modules such as: (i) integration of windmill energy generation system, (ii) an energy exchange model among several micro smart grids, (iii) cloud-based infrastructure to connect micro smart grids, (iv) join the grid using Blockchain technology, (v) efficient storage system and (vi) facilitating electrical vehicles.

Author Contributions: Conceptualization, I.U. and M.B.R.; formal analysis, M.B.R. and T.A.; funding acquisition, M.B.R. and T.A.; investigation, I.U., M.B.R. and T.A.; methodology, I.U., M.B.R. and T.A.; project administration, M.B.R. and T.A.; software, I.U. and M.B.R.; supervision, M.B.R., T.A. and S.T.; validation, I.U., M.B.R. and T.A.; writing—original draft, I.U. and M.B.R.; writing—review and editing, I.U., M.B.R. and S.T. All authors have read and agreed to the published version of the manuscript.

Funding: This Project was funded by the Deanship of Scientific Research (DSR), King Abdulaziz University, Jeddah, under grant No. (DF-397-135-1441). The authors, therefore, gratefully acknowledge DSR technical and financial support.

Conflicts of Interest: The authors have no conflict of interest.

Abbreviations

SG	Smart grid
PV	Photovoltaic
LSM	Least square method
GA	Genetic algorithm
OPRA	Optimal pattern recognition algorithm
ICT	Information and communication technology
EMS	Energy management system
LP	Linear programming
HEM	Home energy management
EMC	Energy management controller
RTP	Real time pricing
CO_2	Carbon-dioxide
ESS	Energy storage system
DA	Dijkstra algorithm
BPSO	Binary particle swarm otpimization
PAR	Peak to average ratio
DR	Demand response
MILP	Mixed integer linear programming
ACO	Ant colony optimization
DSEM	Demand side energy management model
WNC	Wireless network card
PPP	Peak power plant

Nomenclature

$P_{a,t}$	Power consumption of load *a* at time *t*
$\sigma_a(t)$	Status of load *a* during time *t*
β_a	Load *a* finish time
γ_a	Operation Start Time of load *a*
ϑ_a	Operation time of load *a*
$\overline{P_a}$	Maximum power consumption of load *a*
κ_{avg}	PV module average output power
n	Number of PV modules
$\varsigma_{i,t}$	Probability of irradiance at time *t*
b	Voltage-temperature coefficient (V/°C)
l_a	LOT of load *a*
α_a	Load *a* starting time
ρ	Load delay
δ_a	Operation Finish Time of load *a*
ψ_t	Electricity price during time slot *t*
$\underline{P_a}$	Minimum power consumption of load *a*
P^{grid}_{max}	Maximum power imported from grid
$P_{i,t}$	Power produced at irradiance *i* and time *t*
M	Number of particles
I_{sc}	Module short circuit current (A)
S	Total Tilt isolation
R_s	Module series resistance (Ω)
T_r	Reference temperature (°C)
$\overline{I}$	Maximum current (A)
$\overline{V}$	Maximum voltage of module (V)
$P_{c_s,t}$	Electrical charge of battery (W)
$\underline{P_{c_s,t}}$	Minimum charge of battery (W)
$\overline{P_{d_s,t}}$	Maximum discharge of the battery (W)
L_s	Storage capacity
η_s	Efficiency of the battery *s*
j_{pv}	Required PV modules for demand *D*
ϕ_g	Efficiency of grid system
P_g	Power imported from grid (W)
$\vartheta_{a,t}$	Operation time of load *a* during time *t*
$\mathcal{B}$	Fair billing
$g_{i,t}$	Velocity of particle *i* during time *t*
λ_i	Position of particle *i*
S_r	Reference Tilt isolation
T_A	Ambient temperature (°C)
δT	Change in Cell temperature (°C)
V	Voltage of module (V)
V_{oc}	Open circuit voltage of module (V)
$\overline{P_{c_s,t}}$	Maximum charge of the battery (W)
$P_{d_s,t}$	Discharge of the battery (W)
$\underline{P_{d_s,t}}$	Minimum discharge of the battery (W)
N	Number of storage batteries
I	Pre-defined storage quantity (W)
ϕ_{pv}	Efficiency of PV modules
P_o	Power imported from single PV module (W)
z	Ratio that defines the PV and grid
a	Current-temperature coefficient (A/°C)
$\mathcal{E}_{cost,t}$	Cost due to losses during time *t*
w	Weight assigned to a particle
$P_{D,t}$	Total power demand over given time *t* (W)

References

1. Albu, M.M.; Sănduleac, M.; Stănescu, C. Syncretic use of smart meters for power quality monitoring in emerging networks. *IEEE Trans. Smart Grid* **2016**, *8*, 485–492. [CrossRef]
2. Karimi, B.; Namboodiri, V.; Jadliwala, M. Scalable meter data collection in smart grids through message concatenation. *IEEE Trans. Smart Grid* **2015**, *6*, 1697–1706. [CrossRef]
3. Ye, F.; Qian, Y.; Hu, R.Q. A real-time information based demand-side management system in smart grid. *IEEE Trans. Parallel Distrib. Syst.* **2015**, *27*, 329–339. [CrossRef]
4. Viswanath, S.K.; Yuen, C.; Tushar, W.; Li, W.T.; Wen, C.K.; Hu, K.; Chen, C.; Liu, X. System design of the internet of things for residential smart grid. *IEEE Wirel. Commun.* **2016**, *23*, 90–98. [CrossRef]
5. Minoli, D.; Sohraby, K.; Occhiogrosso, B. IoT considerations, requirements, and architectures for smart buildings—Energy optimization and next-generation building management systems. *IEEE Internet Things J.* **2017**, *4*, 269–283. [CrossRef]
6. Forouzandehmehr, N.; Han, Z.; Zheng, R. Stochastic dynamic game between hydropower plant and thermal power plant in smart grid networks. *IEEE Syst. J.* **2014**, *10*, 88–96. [CrossRef]
7. Pahasa, J.; Ngamroo, I. Coordinated control of wind turbine blade pitch angle and PHEVs using MPCs for load frequency control of microgrid. *IEEE Syst. J.* **2014**, *10*, 97–105. [CrossRef]
8. Abdrabou, A. A wireless communication architecture for smart grid distribution networks. *IEEE Syst. J.* **2014**, *10*, 251–261. [CrossRef]
9. Shang, Y. Consensus of hybrid multi-agent systems with malicious nodes. *IEEE Trans. Circuits Syst. II Express Briefs* **2019**. [CrossRef]
10. Eid, B.M.; Rahim, N.A.; Selvaraj, J.; El Khateb, A.H. Control methods and objectives for electronically coupled distributed energy resources in microgrids: A review. *IEEE Syst. J.* **2014**, *10*, 446–458. [CrossRef]
11. Kahrobaeian, A.; Mohamed, Y.A.R.I. Interactive distributed generation interface for flexible micro-grid operation in smart distribution systems. *IEEE Trans. Sustain. Energy* **2012**, *3*, 295–305. [CrossRef]
12. Kumar, S.; Lee, G.K.F.; Ozturk, Y. Residential and Commercial Energy Management System. U.S. Patent 13/660,876, 6 June 2013.
13. Muralitharan, K.; Sakthivel, R.; Shi, Y. Multiobjective optimization technique for demand side management with load balancing approach in smart grid. *Neurocomputing* **2016**, *177*, 110–119. [CrossRef]
14. Rahman, A.; Liu, X.; Kong, F. A survey on geographic load balancing based data center power management in the smart grid environment. *IEEE Commun. Surv. Tutor.* **2013**, *16*, 214–233. [CrossRef]
15. Hassan, N.; Pasha, M.; Yuen, C.; Huang, S.; Wang, X. Impact of scheduling flexibility on demand profile flatness and user inconvenience in residential smart grid system. *Energies* **2013**, *6*, 6608–6635. [CrossRef]
16. Zhang, S.; Liu, J.; Zhao, B.; Cao, J. Cloud computing-based analysis on residential electricity consumption behavior. *Power Syst. Technol.* **2013**, *37*, 1542–1546.
17. Li, H.; Lin, X.; Yang, H.; Liang, X.; Lu, R.; Shen, X. EPPDR: An efficient privacy-preserving demand response scheme with adaptive key evolution in smart grid. *IEEE Trans. Parallel Distrib. Syst.* **2013**, *25*, 2053–2064. [CrossRef]
18. Mohamed, M.A.; Eltamaly, A.M.; Farh, H.M.; Alolah, A.I. Energy management and renewable energy integration in smart grid system. In Proceedings of the 2015 IEEE International Conference on Smart Energy Grid Engineering (SEGE), Oshawa, ON, Canada, 17–19 August 2015; pp. 1–6.
19. Al-Aboosi, F.; El-Halwagi, M. An integrated approach to water-energy nexus in shale-gas production. *Processes* **2018**, *6*, 52. [CrossRef]
20. Tushar, W.; Chai, B.; Yuen, C.; Smith, D.B.; Wood, K.L.; Yang, Z.; Poor, H.V. Three-party energy management with distributed energy resources in smart grid. *IEEE Trans. Ind. Electron.* **2014**, *62*, 2487–2498. [CrossRef]
21. Chen, D.; Kalra, S.; Irwin, D.; Shenoy, P.; Albrecht, J. Preventing occupancy detection from smart meters. *IEEE Trans. Smart Grid* **2015**, *6*, 2426–2434. [CrossRef]
22. Gomez-Vilardebo, J.; Gündüz, D. Smart meter privacy for multiple users in the presence of an alternative energy source. *IEEE Trans. Inf. Forensics Secur.* **2014**, *10*, 132–141. [CrossRef]
23. Tan, O.; Gunduz, D.; Poor, H.V. Increasing smart meter privacy through energy harvesting and storage devices. *IEEE J. Sel. Areas Commun.* **2013**, *31*, 1331–1341. [CrossRef]

24. Rogers, J.; Averyt, K.; Clemmer, S.; Davis, M.; Flores-Lopez, F.; Frumhoff, P.; Kenney, D.; Macknick, J.; Madden, N.; Meldrum, J.; et al. *Water-Smart Power: Strengthening the US Electricity System in a Warming World*; Union of Concerned Scientists: Cambridge, MA, USA, 2013.
25. Averyt, K.; Macknick, J.; Rogers, J.; Madden, N.; Fisher, J.; Meldrum, J.; Newmark, R. Water use for electricity in the United States: An analysis of reported and calculated water use information for 2008. *Environ. Res. Lett.* **2013**, *8*, 015001. [CrossRef]
26. Kanyerere, T.; Tramberend, S.; Levine, A.D.; Mokoena, P.; Mensah, P.; Chingombe, W.; Goldin, J.; Fatima, S.; Prakash, M. Water futures and solutions: Options to enhance water security in sub-Saharan Africa. In *Systems Analysis Approach for Complex Global Challenges*; Springer: Cham, Switzerland, 2018; pp. 93–111.
27. Miceli, R. Energy management and smart grids. *Energies* **2013**, *6*, 2262–2290. [CrossRef]
28. Dodder, R.S.; Barnwell, J.T.; Yelverton, W.H. Scenarios for low carbon and low water electric power plant operations: Implications for upstream water use. *Environ. Sci. Technol.* **2016**, *50*, 11460–11470. [CrossRef] [PubMed]
29. Marler, R.T.; Arora, J.S. Survey of multi-objective optimization methods for engineering. *Struct. Multidiscip. Optim.* **2004**, *26*, 369–395. [CrossRef]
30. Das, I.; Dennis, J.E. A closer look at drawbacks of minimizing weighted sums of objectives for Pareto set generation in multicriteria optimization problems. *Struct. Optim.* **1997**, *14*, 63–69. [CrossRef]
31. Armstrong, N.R.; Shallcross, R.C.; Ogden, K.; Snyder, S.; Achilli, A.; Armstrong, E.L. Challenges and opportunities at the nexus of energy, water, and food: A perspective from the southwest United States. *MRS Energy Sustain.* **2018**, *5*. [CrossRef]
32. Javaid, N.; Ullah, I.; Akbar, M.; Iqbal, Z.; Khan, F.A.; Alrajeh, N.; Alabed, M.S. An intelligent load management system with renewable energy integration for smart homes. *IEEE Access* **2017**, *5*, 13587–13600. [CrossRef]
33. Clauser, C.; Ewert, M. The renewables cost challenge: Levelized cost of geothermal electric energy compared to other sources of primary energy–review and case study. *Renew. Sustain. Energy Rev.* **2018**, *82*, 3683–3693. [CrossRef]
34. Parra, D.; Norman, S.A.; Walker, G.S.; Gillott, M. Optimum community energy storage for renewable energy and demand load management. *Appl. Energy* **2017**, *200*, 358–369. [CrossRef]
35. Smith, J.C.; Milligan, M.R.; DeMeo, E.A.; Parsons, B. Utility wind integration and operating impact state of the art. *IEEE Trans. Power Syst.* **2007**, *22*, 900–908. [CrossRef]
36. Bird, L.; Milligan, M. *Lessons from Large-Scale Renewable Energy Integration Studies*; Technical Report; National Renewable Energy Lab. (NREL): Golden, CO, USA, 2012.
37. Brouwer, A.S.; Van Den Broek, M.; Seebregts, A.; Faaij, A. Impacts of large-scale Intermittent Renewable Energy Sources on electricity systems, and how these can be modeled. *Renew. Sustain. Energy Rev.* **2014**, *33*, 443–466. [CrossRef]
38. Wang, Y.; Lin, X.; Pedram, M. A near-optimal model-based control algorithm for households equipped with residential photovoltaic power generation and energy storage systems. *IEEE Trans. Sustain. Energy* **2015**, *7*, 77–86. [CrossRef]
39. Al Busaidi, A.S.; Kazem, H.A.; Al-Badi, A.H.; Khan, M.F. A review of optimum sizing of hybrid PV–Wind renewable energy systems in oman. *Renew. Sustain. Energy Rev.* **2016**, *53*, 185–193. [CrossRef]
40. Pereira, M.; Limon, D.; de la Peña, D.M.; Valverde, L.; Alamo, T. Periodic economic control of a nonisolated microgrid. *IEEE Trans. Ind. Electron.* **2015**, *62*, 5247–5255. [CrossRef]
41. Pereira, M.; de la Peña, D.M.; Limón, D. Robust economic model predictive control of a community micro-grid. *Renew. Energy* **2017**, *100*, 3–17. [CrossRef]
42. Zhang, X.; Bao, J.; Wang, R.; Zheng, C.; Skyllas-Kazacos, M. Dissipativity based distributed economic model predictive control for residential microgrids with renewable energy generation and battery energy storage. *Renew. Energy* **2017**, *100*, 18–34. [CrossRef]
43. Sechilariu, M.; Wang, B.; Locment, F. Building integrated photovoltaic system with energy storage and smart grid communication. *IEEE Trans. Ind. Electron.* **2012**, *60*, 1607–1618. [CrossRef]
44. Lee, C.K.; Hui, S.Y.R. Reduction of energy storage requirements in future smart grid using electric springs. *IEEE Trans. Smart Grid* **2013**, *4*, 1282–1288. [CrossRef]
45. Ding, Y.M.; Hong, S.H.; Li, X.H. A demand response energy management scheme for industrial facilities in smart grid. *IEEE Trans. Ind. Inform.* **2014**, *10*, 2257–2269. [CrossRef]

46. Borghetti, A.; D'Ambrosio, C.; Lodi, A.; Martello, S. An MILP approach for short-term hydro scheduling and unit commitment with head-dependent reservoir. *IEEE Trans. Power Syst.* **2008**, *23*, 1115–1124. [CrossRef]
47. Pham, T.H.; Wurtz, F.; Bacha, S. Optimal operation of a PV based multi-source system and energy management for household application. In Proceedings of the 2009 IEEE International Conference on Industrial Technology, Gippsland, Australia, 10–13 February 2009; pp. 1–5.
48. Lu, B.; Shahidehpour, M. Short-term scheduling of battery in a grid-connected PV/battery system. *IEEE Trans. Power Syst.* **2005**, *20*, 1053–1061. [CrossRef]
49. Rasheed, M.; Javaid, N.; Ahmad, A.; Jamil, M.; Khan, Z.; Qasim, U.; Alrajeh, N. Energy optimization in smart homes using customer preference and dynamic pricing. *Energies* **2016**, *9*, 593. [CrossRef]
50. Margaret, V.; Rao, K.U.; Ganeshprasad, G. Intelligent load shedding using ant colony algorithm in smart grid environment. In *Power Electronics and Renewable Energy Systems*; Springer: New Delhi, India, 2015; pp. 1149–1162.
51. Stoppato, A.; Cavazzini, G.; Ardizzon, G.; Rossetti, A. A PSO (particle swarm optimization)-based model for the optimal management of a small PV (Photovoltaic)-pump hydro energy storage in a rural dry area. *Energy* **2014**, *76*, 168–174. [CrossRef]
52. Marini, F.; Walczak, B. Particle swarm optimization (PSO). A tutorial. *Chemom. Intell. Lab. Syst.* **2015**, *149*, 153–165. [CrossRef]
53. Phuangpornpitak, N.; Tia, S. Opportunities and challenges of integrating renewable energy in smart grid system. *Energy Procedia* **2013**, *34*, 282–290. [CrossRef]
54. Samadi, P.; Mohsenian-Rad, H.; Schober, R.; Wong, V.W. Advanced demand side management for the future smart grid using mechanism design. *IEEE Trans. Smart Grid* **2012**, *3*, 1170–1180. [CrossRef]
55. Li, N.; Chen, L.; Low, S.H. Optimal demand response based on utility maximization in power networks. In Proceedings of the 2011 IEEE Power and Energy Society General Meeting, Detroit, MI, USA, 24–28 July 2011; pp. 1–8.
56. Basit, A.; Sidhu, G.A.S.; Mahmood, A.; Gao, F. Efficient and autonomous energy management techniques for the future smart homes. *IEEE Trans. Smart Grid* **2015**, *8*, 917–926. [CrossRef]
57. Erol-Kantarci, M.; Mouftah, H.T. Wireless sensor networks for cost-efficient residential energy management in the smart grid. *IEEE Trans. Smart Grid* **2011**, *2*, 314–325. [CrossRef]
58. Ma, J.; Chen, H.H.; Song, L.; Li, Y. Residential load scheduling in smart grid: A cost efficiency perspective. *IEEE Trans. Smart Grid* **2015**, *7*, 771–784. [CrossRef]
59. Saber, A.Y.; Venayagamoorthy, G.K. Resource scheduling under uncertainty in a smart grid with renewables and plug-in vehicles. *IEEE Syst. J.* **2011**, *6*, 103–109. [CrossRef]
60. Yang, J.S.; Choi, J.Y.; An, G.H.; Choi, Y.J.; Kim, M.H.; Won, D.J. Optimal scheduling and real-time state-of-charge management of energy storage system for frequency regulation. *Energies* **2016**, *9*, 1010. [CrossRef]
61. Javaid, N.; Ahmed, F.; Ullah, I.; Abid, S.; Abdul, W.; Alamri, A.; Almogren, A. Towards cost and comfort based hybrid optimization for residential load scheduling in a smart grid. *Energies* **2017**, *10*, 1546. [CrossRef]
62. Rasheed, M.B.; Javaid, N.; Malik, M.S.A.; Asif, M.; Hanif, M.K.; Chaudary, M.H. Intelligent multi-agent based multilayered control system for opportunistic load scheduling in smart buildings. *IEEE Access* **2019**, *7*, 23990–24006. [CrossRef]
63. Hung, D.Q.; Mithulananthan, N.; Lee, K.Y. Determining PV penetration for distribution systems with time-varying load models. *IEEE Trans. Power Syst.* **2014**, *29*, 3048–3057. [CrossRef]
64. Khatod, D.K.; Pant, V.; Sharma, J. Evolutionary programming based optimal placement of renewable distributed generators. *IEEE Trans. Power Syst.* **2012**, *28*, 683–695. [CrossRef]
65. Teng, J.H.; Luan, S.W.; Lee, D.J.; Huang, Y.Q. Optimal charging/discharging scheduling of battery storage systems for distribution systems interconnected with sizeable PV generation systems. *IEEE Trans. Power Syst.* **2012**, *28*, 1425–1433. [CrossRef]
66. Salameh, Z.M.; Borowy, B.S.; Amin, A.R. Photovoltaic module-site matching based on the capacity factors. *IEEE Trans. Energy Convers.* **1995**, *10*, 326–332. [CrossRef]
67. Joo, I.Y.; Choi, D.H. Distributed optimization framework for energy management of multiple smart homes with distributed energy resources. *IEEE Access* **2017**, *5*, 15551–15560. [CrossRef]
68. Paul, S.; Padhy, N.P. Resilient Scheduling Portfolio of Residential Devices and Plug-in Electric Vehicle by Minimizing Conditional Value at Risk. *IEEE Trans. Ind. Inform.* **2018**, *15*, 1566–1578. [CrossRef]

69. Jordehi, A.R. Binary particle swarm optimisation with quadratic transfer function: A new binary optimisation algorithm for optimal scheduling of appliances in smart homes. *Appl. Soft Comput.* **2019**, *78*, 465–480. [CrossRef]
70. Ma, K.; Hu, S.; Yang, J.; Xu, X.; Guan, X. Appliances scheduling via cooperative multi-swarm PSO under day-ahead prices and photovoltaic generation. *Appl. Soft Comput.* **2018**, *62*, 504–513. [CrossRef]
71. Hossain, E.; Khan, I.; Un-Noor, F.; Sikander, S.S.; Sunny, M.S.H. Application of Big Data and Machine Learning in Smart Grid, and Associated Security Concerns: A Review. *IEEE Access* **2019**, *7*, 13960–13988. [CrossRef]
72. Zhao, Z.; Lee, W.C.; Shin, Y.; Song, K.B. An optimal power scheduling method for demand response in home energy management system. *IEEE Trans. Smart Grid* **2013**, *4*, 1391–1400. [CrossRef]
73. Javaid, N.; Javaid, S.; Abdul, W.; Ahmed, I.; Almogren, A.; Alamri, A.; Niaz, I. A hybrid genetic wind driven heuristic optimization algorithm for demand side management in smart grid. *Energies* **2017**, *10*, 319. [CrossRef]
74. Iqbal, M.M.; Sajjad, I.A.; Amin, S.; Haroon, S.S.; Liaqat, R.; Khan, M.F.N.; Waseem, M.; Shah, M.A. Optimal Scheduling of Residential Home Appliances by Considering Energy Storage and Stochastically Modelled Photovoltaics in a Grid Exchange Environment Using Hybrid Grey Wolf Genetic Algorithm Optimizer. *Appl. Sci.* **2019**, *9*, 5226. [CrossRef]

sustainability

Article

An Intelligent Approach to Active and Reactive Power Control in a Grid-Connected Solar Photovoltaic System

Ibrahim Alsaidan [1], Priyanka Chaudhary [2,3,*], Muhannad Alaraj [1] and Mohammad Rizwan [1,2]

1 Department of Electrical Engineering, College of Engineering, Qassim University, Buraydah 52571, Qassim, Saudi Arabia; Alsaidan@qu.edu.sa (I.A.); Muhannad@qu.edu.sa (M.A.); MR.Khan@qu.edu.sa (M.R.)
2 Department of Electrical Engineering, Delhi Technological University, Delhi 110042, India
3 Department of Electrical Engineering, SET, Noida International University, Noida 203201, India
* Correspondence: priyankach.iilm@gmail.com

Abstract: The increasing demand of electrical energy and environmental concerns are invigorating the use of renewable energy resources for power generation. Renewable energy resources can provide an attractive solution for present and future energy requirements. In this scenario, solar photovoltaic systems are becoming prominent and sustainable solutions with numerous advantages. However, the utilization of solar photovoltaic systems in distribution generation makes it mandatory to deploy efficient and organized control measures for integrating solar photovoltaic plants with the grid. In this paper, the control of grid-tied solar photovoltaic systems using a Kalman filter-based generalized neural network is presented with a variable step size perturb and observe-based maximum power point tracking controller to extract the maximum power from a solar photovoltaic plant. The presented system provides power-quality enhancement and supports a three-phase AC grid. The proposed approach extracts the load currents' primary components for efficient harmonics elimination, synchronizes the system with the grid and provides a fast response during rapidly changing conditions. The results of the proposed control technique are also compared with the artificial neural network-based control technique for validation purposes. The proposed algorithm is found more suitable for using a smaller number of unknown weights and training patterns with reduced computational time.

Keywords: renewable energy resources; grid integrated solar PV systems; sustainable power generation; maximum power point tracking; grid reliability and voltage source converter

Citation: Alsaidan, I.; Chaudhary, P.; Alaraj, M.; Rizwan, M. An Intelligent Approach to Active and Reactive Power Control in a Grid-Connected Solar Photovoltaic System. *Sustainability* **2021**, *13*, 4219. https://doi.org/10.3390/su13084219

Academic Editors: M. Dolores Esteban, José-Santos López-Gutiérrez and Vicente Negro

Received: 6 March 2021
Accepted: 30 March 2021
Published: 10 April 2021

1. Introduction

The increasing concern regarding greenhouse gas emissions and the depleting nature of conventional fuels has invigorated the use of renewable energy resources as an alternative and sustainable solutions for the power sector. Solar, wind, biomass and small hydro power are the main renewable energy-based resources which can fulfill the future energy requirements. The solar photovoltaic system (SPV) is more encouraged due to various advantages such as abundance of availability, less environmental pollution, reduced cost and many others. The progress of an SPV system in the existing power system is witnessed for its progress [1]. The integration of an SPV system into the grid causes positive effects, i.e., generating more power, along with some negative effects, i.e., the violation of voltage limitations at common coupling, frequencies disruption and grid stabilization problems, etc. There are set guidelines, codes and standards for grid-connected SPV systems. These standards include IEEE 1547, IEC 61727 and VDE-AR-N4105 [2]. The use of these standards is encouraged and imposed to maintain the stability and power quality related to the grid. The large-scale exploitation of distributed generation make it mandatory to deploy efficiency and organized control measures for integrating and measuring problems. The control algorithms are helpful for the accommodation and facilitation of the integration of an SPV in the distribution grid. Single-stage and two-stage systems have been used for the integration of SPV systems into the grid, as shown in previous works. It has also been shown by various researchers that for a three-phase system, a single-stage configuration is more advantageous [3]. However, conventional single-stage SPV systems suffer from the drawback of converters being idle during the unavailability of power from the SPV system. To overcome the abovementioned problems, the proposed system is designed in such a manner so that it operates as a distribution static compensator (DSTATCOM) while SPV power is not available and restores the operation after recovering the SPV power generation.

The electrical power distribution system has been designed to distribute active power pertaining a smaller number of harmonics to the consumers. DSTATCOM integrated as a shunt compensator with a distribution system effectively controls harmonics, fulfils the demand of reactive power of consumers and improves voltage regulation with balanced and unbalanced loads [4]. The researchers have reported various configurations of DSTATCOM in the literature [5–8]. For efficiently integrating distributed a generation system that compensates for the reactive power and harmonics for utility grids at common coupling points, several converter topologies [9–11] and control strategies [12–18] were reported in the previous work. Additionally, Jain et al. [19] presented a grid-integrated solar energy conversion setup with a movable direct current link voltage to vary the voltages at common pairing points with two-stage circuitry methodology and an associated double-frequency second-order generalized integrator (DFSOGI)-dependent control strategy for controlling the multi-functional voltage source converter (VSC) in abrupt load modification at common connecting points.

Varma et al. [19] have proposed a solar photovoltaic static compensator for improving the power transfer capabilities of the system to transmit real power to utilities with an existing converter system. Humid et al. [20] proposed a strategy that depends on a power conditioning unit located parallel to the plant that works in feed-forward mode to compensate the distorted current photovoltaic output to decrease the harmonic of current from a PV system. Kannan et al. [21] presented icos Φ control technique for a distribution static compensator that gives uninterrupted harmonic degradation, compensates re-active power and compensates the loads along with comparing performances of fuzzy-logic controllers to ordinary proportional integral (PI) controllers. Mishra et al. [22] implemented an optimistic control technique which optimizes the proportional integral's coefficients and photovoltaic fed distribution static compensator's filtration parameters.

To enhance the performance of grid the integrated SPV plant, various control techniques were adopted by the researchers. There are some performance indicators such as less computational time, high accuracy and less complexity, etc. A huge number of techniques depending upon artificial neural networks (ANN) have been widely adopted by a large number of journalists for distribution static compensators implementing dynamic loading [23–27], whereas photovoltaic fed distribution static compensator systems have been adopted by very few. Singh et al. [28] presented a grid interfaced SPV generation plant implementing a neural network-based control technique that utilizes the least mean square (LMS) algorithm, termed adaline (adaptive linear element), for estimating the reference source current. Artificial neural network-based approaches are complex and need large computational times. Further, generalized neural network (GNN)-based control strategies can be used to overcome the abovementioned issues. Applications of GNN are widely available in load forecasting, solar irradiance/energy forecasting, load frequency control in power systems, power system stabilizers, electrical machines and control system-related problems [29–36]. There are few works are reported in the literature related to the application of GNN as a control strategy for grid-tied SPV systems. GNN-based models reduce the training time as well as improve the performance of the system. Considering this, a GNN-based control strategy has been developed for a PV DSTATCOM system in the present work.

The converting capability of a solar photovoltaic plant is comparatively small; thus, to increase the capability of solar photovoltaic plants, it is essential to keep a track of the maximum power point. Maximum Power Point Tracking (MPPT) is complicated because of the non-linear character of SPV plants due to changing meteorological factors, i.e., solar irradiance, ambient temperature, wind velocity and relative humidity, etc. A lot work has been reported in the literature related to MPPT, including techniques such as perturb and observe, incremental conductance, constant voltage, open-circuit voltage, short-circuit current, extremum seeking control and hybrid, etc. Further, there are a number of intelligent techniques also available for MPPT such as artificial neural networks (ANN), fuzzy logic, genetic algorithms, etc. [37–39]. A changeable step size perturb and observe (P&O) MPPT technique is utilized in this research to track the MPP of a solar photovoltaic plant.

Further, to improve the performance of the system, an extended Kalman filter (EKF) [40–42]-based method is adopted to estimate and update the weights of the GNN model. The main contributions of this work are:

- Active power feeding to the connected loads and grid with mitigation of power quality issues. A generalized neural network (GNN)-based approach plays the role of primary control strategy and decides the switching pattern of the voltage source converter (VSC).
- Further, the performance of the proposed algorithm has been improved with the help of EKF for GNN weight estimations.
- The performance of the proposed setup is validated using simulation results implemented in the MATLAB/ Simulink platform.
- The developed system obtains acceptable limits of harmonics in utility currents and voltage fluctuations according to the IEEE-519 and IEEE-1547 standards.

The developed system functions very well with an EKF GNN-based approach and gives a very fast response. Moreover, single-stage topology is able to reduce the losses in semiconductor devices and increase the overall efficiency. The proposed control approach performs with more flexibility in training the network under dynamic conditions. The proposed technique possesses various advantages such as its speed, using a single layer, less mathematical calculations and easy implementation on hardware.

The organization of paper is as follows: Section 2 provides details about the system configuration, followed by Section 3 which explains the developed control technique. The simulation and hardware results are provided in Section 4. The conclusion is given in Section 5, followed by acknowledgements and references.

2. System Description

The proposed system deploys a single-stage circuit topology and consists of a solar PV array, voltage source converter (VSC), ripple filter, loads and a three-phase utility grid, as shown in Figure 1. A solar PV array is a combination of several PV modules in a series and parallel according to the requirements. A PV cell serves as the basic unit for a solar PV module. A single diode solar PV cell is considered for this system, and mathematical modeling is given [43]. The developed system has been implemented with a minimum number of sensors with a variable step size perturb and observe (P&O) MPPT algorithm to achieve maximum power point operation of the solar PV system for different meteorological parameters. An extended Kalman filter (EKF)-based GNN methodology is used for controlling the VSC switching to synchronize the system with the grid and the main task of separating the weights of the primary real and reactive parts of three-phase load currents to obtain reference grid currents. The developed system design specifications are given below:

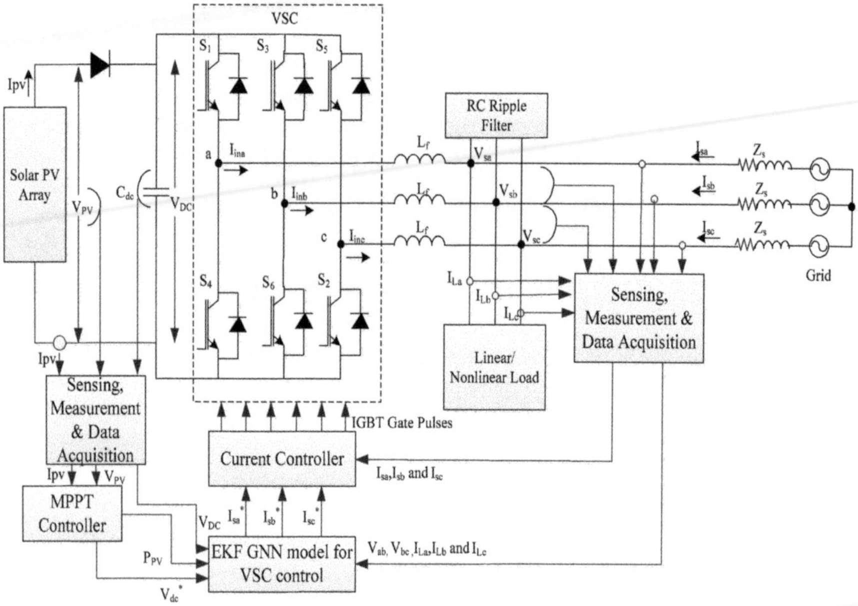

Figure 1. Schematic representation of the system used in this research.

Grid description = three-phase, 415 Volts, 50 Hz, solar PV array ratings = (V_{MPP}) = 700 Volts, current (IMPP) = 13.5 Amp, solar photovoltaic power at maximum power point (PMPP) = 10 kW, rating of interfacing inductors (L_{fa} = L_{fb} = L_{fc}) = 2.6 mH, rating of DC link capacitor (C_{DC}) = 10 mF, DC link voltage (VDC) = 700 V.

3. Extended Kalman Filter-Based GNN Control Algorithm

3.1. Maximum Power Point Tracking Control

A variable step size perturb and observe (P&O)-based MPPT algorithm is employed in this research for the extraction of maximum power from the SPV system. The main operation of the given algorithm is dividing the SPV system's dP_{pv}/dV_{pv} curve into three independent regions. Region 0 denotes the SPV system power's closeness to the highest power point. Region 0 uses the normal value of tracking step size, whereas region 1 and region 2 needed a larger value of step size as compared to region 0 for achieving the high tracking speed. Figure 2 explains the performed functions of the proposed technique. ΔVref0, ΔVref1 and ΔVref2 represent the step size of tracking for region 0, region 1 and region 2, respectively.

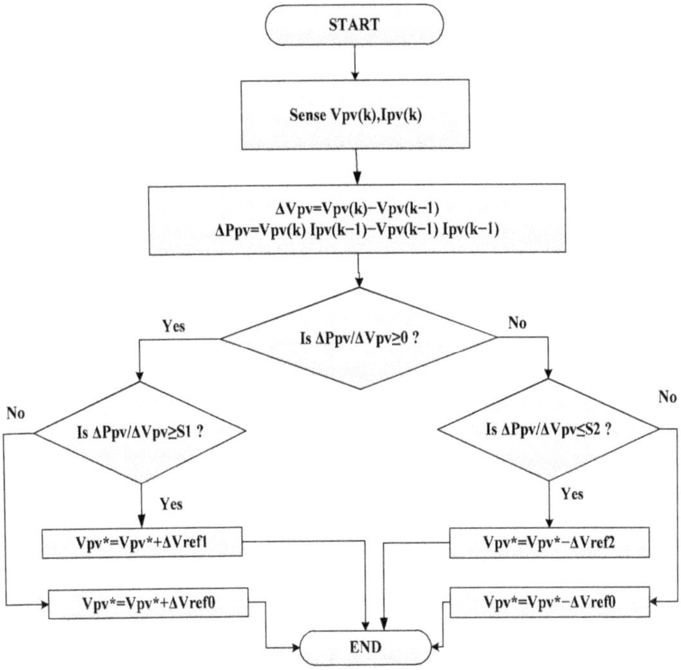

Figure 2. Flowchart of variable step size perturb and observe (P&O) algorithm.

3.2. *Extended Kalman Filter-Based GNN Control Algorithm*

A generalized neural network works like a multi-layer feed-forward network, where every node implements a precise function on all signals that come to the node, and the parameters are set referring to the node. Figure 3 shows an aggregate-type format of a GNN model, with Σ and Π being the aggregated functions. Σ is an aggregated function, which is adopted with the sigmoidal characteristic function f_1, whereas the Π aggregation function is adopted with the Gaussian function f_2. The derivation of active and re-active load current components is performed by assuming load current (i_{La}, i_{Lb}, i_{Lc}) as the input to the summation (Σ_A) and product (Π) neurons having undefined weights of ($W_{\Sigma i}$) and ($W_{\Pi i}$), respectively.

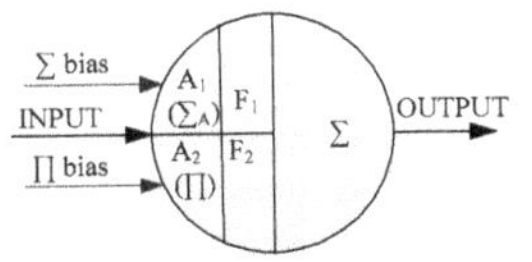

(a)

Figure 3. *Cont.*

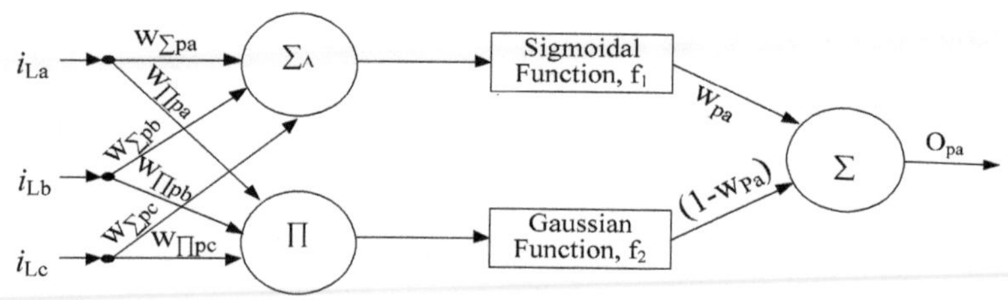

(b)

Figure 3. (**a**) Summation-type generalized neural network (GNN) model, (**b**) a summation-type GNN structure to determine the fundamental active element of a load current.

3.2.1. Estimation of Amplitude of Terminal Voltage and Unit Templates

With the help of sensed line voltages (v_{ab}, v_{bc}, v_{ca}) at the point of common coupling (PCC), the amplitude of phase voltages (v_a, v_b and v_c) are calculated as follows:

$$v_a = \frac{2v_{ab} + v_{bc}}{3}, \; v_b = \frac{-v_{ab} + v_{bc}}{3}, \; v_c = \frac{-v_{ab} - 2v_{bc}}{3} \Sigma \tag{1}$$

The terminal voltage amplitude at PCC can be estimated as

$$V_t = \sqrt{\left[\frac{2({v_a}^2 + {v_b}^2 + {v_c}^2)}{3}\right]} \tag{2}$$

$$u_{pa} = \frac{v_a}{V_t}, u_{pb} = \frac{v_b}{V_t}, \; u_{pc} = \frac{v_c}{V_t} \tag{3}$$

Further, quadrature unit templates can be calculated using in-phase unit templates as

$$u_{qa} = \frac{u_{pc}}{\sqrt{3}} - \frac{u_{pb}}{\sqrt{3}}, u_{qb} = \frac{\sqrt{3}u_{pa}}{2} + \frac{u_{pb} - u_{pc}}{2\sqrt{3}}, u_{qc} = \frac{u_{pb} - u_{pc}}{2\sqrt{3}} - \frac{\sqrt{3}u_{pa}}{2} \tag{4}$$

3.2.2. Terminal Voltage Amplitude and Unit Templates

The DC link voltage (V_{dc}) is sensed and compared with the reference DC link voltage (V_{dc}^*) in order to determine the active loss component.

$$V_{dce}(k) = V_{dc}^*(k) - V_{dc}(k) \tag{5}$$

k is the number of iterations. Further, the error (V_{dce}) of reference V_{dc}^* and sensed V_{dc} are processed through a PI controller, the output of which is an active current component $\left(w_{pdc}\right)$ used to regulate the DC link. The controller output at k^{th} iteration is estimated as

$$w_{pd}(k) = w_{pd}(k-1) + k_{pd}\{V_{dce}(k) - V_{dce}(k-1)\} + k_{id}V_{dce}(k). \tag{6}$$

During no sunshine conditions ($P_{PV} = 0$), the set point DC link voltage is set to reference the DC link voltage of DSTATCOM so that the system operates in the power quality improvement mode (as a DSTATCOM).

The error V_{te} has been considered between actual and set terminal voltage, V_t and V_t^*, at PCC to calculate the reactive loss component. Further, this error is passed through a PI controller for minimization. The calculated reactive loss component is considered to maintain AC terminal voltage constant and close to its reference value.

$$V_{te}(k) = V_t^*(k) - V_t(k) \tag{7}$$

The controller output at k^{th} instant is

$$w_{qt}(k) = w_{qt}(k-1) + k_{pt}\{V_{te}(k) - V_{te}(k-1)\} + k_{it}V_{te}(k) \tag{8}$$

where w_{qt} is a part of the reactive loss current component, and k_{pt} and k_{it} are proportional and integral gains, respectively.

A feed-forward weight is calculated and incorporated into the controller to achieve a fast dynamic response, which can be written as:

$$w_{PV}(k) = \frac{2P_{PV}(k)}{3V_t} \tag{9}$$

where P_{PV} is solar power.

3.2.3. Fundamental Active and Reactive Component of Load Current

A generalized neural network (GNN) of a summation-type network is used in the present work for the estimation of the fundamental active and reactive components of the load current. The output calculations have been divided into two sections: forward calculations and reverse calculations.

With the help of Figure 3, the below mentioned equation is used to obtain the output of the summation part of the GNN:

$$O_\Sigma = f_1(\Sigma W_{\Sigma i}X_i + X_{o\Sigma}) \tag{10}$$

Similarly, the output of the product part of the GNN can be calculated as

$$O_\Pi = f_2(\Pi W_{\Pi i}X_i + X_{o\Pi}) \tag{11}$$

The output of the final GNN will be the sum of the summation part and product part which can be expressed as

$$\boldsymbol{O_i = O_\Sigma * W_\Sigma + O_\Pi(1 - W_\Sigma)} \tag{12}$$

In the above equation, O_Σ denotes the output of the summation part of the neuron, O_Π shows the output of the product part of the neuron, and W depicts the weights. The proposed control strategy using GNN utilizes a GNN model to determine the fundamental active current components of the sensed load currents (i_{La}, i_{Lb}, i_{Lc}). The sensed load currents are the input to the proposed model which are further multiplied with weights $(w_{\Sigma pa}, w_{\Sigma pb}, w_{\Sigma pc})$ and $(w_{\Pi pa}, w_{\Pi pb}, w_{\Pi pc})$ at summation (Σ_A) and product (Π) neurons, respectively. At the initial level, these weights are referred as unknown weights and calculated by considering the in-phase unit templates (u_{pa}, u_{pb}, u_{pc}) as the reference weights. A summation-type GNN structure for the calculation of fundamental active current component of a phase is given in Figure 3b.

The output calculations can be divided into two parts: forward calculations and reverse calculations. The aggregation function of Σ_A and Π of the forward mode calculations for phase "a" is shown in Equations (8) and (9), respectively. Similarly, the aggregation function of Σ_A and Π of the forward mode calculations of another two phases, "b" and "c", can be calculated.

$$\Sigma_{A_{Pa}} = i_{La}w_{\Sigma pa} + i_{Lb}w_{\Sigma pb} + i_{Lc}w_{\Sigma pc} + \Sigma_{bias} \tag{13}$$

$$\Pi_{pa} = i_{La}w_{\Pi pa} * i_{Lb}w_{\Pi pb} * i_{Lc}w_{\Pi pc} * \Pi_{bias} \tag{14}$$

Σ_{bias} and Π_{bias} represent the initial bias of the Σ_A and Π part of the structure, respectively. The proposed developed neuron has both Σ and Π aggregation functions. The Σ_A aggregation function has been used with the sigmoidal characteristic function f_1, while the Π aggregation function has been used with the Gaussian function f_2 as a characteristic function. The output of the Σ_A part with a sigmoidal characteristic transfer function after threshold can be calculated as

$$O_{\Sigma_{Apa}} = f_1(\Sigma_{Apa}) = \frac{1}{1 + e^{-\lambda_{\Sigma p} * \Sigma_{A_{Pa}}}} \tag{15}$$

The output of the Π part is threshold by using Gaussian transfer function and can be written as

$$O_{\Pi pa} = f_2(\Pi_{pa}) = e^{-\lambda_{\Pi p} * \Pi_{pa}^2} \tag{16}$$

where $\lambda_{\Sigma p}$ and $\lambda_{\Pi p}$ are the gain scaling parameters of the Σ_A and Π part of the network, respectively, and are considered as unity here to avoid complexity. Similarly, output of the Σ_A and Π part of the forward mode calculations of other phases, "b" and "c", are calculated as

$$O_{\Sigma_{Apb}} = f_1\left(\Sigma_{Apb}\right) = \frac{1}{1 + e^{-\lambda_{\Sigma p} * \Sigma_{A_{Pb}}}} \tag{17}$$

$$O_{\Pi pb} = f_2\left(\Pi_{pb}\right) = e^{-\lambda_{\Pi p} * \Pi_{pb}^2} \tag{18}$$

$$O_{\Sigma_{Apc}} = f_1(\Sigma_{Apc}) = \frac{1}{1 + e^{-\lambda_{\Sigma p} * \Sigma_{A_{Pc}}}} \tag{19}$$

$$O_{\Pi pc} = f_2(\Pi_{pc}) = e^{-\lambda_{\Pi p} * \Pi_{pc}^2} \tag{20}$$

The final output of the GNN will be the function of two outputs and related to the weights W and $(1 - W)$, respectively, through the linear transfer function which can be written as

$$O_{pa} = O_{\Pi pa}(1 - W_a) + O_{\Sigma Apa} W_a \tag{21}$$

$$O_{pb} = O_{\Pi pb}(1 - W_b) + O_{\Sigma Apb} W_b \tag{22}$$

$$O_{pc} = O_{\Pi pc}(1 - W_c) + O_{\Sigma Apc} W_c \tag{23}$$

where W_a, W_b, W_c are weights associated with the phase "a", "b" and "c", respectively. The mean active component of load currents (w_{Lp}) is obtained by averaging the final output of the GNN for each phase and is given as

$$w_{Lp} = \frac{\left(O_{pa} + O_{pb} + O_{pc}\right)}{3} \tag{24}$$

However, the estimation of the fundamental reactive current component under the load current has been calculated by implementing the proposed GNN model. The sensed load currents (i_{La}, i_{Lb}, i_{Lc}) with their unknown weights, $\left(w_{Aqa}, w_{Aqb}, w_{Aqc}\right)$ and $\left(w_{\Pi qa}, w_{\Pi qb}, w_{\Pi qc}\right)$, are processed as inputs to Σ_A and Π neurons. The aggregation function of Σ_A and Π of the forward mode calculations for phase "a" is given in Equations (23) and (24), respectively. Similarly, the aggregation function of Σ_A and Π of the forward mode calculations of the other remaining phases "b" and "c" are calculated.

$$\Sigma_{A_{qa}} = i_{La} w_{\Sigma qa} + i_{Lb} w_{\Sigma qb} + i_{Lc} w_{\Sigma qc} + \Sigma_{bias} \tag{25}$$

$$\Pi_{qa} = i_{La} w_{\Pi qa} * i_{Lb} w_{\Pi qb} * i_{Lc} w_{\Pi qc} * \Pi_{bias} \tag{26}$$

where Σ_{bias} and Π_{bias} denote the initial bias of the Σ_A and Π part of the network, respectively. Similarly, as explained above while calculating the fundamental active load current component, here also the Σ_A aggregation function has been adopted along with the sigmoidal characteristic function f_1, whereas the Π aggregation function has been adopted along with the Gaussian function f_2, taken as a characteristic function. The output of the Σ_A part with a sigmoidal characteristic transfer function after threshold can be obtained as

$$O_{\Sigma_{Aqa}} = f_1(\Sigma_{Aqa}) = \frac{1}{1+e^{-\lambda_{\Sigma q} * \Sigma_{Aqa}}} \tag{27}$$

The output of the Π part becomes threshold by using the Gaussian transfer function and can be written as given:

$$O_{\Pi qa} = f_2(\Pi_{qa}) = e^{-\lambda_{\Pi q} * \Pi_{qa}^2} \tag{28}$$

where $\lambda_{\Sigma q}$ and $\lambda_{\Pi q}$ are the gain scaling parameters of the Σ_A and Π part of the network, respectively, and are considered as unity here to avoid complexity. Similarly, the output of the Σ_A and Π part of the forward mode calculations of the other two phases "b" and "c" are calculated as

$$O_{\Sigma_{Aqb}} = f_1\left(\Sigma_{Aqb}\right) = \frac{1}{1+e^{-\lambda_{\Sigma q} * \Sigma_{Aqb}}} \tag{29}$$

$$O_{\Pi qb} = f_2\left(\Pi_{qb}\right) = e^{-\lambda_{\Pi q} * \Pi_{qb}^2} \tag{30}$$

$$O_{\Sigma_{Aqc}} = f_1(\Sigma_{Aqc}) = \frac{1}{1+e^{-\lambda_{\Sigma q} * \Sigma_{Aqc}}} \tag{31}$$

$$O_{\Pi qc} = f_2(\Pi_{qc}) = e^{-\lambda_{\Pi q} * \Pi_{qc}^2} \tag{32}$$

The output of the GNN model as a function of weights W and $(1 - W)$ can be written as

$$O_{qa} = O_{\Pi qa}(1 - W_{a1}) + O_{\Sigma Apa} W_{a1} \tag{33}$$

$$O_{qb} = O_{\Pi qb}(1 - W_{b1}) + O_{\Sigma Aqb} W_{b1} \tag{34}$$

$$O_{qc} = O_{\Pi qc}(1 - W_{c1}) + O_{\Sigma Aqc} W_{c1} \tag{35}$$

where W_{a1}, W_{b1} and W_{c1} are the weights associated with the Σ_A and Π part of the developed GNN model for the estimation of the fundamental reactive load current components. The fundamental reactive current component of the load current (w_{Lq}) is determined by taking the average of output of the GNN model for the reactive current component of the load currents and given as:

$$w_{Lq} = \frac{\left(O_{qa} + O_{qb} + O_{qc}\right)}{3} \tag{36}$$

3.2.4. GNN Weight Prediction and Updating Using Extended Kalman Filter (EKF)

The Kalman Filter (KF) is a recursive algorithm used for estimating the state of a dynamic system in the case of less availability of data because of the presence of noise, etc. The KF algorithm utilizes the prior knowledge to predict the past, present and future state of the given system. The main advantage of the KF-based approach is less memory space requirements because the data are updated in each and every iteration. The basic KF theory is based on the probability of the hypothesis of the predicted state of the system under consideration by hypothesis of prior state and then using the available data from measurement sensors to correct the hypothesis to obtain the best estimation for each iteration. Two basic assumptions are made to derive the basic equations for KF to be optimal in the sense of mean square error, which should be described by a model of linear state space; the noises are white and Gaussian with zero mean, uncorrelated with each other.

$$x_{k+1} = F_{k+1,\,k}\, x_k + q_k \tag{37}$$

$$y_{k+1} = H_{k+1}\, x_{k+1} + r_{k+1} \tag{38}$$

Equation (37) is known as a process equation. x_k is a system state vector, a minimal set of data that uniquely defines the behavior of a system, and k depicts discrete time. $F_{k+1,\,k}$ is the transition matrix to take the state x_k from time k to $k + 1$, and q_k is the additive process noise, white and Gaussian, with a zero mean and possessing a covariance matrix Q_k. The measurement step is shown in Equation (38), where H_{k+1} is the measurement matrix, y_{k+1} is observable at time $k + 1$, and r_{k+1} is the additive process noise, white and Gaussian, with a zero mean and possessing a covariance matrix R_k. Both noises are uncorrelated with each other. The KF algorithm works in two repeated functional steps:

The prediction step (time update): This step is to compute the estimation of state and error covariance.

$$\hat{x}^{-}{}_{k+1} = F_{k+1,\,k}\,\hat{x}_k \tag{39}$$

$$P_{k+1} = F_{k+1,\,k}\,P_k\,F^{T}{}_{k+1,\,k} + Q_k \tag{40}$$

where P_k is there error covariance matrix.

Correction step (measurement update): This step is to correct the estimated state according to the previous step with the help of y_{k+1}

$$K_{k+1} = P^{-}{}_{k+1}\,H^{T}{}_{k+1}\left[H_{k+1}P^{-}{}_{k+1}\,H^{T}{}_{k+1} + R_{k+1}\right]^{-1} \tag{41}$$

$$\hat{x}_{k+1} = \hat{x}_k^{-}\,K_{k+1}\left(y_{k+1} - H_{k+1}\,\hat{x}^{-}{}_k\right) \tag{42}$$

$$P_{k+1} = (I - K_{k+1}\,H_{k+1})P_{k+1}{}^{-} \tag{43}$$

where K_{k+1} is the Kalman gain matrix.

The GNN model is a nonlinear system, so the basic KF approach should be extended by using the linearization process and is known as the extended Kalman Filter (EKF). The basic difference between KF and EKF is the linearization of the nonlinear system function performed by using the Jacobian matrix in EKF, and then rest of the KF steps can be applied. The nonlinear dynamic system can be defined by using the following equations:

$$x_{k+1} = x_k + q_k \tag{44}$$

$$y_{k+1} = h(x_{k+1},\,u_{k+1}) + r_{k+1} \tag{45}$$

Equation (44) depicts the state of a stationary process corrupted with process noise q_k, and state x_k consists of network weights. x_{k+1} is the weight vector matrix and input vector depicted by u_{k+1}. The noise covariance matrix can be written as $R_{k+1} = E\left[r_{k+1}r^{T}{}_{k+1}\right]$ and $Q_{k+1} = E\left[q_{k+1}q^{T}{}_{k+1}\right]$

In EKF, the linearization of measurement equation is carried out at each time step around the newest state estimation by using the first-order Taylor approximation. The GNN training problem is taken as a problem to find the state estimation of x_{k+1} for minimizing the least square errors by implementing the previous calculations. K_{k+1} Kalman gain can be written as

$$K_{k+1} = P_{k+1}H^{T}{}_{k+1}\left[H_{k+1}\,P_{k+1}H^{T}{}_{k+1} + R_{k+1}\right]^{-1} \tag{46}$$

$$\hat{x}_{k+1} = \hat{x}_{k+1} + K_{k+1}[\,y_{k+1} - h(\,\hat{x}_{k+1} + u_{k+1})] \tag{47}$$

$$P_{k+1} = P_k - K_k H_k\,P_k + Q_k \tag{48}$$

3.2.5. Reference Current Calculation

The fundamental real power current component of the load current can be obtained as

$$I_{Lp} = w_{pd} + w_{Lp} - w_{PV} \tag{49}$$

The reference active components of the grid currents can be written as

$$i_{psa} = I_{Lp} * u_{pa},\; i_{psb} = I_{Lp} * u_{pb},\, i_{psc} = I_{Lp} * u_{pc} \tag{50}$$

Similarly, the fundamental reactive current component of a load current can also be obtained as

$$I_{Lq} = w_{qt} + w_{Lq} \tag{51}$$

The grid currents' active reference components can be written as

$$i_{qsa} = I_{Lq} * u_{qa},\ i_{qsb} = I_{Lq} * u_{qb}, i_{qsc} = I_{Lq} * u_{qc} \tag{52}$$

The net fundamental reference grid currents (i^*_{sa}, i^*_{sb}, i^*_{sc}) can be obtained as

$$i^*_{sa} = i_{psa} + i_{qsa},\ i^*_{sb} = i_{psb} + i_{qsb},\ i^*_{sc} = i_{psc} + i_{qsc} \tag{53}$$

VSC gating signals have been generated by comparing reference (i^*_{sa}, i^*_{sb}, i^*_{sc}) and actual (i_{sa}, i_{sb}, i_{sc}) grid currents for an individual phase, and the error is processed through the PI current regulator. A hysteresis current regulator is used for indirect current control.

4. Results

The SPV array is designed for a maximum power rating of 10 kW. The given system is designed, developed and simulated using the MATLAB/Simulink platform. The performance evaluation of the proposed system is carried out under a linear and nonlinear load with dynamic changes and a changing solar irradiance scenario.

4.1. *Performance Analysis of Proposed Controller for Linear Load for PFC*

The performance of the proposed EKF-based GNN controller is evaluated with a combined structure of the summation and product neurons collectively in single layer. In the proposed GNN network, each processed neuron consists of different weights with the inputs as measured load currents (i_{La}, i_{Lb}, i_{Lc}). With the help of this, the summation weights ($w_{\Sigma pa}, w_{\Sigma pb}, w_{\Sigma pc}$) and product weights ($w_{\Pi pa}, w_{\Pi pb}, w_{\Pi pc}$) are estimated. The proposed GNN is trained to calculate the unspecified weights and EKF filter which is used to predict and update the weights of the GNN network. The output of the proposed system is evaluated under the influence of several performance parameters, i.e., grid voltage (v_s), grid current (i_s), load current (i_L), voltage source converter current (i_{vsc}), grid real power (P_g), grid re-active power (Q_g), PCC terminal voltage (V_t), SPV power (P_{pv}), SPV array current (I_{pv}) and voltage at direct current link (V_{dc}), provided in Figure 4a,b in steady-state linear load (at t = 0.2 to 0.3 s) conditions. Intermediate signals are given in Figure 4b, which depicts the performance parameters and corresponding weights with a variation in error between the active fundamental components of the load currents. The SPV system remains in operation for feeding the maximum value of power onto the load and grid at unity power factor (UPF).

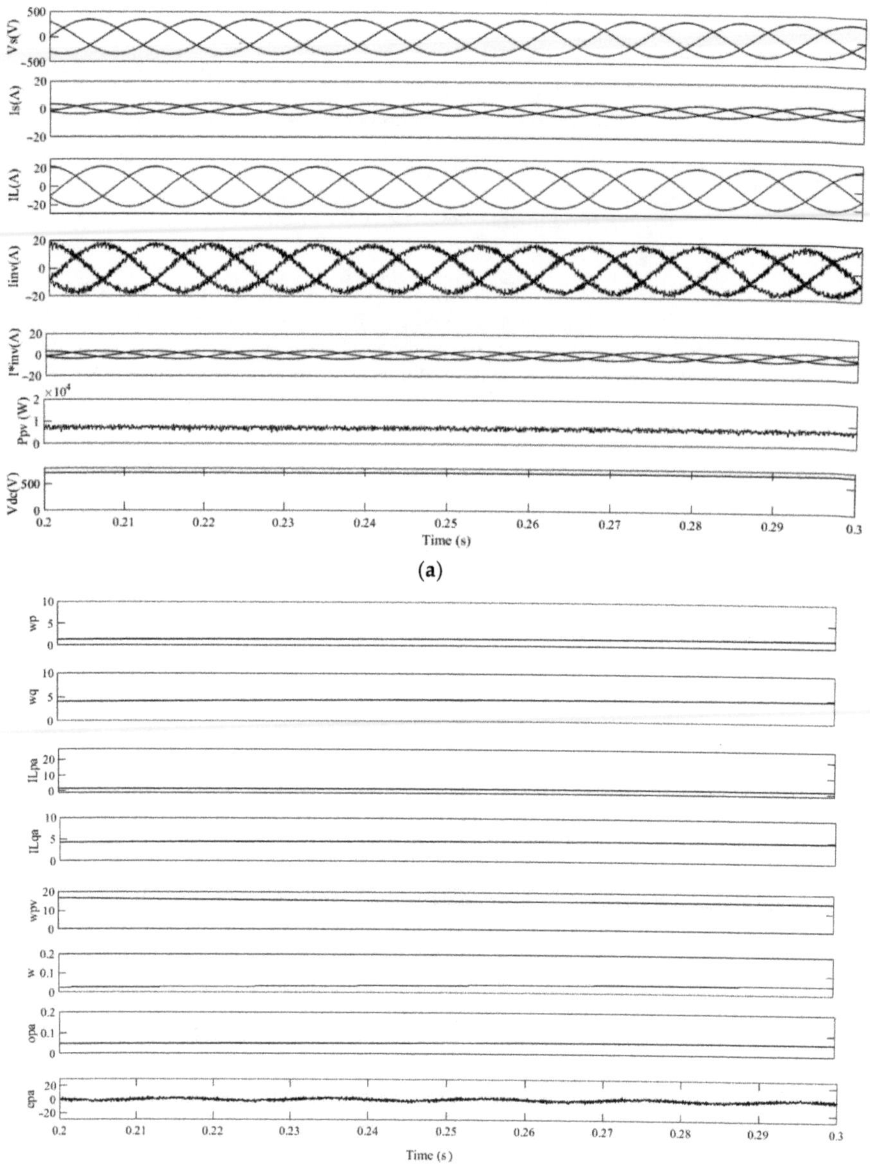

Figure 4. (**a**) Performance parameters, (**b**) estimated weights under steady-state linear load conditions in PFC mode.

4.2. Performance Analysis of Proposed Controller Considering Dynamic Linear Load for Zero Voltage Regulation (ZVR)

If the load is drawing reactive power with unbalanced behavior, then a change in the terminal voltage (V_t) occurs at the point of common coupling. Figure 5a,b shows the behavior of the given system for dynamic linear loading conditions under a zero voltage regulation (ZVR) mode. It can be observed from the results after phase "c" of the load is taken out from the supply at 1.1 s, the grid currents (i_s) remain sinusoidal with the help of a voltage source converter. The PCC voltage (V_t) and voltage at the DC junction are maintained at set values, which are 415 V and 700 V, respectively, without any fluctuations. Intermediate signals are provided in Figure 5b, which shows the performance parameters and corresponding weights with a variation in error between the active fundamental components of the load currents. The reactive power (Qg) taken from the utility grid is approximately equivalent to 0, for compensating the linear loads using reactive power.

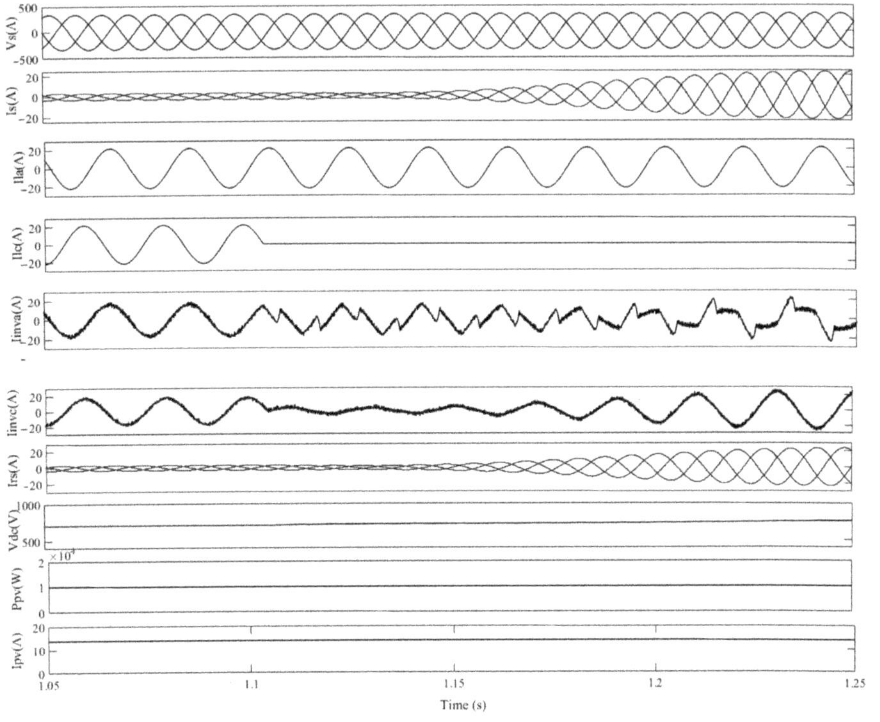

(a)

Figure 5. *Cont.*

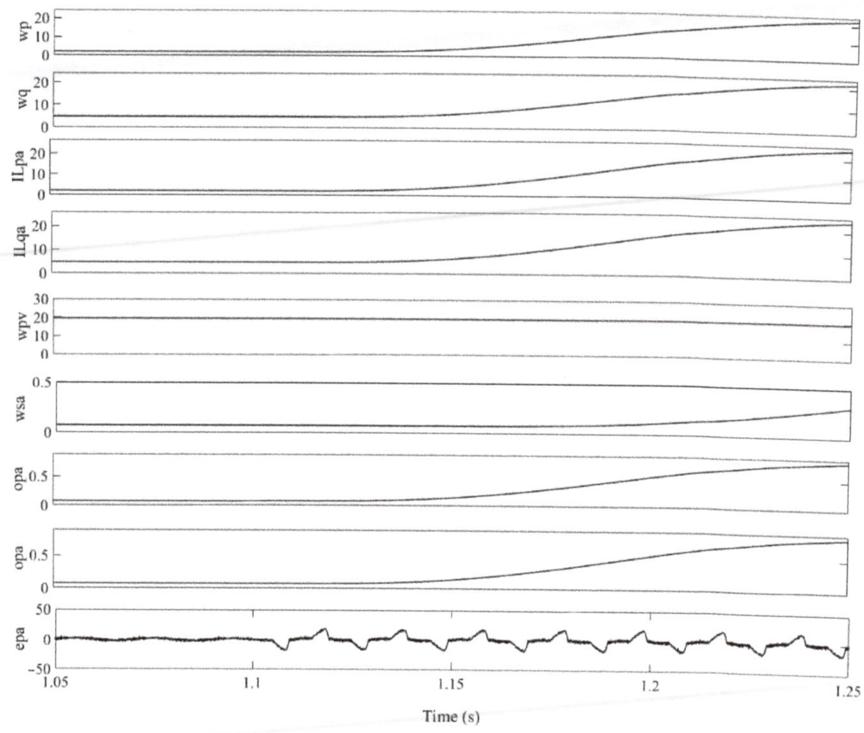

(b)

Figure 5. (**a**) Performance parameters, (**b**) estimated weights under dynamic linear load conditions for zero voltage regulation (ZVR).

4.3. *Performance under a Nonlinear Load*

Figure 6a,b show the behavior of a developed system for dynamic nonlinear loads under the zero voltage regulation (ZVR) mode. By using the given control algorithm, the grid current (i_s) is sinusoidal in nature, when the load of phase "*c*" is extracted at 1.1 s. Intermediate signals are given in Figure 6b for corresponding weight signals with a variation in error between the active fundamental components of the load currents. W_{pa} of the linear transfer function Σ and the estimated GNN output of the active current component of the load current are represented by O_{pa}, and its actual value is achieved by scaling and depicted by I_{Lpa}. This indicates that the calculated weights change according to the load condition. Figure 7a,c show the total harmonic distortion (THD) at the point of common coupling for phase "*a*" of a load current, grid current and grid voltage, respectively, and obtained as 37.48%, 1.04% and 1.07%, respectively, which is within acceptable limits of harmonics in utility currents and voltage fluctuations according to the IEEE-519 and IEEE-1547 standards. THD analysis of the given system is provided in Table 1 for a nonlinear load.

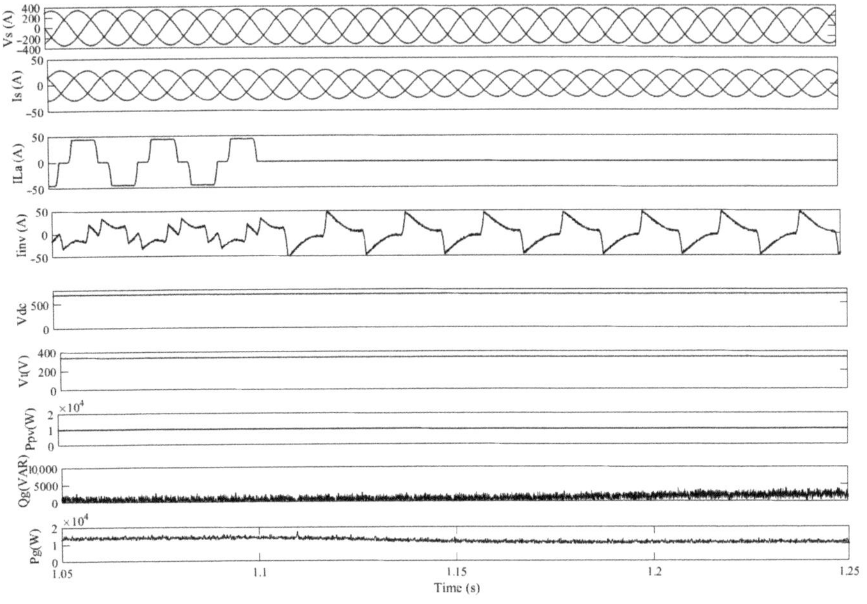

(**a**)

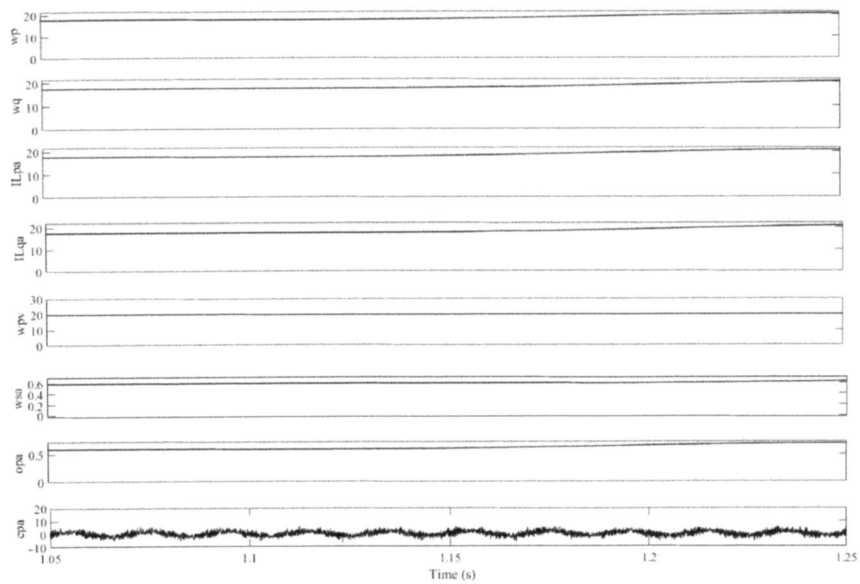

(**b**)

Figure 6. (**a**) Performance indices, (**b**) weights with training parameters for phase "*a*" with steady-state loading in PFC mode.

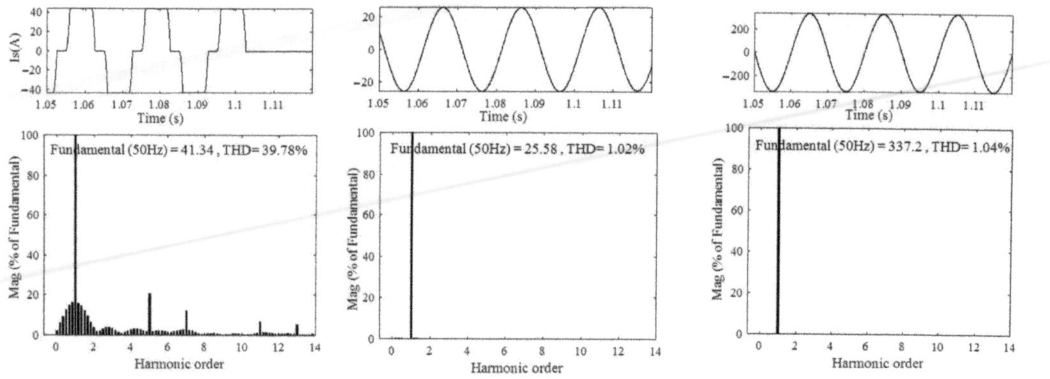

(a) **(b)** **(c)**

Figure 7. Total harmonic distortion (THD) for (**a**) load current of phase "*a*", (**b**) grid current and (**c**) grid voltage in ZVR mode.

Table 1. Parameters of the developed system for the proposed controller.

Operating Mode	Parameters	GNN Based Control Algorithm	EKF GNN Based Control Algorithm
ZVR	Grid voltage (V), %THD at PCC	333.02 V, 1.89%	337.2 V, 1.04%
	Grid current (A), %THD at PCC	23.17 A, 2.54%	25.58 A, 1.02%
	Load current (A), %THD at PCC	40.58 A, 40.62%	41.34 A, 39.78%

4.4. Performance Analysis at Varying Solar Irradiance

Dynamic behavior of the developed system is also observed under varying solar irradiance conditions. The solar irradiance (S) is raised to 1000 W/m^2 from 600 W/m^2 at 0.5 s. The SPV feeds the connected load and excess amount of power output provided to the utility grid. As shown in the results, the grid power (P_g) decreases after 0.5 s because of an increase in power (P_{pv}) output from SPV, given in Figure 8. The proposed system under variable solar irradiance remains operating at the maximum power point of the SPV array and works at unity power factor. The grid side current is maintained to be sinusoidal, and the voltage at DC link is also maintained at a set point.

4.5. Comparative Study of Developed Algorithm with Other Conventional Approaches

The proposed algorithm is analyzed for different scenarios and found accurate as per the desired performance. The parameters of proposed EKF GNN-based controller is presented in Table 1 and compared with the GNN approach. The developed EKF GNN based control approach is found better than ANN conventional approaches such as ADALINE and multilayer perceptron neural network (MLPN). A comparative analysis is made and presented in Table 2 under dynamic nonlinear load conditions on the basis of various performance parameters. The smaller number of unknown weights required and a single layer enhanced the performance of the controller. Additionally, the weight update time, settling period and maximum change in DC voltage is less for the proposed technique.

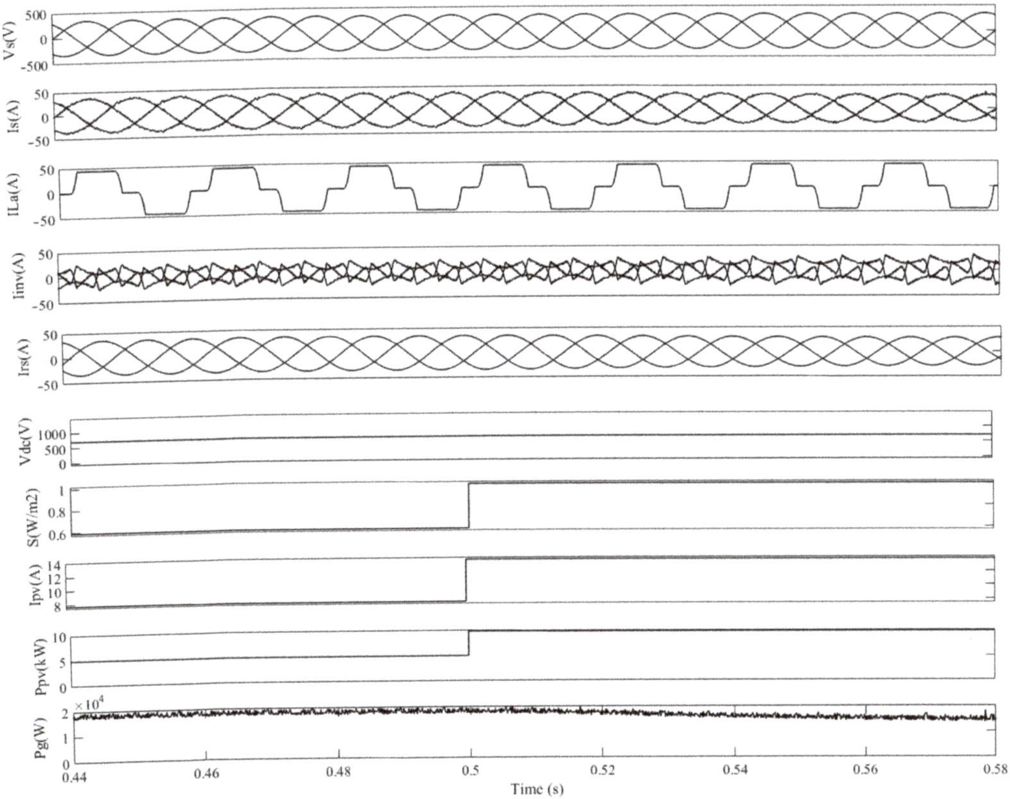

Figure 8. Analysis of performance for a non-linear load with variable solar irradiance.

Table 2. Comparative performance analysis of the proposed technique.

Performance Parameters	ADALINE	MLPN	GNN
Training	Least Mean Square	Back Propagation	EKF
Learning	Gradient decent (GD)	GD/GDM	GDM
Layers	Two	Three	One
Training pattern	Online Training	Offline stochastic training	Online training
Estimation nature	Linear	Nonlinear	Both types
Transfer function	Linear	Sigmoidal	All
Weight update Time	15.7 µs	82 µs	6 µs
Settling period of DC link	1 cycle	$2\frac{1}{2}$ cycle	1 cycle
Max change in DC link voltage	4.5 V	10 V	3.7 V

Figure 9 shows the performances by experimenting the given system in power factor correction mode by implementing linear and nonlinear loads. It depicts the voltage source current (I_{vsc}), load current (I_{La}), grid current (I_g) and direct current link voltage (V_{dc}). Phase "*a*" gid voltage and grid current are shown in Figure 10; it also shows the performances of the controller under PFC mode.

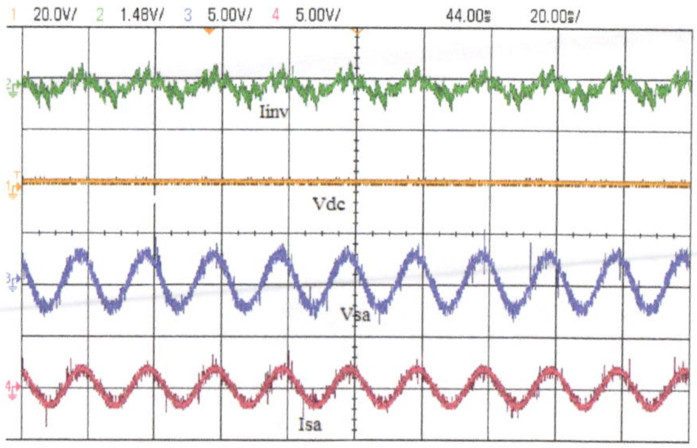

Figure 9. Performance parameters under linear load for the extended Kalman filter (EKF) GNN.

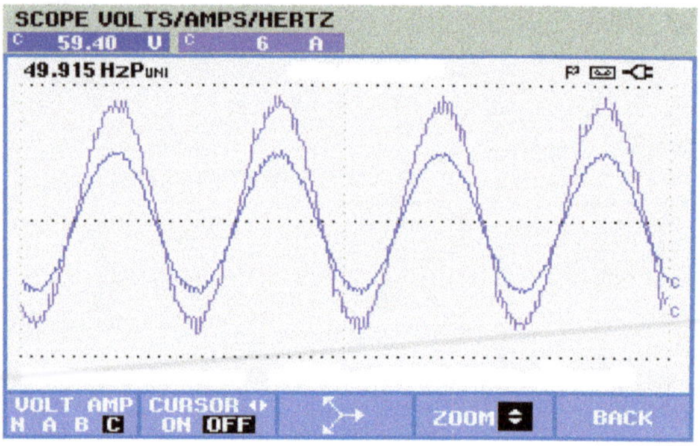

Figure 10. Source side voltage and current of phase "*a*" under the power factor correction mode.

Figure 11 provides the performance of the SPV plant for nonlinear loads, whereas Figure 12a depicts non-linear current and Figure 12b depicts the source current and voltages for phase "*a*". The THD of the load current is given in Figure 13.

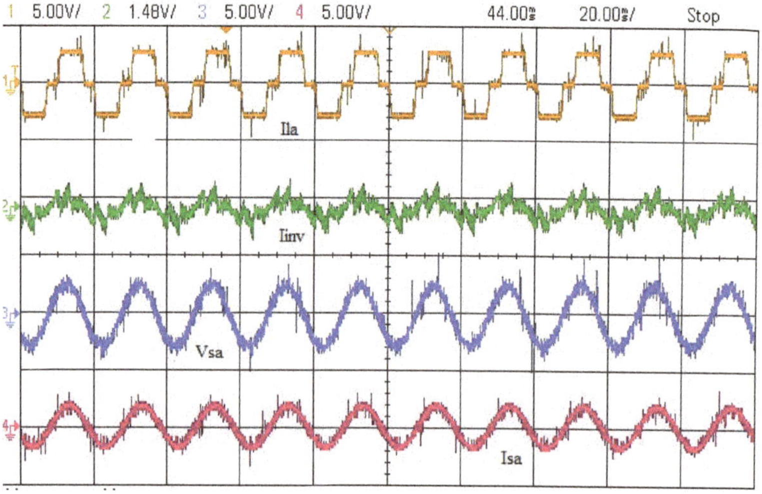

Figure 11. Performance analysis considering non-linear load conditions for EKF GNN.

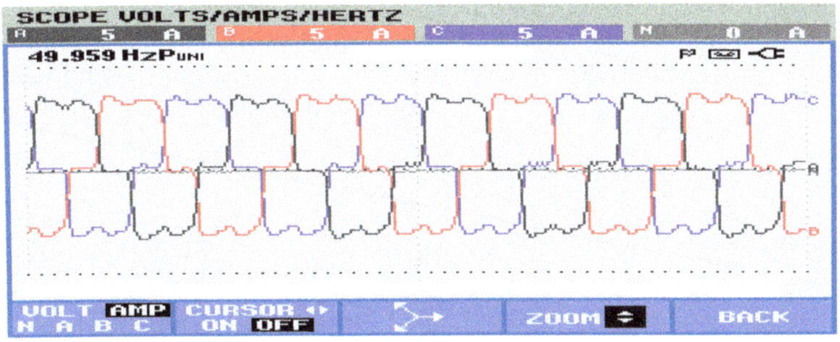

(**a**)

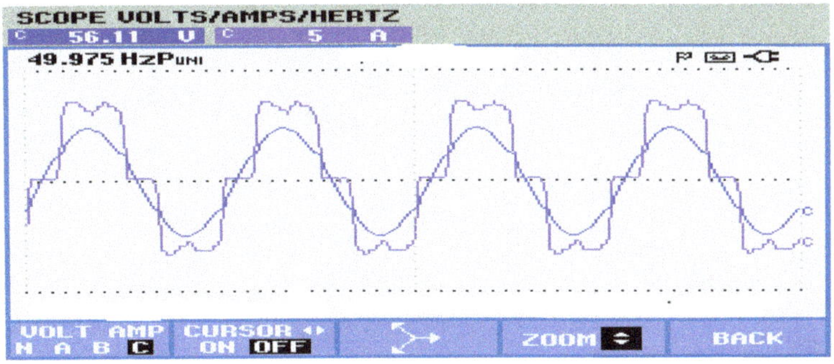

(**b**)

Figure 12. (**a**) Non-linear current, (**b**) current and voltage of phase "*a*".

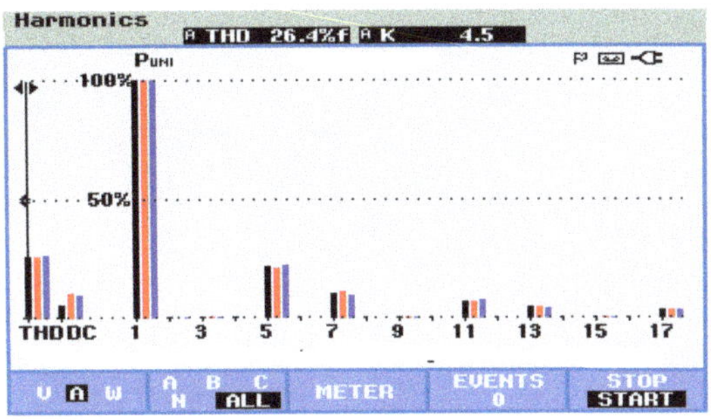

(**a**)

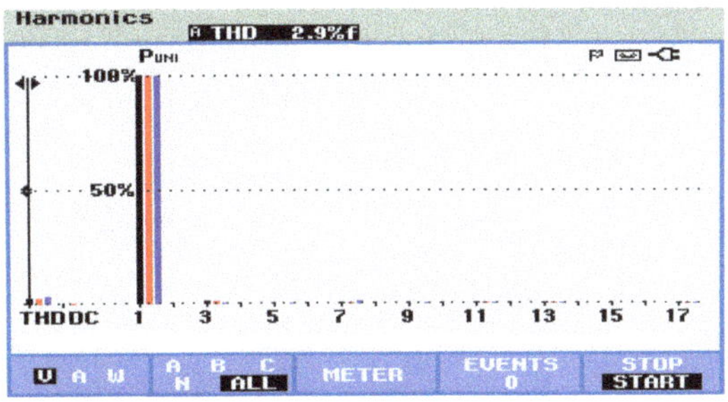

(**b**)

Figure 13. (**a**) THD spectrum of nonlinear load, (**b**) improved THD for grid current under nonlinear load.

The performance of the proposed controller under nonlinear dynamic load is shown in Figure 14; when load of phase "*b*" is removed, the grid current is reduced. Further, the performance of intermediate signals for different weights is also shown in Figure 15. Figures 16 and 17 show the controller performance under variable solar irradiance.

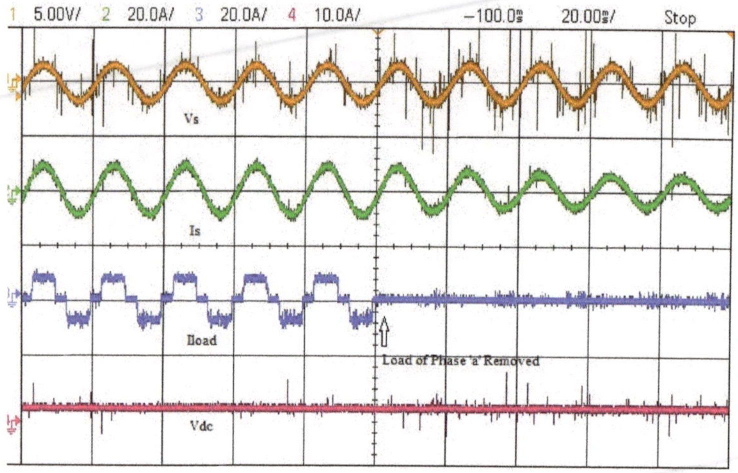

Figure 14. Performance parameters under dynamic nonlinear load for EKF GNN.

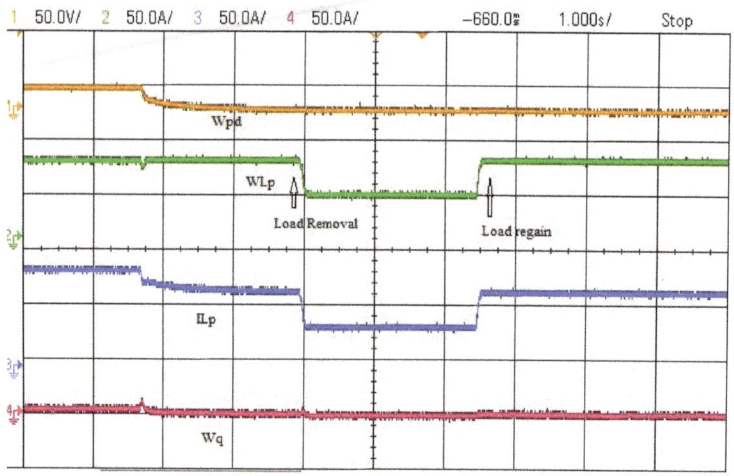

Figure 15. Performance of intermediate weight signals under dynamic nonlinear load for EKF GNN.

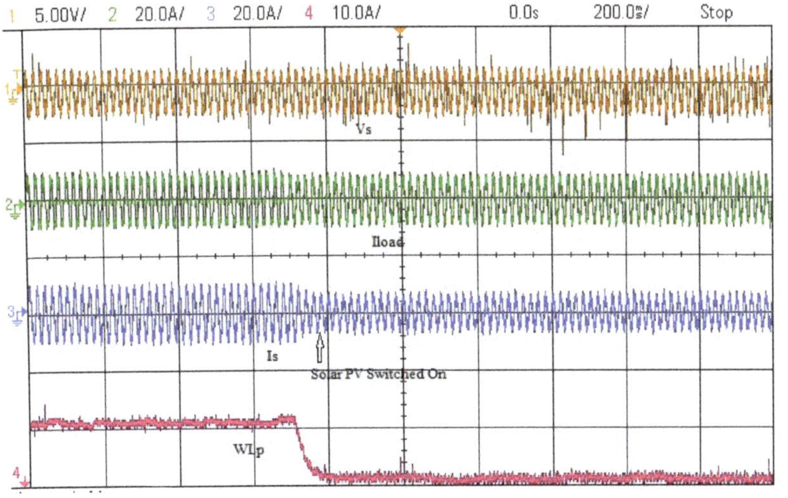

Figure 16. Performance parameters under nonlinear load with solar PV ON for EKF GNN.

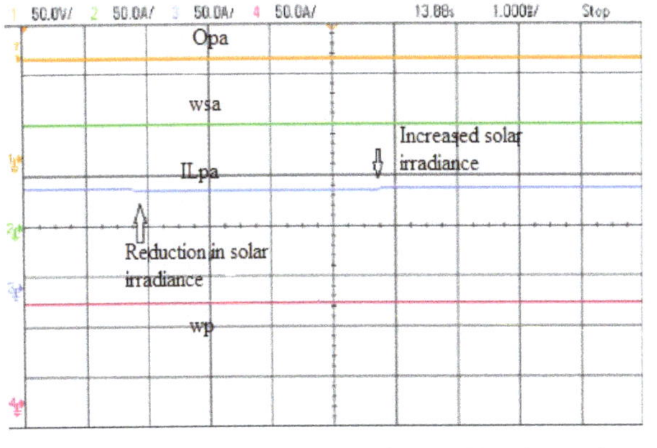

Figure 17. Performance parameters under variable solar irradiance for EKF GNN.

5. Conclusions

The proposed system was developed and tested using MATLAB simulation environment, and the performance of the EKF GNN-based VSC control approach was validated under linear and nonlinear loads for static and dynamic scenarios. The proposed EKF GNN approach has successfully improved the function of the developed SPV system. The performance of the developed system was found satisfactory under load unbalancing and varying solar irradiance. Additionally, it continues to operate in a UPF mode of operation, providing reactive power compensation, load balancing, MPP extraction and harmonics compensation. A changeable step size perturb and observe (P&O) MPPT approach was utilized in this research and enabled fast tracking of MPP and convergence. Moreover, single-stage topology was able to reduce the losses in semiconductor devices and increase the overall efficiency.

The proposed control approach performs with more flexibly in training the network under dynamic conditions and provides improved online learning. This control technique needs a smaller number of training patterns and unknown weights and hence reduces the complexity and the computational time. The developed system obtains acceptable limits of harmonics in utility currents and voltage fluctuations according to the IEEE-519 and IEEE-1547 standards.

Author Contributions: Conceptualization, P.C., M.R. and I.A.; methodology, P.C., M.A. and M.R.; validation, P.C. and I.A.; formal analysis, P.C., I.A., M.A. and M.R.; resources, M.R. and I.A.; writing—original draft, P.C. and M.R.; writing—review and editing, P.C., I.A., M.R. and M.A.; visualization, P.C.; supervision, M.R.; funding acquisition, I.A., M.A. and M.R. All authors have read and agreed to the published version of the manuscript.

Institutional Review Board Statement: Not applicable.

Informed Consent Statement: Not applicable.

Data Availability Statement: No data are used here.

Acknowledgments: Researchers would like to thank the Deanship of Scientific Research, Qassim University for funding publication of this project.

Conflicts of Interest: The authors declare no conflict of interest.

Nomenclature

C	DC link capacitor (μF)
f_1	Sigmoidal characteristic function
f_2	Gaussian characteristic function
I_{inv}	Inverter output current
I^*_{inv}	Reference Inverter output current
i_{La}, i_{Lb}, i_{Lc}	Load currents of phase 'a', 'b' and 'c' respectively (Ampere)
I_{Lp}	Fundamental active current component
I_{Lpa}	Fundamental active current component for phase 'a'
I_{Lq}	Fundamental reactive current component
I_{Lqa}	Fundamental reactive current component for phase 'a'
I_{MPP}	PV current at maximum power point (Ampere)
i_{psa}	Active reference component of grid current
I_{PV}	PV array output current (Ampere)
i_{qsa}	Reactive reference component of grid current
I_{rs}	Reference component of grid current
i_{sa}, i_{sb}, i_{sc}	Grid currents of phase 'a', 'b' and 'c' respectively (Ampere)
$i^*_{sa}, i^*_{sb}, i^*_{sc}$	Reference currents of phase 'a', 'b' and 'c' respectively (Ampere)
K_i	Integral gain of PI controller
K_p	Proportional gain of PI controller
L_f	Interfacing inductor (mH)
O_i	Final output of generalized neuron
O_{pa}	output of the GNN model for active component
O_{qa}	output of the GNN model for reactive component
O_Σ	Output of Σ_A part network
O_Π	Output of Π part network
P_{pv}	PV power (W)
P_g	Grid Active power (W)
Q_g	Grid Reactive power (vAR)
S	Solar irradiance (W/m^2)
u_{pa}, u_{pb}, u_{pc}	In phase unit templates of phase voltages
u_{qa}, u_{qb}, u_{qc}	Quadrature unit templates of phase voltages
V_a, V_b, V_c	Phase voltage of utility grid (Volts)
v_{ab}, v_{bc}, v_{ca}	Line voltage of utility grid (Volts)
V_{dc}	DC link voltage (Volts)
V^*_{dc}	Reference DC link voltage (Volts)
V_{MPP}	PV voltage at maximum power point (Ampere)
V_{pv}	PV output voltage (Volts)
V_t	Voltage at point of common coupling (Volts)

V^*_t	Reference voltage at point of common coupling (Volts)
V_{te}	Error between sensed and reference voltage at point of common coupling (Volts)
w	Weights of GNN
w_{ap}, w_{bp}, w_{cp}	Updated weights for hidden layer of active components
w_{aq}, w_{bq}, w_{cq}	Updated weights for hidden layer of reactive components
w_{Lp}	Mean active component of load current
w_{Lq}	Mean reactive component of load current
w_p	Active current component
w_{pdc}	Active current component phase c
w_{pv}	Feed forward weight function of solar power
w_q	Reactive current component
w_{qt}	Function of reactive loss current component
w_{sa}	weight of summation neuron
W_Σ	Weights of summation part of GNN
ΔW	Change in weights of GNN
Greek Symbols	
α	Learning rate
μ	Micro
Ω	Ohm
ϕ	Phase
Σ_A	Aggregation function used with sigmoidal characteristic function
Π	Aggregation function used with Gaussian characteristic function
η	Learning rate

References

1. Hudson, R.; Heilscher, G. PV Grid Integration—System Management Issues and Utility Concerns. *Energy Procedia* **2012**, *25*, 82–92. [CrossRef]
2. IEEE. *IEEE Recommended Practices and Requirement for Harmonic Control on Electric Power System*; IEEE Std.: New York, NY, USA, 1992; p. 519.
3. Wu, T.-F.; Chang, C.-H.; Lin, L.-C.; Kuo, C.-L. Power loss comparison of single and two-stage grid-connected photovoltaic systems. *IEEE Trans. Energy Convers.* **2011**, *26*, 707–715. [CrossRef]
4. Arya, S.R.; Niwas, R.; Bhalla, K.K.; Singh, B.; Chandra, A.; Al-Haddad, K. Power quality improvement in isolated distributed power generating system using DSTATCOM. *IEEE Trans. Ind. Appl.* **2015**, *51*, 4766–4774. [CrossRef]
5. Singh, B.; Jayaprakash, P.; Kothari, D.P.; Chandra, A.; Al-Haddad, K. Comprehensive study of DSTATCOM configurations. *IEEE Trans. Ind. Inform.* **2014**, *10*, 854–870. [CrossRef]
6. Singh, B.; Solanki, J. A comparison of control algorithms for DSTATCOM. *IEEE Trans. Ind. Electron.* **2009**, *56*, 2738–2745. [CrossRef]
7. Kumar, C.; Mishra, M.K. A Multifunctional DSTATCOM Operating Under Stiff Source. *IEEE Trans. Ind. Electron.* **2013**, *61*, 3131–3136. [CrossRef]
8. Ahmad, M.T.; Kumar, N.; Singh, B. Generalised neural network-based control algorithm for DSTATCOM in distribution systems. *IET Power Electron.* **2017**, *10*, 1529–1538. [CrossRef]
9. Bag, A.; Subudhi, B.; Ray, P.K. A combined reinforcement learning and sliding mode control scheme for grid integration of a PV system. *CSEE J. Power Energy Syst.* **2019**, *5*, 498–506.
10. He, J.; Li, Y.W. Hybrid Voltage and Current Control Approach for DG-Grid Interfacing Converters with *LCL* filters. *IEEE Trans. Ind. Electron.* **2012**, *60*, 1797–1809. [CrossRef]
11. Tomar, A.; Mishra, S.; Bhende, C.N. Modified MISO DC-DC converter based PV water pumping system. In Proceedings of the 2016 IEEE 7th Power India International Conference (PIICON), Bikaner, India, 25–27 November 2016; Institute of Electrical and Electronics Engineers (IEEE): New York, NY, USA, 2016; pp. 1–6.
12. Alfaris, F.E.; Bhattacharya, S. Control and Real-Time Validation for Convertible Static Transmission Controller Enabled Dual Active Power Filters and PV Integration. *IEEE Trans. Ind. Appl.* **2019**, *55*, 4309–4320. [CrossRef]
13. Rengasamy, M.; Gangatharan, S.; Elavarasan, R.M.; Mihet-Popa, L. The Motivation for Incorporation of Microgrid Technology in Rooftop Solar Photovoltaic Deployment to Enhance Energy Economics. *Sustainability* **2020**, *12*, 10365. [CrossRef]
14. Jain, C.; Singh, B. A Three-Phase Grid Tied SPV System with Adaptive DC Link Voltage for CPI Voltage Variations. *IEEE Trans. Sustain. Energy* **2016**, *7*, 337–344. [CrossRef]
15. Agarwal, R.K.; Hussain, I.; Singh, B. Three-phase single-stage grid tied solar PV ECS using PLL-less fast CTF control technique. *IET Power Electron.* **2017**, *10*, 178–188. [CrossRef]
16. Singh, B.; Jain, C.; Goel, S.; Chandra, A.; Al-Haddad, K. A multifunctional grid-tied solar energy conversion system with ANF-based control approach. *IEEE Trans. Ind. Appl.* **2016**, *52*, 3663–3672. [CrossRef]
17. Vidal, H.; Rivera, M.; Wheeler, P.; Vicencio, N. The Analysis Performance of a Grid-Connected 8.2 kWp Photovoltaic System in the Patagonia Region. *Sustainability* **2020**, *12*, 9227. [CrossRef]

18. Campanhol, L.B.G.; da Silva, S.A.O.; de Oliveira, A.A.; Bacon, V.D. Single-stage three-phase grid-tied PV system with universal filtering capability applied to DG systems and AC microgrids. *IEEE Trans. Power Electron.* **2017**, *32*, 9131–9142. [CrossRef]
19. Tomar, A.; Mishra, S. CMPVI-Based MIDO Scheme under SSE for Optimum Energy Balance and Reduced ROI. *IEEE Trans. Sustain. Energy* **2017**, *9*, 1318–1327. [CrossRef]
20. Huang, P.; Zhang, X.; Copertaro, B.; Saini, P.; Yan, D.; Wu, Y.; Chen, X. A Technical Review of Modeling Techniques for Urban Solar Mobility: Solar to Buildings, Vehicles, and Storage (S2BVS). *Sustainability* **2020**, *12*, 7035. [CrossRef]
21. Varma, R.K.; Rahman, S.A.; Vanderheide, T. New control of PV solar farm as STATCOM (PV-STATCOM) for increasing grid power transmission limits during night and day. *IEEE Trans. Power Deliv.* **2014**, *30*, 755–763. [CrossRef]
22. Hamid, M.I.; Jusoh, A.; Anwari, M. Photovoltaic plant with reduced output current harmonics using generation-side active power conditioner. *IET Renew. Power Gener.* **2014**, *8*, 817–826. [CrossRef]
23. Tomar, A.; Mishra, S.; Bhende, C.N. AOMH–MISO based PV–VCI irrigation system using ASCIM pump. *IEEE Trans. Ind. Appl.* **2018**, *54*, 4813–4824. [CrossRef]
24. Kannan, V.K.; Rengarajan, N. Investigating the performance of photovoltaic based DSTATCOM using I cos Φ algorithm. *Int. J. Electr. Power Energy Syst.* **2014**, *54*, 376–386. [CrossRef]
25. Panigrahi, R.; Mishra, S.K.; Srivastava, S.C.; Srivastava, A.K.; Schulz, N.N. Grid Integration of Small-Scale Photovoltaic Systems in Secondary Distribution Network—A Review. *IEEE Trans. Ind. Appl.* **2020**, *56*, 3178–3195. [CrossRef]
26. Mazumdar, J.; Harley, R.G. Recurrent neural networks trained with backpropagation through time algorithm to estimate nonlinear load harmonic currents. *IEEE Trans. Ind. Electron.* **2008**, *55*, 3484–3491. [CrossRef]
27. Mukundan, N.; Singh, Y.; Naqvi, S.B.Q.; Singh, B.; Pychadathil, J. Multi-Objective Solar Power Conversion System with MGI Control for Grid Integration at Adverse Operating Conditions. *IEEE Trans. Sustain. Energy* **2020**, *11*, 2901–2910.
28. Singh, B.; Arya, S.R. Back-Propagation Control Algorithm for Power Quality Improvement Using DSTATCOM. *IEEE Trans. Ind. Electron.* **2013**, *61*, 1204–1212. [CrossRef]
29. Jayachandran, J.; Sachithanandam, R.M. Neural Network-Based Control Algorithm for DSTATCOM Under Nonideal Source Voltage and Varying Load Conditions. *Can. J. Electr. Comput. Eng.* **2015**, *38*, 307–317. [CrossRef]
30. Janpong, S.; Areerak, K.L.; Areerak, K.N. A literature survey of neural network applications for shunt active power filters. *World Acad. Sci. Eng. Technol.* **2011**, *5*, 273–279.
31. Chaudhary, P.; Rizwan, M. QNBP NN-based I cos ϕ algorithm for PV systems integrated with LV/MV grid. *Soft Comput.* **2021**, *25*, 2599–2614. [CrossRef]
32. Chaudhary, P.; Rizwan, M. Intelligent approach-based hybrid control algorithm for integration of solar photovoltaic system in smart grid environment. *IET Smart Grid* **2019**, *2*, 445–454. [CrossRef]
33. Chaturvedi, D.K.; Malik, O.P.; Kalra, P.K. Generalised neuron-based adaptive power system stabiliser. *IEEE Proc. Gener. Transm. Distrib.* **2004**, *151*, 213–218. [CrossRef]
34. Chaturvedi, D.K. *Soft Computing Techniques and Its Applications in Electrical Engineering*; Springer: Berlin/Heidelberg, Germany, 2008.
35. Rizwan, M.; Jamil, M.; Kothari, D.P. Generalized Neural Network Approach for Global Solar Energy Estimation in India. *IEEE Trans. Sustain. Energy* **2012**, *3*, 576–584. [CrossRef]
36. Kulkarni, R.V.; Venayagamoorthy, G.K. Generalized neuron: Feedforward and recurrent architectures. *Neural. Netw.* **2009**, *22*, 1011–1017. [CrossRef]
37. Reisi, A.R.; Moradi, M.H.; Jamasb, S. Classification and comparison of maximum power point tracking techniques for photovoltaic system: A review. *Renew. Sustain. Energy Rev.* **2013**, *19*, 433–443. [CrossRef]
38. Almutairi, A.; Abo-Khalil, A.; Sayed, K.; Albagami, N. MPPT for a PV Grid-Connected System to Improve Efficiency under Partial Shading Conditions. *Sustainability* **2020**, *12*, 10310. [CrossRef]
39. Salah, C.B.; Ouali, M. Comparison of fuzzy logic and neural network in maximum power point tracker for PV systems. *Electr. Power Syst. Res.* **2011**, *81*, 43–50. [CrossRef]
40. Williams, R.J. Training recurrent networks using the extended Kalman filter. In Proceedings of the 1992 IJCNN International Joint Conference on Neural Networks, Baltimore, MD, USA, 7–11 June 1992; IEEE: New York, NY, USA, 2003; pp. 241–246.
41. de Oliveira, M.A. An application of neural networks trained with kalman filter variants (ekf and ukf) to heteroscedastic time series forecasting. *Appl. Math. Sci.* **2012**, *6*, 3675–3686.
42. Bishop, G.; Welch, G. An introduction to the kalman filter. *Proc. SIGGRAPH Course* **2001**, *8*, 23175–27599.
43. Chaudhary, P.; Rizwan, M. Hybrid control approach for PV/FC fed voltage source converter tied to grid. *Int. J. Hydrogen Energy* **2018**, *43*, 6851–6866. [CrossRef]

sustainability

MDPI

Article

Accurate Sizing of Residential Stand-Alone Photovoltaic Systems Considering System Reliability

Eduardo Quiles [1,*], Carlos Roldán-Blay [2], Guillermo Escrivá-Escrivá [2] and Carlos Roldán-Porta [2]

1 Instituto de Automática e Informática Industrial, Universitat Politècnica de València, Camino de Vera, s/n, 46022 Valencia, Spain

2 Institute for Energy Engineering, Universitat Politècnica de València, Camino de Vera, s/n, edificio 8E, escalera F, 5a planta, 46022 Valencia, Spain; carrolbl@die.upv.es (C.R.-B.); guieses@die.upv.es (G.E.-E.); croldan@die.upv.es (C.R.-P.)

* Correspondence: equiles@isa.upv.es; Tel.: +34-963-877-007

Received: 10 January 2020; Accepted: 6 February 2020; Published: 10 February 2020

Abstract: In rural areas or in isolated communities in developing countries it is increasingly common to install micro-renewable sources, such as photovoltaic (PV) systems, by residential consumers without access to the utility distribution network. The reliability of the supply provided by these stand-alone generators is a key issue when designing the PV system. The proper system sizing for a minimum level of reliability avoids unacceptable continuity of supply (undersized system) and unnecessary costs (oversized system). This paper presents a method for the accurate sizing of stand-alone photovoltaic (SAPV) residential generation systems for a pre-established reliability level. The proposed method is based on the application of a sequential random Monte Carlo simulation to the system model. Uncertainties of solar radiation, energy demand, and component failures are simultaneously considered. The results of the case study facilitate the sizing of the main energy elements (solar panels and battery) depending on the required level of reliability, taking into account the uncertainties that affect this type of facility. The analysis carried out demonstrates that deterministic designs of SAPV systems based on average demand and radiation values or the average number of consecutive cloudy days can lead to inadequate levels of continuity of supply.

Keywords: renewable energy; photovoltaic generation; battery storage; reliability evaluation; Monte Carlo Simulation

1. Introduction

The need to reduce dependency in fossil fuels has promoted the use of renewable energy sources. The appearance of renewable energy sources such as wind power, solar power, or small hydro plants in the electrical market is increasing every day. Many of these renewable energy plants are customer owned and have small unitary power.

These renewable microgeneration systems can be isolated from the distribution network (off-grid) or connected to it exporting their energy surpluses [1–3]. In rural areas or in isolated communities in developing countries it is increasingly common to install micro-renewables sources by residential consumers without access to the utility distribution network (DN). In Spain these micro-renewables are mainly off-grid photovoltaic (PV) systems for electrification of single residential households. These stand-alone photovoltaic (SAPV) energy systems generally include batteries for energy storage [4–6].

The Spanish Administration has recently regulated self-consumption of electricity [7]. This regulation was issued as an urgent measure for renewable energy promotion and consumer protection. It has reduced the administrative hurdles suffered by small-scale energy plants and it allows collective self-consumption for the first time.

This new regulation defines the concept of "neighbor facility" in order to regulate collective self-consumption and confirms the elimination of the charge to self-consumed energy. It creates a simplified compensation mechanism to compensate self-consumers with surplus energy that export to the network and facilitates the installation of energy storage elements without more requirements than to comply with safety and industrial quality regulations. This regulation is expected to incentive the installation of more SAPV generation systems for residential customers.

To design a SAPV system it is necessary to determine the PV panels' rated power and battery storage capacity. PV generation has uncertainty associated to its energy output which depends on the irradiation level [8,9]. This irradiation level depends on the weather condition, varying widely between sunny and cloudy days, in the short term (hour to hour) and in the long term (seasonal variations) [10,11]. Even for a typical clear-sky day, fluctuations of PV power are caused by passing clouds. Getting a good prediction of the reliability of a SAPV is mandatory in order to improve its sustainability. A detailed analysis of the characteristics of solar radiation for the area where a SAPV will be installed is convenient. Uncertainty associated to PV generation must be modelled to predict average performance in the future. Time-series measurements of solar radiation data from near-site weather stations are required to estimate expected generation [12–14].

Simultaneity between PV generation and energy consumption in residential households is limited. PV generation is maximum at noon while for a typical residential customer the peak period occurs in the evening. The battery system allows to storage PV energy surpluses produced during the day. As long as the battery does not reach its maximum state of charge (SOC), the PV energy surpluses will be used to charge the battery. Another option to improve self-sufficiency of PV systems is shifting the consumption of deferrable loads by demand-side management to periods with PV-surpluses [15,16]. The creation of cooperative microgrids with different SAPV generators has also been proposed [17,18] to address this problem.

To make a realistic design of the SAPV system an accurate model of load demand is required. This is especially difficult for individual residential customers that usually have a variable load profile. Residential load has a time-varying nature and it changes depending on the time of the day, day of the week and season of the year. To simulate the energy flows between PV unit, battery and load, time series data of PV generation and load demand with a high temporal resolution are required.

Many studies have been carried out to determine the feasibility of SAPV systems [19]. There are some studies [20–23] that evaluate residential PV plants connected to the DN. In this way they can sell energy surpluses to the utility and import energy from the DN when needed. Some other studies [24,25] consider SAPV systems, autonomous and isolated from the DN. This is usually the only available option in many rural areas. Different solutions have been proposed for the design of such SAPV systems. The main objective is to determine the most reliable and cost-effective configuration of PV units for energy generation and batteries for energy storage.

Some authors propose analytical solutions based on energy balance equations [26]. Some of them include statistical approaches to consider solar radiation fluctuations [12]. A review of SAPV systems sizing methodologies can be found in [1,27].

The SAPV design considering reliability of involved systems requires a realistic model of the energy resource, energy demand, and system components faults. Several studies usually consider average solar irradiance levels for a broad area. These values are monthly averages obtained from databases of meteorological services [28,29] In this study local hourly irradiance levels are measured and included in the design process.

As previously stated, residential demand is highly variable between different users and different days. The simplified approach of considering an average daily peak demand clearly underestimates demand fluctuation [1]. This work considers actual demand data measured in an hourly basis from average customers.

Another factor that influences the applicability of simulation results is the temporal resolution of the PV generation and load demand. At least an hourly resolution is required to reflect the power balance between PV generation and energy demand [30–32]. Some authors [1] propose smaller temporal resolutions (10-min sampled data) in order to evaluate energy flows between PV units, batteries, and loads.

Several studies conclude that SAPV systems are an economic and profitable solution for residential customers without access to the DN [33]. These studies show that the initial investments in PV panels, batteries, and installation costs have an amortization period smaller than the PV plant useful life [20,34]. Therefore, for many residential customers the reliability of the electrical supply is the key factor when planning the installation of a SAPV system, rather than the investment costs. Many individual residential investors are willing to afford the investment costs of the PV generation system as far as a satisfactory electrical supply reliability level is warranted [35].

SAPV design must take into account faults that unexpectedly occur in the system. Assuming no component faults can result in an over-optimistic performance prediction and in the subsequent infra-sizing of PV units and batteries [36].

Monte Carlo simulation (MCS) can be performed in a sequential or non-sequential manner [37]. In the sequential MCS the states of the components are sequentially sampled simulating the chronology of the stochastic process of the system operation. For renewable-energy systems with energy storage, the state of the system depends on previous states, i.e., battery SOC level. PV generation and residential demand are not usually correlated. The complexity of this reliability analysis can better be dealt with a sequential MCS. Other approaches as reliability evaluation based on analytical models or Markov models require modeling simplifications that are not suitable for a realistic assessment of SAPV systems [38].

This study addresses the evaluation of component faults using a sequential Monte Carlo simulation methodology. The objective of the paper is to obtain an optimal sizing of the SAPV system from the economical point of view, but imposing some constraints related to the desired reliability of the system. The novelty of this work compared with existing studies is the consideration of actual PV generation and load demand time series data, and the simultaneous evaluation of uncertainties associated to PV generation, load demand, and system component faults. To the authors' best knowledge this is the first work where these three sources of uncertainty are simultaneously taken into account in the design of SAPV systems. The objective is to guarantee a desired reliability in the continuity of the supply in the design of a SAPV generation system.

To validate the results of the proposed method, the sizing of a SAPV system is firstly approached using a deterministic worst case procedure. Then, a reliability evaluation method is used, considering PV generation uncertainty, demand uncertainty and unexpected faults performing a sequential Monte Carlo simulation. The results of both approaches are compared to extract relevant conclusions about the design process of SAPV systems.

2. Materials and Methods

2.1. Stand-Alone Photovoltaic Energy System

Many residential households have installed PV renewable generation to satisfy its own energy requirements. This generation is accompanied by batteries to storage energy when surpluses are available and to supply energy when the PV output is insufficient. Figure 1 shows a scheme of the installation of PV panels and batteries to supply a residential load without connection to the grid. The battery has a regulator or battery controller (BC) to control the SOC and its maximum current, both in charge and discharge operations. The BC decides the power flows between the PV panel array and the battery.

The SAPV system has been divided into two sections:

- The generation section: PV panel array, BC and batteries.
- The load section: inverter and loads.

The reliability analysis performed in this work will take into account the possible failures in the PV panel array and the BC considering the failure rate per year of these elements (λ_c). These failures can be covered by the batteries, that also present a specific failure rate (λ_b). The possible failures in the inverter (λ_i) cannot be supported by other elements in the proposed scheme. Therefore, to include these failures in the analysis, this element should be considered in series with the generation section and the overall reliability would be the product of the reliability of both systems.

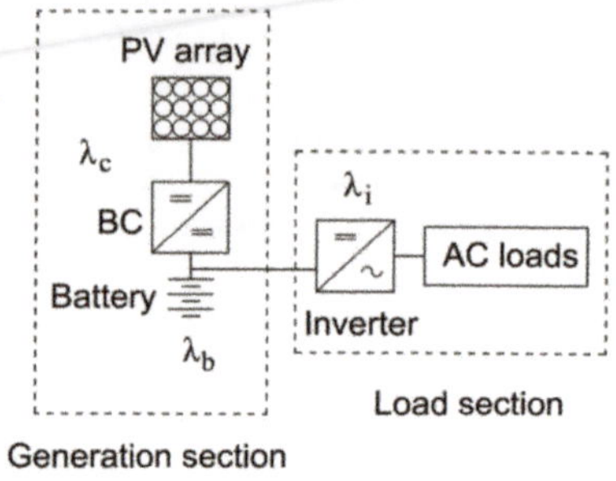

Figure 1. Stand-alone photovoltaic (SAPV) system electric scheme.

To model a typical household energy demand, the authors of this study have registered time series of data with a temporal resolution of an hour. Figure 2 shows typical values of the daily demand from a residential customer in different seasons.

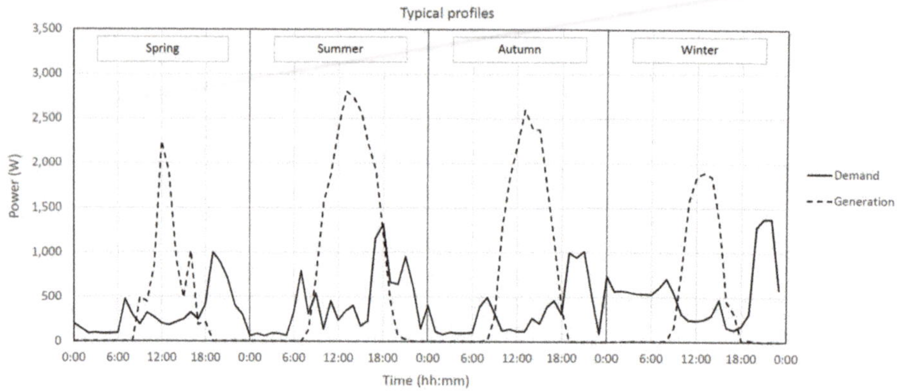

Figure 2. Daily load profile for residential customers and hourly generation curve in different seasons.

PV generation used in this work is based in the actual measurements obtained in the laboratory of Distributed Energy Resources (LABDER) at the Institute for Energy Engineering of the Universitat Politècnica de València, Spain [39]. This laboratory enables the assembly of Hybrid Renewable Energy Systems (HRES) combining different renewable sources: photovoltaic, biogas, wind power and hydrogen fuel cells, interconnected by a controlled microgrid that supplies a specific load. Additionally, the laboratory includes the capability to store energy, both in batteries and hydrogen, to cover most of the possible HRES configurations [40].

All the systems are working in the 10 kW range. Specifically, the photovoltaic generator in the LABDER is made up of monocrystalline and polycrystalline silicon modules mounted on the roof of the laboratory, facing south with tilt angle of 30 degrees to produce maximum annual energy. The total power of the photovoltaic generator now installed in this lab is 2.1 kWp and the panels are connected to a single phase grid inverter.

The operating point of the panels and the inverter, currents, voltages, power and energy injected to the grid are also monitored using a power meter installed at the AC output of the inverter and a data logger to storage the data. This information enables the management system to check the correct operation of the system and to know the energy produced at any time.

The registered power produced during a year has been used to obtain the PV power generation in this work, as depicted in Figure 2. This register allows to use actual data and to obtain reliable values in the different simulations performed later.

To provide a deterministic design (DD) of the size of the SAPV system, the average demand recorded in this study, $P_d(t)$, at the most unfavorable time of the year (winter) is considered. The average daily energy consumption in this period E_d (Wh/day) will be:

$$E_d = \int_0^{24} P_d(t)dt. \quad (1)$$

This energy is obtained from the PV array. The output power of the PV system depends on the irradiance level. This fact introduces uncertainty in the energy resource.

The battery must be designed to secure the energy supply to the loads when the PV output is not enough (at nights and in cloudy days). From the values of the instantaneous PV energy generation $P_S(t)$ and the energy demand $P_d(t)$ the useful energy contributed by the battery in one day is obtained, as depicted in Figure 3.

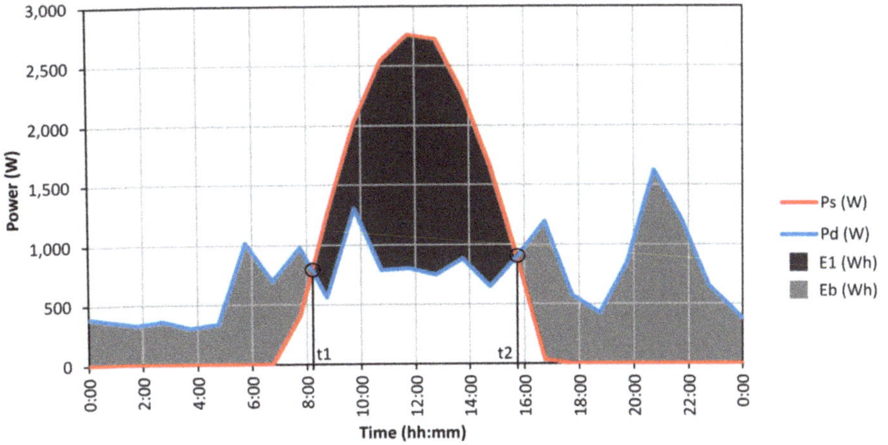

Figure 3. Useful energy contributed by the battery in one day.

As shown in Figure 3, the power produced by the array on a sunny day, $P_S(t)$, must exceed the demand during the time interval t1–t2. Then, the energy available to be stored in the battery, E_1, will be:

$$E_1 = \int_{t_1}^{t_2} (P_s(t) - P_d(t))dt \text{ when } P_S(t) > P_d(t). \quad (2)$$

To keep the system stable, the energy that the battery can return in a daily discharge cycle, E_b will be computed according to Equation (3), where η_c and η_d are the battery charge and discharge performance and η_b the overall efficiency of the energy storage and recovery process.

$$E_b = (\eta_c \cdot \eta_d) \cdot E_1 = \eta_b \cdot E_1. \quad (3)$$

From the adjustment to the balance of the previous data, Equation (4) must be fulfilled:

$$\int_0^{24} P_s(t)dt = E_d + E_1 \cdot (1 - \eta_b). \quad (4)$$

The minimum rated power required for the PV array (PV_{peak}) is then obtained from Equation (4).

For a set of n_c cloudy days, the energy production would be much less than the theoretical one with good weather. For example, admitting that 15% of the theoretical value is obtained, since the irradiance can be between 10% and 20% of a sunny day [41], it would result:

$$n_c \cdot \left(E_d - 0.15 \cdot \int_0^{24} P_s(t)dt \right) = E\prime_b, \quad (5)$$

$$Q_b = \frac{E\prime_b \cdot 100}{\eta_d \cdot (100 - SOC_{min})} \quad (6)$$

where $E\prime_b$ is the energy contributed by the battery the n_c cloudy days (it has been assumed that $P_d(t) > 0.15 \cdot P_S(t)$ during all the time those days), SOC_{min} is the minimum admissible value of SOC after those n_c days (for example 15%) and Q_b is the value of the necessary battery capacity.

After these days it is necessary to recover the normal state of charge in the batteries in a not very long period, for this reason it is necessary to oversize the value PV_{peak} (for example 20%).

In this way, the batteries are designed for a sufficient capacity to supply the load for n_c cloudy days.

2.2. *Reliability Assessment Methodology*

To evaluate the reliability of the SAPV system two questions must be addressed. An interruption of supply will occur if the instant power provided by the PV arrays plus the energy stored in the batteries is not enough to satisfy the demand. The second cause for an interruption of supply is when unexpected faults occur in the system.

The reliability model of a PV plant is a complex issue [42–44]. For the objectives of this paper an aggregated reliability model of the PV plant (panels and BC) is used (λ_c). The reliability model of the PV plant is assumed to be a two state model, with total power output in the up state and zero power output in the down state (Figure 4). Faults in the batteries are not considered as they are assumed to be maintained and substituted before ending its useful life period (so it is considered $\lambda_b = 0$).

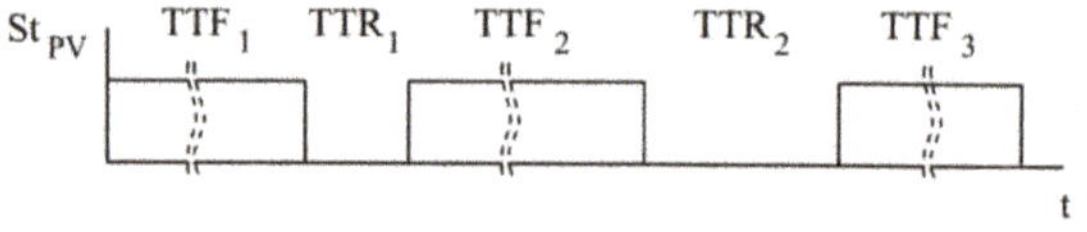

Figure 4. Up and down sequence for the photovoltaic (PV) generation unit.

Time to failure (TTF) of the PV plant is modeled using an exponential distribution and the mean time to repair (TTR) is modeled following a Rayleigh distribution [45]. TTF and TTR are randomly generated using the inverse transform method [46].

The PV generation curve is combined with the sequence of failures obtained, making it zero when there is a PV panel array or BC failure. In this way the generating capacity sequence (GCS) is obtained. The SOC of the battery is then calculated by combining the GCS with the demand sequence.

Considering the instantaneous power demand $P_d(t)$ and $GCS(t)$, SOC (t) is obtained from Equation (7):

$$\mathrm{SOC(t)} = \mathrm{SOC(t-1)} + \alpha(\mathrm{GCS(t)} - \mathrm{P_d(t)}) \cdot \Delta \mathrm{t}, \quad (7)$$

where $\alpha = \eta_c$, if $GCS(t) - P_d(t) > 0$ and $\alpha = 1/\eta_d$, if $GCS(t) - P_d(t) < 0$.

An interruption of supply due to generation inadequacy occurs when:

$$SOC(t) \leq SOC_{min} \text{ and } GCS(t) < Pd(t). \tag{8}$$

With Equations (7) and (8) during PV failure periods, the energy provided by the battery is evaluated to reduce the failure time totally or partially until the battery is discharged to SOC_{min}. If a state in which P_d cannot be supplied is reached, the frequency of interruptions (FOI in number of interruptions/yr) is increased and the loss of energy expectation (LOEE in *Wh/yr*), i.e., energy not supplied expectation, is evaluated as indicated in the description of the reliability analysis method.

If SOC(t) = SOC_{max} and $GCS(t) > P_d(t)$ there is an energy production capacity that cannot be used. This energy will correspond to energy not used (ENU).

To apply the analysis method numerically, variables $P_s(t)$, and $P_d(t)$ must be discretized. For each hour h of the year, the energy produced by the PV array is calculated and its value is assigned to the discrete variable $P_s(h)$. Since its value corresponds to the average hourly power it can be expressed both in kW or in kWh equivalently. The *GCS(t)* is automatically discretized as *GCS(h)* if it is calculated with the discrete values of $P_s(h)$. Similarly, the energy demanded in each hour is assigned to the discrete variable $P_d(h)$.

Description of the Reliability Analysis Method

A sequential Monte Carlo Simulation (MCS) is performed where the hourly behavior of the system for a series of years is simulated (400 years in this work). The randomized irradiance level for every hour of the day for every month of the year is considered, together with the hourly demand to construct chronological generation and demand random sequences.

The reliability evaluation method is performed as follows (Figure 5):

- BEGIN: Initialize counter: n = 1 (number of years). Obtain initial system parameters.
- FOR n = 1 to 400 DO//Consider a possible convergence criterion (*).
    - Initialize counters: h = 1 (number of simulated hours of the year); i = 0 (counter of interruptions); H = 0 (hours of interruption); LOEE = 0; ENU = 0, SOC = 80% (battery state of charge).
    - Simulate TTF and TTR consecutively to generate the annual failure sequence.
    - Obtain randomized hourly PV generation time series data $P_s(h)$ from the historical record.
    - Generate the hourly chronological curve of annual demand $P_d(h)$ from the historical record.
    - Combine $P_s(h)$ and the annual failure sequence to get the generating capacity sequence GCS(h) for the simulated year.
    - FOR h = 1 to 8760 DO:
        - Using GCS(h) and $P_d(h)$, obtain SOC(h) with Equation (7).
        - Update the number of interruptions i and evaluate the duration in hours of each interruption H_i
        - If SOC = SOC_{min} and GCS(h)<$P_d(h)$, update LOEE: LOEE = LOEE+Pd(h)-GCS(h)
        - If SOC = SOC_{max} and GCS(h)>$P_d(h)$, update ENU: ENU = ENU+GCS(h)−$P_d(h)$
    - Evaluate FOI index: FOI = i.
    - Evaluate the loss of load expectation (LOLE) index: LOLE = $\sum$Hi (*h/yr*).
    - Evaluate the loss of load probability (LOLP) index: LOLP = 100·LOLE/8760.
    - Calculate average values of the indices for the n simulated years//Consider a possible convergence criterion (*).
- Calculate frequency histograms for the reliability indices per year.

(*) A possible convergence criterion is to stop the process if for 10 consecutive years, for example, the average values of the indices vary less than a fixed value.

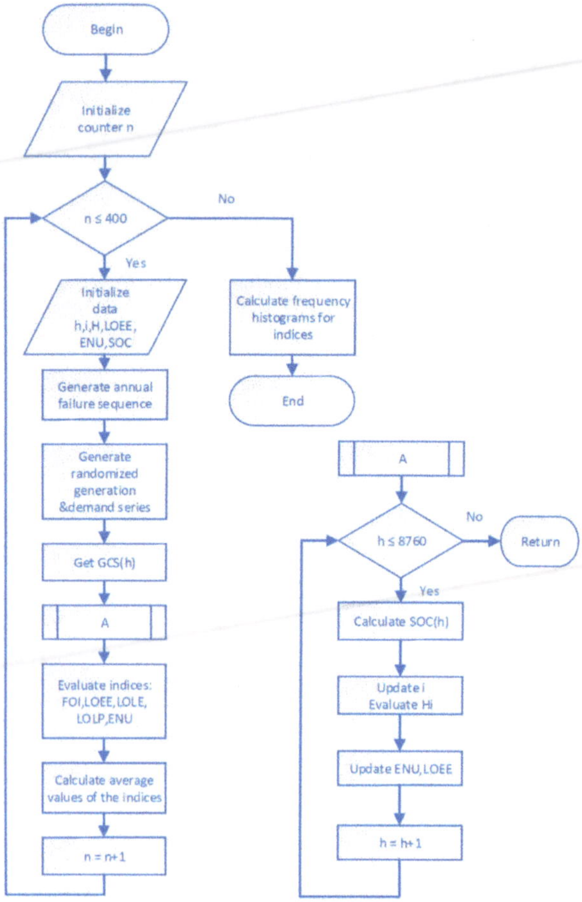

Figure 5. Flowchart of the reliability evaluation method for the SAPV system.

3. Results

In this section, the reliability of supply is evaluated for a real SAPV system, using the proposed methodology.

On the one hand, using the registered demand data of a real consumer during a whole year, a deterministic design (DD) using equations (1) to (6) provides a design of PV_{peak} = 4 kW. Assuming n_c = 4 days, it results in Q_b = 35 kWh (Table 1). The selected case study developed corresponds to a dwelling, but the proposed method can be applied to larger SAPV systems.

Table 1. Generation and demand data.

Method	Variable	Value	Comments
DD	P_d(peak)	3 kW	Maximum demanded power
DD	η_c, η_d	0.9	Efficiency of Li-ion battery (charge and discharge)
DD	n_c	4	Consecutive cloudy days
DD	SOC_{min}	15%	SOC min considered
DD	PV_{peak}	4 kW	Obtained rated power in PV panels (24 m^2)
DD	Q_b	35 kWh	Obtained battery capacity
MCS	Failure rate λ_c	2 f/yr	Exponential distribution
MCS	Repair time	24 h	Rayleigh distribution

On the other hand, using the MCS proposed several situations have been evaluated and different cases are obtained for the considered system.

First, depending on the load demand, PV generation and battery capacity, some days present a generation surplus, as represented in Figure 6a. The PV generation capacity exceeds the load demand and the SOC of the battery has reached SOC_{max}. This results in surplus power not used (PNU), due to a lack of energy storage capacity in the batteries, that would lead to surplus ENU.

Secondly, there are days when PV generation is low and the battery capacity is exhausted resulting in load curtailments and interruptions of supply to the user, as depicted in Figure 6b. These events result in a loss of energy expectation, or its instantaneous equivalent loss of power expectation (LOPE).

Lastly, in addition to the variability in PV generation, unexpected faults occur in the system that cause PV energy generation interruptions. These faults may result in interruptions of supply depending on the duration of the power interruption, the energy demand during that time and the energy stored in the battery (SOC). Figure 6c shows an example where the fault is successfully covered by the batteries and Figure 6d shows an example where the fault is not covered by the batteries.

The proposed MCS method enables the evaluation of reliability indices taking into account the uncertainty associated to variable PV generation, variable load demand and unexpected failures. The computed LOLE is shown in Figure 7 with an average value of 791.6 h/yr. This continuity of supply corresponds to an average LOLP of 9% and supposes that LOEE equals 450 kWh/yr (Figure 8). ENU is 1634 kWh/yr occurring during 1065 h when there is surplus non-used PV generation.

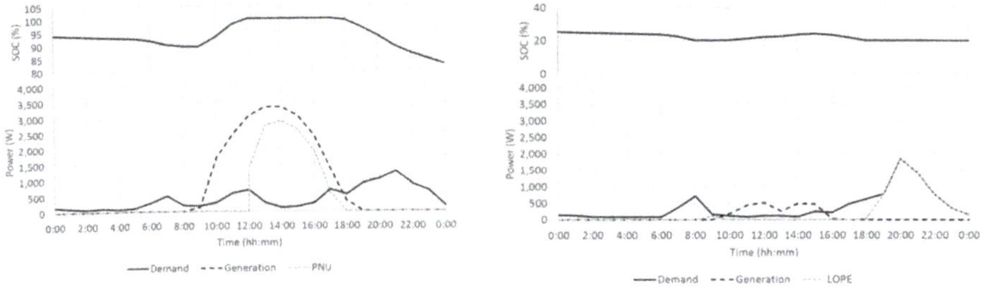

(a) (b)

Figure 6. *Cont.*

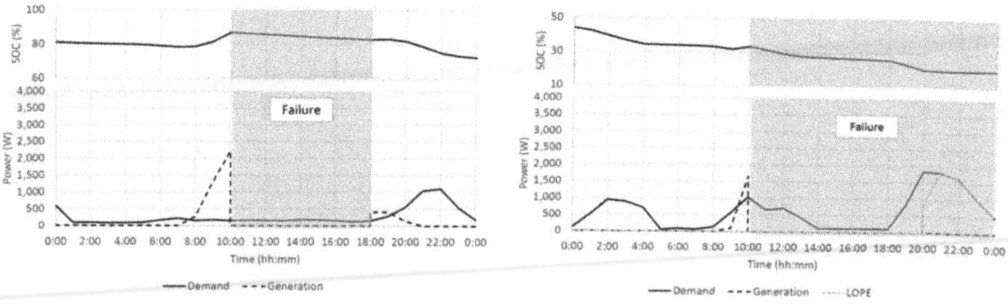

(c) **(d)**

Figure 6. (**a**) (15/4/2018) Day without failures and surplus energy not used (ENU) (ENU > 0); (**b**) (12/11/2018) Day without failures with an interruption due to a deficit of generation which is not supplied by the battery (state of charge (SOC) < SOC_{min} and LOEE > 0); (**c**) (20/4/2018) day with a failure that is supplied by the battery (SOC > SOC_{min} and LOEE = 0); (**d**) (11/11/2018) day with a failure which is not covered by the battery (SOC < SOC_{min} and LOEE > 0).

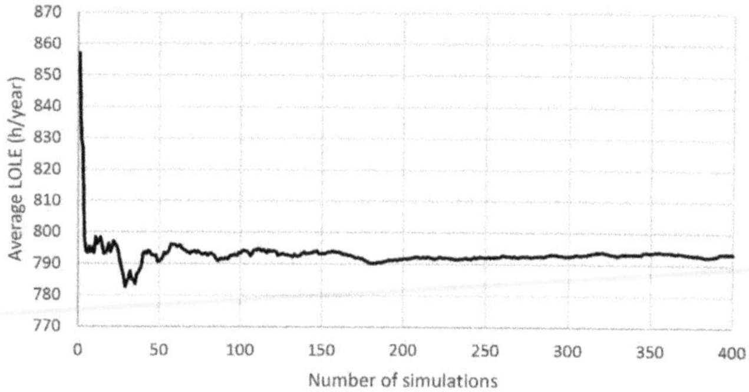

Figure 7. Evolution of average loss of load expectation (LOLE) along 400 simulations.

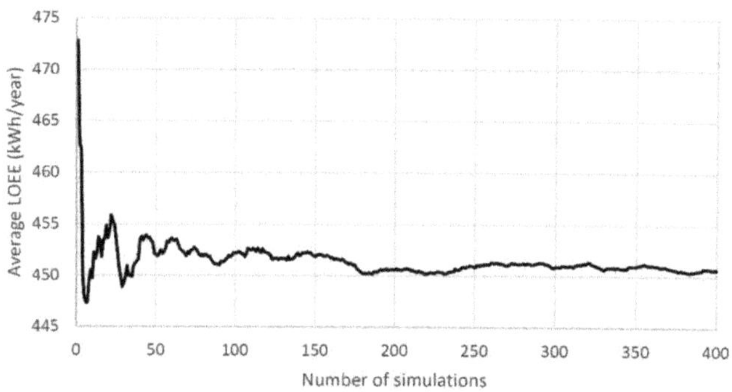

Figure 8. Evolution of average loss of energy expectation (LOEE) along 400 simulations.

This MCS method provides not only long-term average values but also the reliability indices distribution, useful to determine possible extreme values. Figure 9—(a) shows the yearly histograms of the number of failures in the generation system, (b) FOI due to a deficit of PV generation and energy storage, including the effect of the system failures, (c) LOLE and (d) ENU.

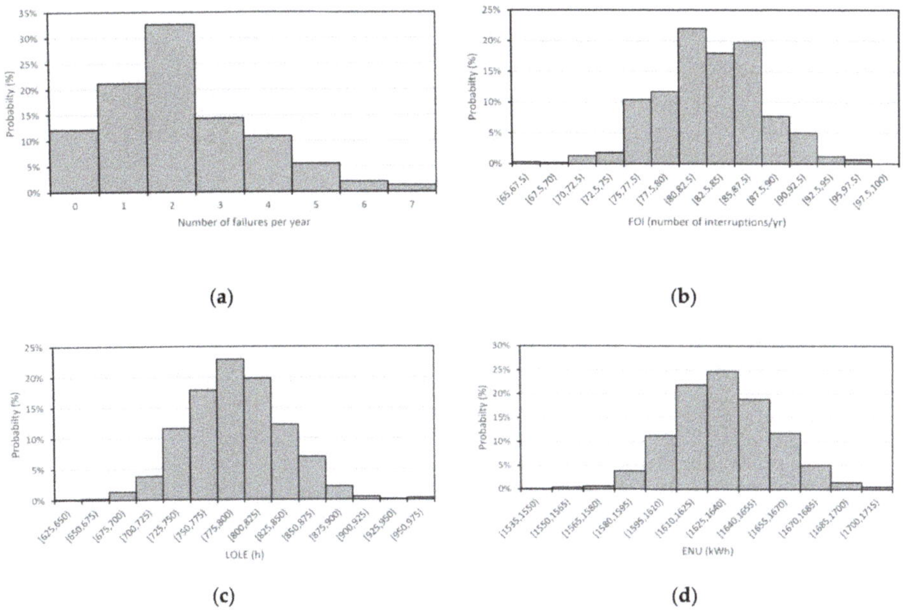

Figure 9. (**a**) Histograms of the number of failures in the generation system, (**b**) number of supply interruptions, (**c**) LOLE and (**d**) ENU.

Although average values of 2 failures per year, 83 energy deficit events, 790 h of LOLE and 1634 kWh of ENU are estimated, more extreme values are not unlikely.

4. Discussion

Results displayed in Figure 6 show that several factors interplay to determine supply interruptions in the system. Insufficient PV generation can lead to power interruptions when the battery energy is exhausted (Figure 6b). Component faults unexpectedly occur that can lead to an interruption of supply depending in the outage duration and the SOC of the battery (Figure 6c,d). The uncertainty associated to solar radiation, demand and component faults and the dependency on previous states of the system (battery SOC) to determine power outages make the evaluation of the system's expected reliability indices complex.

A method based on sequential MCS is then convenient as proposed in this work. Figures 7 and 8 show that when the number of simulations increases, the reliability indices converge to a steady state value.

One advantage of the proposed method is that the distribution of the reliability indices can be easily obtained (Figure 9). These distributions allow considering extreme values in the design of the SAPV system that are unlikely to happen.

A sensitivity analysis is performed to evaluate the impact of different parameters of the SAPV generation system in the reliability indices. The average failure tax and repair time will affect the number and duration of the interruptions. The capacity of the batteries and the installed peak power of the PV panels will affect the reliability indices experienced by the user in stand-alone installations.

The continuity of supply depends on the failure tax of the system. Figure 10a,b show, respectively, how the LOLP and LOEE indices increase linearly when the failure rate λ_c increases from 0.1 f/yr to 20 f/yr for a fixed average repair time of 24 h.

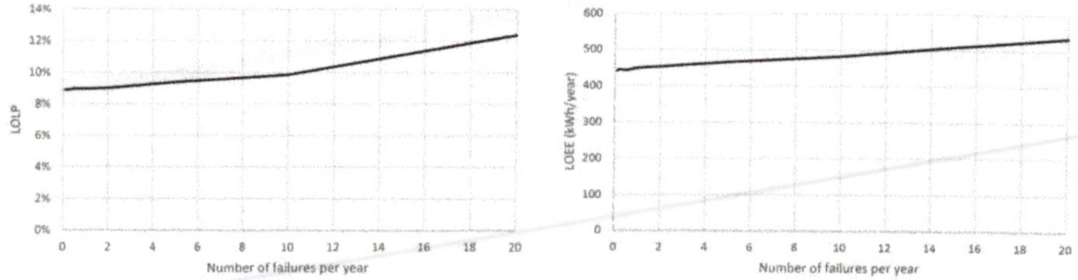

(a) **(b)**

Figure 10. (**a**) loss of load probability (LOLP) and (**b**) LOEE evolution when λ_c varies from 0.1 to 20 f/yr.

In the proposed method system failures are just one of the three different sources of uncertainty considered (PV generation, power demand, and component faults). In fact, the main contribution to the outage time is due to deficits in the energy balance. As shown in Figure 10 for negligible failure rates, resulting in years with no faults in the PV components, power interruptions occur due to insufficient PV generation and stored energy in the batteries. Reliability indices for years with no component faults are LOLP = 9% and LOLE = 440 kWh for the installed PV power and battery storage resulting from the deterministic design. These results show that the design method based on worst case scenarios can lead to inadequate continuity of supply levels.

The average repair time after any failure also affects the reliability indices. Figure 11a,b show, respectively, how the LOLP and LOEE indices increase linearly when the average TTR increases from 6 h to 7 days for a fixed failure tax of 2 f/yr.

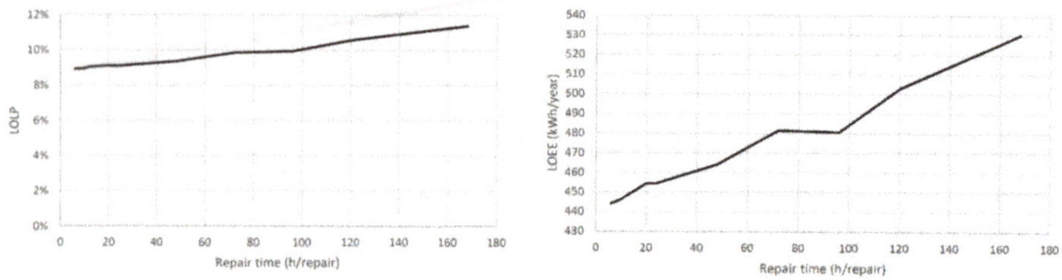

(a) **(b)**

Figure 11. LOLP (**a**) and LOEE (**b**) evolution when time to repair (TTR) varies from 6 h to 7 days.

Higher installed peak power (PV_{peak}) in the PV panels will suppose increased self-sufficiency of the SAPV system. A sensitivity analysis is performed for variable PV_{peak} with fixed failure rate (2 f/yr), repair time (24 h) and for several values of Q_b, from 10 to 50 kWh. Figure 12 shows how increasing PV_{peak} from 1 kWp to 8 kWp improves exponentially the LOLP, while further increases result in marginal improvements. The same can be said about the LOEE.

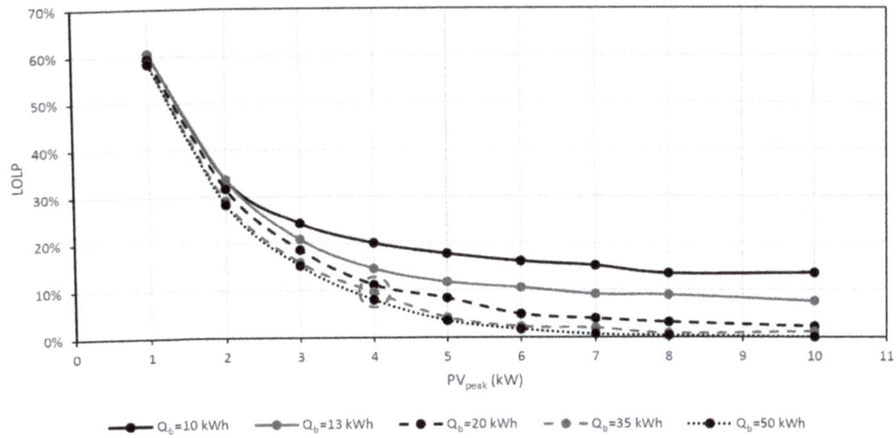

Figure 12. LOLP evolution when PV_{peak} varies from 1 to 8 kWp for several values of Q_b.

Another sensitivity analysis is performed for variable Q_b and fixed λ_c (2 f/yr), repair time (24 h) and varying PV_{peak} from 4 to 10 kW. Figure 13 shows how increasing Q_b from 1 kWh to 25 kWh improves exponentially the LOLP, while further increases result in a much slower decrease. The same can be said about the LOEE. Similarly, the ENU is reduced as Q_b increases up to the same limit of Q_b = 25 kWh where further reductions are marginal.

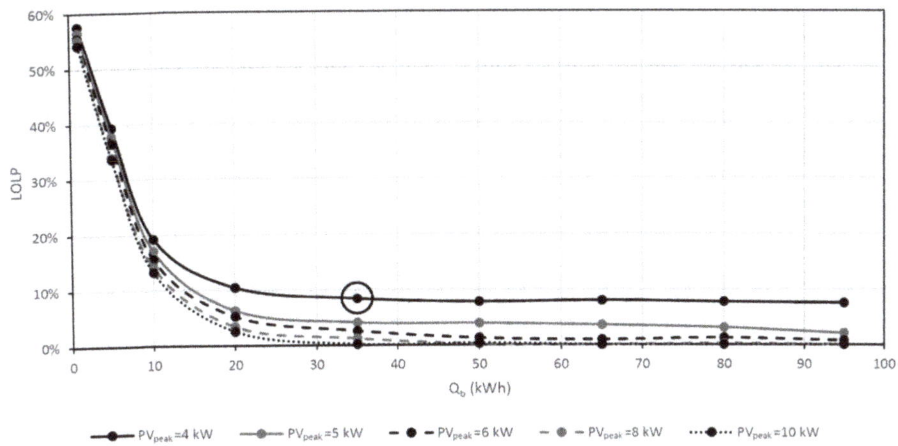

Figure 13. LOLP evolution when Q_b varies from 1 to 95 kWh for several values of PV_{peak}.

The results shown demonstrate the usefulness of reliability analysis based on MCS to improve the design of SAPV systems. It is observed that the optimal values of Q_b are between 20 and 30 kWh, while higher values provide very little improvements to the system. Likewise, it is observed that the installed power PV_{peak} must be increased to values between 6 and 8 kW in order to obtain a LOLP around 2%. Comparing to the initial deterministic design (PV_{peak} = 4 kW and Q_b = 35 kWh are shown as circled points in Figures 12 and 13), this means that it is more convenient to increase PV_{peak} and slightly reduce Q_b (to a value of 25–30 kWh). This type of design is the only one that can guarantee continuity in the supply, based on statistical values.

The main contributions of this work are:

- Development and implementation of a reliability evaluation method in an SAPV generation system with energy storage.
- Consideration of the uncertainty associated with the generation, demand and system failures simultaneously.
- Parametric analysis of the influence of TTF and TTR on the operation of the system.
- Use of a local weather model of PV generation and demand for each day of the year, to achieve realistic results.
- Demonstration of the advantages offered by the sequential Monte Carlo simulation versus deterministic methods to achieve a design of an SAPV generation system with energy storage based on required continuity of supply values.
- Although the case study has been carried out for a domestic residence, the method is directly applicable to any other installation if adequate generation and consumption data are available. As an example, it could be applied to small residential communities, agricultural facilities or others.

5. Conclusions

The application of the developed probabilistic method enables the evaluation of reliability indices according to the system parameters. From the experimental results it can be concluded that, as one would expect, the values of loss of load probability (LOLP) or loss of energy expectation (LOEE) indices increase as the average failure rate per year and the average repair time increase. However, this increase is very small for moderate increases in the failure rate in the panels and battery controller (between 1 failure every 10 years and 3 failures per year) and in the average time to repair (between 6 and 24 h). This analysis makes it possible to evaluate the utilization of solar panels with greater robustness and an assistance service with a limited response time.

From the sensitivity analysis carried out in the presented case study, it is concluded that using a standard deterministic method based in worst case scenarios, as stated in Section 2, for the initial design of the installed power of photovoltaic panels and the battery capacity criteria results in inadequate levels of reliability. The developed method allows the determination of the expected continuity of supply values based on the energy parameters of the installation and the probability distributions associated with failure and repair times. It is concluded that the deterministic design results in an undersized installed power in PV panels and an oversized battery capacity.

The reliability assessment method proposed by the authors in Section 3 enables an optimal design of the installed power and the energy storage for the desired reliability in the system supply (a LOLP around 2%).

The use of renewable energy sources, such as solar energy, is one of the keys to improving energy sustainability. However, the intermittent nature of these sources and their difficult management are great disadvantages for their use. To improve these aspects, energy storage systems are used. As demonstrated in this paper, if the design of renewable energy facilities is not carried out with appropriate criteria, supply interruption periods may be inadmissible. Very few consumers would be willing to rely on energy systems that cause frequent interruptions. The article demonstrates that a design based on reliability parameters is possible, providing higher security to users and improving their confidence in these sustainable energy systems.

The obtained results can also be useful to guide the design of new installations in the same geographical area. In addition, the presented analysis method enables to draw conclusions about the modifications needed in a facility in which the admissible reliability criteria are not met. The results prove which measures will be more effective to improve reliability: increase batteries capacity, installed peak power or both at once. Without this kind of analysis, some extensions of facilities could be ineffective, have a very low impact or represent a very high and unjustified cost.

The presented case study and the discussion of the results show that in this type of stand-alone systems there are situations of lack of energy as well as moments with ENU due to an excess of generation. The diversity of consumers and their energy systems suggests studying the creation of small microgrids formed by several facilities such as the one described in this paper, interconnected by the grid, so that these facilities can be supported by energy transfers between them. Reliability analysis of these small communities linked in microgrids is now an objective for future research of the authors.

Author Contributions: Conceptualization and methodology, E.Q. simulation software design C.R.-B. experiments design and validation, all authors; formal analysis, E.Q. and C.R.-P. data curation C.R.-B., G.E.-E., and E.Q. writing-original draft preparation, E.Q., C.R.-B., and G.E.-E. writing-review and editing, all authors; supervision E.Q. and C.R.-P. funding acquisition, all authors. All authors have read and agreed to the published version of the manuscript.

Funding: This work has been supported by research funds of the Universitat Politècnica de València.

Acknowledgments: This work has been supported by the Universitat Politècnica de València.

Conflicts of Interest: The authors declare no conflict of interest.

Abbreviations

AC	alternating current
BC	battery controller
DC	direct current
DN	distribution network
E_b	energy contributed by the battery
E_d	daily energy consumption
E_1	excess energy produced by the photovoltaic panels
ENU	energy not used
FOI	frequency of interruptions
GCS	generating capacity sequence
HRES	hybrid renewable energy systems
LOEE	loss of energy expectation index
LOLP	loss of load probability index
LOLE	loss of load expectation index
LOPE	loss of power expectation index
MCS	Monte Carlo simulation
n_c	consecutive cloudy days
$P_d(t)$	instantaneous power demand
$P_{d(peak)}$	maximum demanded power
PNU	power not used
PV_{peak}	rated power installed in the photovoltaic panels
$P_S(t)$	power produced by the photovoltaic array
PV	photovoltaic array generation system
Q_b	battery capacity
SAPV	stand-alone photovoltaic system
SOC	state of charge of battery
SOC_{max}	maximum admissible value of SOC
SOC_{min}	minimum admissible value of SOC
TTF	time to failure
TTR	time to repair
λ_b	battery failure rate per year
λ_c	photovoltaic panel array and battery controller failure rate per year
λ_i	inverter failure rate per year
η_b	li-ion battery efficiency
η_c	battery charging efficiency
η_d	battery discharging efficiency

References

1. Twaha, S.; Ramli, M.A. A review of optimization approaches for hybrid distributed energy generation systems: Off-grid and grid-connected systems. *Sustain. Cities Soc.* **2018**, *41*, 320–331. [CrossRef]
2. IEA PVPS. *Trends 2018 in Photovoltaic Applications*; T1-34:2018; IEA PVPS: Paris, France, 2018; Available online: http://www.iea-pvps.org/fileadmin/dam/intranet/task1/IEA_PVPS_Trends_2018_in_Photovoltaic_Applications.pdf (accessed on 3 February 2020).
3. Masson, G.; Latour, M. Self-consumption as the new Holy Grail of the PV industry: From theory to reality. In *Photovoltaics International*, 17th ed.; PV Tech: London, UK, 2012; pp. 166–169.
4. Mandelli, S.; Barbieri, J.; Mereu, R.; Colombo, E. Off-grid systems for rural electrification in developing countries: Definitions, classification and a comprehensive literature review. *Renew. Sustain. Energy Rev.* **2016**, *58*, 1621–1646. [CrossRef]
5. Luthander, R.; Widén, J.; Nilsson, D.; Palm, J. Photovoltaic self-consumption in buildings: A review. *Appl. Energy* **2015**, *142*, 80–94. [CrossRef]
6. Evans, A.; Strezov, V.; Evans, T.J. Assessment of utility energy storage options for increased renewable energy penetration. *Renew. Sustain. Energy Rev.* **2012**, *16*, 4141–4147. [CrossRef]
7. Decree, R.; del Estado, B.O. Available online: https://www.boe.es/diario_boe/txt.php?id=BOE-A-2019-5089 (accessed on 23 December 2019).
8. Bugała, A.; Zaborowicz, M.; Boniecki, P.; Janczak, D.; Koszela, K.; Czekała, W.; Lewicki, A. Short-term forecast of generation of electric energy in photovoltaic systems. *Renew. Sustain. Energy Rev.* **2018**, *81*(Part 1), 306–312. [CrossRef]
9. Abuagreb, M.; Allehyani, M.; Johnson, B.K. Design and Test of a Combined PV and Battery System under Multiple Load and Irradiation Conditions. In Proceedings of the 2019 IEEE Power & Energy Society Innovative Smart Grid Technologies Conference (ISGT), Washington, DC, USA, 17–20 February 2019; pp. 1–5. [CrossRef]
10. Moharil, R.M.; Kulkarni, P.S. Reliability analysis of solar photovoltaic system using hourly mean solar radiation data. *Sol. Energy* **2010**, *84*, 691–702. [CrossRef]
11. Dissawa, D.M.L.H.; Godaliyadda, G.M.R.I.; Ekanayake, M.P.B.; Ekanayake, J.B.; Agalgaonkar, A.P. Cross-correlation based cloud motion estimation for short-term solar irradiation predictions. In Proceedings of the 2017 IEEE International Conference on Industrial and Information Systems (ICIIS), Peradeniya, Sri Lanka, 15–16 December 2017; pp. 1–6. [CrossRef]
12. Kaplani, E.; Kaplanis, S. A stochastic simulation model for reliable PV system sizing providing for solar radiation fluctuations. *Appl. Energy* **2012**, *97*, 970–981. [CrossRef]
13. Benmouiza, K.; Tadj, M.; Cheknane, A. Classification of hourly solar radiation using fuzzy c-means algorithm for optimal stand-alone PV system sizing. *Int. J. Electr. Power Energy Syst.* **2016**, *82*, 233–241. [CrossRef]
14. Ozoegwu, C.G. Artificial neural network forecast of monthly mean daily global solar radiation of selected locations based on time series and month number. *J. Clean. Prod.* **2019**, *216*, 1–13. [CrossRef]
15. Palensky, P.; Dietrich, D. Demand Side Management: Demand Response, Intelligent Energy Systems, and Smart Loads. *IEEE Trans. Ind. Inform.* **2011**, *7*, 381–388. [CrossRef]
16. Roldán-Blay, C.; Escrivá-Escrivá, G.; Roldán-Porta, C. Improving the benefits of demand response participation in facilities with distributed energy resources. *Energy* **2019**, *169*, 710–718. [CrossRef]
17. Roldán-Porta, C.; Roldán-Blay, C.; Escrivá-Escrivá, G.; Quiles, E. Improving the Sustainability of Self-Consumption with Cooperative DC Microgrids. *Sustainability* **2019**, *11*, 5472. [CrossRef]
18. Huang, Y.; Yang, L.; Liu, S.; Wang, G. Cooperation between Two Micro-Grids Considering Power Exchange: An Optimal Sizing Approach Based on Collaborative Operation. *Sustainability* **2018**, *10*, 4198. [CrossRef]
19. Goel, S.; Sharma, R. Performance evaluation of stand alone, grid connected and hybrid renewable energy systems for rural application: A comparative review. *Renew. Sustain. Energy Rev.* **2017**, *78*, 1378–1389. [CrossRef]
20. Weniger, J.; Tjaden, T.; Quaschning, V. Sizing of residential PV battery systems. *Energy Procedia* **2014**, *46*, 78–87. [CrossRef]
21. Maleki, A.; Rosen, M.; Pourfayaz, F. Optimal operation of a grid-connected hybrid renewable energy system for residential applications. *Sustainability* **2017**, *9*, 1314. [CrossRef]
22. Cao, S.; Hasan, A.; Sirén, K. Matching analysis for on-site hybrid renewable energy systems of office buildings with extended indices. *Appl. Energy* **2014**, *113*, 230–247. [CrossRef]

23. Ren, H.; Wu, Q.; Gao, W.; Zhou, W. Optimal operation of a grid-connected hybrid PV/fuel cell/battery energy system for residential applications. *Energy* **2016**, *113*, 702–712. [CrossRef]
24. Ghafoor, A.; Munir, A. Design and economics analysis of an off-grid PV system for household electrification. *Renew. Sustain. Energy Rev.* **2015**, *42*, 496–502. [CrossRef]
25. Maleki, A.; Hajinezhad, A.; Rosen, M.A. Modeling and optimal design of an off-grid hybrid system for electricity generation using various biodiesel fuels: A case study for Davarzan, Iran. *Biofuels* **2016**, *7*, 669–712. [CrossRef]
26. Castillo-Cagigal, M.; Caamano-Martín, E.; Matallanas, E.; Masa-Bote, D.; Gutiérrez, A.; Monasterio-Huelin, F.; Jiménez-Leube, J. PV self-consumption optimization with storage and Active DSM for the residential sector. *Sol. Energy* **2011**, *85*, 2338–2348. [CrossRef]
27. Zhou, W.; Lou, C.; Li, Z.; Lu, L.; Yang, H. Current status of research on optimum sizing of stand-alone hybrid solar–wind power generation systems. *Appl. Energy* **2010**, *87*, 380–389. [CrossRef]
28. Yadav, A.K.; Chandel, S.S. Solar radiation prediction using Artificial Neural Network techniques: A review. *Renew. Sustain. Energy Rev.* **2014**, *33*, 772–781. [CrossRef]
29. JPW Stackhouse. Surface meteorology and Solar Energy. Atmospheric Science Data Center. 2011. Available online: https://eosweb.larc.nasa.gov/ (accessed on 3 February 2020).
30. Roldán-Blay, C.; Escrivá-Escrivá, G.; Roldán-Porta, C.; Álvarez-Bel, C. An optimisation algorithm for distributed energy resources management in micro-scale energy hubs. *Energy* **2017**, *132*, 126–135. [CrossRef]
31. Hoevenaars, E.J.; Crawford, C.A. Implications of temporal resolution for modeling renewables-based power systems. *Renew. Energy* **2012**, *41*, 285–293. [CrossRef]
32. Cao, S.; Sirén, K. Impact of simulation time-resolution on the matching of PV production and household electric demand. *Appl. Energy* **2014**, *128*, 192–208. [CrossRef]
33. Cucchiella, F.; D'Adamo, I.; Gastaldi, M.; Stornelli, V. Solar Photovoltaic Panels Combined with Energy Storage in a Residential Building: An Economic Analysis. *Sustainability* **2018**, *10*, 3117. [CrossRef]
34. Kosmadakis, I.E.; Elmasides, C.; Eleftheriou, D.; Tsagarakis, K.P. A Techno-Economic Analysis of a PV-Battery System in Greece. *Energies* **2019**, *12*, 1357. [CrossRef]
35. Werner, C.; Breyer, C.; Gerlach, A.; Beckel, O. Photovoltaic with Energy Storage: An Overview on Economics, System Design and Politics. In Proceedings of the 27th European Photovoltaic Solar Energy Conference, Frankfurt, Germany, 24–28 September 2012.
36. Faza, A. A probabilistic model for estimating the effects of photovoltaic sources on the power systems reliability. *Reliab. Eng. Syst. Saf.* **2018**, *171*, 67–77. [CrossRef]
37. Borges, C.L.T. An overview of reliability models and methods for distribution systems with renewable energy distributed generation. *Renew. Sustain. Energy Rev.* **2012**, *16*, 4008–4015. [CrossRef]
38. Billinton, R. Reliability considerations in the utilization of wind energy, solar energy and energy storage in electric power systems. In Proceedings of the 2006 International Conference on Probabilistic Methods Applied to Power Systems, Stockholm, Sweden, 11–15 June 2006; IEEE: Piscataway, NJ, USA, 2006; pp. 1–6.
39. Roldán-Blay, C.; Roldán-Porta, C.; Peñalvo-López, E.; Escrivá-Escrivá, G. Optimal Energy Management of an Academic Building with Distributed Generation and Energy Storage Systems. *IOP Conf. Ser. Earth Environ. Sci.* **2017**, *78*, 012018. [CrossRef]
40. Pérez-Navarro, A.; Alfonso, D.; Ariza, H.E.; Cárcel, J.; Correcher, A.; Escrivá-Escrivá, G.; Hurtado, E.; Ibáñez, F.; Peñalvo, E.; Roig, R.; et al. Experimental verification of hybrid renewable systems as feasible energy sources. *Renew. Energy* **2016**, *86*, 384–391. [CrossRef]
41. Wang, J.Y.; Qian, Z.; Zareipour, H.; Wood, D. Performance assessment of photovoltaic modules based on daily energy generation estimation. *Energy* **2018**, *165 Pt B*, 1160–1172. [CrossRef]
42. Eltawil, M.A.; Zhao, Z. Grid-connected photovoltaic power systems: Technical and potential problems: A review. *Renew. Sustain. Energy Rev.* **2010**, *14*, 112–129. [CrossRef]
43. Zhang, P.; Li, W.; Li, S.; Wang, Y.; Xiao, W. Reliability assessment of photovoltaic power systems: Review of current status and future perspectives. *Appl. Energy* **2013**, *104*, 822–833. [CrossRef]
44. Collins, E.; Dvorack, M.; Mahn, J.; Mundt, M.; Quintana, M. Reliability and availability analysis of a fielded photovoltaic system. In Proceedings of the 34th IEEE Photovoltaic Specialists Conference (PVSC), Philadelphia, PA, USA, 7–12 June 2009.

45. Billinton, R.; Jonnavithula, A. Application of sequential monte carlo simulation to evaluation of distributions of composite system indices. *IEEE Proc. Gener. Transm. Distrib.* **1997**, *144*, 87–90. [CrossRef]
46. Billinton, R.; Allan, R.N. *Reliability Evaluation of Power Systems*; Springer: Boston, MA, USA, 1984.

sustainability

MDPI

Perspective

Assessment on Global Urban Photovoltaic Carrying Capacity and Adjustment of Photovoltaic Spatial Planning

Siyuan Chen, Yukun Zhang and Jie Zheng *

School of Architecture of Tianjin University, Tianjin 300000, China; miracle1987@126.com (S.C.); zyk.tj@163.com (Y.Z.)
* Correspondence: tjuzhengjie@163.com

Abstract: To promote the effective combination of photovoltaic (PV) utilization and urban development, this study proposes that solar PV generation should be taken as an important resource and environmental carrying capacity factor, which is defined as "Photovoltaic Carrying Capacity (PVCC)", to be integrated into future urban planning. According to the PVCC assessment on global cities, the sensitivity of PVCC to different influence factors is analyzed and the benefits of different optimization strategies on PVCC are also discussed. Additionally, in an equilibrium analysis of PVCC distribution in six sample countries, we discuss both the causes and impact of their distribution gap. That analysis shows that the average PVCC in global urban areas can reach 23.13%, which approaches the anticipated needs of PV development in cities by 2050.Though different optimization strategies may be feasible to promote urban PVCC, they might only make significant changes in medium or large cities. When considering the high-efficient utility of local energy, dispersed layout of cities and population is necessary to implement PV spatial planning. According to the assessment and analysis results, the adjustment suggestions of PV spatial planning in each sample country are also discussed. Finally, it is pointed out that the proactive PV spatial may be of great significance to achieve higher solar energy supply and PVCC will be an available cognition in guiding this planning in the future.

Keywords: photovoltaic spatial planning; photovoltaic carrying capacity; influence factors; optimization strategies; carrying capacity distribution; planning adjustment

check for updates

Citation: Chen, S.; Zhang, Y.; Zheng, J. Assessment on Global Urban Photovoltaic Carrying Capacity and Adjustment of Photovoltaic Spatial Planning. *Sustainability* **2021**, *13*, 3149. https://doi.org/10.3390/su13063149

Academic Editors: M. Dolores Esteban and Maria Malvoni

Received: 20 January 2021
Accepted: 10 March 2021
Published: 12 March 2021

1. Introduction

With the depletion of fossil energy, countries all over the world have put forward their own energy transition plans. Renewable energy, mainly solar photovoltaic (PV) have increasingly become the mainstream development trend of global energy transition due to its more general applicability and sustainability. By the end of 2018, the world's total installed photovoltaic capacity had reached to about 480 GW, which had contributed 2% of the world's total energy consumption [1]. Meanwhile, the cost of PV generation per kilowatt hour has also shown a trend of rapid decline. Taking utility scale PV as an example, the average cost per kilowatt hour of the electricity has been reduced from 371 USD/MWh in 2010 to 85 USD/MWh in 2018, dropping by 77% in only eight years [2]. Although there is still a cost gap compared with the traditional energy, such as thermal power generation, solar PV is catching up with traditional energy to achieve connection to grid at an equal price and its competitive advantage constantly improves. An estimate from the International Renewable Energy Agency (IRENA) shows that, the total global installed PV capacity will reach 8519 GW by 2050, including about 60% utility scale and 40% distributed rooftop and the power output will take up 25% of the world's total electrical energy consumption [1], becoming a significant part of the energy consumption structure.

The energy transition to PV (and many other renewable energies) is not in doubt, but compared with traditional fossil energy, solar PV has particular characteristics, such as intermittence and lower energy density, which will necessitate a significantly different planning mode. Currently, comprehensive assessments of its sustainability (including energy production, transportation and consumption) are not generally carried out [3]. To date, for the most part, spatial planning and energy planning have been treated as two separate domains. So-called "Energy Planning" is usually just some target-based guidelines which are largely concerned with the security of supply and energy demand. However, without resource distribution analysis and spatial construction planning, there will be obvious deficiencies in the guidance of local investment in the construction of renewable energy facilities. As a result, renewable energy planning is still carried out in the same way as for traditional fossil energy, becoming thereby subordinated to urban planning in the form of power distribution only. With regard to this background, many researchers have proposed the application of "Landscape: into energy planning, putting forward the concept of "Energy Landscape" [3–5]. According to the spatio-temporal modeling of renewable energy carriers' distribution, a comprehensive optimized utilization strategy could be achievable through methods of planning adjustment, policy incentive and financial support, etc. In that case, the goal of future renewable energy planning will no longer be simply limited to the development expectations of a single increasing of energy consumption proportion, but will be more inclined to the deployment of both regional energy distribution and urban development in a view of comprehensive spatial planning.

Energy is of great significance in the process of urban development. It is not only an indispensable natural resource, but also a necessary condition for the transformation of other natural resources into economic output. From the economic perspective, the essence of urban development is just a process of population accumulation during which natural resources are exhausted for economic production by converting their economic mode [6]. So, like many other natural resources, energy adequacy is usually reflected as an important "Resource and Environment Carrying Capacity (RECC)", which is initially used to assess the environmental constraints on population growth [7]. RECC has been defined by FAO and UNESCO as the maximum population that a country or region can afford by using local natural resources, energy, technology and technology [8]. However, with the development of worldwide research, its measurement has been gradually extended to include many other criteria, such as "Ecological Footprint", "Net Primary Productivity" and "Emergy" [9–11]. Although there is not yet a universal definition, the connotation of RECC allows the measurement of the resource endowment and environmental capacity of a certain regional space and the population and economic scale it could afford so as to maintain a good ecosystem without doing any harm to the natural ecological environment [12].

A great quantity of researches of RECC can be found on fossil energies like coal or gas-fired power generation [13–16], but there is as yet none on solar PV. The obvious reason is that these energy forms are generally considered as non-renewable. They are also affected to the stability and security of their energy supply. Compared with fossil energy forms, solar PV has not yet had any serious impact on either urban or economic development. That said, the proportion of PV power consumption will increase significantly by 2050. Solar PV will play the same role as other power generation technologies, providing important energy guarantees in the process of urban development in the future. Furthermore, according to the goal of "Energy Landscape" planning, solar PV will also become a key part of future urban spatial planning. We therefore propose, in this study, to take solar PV as a latent but necessary resource and environment carrying capacity factor that should be integrated into the comprehensive carrying capacity index system of future urban planning. On this basis, the distribution of PVCC in cities around the world is evaluated, then the deployment and further optimization strategy of future PV energy in urban spatial planning are also discussed in this study. These will help to make a tentative discussion on how PVCC would affect the strategy determination of solar PV spatial planning in a range that from urban planning to national geospatial planning.

2. Method

2.1. Photovoltaic Carring Capacity

According to the connotation of RECC, the "Photovoltaic Carrying Capacity (PVCC)" here in this paper is defined as the capacity of a city to use solar energy resource in its region for PV power generation to meet its own demand, as is shown in Equation (1). By the differences of PV system construction forms, the evaluation objects include two parts of power output: distributed rooftop PV system in urban built-up areas and utility scale PV stations in non-built-up areas.

$$PVCC = \frac{E}{C} \tag{1}$$

where $PVCC$ is the photovoltaic carrying capacity of a city, measured in %; E is the annual power output that can be achieved by a city after PV installation, measured in kWh/year; C is the annual power consumption of a city, measured in kWh/year.

2.2. Influence Factors

According to Equation (1), the influence factors of PVCC can be generally divided into two main aspects: power generation and power consumption. On the basis of some existing researches, the generation aspect includes natural conditions such as solar irradiation, effective sunshine time, latitude, slope, geology, natural disasters, land usage, protected area, temperature, snow, dust, etc. and other spatial morphology factors like building density, spatial layout, building style, urban built up area, population density, population distribution, infrastructure (roads, transmission lines, substations) distribution, etc. [17–20]. The consumption aspect is often related to the living standard or urban development condition of a city, such as urbanization rate, industrial structure, power consumption level, national economy and spot price [21–24]. The assessment results of PVCC are closely related to the influencing factors selected. However, this does not mean that the PVCC of a city is always immutable. For instance, the change of urban spatial morphology by adjusting its planning strategy can significantly affect the city's final PVCC. In addition, considering the convenience and feasibility of the assessment, it is also necessary to define the range of influence factors selected according to different spatial assessment scales. Since the objective discussing spatial scale of this study is set as the range from urban planning to national geospatial planning, considering the replication of these assessments on such geographical scope (worldwide), we take only six main factors into consideration in this study, each of which is supposed to make more significant impact on final PVCC result than the other detailed factors. They are solar irradiation, urban population density, urban built-up area, annual electricity consumption per capita, available land area for PV installation and transmission distance. Moreover, each of these six factors could correspond to at least one independent optimization strategy (without causing any change of other influence factors) directly or indirectly, as shown in Table 1. In addition, these strategies will be later used to underpin a theoretical discussion on the effect that it would make through different methods of planning adjustment. As for the assessment of power consumption, considering the real situation of urban economic production, built-up area will be the main part where energy is exhausted (Almost 80% of total social electricity consumption) [25], so we only take this part of consumption into account for the PVCC assessment.

2.3. Urban Geographical Division

A key quality of future energy systems is low transmission distance to realize a high-efficient utility of local energy [3]. Therefore, it is necessary to delimit the boundary of "local" region first to determine to which city the PV power output (Especially for utility scale) will be mainly supplied. On consideration of transmission line loss and infrastructure investment, PV power output is usually preferentially transmitted to its nearest urban load area to minimize extra consumption, therefore, the urban region division here in this study is determined by the geographical distance between the land location and its nearest urban center by using the cost allocation function in Arcgis, instead of a traditional administrative division. According to the quality of low transmission distance, this method can ensure that the distance between any land pixel in a certain geographical division and its subordinate urban center is shorter than that from any other urban centers. Then, by extracting the average value of each influence factor in a certain geographical division (by zonal statistics function in Arcgis), the local condition of a city could be further described in brief.

Table 1. Influence factors of PVCC and its corresponding optimization strategies.

Influence Factors	Corresponding Optimization Strategies
Solar irradiation	Choosing PV panels of higher conversion efficiency, PV installation on building facades or using inclined roof to gain more irradiation in the same land area
Urban population density	When the urban built-up area is constant, population density will change with the variation of building density
Built-up area	Urbanization in spatial dimension, urbanization or counter-urbanization, make more land into built-up area or in the opposite
Electricity consumption per capita	Reduce building energy consumption, change people's electricity consumption habits or develop low-carbon economy
Available land area	Exploit more land to achieve more installation area (for utility scale photovoltaic stations)
Transmission distance	Closer site selection to urban load to decrease transmission line loss

2.4. Assessment of PV Generation Potential

Different assessment methods of generation potential are adopted in two forms of PV system. As for distributed rooftop, the key point is to evaluate the PV installation area on the roof of a city. In this regard, the International Energy Agency (IEA)has given out an exponential function to describe the relationship between the installation area per capita and the urban population density [26], in order to meet the needs of PV potential assessment in a large range of geographical area, as is shown in Equation (2). Applying this, the annual power generation of distributed rooftop PV can be further calculated by Equation (3).

$$A_a = \alpha \cdot p^{-\beta} \quad (2)$$

$$E_d = A_a \cdot A_b \cdot p \cdot GTI \cdot i \cdot K \quad (3)$$

where A_a is the rooftop PV installation area per capita, measured in m^2/person; p is the urban population density, measured in persons/km^2; α and β are theadjustment coefficients, unless otherwise specified, α = 172.3, β = 0.352; E_d is the annual power generation of distributed rooftop PV, measured in kWh/year; A_b is the urban built-up area, measured in km^2; GTI is annual solar irradiation on tilted surface, measured in kWh/m^2/year; i is the conversion efficiency of PV panel, measured in %; K is the total efficiency of PV system, K = 75% in general.

As for utility scale PV power stations, the key factors include two main parts, the evaluations of the available land area and the transmission distance. According to a standard regulation of China [27] and our field investigation, four kinds of land usage are selected to be available for PV installation and each these four can be further reflected as a corresponding land cover type on basis of the "Land Cover Classification System (LCCS)" [28], including bare areas (LCCS-200), sparse vegetation areas (LCCS-150), mosaic vegetation/cropland areas (LCCS-30) and closed to open herbaceous vegetation areas (LCCS-140). The annual power output can be then further estimated after considering an average 40% ratio of panels area to land occupation and the adjustment coefficient of each available land type (to avoid vegetation occupation, see appendix II in [28]). Meanwhile, considering the lack of or absence of feasibility of installing PV over steep terrains, land with an average slope larger than 5° is excluded by referring to the value in [29] and the pixel size of the DEM data used in this study (2397 × 2397 m). The assessment of transmission line loss is derived by extracting an average value of the distances between each available land pixel and its subordinate urban center. Moreover, an average line loss rate of 0.4%/km is considered in assumption that all the transmission lines are constructed in the system form of 10 kV three-phase four wires. Then, the annual PV power received by each urban built-up area from its surrounding utility scale PV stations could be further estimated as follows:

$$E_u = 40\% \cdot \left(\sum_i^n Au_i \cdot R_i \right) \cdot GTI \cdot i \cdot K \cdot (1 - 0.4\% \cdot D) \quad (4)$$

where E_u is the annual power received from utility scale PV, measured in kWh/year; Au_i is the total area of each type of land usage (I = 30, 140, 150, 200), measured in m^2; R_i is the adjustment coefficient of each available land type (R_{30} = 50%, R_{140} = 70%, R_{150} = 90%, R_{200} = 100%); D is the average transmission distance, measured in km. GTI, i and K see Equation (2).

2.5. PVCC Assessment

Through methods above, it can be estimated that the global PV installed capacity could reach a level of 8.58 TW for distributed rooftop and 3704 TW for utility scale, which has far exceeded the development goal in 2050 (3470.6 GW for distributed rooftop, 40%; 5111.4 GW for utility scale, 60% [1]). This just reflects a theoretical installed capacity. While considering the feasibilities of technical condition, development cost and industry capacity, such amount of installation would be hardly achieved, therefore, a hypothetical restrict of maximum installed capacity is further needed to reflect the realistic situation. By comparing the ratio of development goal to theoretical installation, on average, merely 0.14% of available land area and 39% of roof area are needed to meet 25% proportion of PV power in 2050. Therefore, the PVCC of each urban geographical division can be further evaluated as following equation:

$$PVCC = \frac{39\% \cdot E_d + 0.14\% \cdot E_u}{A_b \cdot p \cdot C_a} \quad (5)$$

where C_a is the annual electricity consumption per capita; The other symbols see Equations (1)–(4).

2.6. Data Sources

The data used in this study include global digital elevation model (DEM) [30], Globcover 2009 [31], Gridded Population of the World, Version 4 (GPWv4) [32], Global irradiation for optimally tilted surface [33] and Urban population gathering points (urban centers). The power consumption is estimated based on the statistics data of "Electricity consumption per capita" from [34,35]. Although they only reflect the electricity consumption of the overall level of a country, while taking into consideration that the citizens of the same country would have certain similarities in their electricity consumption behaviors, habits or customs, these data are still considered as valid. Meanwhile, to ensure the consistency of assessment results, the conversion efficiency of PV panel is unified as 16.54% by referring to Jkm270pp-60-dv poly-silicon PV module [36].

3. Results and Analysis

3.1. Results of PVCC Assessment

The assessment result of each urban division worldwide is shown in Figure 1. In addition, the annual PV power generation from all urban divisions would take up in a total ratio of 43.22% to the global electricity demand. Considering that over 30% of this power output might not be used directly or even wasted because of intermittence and electrical-surplus in off-peak period, this assessment result is relatively reasonable. In addition, it can be seen from Figure 1 that the vast majority of PVCC are provided by deserts, tundra and other sparsely populated areas in the form of ground utility scale PV station. Unless absolutely necessary, these sparsely populated areas are always suggested to better remain unexploited to protect their ecology and biodiversity. Generally, it is not essential to discuss the PVCC of these areas in priority. Therefore, by utilizing a regulation applicable in China [37], the standard of urban scale with a population of 50,000 is adopted in this study to distinguish urban and non-urban areas. On this basis, the average level of PVCC in urban areas worldwide could still reach up to 23.13%, nearly approaching the target of 25% by 2050. In other words, the global PV target of 2050 could be almost achieved only by PV installation in urban areas themselves together with methods such as low-priced energy storage and electric vehicles recharging (to avoid electrical-surplus), though this is just on the basis of global average level.

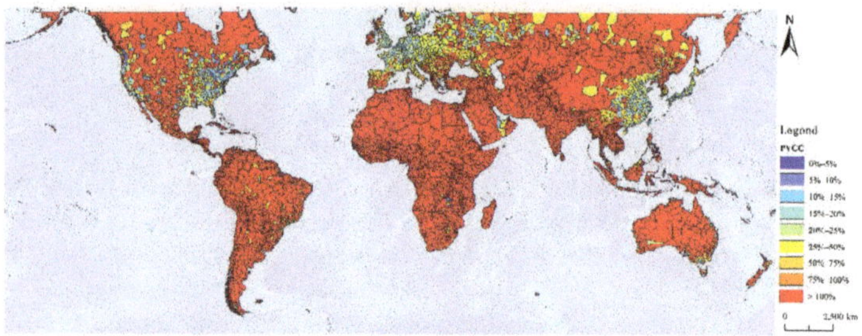

Figure 1. Assessment on PVCC of global urban divisions.

3.2. *Sensitivity Analysis of PVCC*

As mentioned above, the PVCC of a city will change with its influence factors. This section analyses the sensitivity of PVCC result to changes in the influence factors inputs. While considering the effects of different urban types to analysis accuracy, these analyses are conducted separately by different population scales based on a classification standard of China [37], specifically including seven scales that range from miniature city to mega city. For each type of urban scale, the median value of each influence factor is extracted as an assumed global average standard input into the PVCC equation. It is then adjusted independently of the others by ±50% (theoretical maximum and minimum) to observe its sensitivity and the results are shown in Figure 2. These analyses will further help to illustrate how different optimization strategies could make changes on the PVCC result, according to Table 1.

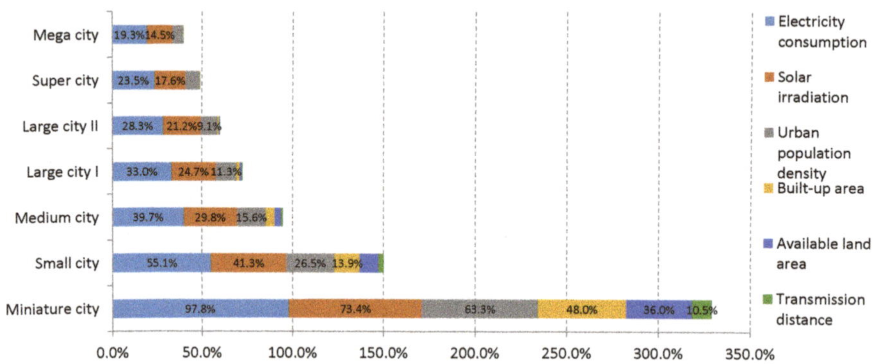

Figure 2. Sensitivity analysis of PVCC to influence factors in different urban scales. Note: The length of each colored bar indicates the range of percentage of PVCC change by adjusting each influence factor to ±50%. In addition, the urban types are classified by its population scale, Miniature city: 50,000–200,000, Small city: 200,000–500,000, Medium city: 500,000–1,000,000, Large city I: 1,000,000–3,000,000, Large city II: 3,000,000–5,000,000, Super city: 5,000,000–10,000,000, Mega city: >10,000,000.

3.3. Equilibrium Analysis of PVCC Distribution

The total PVCC of 23.13% in urban areas worldwide would approximately meet the development needs in 2050, but this is just from the view of total average. As is shown in Figure 1, the PVCC of different urban divisions are actually changing in varying degrees. That is also to say, in order to achieve the appropriate proportion of PV energy consumption in each urban division, long distance energy allocations are inevitable, but this would definitely break the key quality of low transmission distance and high-efficient local utility of future energy systems. In order to describe this imbalance distribution of PVCC, six countries (or regions) are extracted as the contrasting analyses in this study, being: China, USA, Europe (partial), Australia, Brazil and India. The equilibrium of the PVCC distribution can be quantified as a "Gini index" which is commonly used as an indicator to measure the income gap of residents in a country or region. As shown in Equation (6), the Gini index is a value between 0 and 1, where 0 indicates the absolute equilibrium distribution of PVCC and 1 indicates the totally unequal distribution and 0.4 is generally used as the warning line to make a distinction between equilibrium and inequilibrium according to the definition of United Nations [38].It is worth noting, however, that the Gini index will change with the selected objects that are taken into calculation and that, according to the results of global assessment, the PVCC would usually decrease with the increase of urban population. Therefore, the equilibrium analysis of each county is conducted in the range of different urban scale sections, such as miniature cites and above and small cities and above (shown in Figure 3), and this would be used to discover in which section the largest distribution gap will occur. By combination then with the average level of PVCC and annual energy acquisition per capita in each urban scale (shown in Table 2), this would help to make further discussions on the adjustment strategies of the PV spatial planning in each country or region.

$$G = \frac{1}{2n^2 \cdot \overline{PVCC}} \sum_{j=1}^{n} \sum_{i=1}^{n} \left| PVCC_j - PVCC_i \right| \quad (6)$$

where G is the Gini index; n is the number of cities selected; $\overline{PVCC}$ is the average value of PVCC in cities selected; $PVCC_j$ is the PVCC of the jth city; $PVCC_i$ is the PVCC of the ith city.

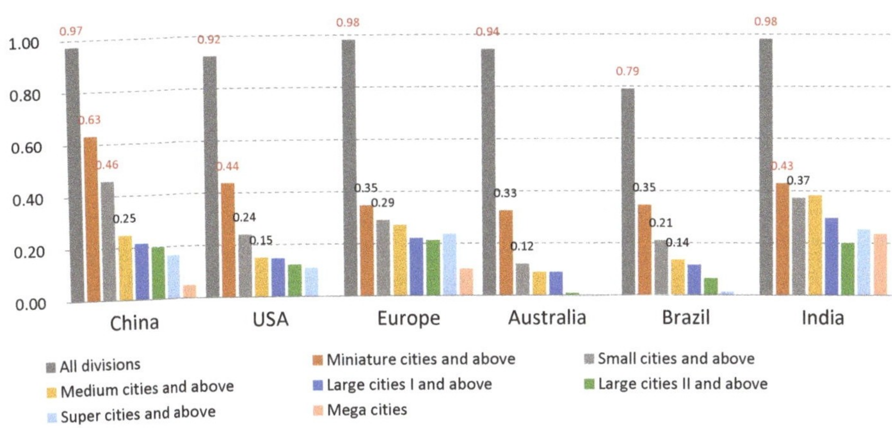

Figure 3. Gini indices of PVCC distribution in each urban scale section of different countries. Note: The definition of Gini index value: 0.0–0.2 Absolutely average, 0.2–0.3 Comparative average, 0.3–0.4 Relatively reasonable, 0.4–0.5 Large disparity, 0.5–1.0 Great disparity. The value of 0.4 is generally considered as the warning line.

Table 2. Average level of PVCC and annual energy acquisition per capita in each urban scale.

Urban Scale	Average PVCC (%)							Annual Average Energy Acquisition Per Capita (kWh)						
	China	USA	Europe	Australia	Brazil	India	World	China	USA	Europe	Australia	Brazil	India	World
Miniature cities	146.8%	38.8%	52.9%	27.1%	118.7%	411.7%	428.6%	5887.2	4948.7	2599.4	2603.4	2879.5	3236.7	4776.9
Small cities	62.1%	18.8%	34.8%	16.2%	64.1%	248.0%	203.2%	2568.3	2287.9	1678.3	1556.2	1513.2	1794.1	2132.2
Medium cities	26.7%	12.3%	29.3%	11.3%	44.8%	257.1%	82.7%	1193.7	1597.9	1302.7	1083.7	1082.3	1690.3	1389.1
Large cities I	20.8%	10.9%	21.4%	14.2%	36.9%	175.0%	88.8%	929.5	1376.6	1088.0	1370.6	891.5	1209.9	1141.4
Large cities II	18.8%	9.0%	17.3%	10.1%	32.3%	99.2%	40.5%	820.1	1198.9	881.6	972.7	780.1	798.1	891.8
Super cities	17.0%	7.3%	19.7%	-	26.3%	99.9%	65.1%	732.4	988.9	936.3	-	635.0	694.5	782.6
Mega cities	13.9%	5.1%	8.9%	-	27.8%	109.9%	34.9%	600.2	670.9	519.5	-	671.9	652.7	611.6

Note: The data in this table are extracted from the global PVCC assessment results of this study, by calculating the average level of each urban scale, according to the scope of each sample country.

4. Discussions

4.1. How Could Different Optimization Strategies Make Changes on the PVCC?

Necessarily, as shown in Figure 2, the PVCC will be sensitive to each influence factor input in different degrees. Therefore, the beneficial effects from different methods of planning adjustment (Table 1) will change as well. For instance, reducing energy consumption will always be the most direct way to improve PVCC; methods of building morphology optimization and decreasing building density will also be positive in promoting irradiation acquiring; and it might not be the most efficient way to improve PVCC by building utility scale stations closer to urban load, because this merely reduces the transmission loss. Different optimization strategies will all help to improve PVCC more or less, but it is also worth noting that the beneficial effects brought by these optimization strategies will decrease with the upgrading of urban scale. In fact, their actual benefits might not always be as optimistic or even be limited. All optimization strategies show greater benefits in smaller urban scale such as miniature cites and small cities, but it can be seen from Table 2 that the original PVCC of these cities are already in a high level and that they are capable of meeting the PV consumption proportion even without any further adjustment. That said, the benefits of such optimization strategies in larger urban scales would probably become very limited. Taking the assumed global standard in this study as an example, by using all these optimization strategies, the PVCC in mega cities and super cities are likely to be improved on average, from 14.4% to 19.9% and 17.6% to 24.5%, respectively. That will, in most cases, fall short of meeting the desired 25%. Considering the difference of original PVCC in each country and the actual implementation difficulty, it could be roughly concluded that these optimization strategies might be effective only in urban scales that range from medium to large cities.

4.2. In Which Section Will the Largest PVCC Distribution Gap Occur?

As shown in Figure 3, the largest gap of PVCC distribution occurs between urban and non-urban areas. This reflects the fact that most of the PVCC are provided by sparsely populated areas and this situation is almost the same for all countries. In addition, this distribution gap will appear obviously between the medium cities and small cities in China and in miniature cities and small cities in USA and India. To illustrate the cause of these differences in PVCC distribution, we also conducted an equilibrium analysis of the distribution of influence factors in each country applying the same method as in Equation (6). If the distribution of certain influence factor tends to be even, then it cannot be the main reason to cause the gap of PVCC distribution. As shown in Table 3, the distribution of solar irradiation in each country does not show obvious imbalance sufficiently to effect the distribution of PVCC. The gap of PVCC distribution is mainly affected by the distribution differences of urban population density, built-up area and available land area. In fact, the gap of PVCC is closely related to the urban spatial layout of a country. As shown in Figure 4, the distribution of cities in the USA, Europe and India is more even, which makes the solar energy acquisition of each urban division relatively equal. Cities in Australia and Brazil are more closely around the harbors, but the only difference is that Brazil's urban distribution is more dispersed, which leads to a higher average PVCC than that in Australia. Compared with countries above, however, the urban layout of China is gradually spreading and gathering from west to east, resulting in a huge gap in the distribution of PVCC. In order to achieve the overall PV development goal, China has to rely on long-distance power transmission, which would definitely break the key quality of high-efficient local utility. So, it can be concluded that the dispersed layout of cities and population is also of great significance for the future PV spatial planning.

Table 3. Gini indices of influence factors distribution in each urban division of different countries.

Influence Factors	China	USA	Europe	Australia	Brazil	India
Electricity consumption per capita	-	-	-	-	-	-
Solar irradiation	0.08	0.07	0.09	0.06	0.04	0.04
Urban population density	0.42	0.33	0.42	0.29	0.40	0.58
Built-up area	0.74	0.70	0.61	0.69	0.74	0.70
Available land area	0.67	0.50	0.42	0.75	0.36	0.87
Transmission distance	0.31	0.22	0.19	0.42	0.28	0.28

Note: The influence factors are arranged in descending order by the sensitivity of PVCC to each. The data of electricity consumption per capita is the overall level of a country, considering the similarity of electricity consumption behavior, the specific impact of its distribution is therefore ignored in this study.

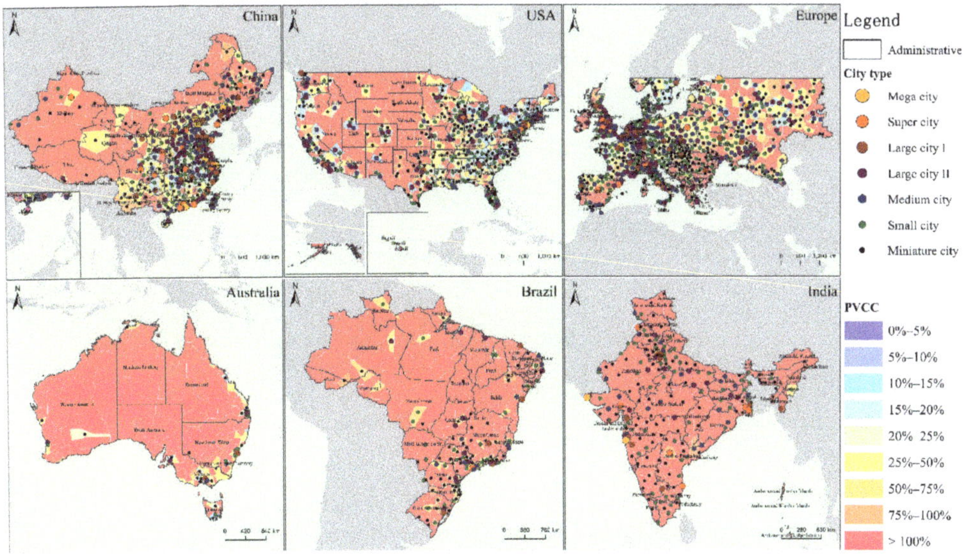

Figure 4. Comparison of urban spatial layout and PVCC distribution in each country or region.

4.3. *What Should Be Done to Adjust the PV Spatial Planning?*

The key point of PV spatial planning adjustment is the crucial means by which to determine the most effective way to improve PVCC. Optimization strategies by transformations of electricity consumption mode, urban planning or building morphology will be still feasible, but as mentioned above, these methods might only be effective in certain urban scales. Realistically, the adjustment of PV spatial planning needs to pay close attention to the actual condition of each country. In the case of Brazil and India, unless higher development target is proposed, it is unnecessary to contemplate their adjustment, because all the PVCC in each urban scale in either country has exceeded the proportion of global average PV consumption (refer to Table 2). For Australia, the average PVCC in urban areas is relatively low, because Australia has relatively few cities, which are each concentrated but mostly distributed some distance from each other, it will be able to meet the needs of future PV planning and development by just establishing a few utility scale PV stations and sufficient lengths of transmission lines. The PVCC in urban areas of USA and Europe is also at a low level, but, in fact, their energy acquisition per capita is even higher than other countries (refer to Table 2). Therefore, the method of controlling energy consumption would be more effective for these countries. As for China, the excessive urban agglomeration would serve to aggravate the gap in PVCC distribution, which might seriously impact the high-efficient utility of local energy and expand its overall transmission grid coverage. In this regard, the equilibrium development of different urban scales (especially scales below small cities and above medium cities) by adjustments of population distribution, urban layout or even industrial deployment, might become a more challenging problem currently faced by China in its overall PV planning.

4.4. What Is the Essence of Solar PV Spatial Planning?

The increasing utilization of renewable energy is essential to sustainable urban development in the future. In 2020, China put forward its guiding task for the development of renewable energy in the next five years [39]. To respond this, 30 provinces had also given out their own energy planning measures, which can be generally summarized into five aspects, including: (1) Increase the construction of renewable energy infrastructures to enhance the power generation capacity. (2) Deepen the reform of smart grid to realize the flexible allocation of power in urban area. (3) Expand the scale and capacity of trans-provincial power transmission to balance the difference of renewable energy power proportion in different regions. (4) Improve the manufacturing capacity and technology of renewable energy facilities to reduce the cost of power generation. (5) Transform the planning of energy storage facilities and energy utilization mode (electric vehicles, electric heating, etc.) to improve local renewable energy consumption. Apparently, all these measures are the continuation of previous cognition of "Power distribution and allocation", in the working mode of which, the city (energy demand side) is still considered as the principal part of planning and the power planning has to be adjusted passively with the change of urban energy demand. An inevitable result of this is that, still we have to make large scale of investment into building thousands miles of transmission lines, merely to deliver solar energy, the ubiquitous resource on earth, from one place to another. However, as repeatedly stated before, the main obstacle to the large-scale utilization of solar PV is more likely to be the inadaptability of existing spatial planning, such as population distribution, urban layout or industrial deployment. Therefore, the PV spatial planning proposed in this study is essentially a kind of urban planning or national geographic spatial planning, which could balance the relation between energy demand and supply, coordinated with regional (or urban) economy, resources, social structure and other comprehensive planning elements. Such "PV spatial planning" requires the active adjustment of urban (and national geographic) planning to adapt the needs of large-scale PV utilization in the future. However, it is not meant that solar PV would become the dominant factor to determine the morphology or deployment of the city. Instead, it is just one factor related to many other planning elements (e.g., water resources, food, transportation capacity, natural disasters, ecological or historical protection). In addition, PV spatial planning does not deny any current attempt of optimization measures, but at least these measures need to be integrated into a hierarchical process to cope with the transformation of geographic and urban spatial planning. For instance: (1) At the very first hierarchy of national geospatial planning, it needs the reasonable deployment of urban location and population scale to reduce the gap of PVCC distribution and maximize the efficient use of local energy (2) Through the assessment of regional solar irradiation, available land conditions and other planning elements, the function target and PV development anticipate of the city (or region) should be then formulated. (3) Furthermore, at the hierarchy of urban planning, it determines the scope of urban land use, building density control standards, building function layout and the optimization strategy of building morphology, etc. Compare the adjustment strategies of each planning hierarchy, to explore the most effective way to achieve the optimal utilization of solar PV energy. Of course, such "PV spatial planning" is a kind of forward-thinking and might be regarded as a dispensable work for now, because, at least at present, the proportion of solar PV utilization is still at a low level, but with the rapid increasing utilization of renewable energy, such proactive adjustment of urban (or even geographic) spatial planning may be of great significance for achieving higher solar energy supply in the future.

5. Conclusions

Out of the demand for energy sustainability, solar PV will gradually play an important role in future energy consumption. PV utilization focuses on the improvement of power consumption proportion, meanwhile, it also needs the reasonable deployment of energy and urban development. Both of these factors are the key elements in the construction of energy systems in the future. As discussed herein, the spatial planning of PV might not only be the planning of solar PV itself, the cognition of PV spatial planning needs to be transformed into a kind of urban or even national geographic spatial planning, in the progress of which solar PV utilization should be taken as an important planning element into consideration. The concept of PVCC described in this study established a link between the PV development anticipation and its close relationship to other spatial planning on the one hand and make people think about PV utilization from a view of resource and environment carrying capacity on the other hand. Such carrying capacity may in turn become a valid intuitive concept for future PV spatial planning, providing analysis capabilities and methods, with which the spatial planners could make comparisons and decisions to plan future courses of action. At the very least, it is suggested that the framework in this study can be considered as a starting point, aiming to stimulate interdisciplinary discussions between renewable energy developers, urban planners and policy makers. The methods mentioned in this study may be still not precise or comprehensive enough to make implementable planning adjustments independently, after all, only six factors were fit into the comparison, but they do provide the logic to figure out optimal solutions in the process of PV spatial planning by analyzing PVCC distribution and comparing the beneficial effects from each optimization strategy. Therefore, we can conclude that these tentative methods are available, but still need to be connected with other planning elements (or detailed influence factors) to give both planners and policy-makers the ability to guide the feasible PV spatial planning in each hierarchy, for the time being and in the foreseeable future.

Author Contributions: Conceptualization, Y.Z. and S.C.; methodology, S.C.; software, S.C.; validation, Y.Z., J.Z. and S.C.; formal analysis, S.C.; investigation, S.C.; resources, S.C.; data curation, S.C.; writing—original draft preparation, S.C.; writing—review and editing, Y.Z., J.Z. and S.C.; visualization, S.C.; supervision, J.Z.; project administration, J.Z.; funding acquisition, Y.Z., J.Z. All authors have read and agreed to the published version of the manuscript.

Funding: This research was funded by National Natural Science Foundation of China, grant number 51978443, National Natural Science Foundation of China, grant number 52078322, National Natural Science Foundation of China, grant number 51708395 and The APC was funded by National Natural Science Foundation of China, grant number 51708395.

Institutional Review Board Statement: Not applicable.

Informed Consent Statement: Not applicable.

Data Availability Statement: Data is contained within the article or supplementary.

Acknowledgments: Not applicable.

Conflicts of Interest: The authors declare no conflict of interest. The funders had no role in the design of the study; in the collection, analyses, or interpretation of data; in the writing of the manuscript, or in the decision to publish the results.

References

1. IRENA. *Future of Solar Photovoltaic: Deployment, Investment, Technology, Grid Integration and Socio-Economic Aspects (A Global Energy Transformation: Paper)*; International Renewable Energy Agency: Abu Dhabi, United Arab Emirates, 2019.
2. IRENA. *Renewable Power Generation Costs in 2017*; International Renewable Energy Agency: Abu Dhabi, United Arab Emirates, 2018.
3. Blaschke, T.; Biberacher, M.; Gadocha, S.; Schardinger, I. 'Energy landscapes': Meeting energy demands and human aspirations. *Biomass Bioenergy* **2013**, *55*, 3–16. [CrossRef] [PubMed]
4. Stremke, S.; van den Dobbelsteen, A. *Sustainable Energy Landscape*; CRC Press: Boca Raton, FL, USA, 2013.
5. Nadaï, A.; Van der Horst, D. Introduction: Landscapes of Energies. *Landsc. Res.* **2010**, *35*, 143–155. [CrossRef]
6. Rees, W.; Wackernagel, M. Urban ecological footprints: Why cities cannot be sustainable—And why they are a key to sustainability. *Environ. Impact Assess. Rev.* **1996**, *16*, 223–248. [CrossRef]
7. Malthus, T.R. *An Essay on the Principle of Population*; St Paul's Church-Yard: London, UK, 1798.
8. UNESCO & FAO. *Carrying Capacity Assessment with a Pilot Study of Kenya: A Resource Accounting Methodology for Exploring National Options for Sustainable Development*; Food and Agriculture Organization of the United Nations: Rome, Italy, 1985.
9. Wackernagel, M.; Rees, B. *Our Ecological Footprint: Reducing Human Impact on the Earth*; New Society Publishers: Philadelphia, PA, USA, 1996.
10. Vitousek, P.M.; Ehrlich, P.R.; Ehrlich, A.H.; Matson, P.A. Human appropriation of the products of photosynthesis. *BioScience* **1986**, *36*, 368–373. [CrossRef]
11. Odum, H.T. *Environmental Accounting: Emergy and Decision Making*, 1st ed.; John Wiley and Sons Inc.: New York, NY, USA, 1996.
12. Feng, Z.M.; Yang, Y.Z.; Yan, H.M.; Pan, T.; Li, P. A review of resources and environment carrying capacity research since the 20th Century from theory to practice. *Resour. Sci.* **2017**, *39*, 379–395.
13. Zhang, L.; Nie, Q.; Chen, B.; Chai, J.; Zhao, Z. Multi-scale evaluation and multi-scenario simulation analysis of regional energy carrying capacity-Case study: China. *Sci. Total Environ.* **2020**, *734*, 139440. [CrossRef] [PubMed]
14. Xue, R. Research on Fuzhou Energy Ecological Footprint and Ecological Carrying Capacity Based on Net Primary Productivity Model. *Anhui Agric. Sci.* **2020**, *48*, 64–67.
15. DeAngelis, D.L.; Zhang, B.; Ni, W.M.; Wang, Y. Carrying Capacity of a Population Diffusing in a Heterogeneous Environment. *Mathematics* **2020**, *8*, 49. [CrossRef]
16. Peng, X.; Zhu, H.; Zhu, E. Evaluation and Analysis of Energy Bearing Capacity in Beijing-Tianjin-Hebei Region: Based on data from 2007–2011. *J. Cap. Univ. Econ. Bus.* **2015**, *17*, 15–22.
17. Mentis, D.; Welsch, M.; Nerini, F.F.; Broad, O.; Howells, M.; Bazilian, M.; Rogner, H. A GIS-based approach for electrification planning-A case study on Nigeria. *Energy Sustain. Dev.* **2015**, *29*, 142–150. [CrossRef]
18. Sabo, M.L.; Mariun, N.; Hizam, H.; MohdRadzi, M.A.; Zakaria, A. Spatial matching of large-scale grid-connected photovoltaic power generation with utility demand in Peninsular Malaysia. *Appl. Energy* **2017**, *191*, 63–88. [CrossRef]
19. Watson, J.J.W.; Hudson, M.D. Regional Scale wind farm and solar farm suitability assessment using GIS-assisted multi-criteria evaluation. *Landsc. Urban Plan* **2015**, *138*, 20–31. [CrossRef]
20. Lee, A.H.; Kang, H.Y.; Lin, C.Y.; Shen, K.C. An Integrated Decision-Making Model for the Location of a PV Solar Plant. *Sustainability* **2015**, *7*, 13522–13541. [CrossRef]
21. Mohajeri, N.; Upadhyay, G.; Gudmundsson, A.; Assouline, D.; Kämpf, J.; Scartezzini, J.L. Effects of urban compactness on solar energy potential. *Renew. Energy* **2016**, *93*, 469–482. [CrossRef]
22. Natanian, J.; Auer, T. Balancing urban density, energy performance and environmental quality in the Mediterranean: A typological evaluation based on photovoltaic potential. *Energy Procedia* **2018**, *152*, 1103–1108. [CrossRef]
23. Sarralde, J.J.; Quinn, D.J.; Wiesmann, D.; Steemers, K. Solar energy and urban morphology: Scenarios for increasing the renewable energy potential of neighbourhoods in London. *Renew. Energy* **2015**, *73*, 10–17. [CrossRef]
24. Lobaccaro, G.; Frontini, F. Solar Energy in Urban Environment: How Urban Densification Affects Existing Buildings. *Energy Procedia* **2014**, *48*, 1559–1569. [CrossRef]
25. Yin, L. *Annual Report on World Smart City (2016–2017)*; Social Sciences Academic Press: Beijing, China, 2017.
26. IEA. *Energy Technology Perspectives 2016: Towards Sustainable Urban Energy Systems*; International Energy Agency: Paris, France, 2016.
27. China Electricity Council. *Code for Design of Photovoltaic Power Station (GB50797-2012)*; China Planning Press: Beijing, China, 2012.
28. Sophie, B.; Pierre, D.; Van Bogaert, E.; Arino, O.; Kalogirou, V.; Perez, J.R. *GLOBCOVER 2009 Products Description and Validation Report*; UC Louvain: Leuven, Belgium; ESA Team: Paris, France, 2011.
29. Hernandez, R.R.; Hoffacker, M.K.; Field, C.B. Efficient use of land to meet sustainable energy needs. *Nat. Clim. Chang.* **2015**, *5*, 353–358. [CrossRef]
30. Available online: http://www.rivermap.cn/ (accessed on 4 November 2019).
31. Available online: http://due.esrin.esa.int/page_globcover.php (accessed on 12 October 2019).
32. Center for International Earth Science Information Network—CIESIN—Columbia University. *Gridded Population of the World, Version 4 (GPWv4): Population Density Adjusted to Match 2015 Revision UN WPP Country Totals*; NASA Socioeconomic Data and Applications Center (SEDAC): New York, NY, USA, 2016.
33. Available online: https://solargis.com/cn/maps-and-gis-data/download (accessed on 4 November 2019).

34. Available online: https://data.worldbank.org/ (accessed on 14 July 2020).
35. Central Intelligence Agency. *The CIA World Facebook 2018–2019*; Skyhorse Publishing: New York, NY, USA, 2018.
36. Available online: https://www.jinkosolar.com/ (accessed on 26 July 2020).
37. The State Council of China. *Notice on Adjusting the Standard of City Scale Division*; The State Council of China: Beijing, China, 2014.
38. Han, J.; Zhao, Q.; Zhang, M. China's income inequality in the global context. *Perspect. Sci.* **2016**, 7, 24–29. [CrossRef]
39. National Energy Administration of China. *Notice on the Preparation of the 14th Five Year Plan for Renewable Energy Development*; Comprehensive Department of National Energy Administration: Beijing, China, 2020.

sustainability

MDPI

Article

Effect of Spatial and Temporal Resolution Data on Design and Power Capture of a Heaving Point Absorber

Tunde Aderinto [1] and Hua Li [2,*]

1 Sustainable Energy Systems Engineering, Texas A&M University-Kingsville, Kingsville, TX 78363, USA; tundeaderintos@yahoo.com

2 Mechanical and Industrial Engineering Department, Texas A&M University-Kingsville, Kingsville, TX 78363, USA

* Correspondence: hua.li@tamuk.edu

Received: 23 October 2020; Accepted: 13 November 2020; Published: 16 November 2020

Abstract: For a heaving point absorber to perform optimally, it has to be designed to resonate to the prevailing ocean wave period. Hence, it is important to make the ocean wave data analysis to be as accurate as possible. In this study, existing wave condition data is used to investigate the effect of the temporal resolution (daily vs. hourly) of wave data on the design of the device and power capture. The temporal resolution effect on the estimation of ocean wave resource theoretical potential is also investigated. Results show that the temporal resolution variation of the ocean wave data affects the design of the device and its power capture, but the theoretical power resource assessment is not significantly affected. The device designed for the Gulf of Mexico is also analyzed with wave condition in Oregon, which has about 40 times the wave resource theoretical potential compared to the Gulf of Mexico. The results confirmed that a device should be designed for a specific location as the device performed better in the Gulf of Mexico, which has much less ocean wave resource theoretical potential. At last, the effect of the design, diameter and season (summer and winter) on the power output of the device is also investigated using statistical hypothesis testing methods. The results show that the power capture of a device is significantly affected by these parameters.

Keywords: wave energy converters; heaving point absorber; design and performance; spatial and temporal variation

---

1. Introduction

Ocean wave energy has continued to see increase in the level of awareness in recent years. Moreover, the last three to five years have seen a lot of research and development efforts into the ocean wave energy industry [1–3]. Others studies focused on specific aspects of the ocean wave energy such as resource characterization have been performed at global [4,5], regional [6–8] and local levels [9,10]. Serious exploration and exploitation of ocean wave energy resources are currently being investigated in U.S. [3], Europe [11–13], China [14], India [15], etc. Other aspects of ocean wave energy such as economics [16], environmental [17], design [18] and efficiency and performance [19] have all being studied by different researchers. Apart from the general aim for ocean wave energy to supply energy into the traditional grid systems, the work done by [20] investigated the potential of using ocean wave energy to supply power for offshore oil rigs and other offshore structures which can broaden the application of this vast but underutilized energy resource.

Wave energy converters (WECs) can be classified based on their working principles. Another classification method is based on the water depth of the WEC's site location (shoreline, nearshore and offshore). WECs are also classified based on the ratio of the wavelength magnitude to the interacting part of the

WEC. For example, WECs can be classified into oscillating water columns, oscillating body systems and overtopping devices (Figure 1) based on their working principles. Under this classification, a point absorber, which is when the WEC interacting part dimensions is considerably smaller than that of the interacting ocean wavelength [21,22], is considered as an oscillating body-based WEC. It is a terminator if the dominant wave direction is perpendicular to the structural extension of the WEC [23], while it is an attenuator if its structural extension is parallel to the interacting wave direction [24].

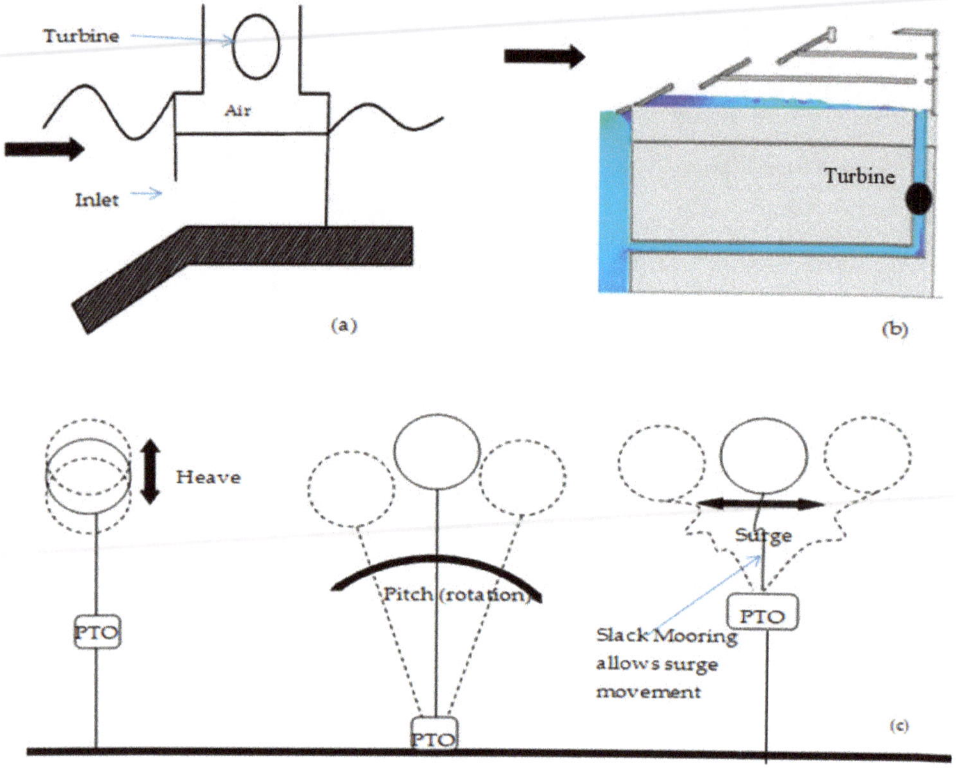

Figure 1. Classification of wave energy converter (WEC) extraction technologies: (**a**) oscillating water column, (**b**) overtopping devices and (**c**) oscillating bodies [2].

One of the promising methods of wave energy capture is the oscillating body system. One major thing that makes the use of these types of converters attractive is because the amount of the energy absorbed by the body can be improved upon significantly under the same wave conditions when the body is at resonance with the incoming waves as illustrated in Figure 2. In fact, the team that won the prestigious ocean wave energy prize offered by the United States Department of Energy developed their concept and design based on a heavingf oscillating buoy [25]. The Pelamis [26], which is one of the most studied converters, is a pitching (rotating) oscillating converter and is also another type of a wave activated body system.

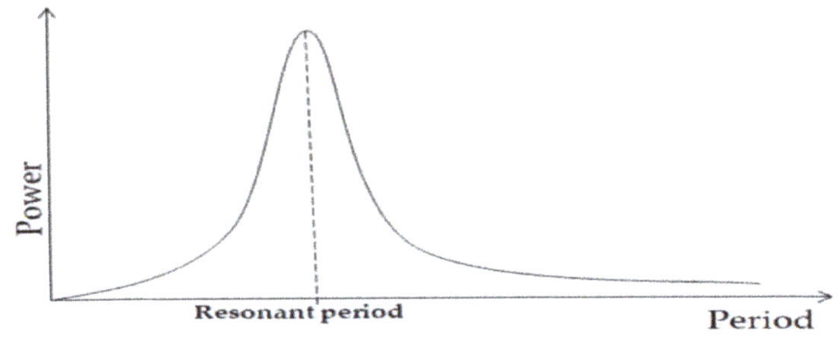

Figure 2. Illustration of theoretical power capture of oscillating body system converters [1,2].

The hydrodynamics of oscillating body systems including heaving systems were independently solved by [27–29], which show the theoretical maximum energy to be captured by an oscillating body system-based wave energy converter. The results confirm that the highest possible capture occurs when the body is at resonance with the incoming waves. This behavior of a floating oscillating body poses a challenge for WEC designers because a typical body has a narrow resonance frequency band, and the body performs poorly outside this band. Hence, one of the many characteristics of a good oscillating WEC design is to make the buoy resonate to the prevailing ocean wave properties [2]. It should be noted that other factors, including survivability and profitability, need to be considered as well when designing a WEC. In order to capture considerable power outside the resonance frequency band, different optimization methods have been proposed and investigated. Some optimization methods include changes made to the shape and dimensions of the buoy [30], while latching and declutching methods [31] are also used in some existing studies. Latching control is achieved by holding the heaving WEC in a fixed position when the velocity is zero and releasing it at the right time so that its velocity can be in phase with the excitation force to achieve resonance [32]. On the other hand, declutching works by alternatively switching the power take off system on and off [31]. Another method is the model predictive control. It is an advanced control strategy [33] compared to the passive control methods, which may employ complex algorithms and simulations to achieve the optimization of power absorption by the WECs. While these methods have theoretical possibilities, there exists very little information reporting their applications in real ocean conditions.

In all these designs and optimization methods, the ocean wave properties have to be properly characterized first in order to have a good and effective WEC design. Although two ocean wave properties (wave height and wave period) guide the estimation of ocean wave resource potential and power capture of a WEC, it is the wave period that determines how the resonance behavior of a floating body will be engineered so it will operate near resonant level with the desired ocean wave period. There are existing studies that focus on general guidelines for designing a WEC. Meanwhile, most available wave condition database or monitoring/forecasting systems are designed for other marine systems instead of WEC design [8,9], so it is important to investigate the possibility of using existing wave condition data on designing a WEC. Instead of optimizing the size of a heaving point absorber WEC, this paper focuses on conducting detailed quantitative analysis on the changes on the power output of a heaving point absorber due to the variation of temporal and spatial resolutions of existing wave condition data. This paper uses available wave condition data from an existing database instead of collecting new wave condition data in the analysis. Section 2 introduces the methodology used including the ocean wave data analysis in the studied regions, the method for estimating the average yearly energy resource potential as well as the design process for the WEC which led to two different WEC designs in terms of dimensions. In Section 3, the annual energy resource potential is estimated under three different scenarios based on existing wave condition data: (1) wave data based on hourly resolution at the selected location in the Gulf of Mexico, (2) wave data based on daily resolution at the selected location in the Gulf of Mexico, and (3) wave data based on hourly resolution in an offshore location in Oregon. The power and annual energy matrices of the two WEC designs based on the hourly and daily resolution data are analyzed in Section 4. The design tailored to the hourly resolution data in the Gulf of Mexico is tested with the hourly data of an offshore location in Oregon, and the results are compared. Section 5 shows the results of a series of statistical hypothesis analysis such as *t*-test using data obtained from the Gulf of Mexico to determine the significance on the power output by different parameters such as diameter, design and seasonal variation (winter and summer). Section 6 is for discussions and conclusions.

2. Methods

2.1. Ocean Wave Data

Existing default data of significant wave height and dominant wave period over a 9-year period was obtained from a buoy operated by National Data Buoy Center at a location in the Gulf of Mexico (GoM) at coordinates 26.968° N and 96.693° W with sea depth of 84 m in a watch circle of radius 138 m (Figure 3a) [34]. The percentage of occurrences were analyzed for two scenarios in terms of the temporal variation of the collected data: data obtained hourly (Table 1) and data obtained daily (Table 2). The daily and hourly significant wave height data represent the average of the highest one-third of waves in the given period of capture. For both datasets, the wave heights ranged from 1 m to 5 m and wave periods ranging from 3 s to 12 s captured about 99% of all data points. Another set of ocean wave data was obtained from a location in Oregon which is close to the PacWave test site at coordinates 44.667° N and 124.515° W with sea depth of 140 m and a watch circle of radius of 230 m (Figure 3b) [34]. Another reason to choose the Oregon site for comparison is that the Oregon site is considered as high wave energy potential site (Table 3) while the GoM site is normally considered as low wave energy potential site. It should be noted that extreme wave conditions exist with a very small occurrence, such that there are 0.0001% of waves with significant wave height larger than 12 m and dominant wave period higher than 22 s in the GoM. However, we didn't include these extreme wave conditions in the data tables and power generation estimation, because the focus of this paper is on the power generation, while the proposed WEC will not generate power during the extreme conditions. The extreme conditions will be considered when investigating the structural reliability of the proposed WEC.

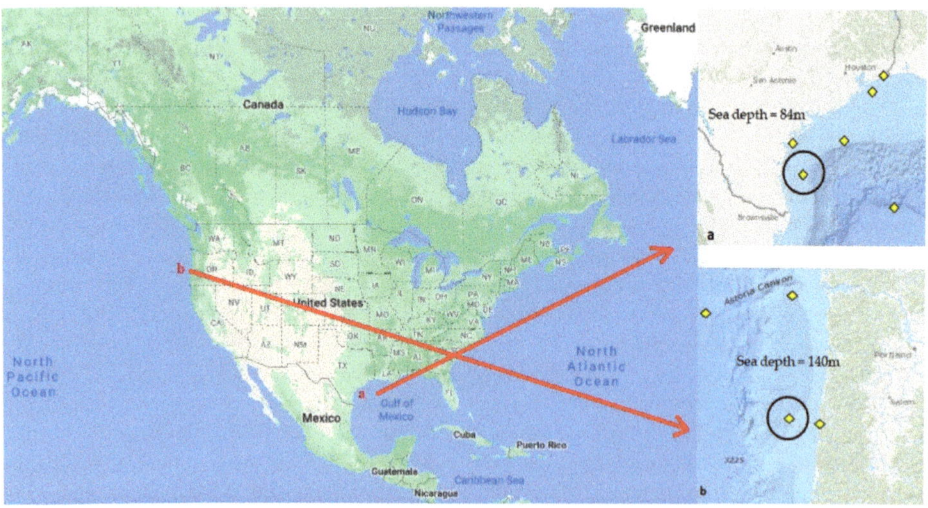

Figure 3. Studying locations: (**a**) location 1: Gulf of Mexico, and (**b**) location 2: Oregon [34].

Table 1. Percentage of ocean wave height and period occurrence based on hourly wave data (location 1: Gulf of Mexico).

		Significant Wave Height (m)				
		0–1	**1–2**	**2–3**	**3–4**	**4–5**
Dominant Wave Period (s)	**0–1**	0.37%	0.00%	0.00%	0.00%	0.00%
	1–2	0.00%	0.00%	0.00%	0.00%	0.00%
	2–3	0.80%	0.06%	0.01%	0.01%	0.00%
	3–4	4.14%	0.31%	0.06%	0.01%	0.00%
	4–5	8.26%	2.86%	0.19%	0.02%	0.00%
	5–6	12.91%	12.55%	0.80%	0.09%	0.03%
	6–7	6.49%	14.54%	2.64%	0.17%	0.01%
	7–8	3.64%	12.47%	5.33%	0.79%	0.03%
	8–9	1.03%	2.60%	2.02%	0.54%	0.04%
	9–10	0.48%	1.09%	0.91%	0.31%	0.06%
	10–11	0.17%	0.37%	0.23%	0.10%	0.04%
	11–12	0.02%	0.02%	0.04%	0.03%	0.01%

Table 2. Percentage of ocean wave height and period occurrence based on daily resolution data (location 1: Gulf of Mexico).

		Significant Wave Height (m)				
		0–1	**1–2**	**2–3**	**3–4**	**4–5**
Dominant Wave Period (s)	**0–1**	0.30%	0.00%	0.00%	0.00%	0.00%
	1–2	0.03%	0.00%	0.00%	0.00%	0.00%
	2–3	0.03%	0.00%	0.00%	0.00%	0.00%
	3–4	2.40%	0.00%	0.00%	0.00%	0.00%
	4–5	11.26%	0.94%	0.00%	0.00%	0.00%
	5–6	13.60%	12.14%	0.00%	0.00%	0.00%
	6–7	7.09%	22.52%	1.77%	0.00%	0.00%
	7–8	1.77%	10.99%	7.06%	0.24%	0.00%
	8–9	0.64%	2.31%	2.50%	0.64%	0.00%
	9–10	0.06%	0.67%	0.40%	0.27%	0.00%
	10–11	0.06%	0.06%	0.03%	0.00%	0.00%
	11–12	0.00%	0.00%	0.00%	0.06%	0.00%

Table 3. Percentage of ocean wave height and period occurrence based on hourly resolution data (location 2: Oregon).

		Significant Wave Height (m)					
		0–1	**1–2**	**2–3**	**3–4**	**4–5**	**5–6**
Dominant Wave Period (s)	**4–5**	0.17%	0.69%	0.00%	0.00%	0.00%	0.00%
	5–6	0.47%	2.58%	0.30%	0.00%	0.00%	0.00%
	6–7	0.55%	3.66%	0.93%	0.08%	0.00%	0.00%
	7–8	0.89%	6.89%	2.00%	0.42%	0.07%	0.01%
	8–9	0.37%	5.67%	1.94%	0.50%	0.18%	0.05%
	9–10	0.49%	7.16%	2.92%	0.62%	0.34%	0.08%
	10–11	0.52%	7.98%	7.57%	2.55%	0.78%	0.36%
	11–12	0.19%	2.16%	2.35%	1.21%	0.38%	0.18%
	12–13	0.46%	4.28%	4.65%	2.78%	1.32%	0.44%
	13–14	0.38%	1.94%	1.80%	1.33%	0.71%	0.27%
	14–15	0.60%	1.77%	1.25%	1.02%	0.66%	0.18%
	15–16	0.00%	0.00%	0.00%	0.00%	0.00%	0.00%
	16–17	0.33%	1.24%	0.81%	0.81%	0.48%	0.16%

2.2. Design of Heaving WEC Dimensions

The initial dimensions of the proposed WEC (Figure 4) were estimated based on the theoretical hydrodynamics of floating bodies. The hydrodynamics describe the motion of a floating body under the action of external forces. The external forces in this case were mainly generated by the ocean waves. A combination of wave data analysis and the theoretical wave hydrodynamics were used to estimate the diameter of the cylindrical buoy, which is the shape of the proposed WEC device. From the wave data analysis shown in Tables 1 and 2, the dominant wave period falls between 5 and 6 s with 26.38% occurrence for the hourly data and between 6 and 7 s with 31.38% occurrence for the daily data. The following sets of equations were used to estimate the initial dimensions of the proposed WEC.

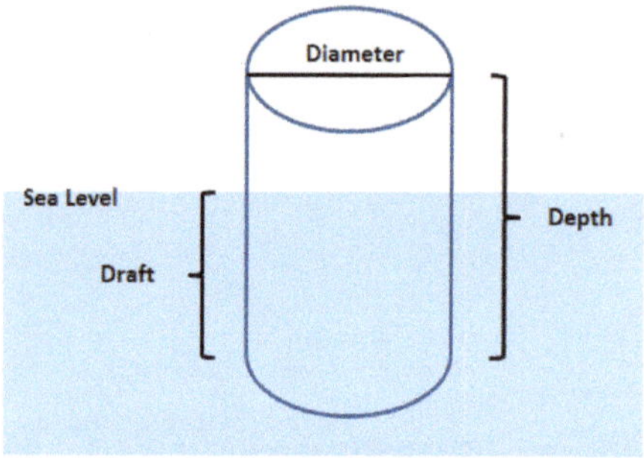

Figure 4. Schematic design of the proposed WEC buoy.

From previous literature [35],

$$L_{max} = \frac{\lambda}{2\pi} \tag{1}$$

$$\lambda = \frac{gT^2}{2\pi} \tag{2}$$

and as recommended by [36],

$$C = \frac{L_{max}}{D} \geq 3 \tag{3}$$

where L_{max} is Maximum capture width, λ is wave length, T is dominant wave period, D is buoy diameter, C is captured width ratio and g is gravitational constant. These equations act as a guide to determine the size of a point absorber at the initial stage. It should be noted that Equation (2) is based on approximation made for deep water condition.

Theoretically, the resonance frequency ω_n of a submerged body is given by Equation (4) below. This simplified equation can be used at an initial stage to estimate the natural frequency of the point absorber system.

$$\omega_n = \sqrt{\frac{\rho g A_w}{M_w + a}} \tag{4}$$

where A_w = water plane area, M_w is mass of displaced water, a is added mass and ω_n is natural period. Added mass at this preliminary stage is given as 0.167 ρD^3 [36]. The effective drafts for different diameters of a buoy whose density is equal to the density of water is given in Table 4. This application is for floating bodies alone. Steel is chosen as the material for the buoy in this paper. For the buoy to oscillate freely in seawater, the buoy must be hollow as the density of steel is greater than that of water. The mass of the steel buoy should be equal to the mass of the displaced water such that

$$M_w = M_s; \tag{5}$$

where, M_s is the mass of the steel cylindrical buoy.

Table 4. Dimensions of initial designs of the proposed WEC. GoM = Gulf of Mexico.

Design	Location	Depth (m)	Draft (m)	Thickness (m)	Diameter (m)
1	GoM	12.9	7.3	0.15	8
2	GoM	18.6	10.5	0.15	8

For any selected diameter and wave period, the corresponding draft and depth under 0.15 m thickness can be obtained in Tables 5 and 6, respectively. It should be noted that the depth of the buoy (Table 6) must be greater than or equal to the draft (Table 5) for a floating cylinder. Thus, the feasible diameter of the buoy is 5 m and above. According to Equation (4), the resonance period is a function of mass of displaced water, water plain area, etc. Thus, for any thickness, the mass of water displaced and the mass of the cylindrical buoy should be constant when choosing the diameter of the buoy, but the depth of the cylinder will change. Therefore, the impact of the buoy thickness on its performance may be minimal in this research. For the initial design of the buoy in this paper, 8 m was chosen as its diameter with a steel of density 7850 kg/m^3. For design 1, whose dimensions were tailored to the most probable wave period of the hourly resolution data in the GoM, the most prevalent wave period lies between 5 and 6 s, so the initial draft and depth of design 1 were 7.3 m and 12.9 m, respectively. Similarly, for design 2, whose dimensions were tailored to the most prevalent wave period between 6 and 7 s of the daily resolution data in the GoM, the initial draft and depth of design 2 were 10.5 m and 18.6 m, respectively. These initial dimensions were further analyzed in ANSYS diffraction module to give more accurate values of the resonance period corresponding to the initial dimensions, which decide the final dimensions used to estimate the power capture of the WEC.

Table 5. Draft under different diameters and wave periods to achieve resonance (density of material = density of sea water).

		Diameter (m)												
		1.3	2	3	4	5	6	7	8	9	10	11	12	
Period (s)	3–4	3.7	3.6	3.3	3.1	2.9	2.7	2.5	2.3	2.1	1.9	1.7	1.5	Draft range (m)
	4–5	5.9	5.8	5.6	5.4	5.2	4.9	4.7	4.5	4.3	4.1	3.9	3.7	
	5–6	8.7	8.5	8.3	8.1	7.9	7.7	7.5	7.3	7.0	6.8	6.6	6.4	
	6–7	11.9	11.7	11.5	11.3	11.1	10.9	10.7	10.5	10.3	10.1	9.9	9.6	
	7–8	15.6	15.5	15.3	15.0	14.8	14.6	14.4	14.2	14.0	13.8	13.6	13.4	
	8–9	19.8	19.7	19.5	19.3	19.1	18.8	18.6	18.4	18.2	18.0	17.8	17.6	

Table 6. Depth under different diameters and wave period to achieve resonance (density of material = density of steel).

		Diameter (m)												
		1.3	2	3	4	5	6	7	8	9	10	11	12	
Period (s)	3–4	1.2	1.7	2.3	2.8	3.3	3.6	3.9	4.1	4.1	4.1	4.0	3.8	Depth (m)
	4–5	1.9	2.7	3.8	4.9	5.8	6.6	7.4	8.0	8.6	9.1	9.5	9.7	
	5–6	2.8	4.0	5.7	7.3	8.8	10.3	11.6	12.9	14.0	15.1	16.1	17.0	
	6–7	3.8	5.5	7.9	10.2	12.5	14.6	16.6	18.6	20.5	22.2	23.9	25.5	
	7–8	5.0	7.3	10.5	13.6	16.6	19.6	22.4	25.2	27.9	30.5	32.9	35.3	
	8–9	6.3	9.3	13.4	17.4	21.4	25.2	29.0	32.7	36.3	39.8	43.2	46.5	

Colored cells represent feasible depth of buoy.

A buoy diameter and associated draft used to form initial dimensions were tested using ANSYS/AQWA 18.1 [37]. ANSYS/AQWA is capable of simulating linearized hydrodynamic fluid wave loading on either floating or fixed rigid structures based on potential flow theory. It employs a three-dimensional diffraction theory in regular waves in the frequency domain. Furthermore, real time motion and force responses of bodies operating in regular or irregular waves can be studied. For each prevailing wave period based on the two temporal resolution data, the optimal dimensions (depth and draft to make buoy resonate with the wave period) for the selected diameters were decided after the analysis in ANSYS. AQWA The process is summarized in Figure 5.

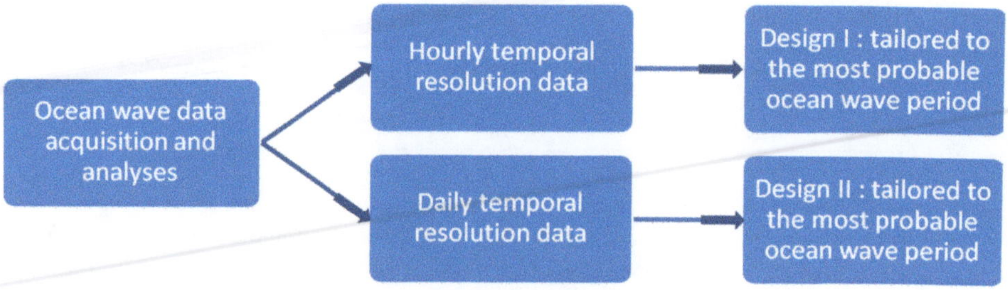

Figure 5. Design process for arriving at each design.

3. Annual Energy Resource Assessment and Estimation

The ocean wave power density in a location is a function of wave height and wave period given by the relation in Equation (6) based on approximation made for deep water condition.

The annual power density can be estimated if information about the percentage occurrence of the wave height/period in the location is available [2].

$$P = \frac{\rho g^2 T_e H^2}{64\pi} \quad (6)$$

where P (KW/m) is power density (power per unit width of wave front), ρ (Kg/m^3) is seawater density, g (m/s^2) is gravitational acceleration, H (m) is significant wave height and T_e (s) is energy wave period. Since the wave period date provided by National Oceanic and Atmospheric Administration (NOAA) is dominant wave period T_p, the energy wave period (T_e) is calculated by multiplying a wave period conversion factor α to dominant wave period (T_p). In this study, the wave conversion factor was considered as 0.9, which is the equivalent of JONSWAP software [38] and has been used in the past by different studies [9,20,38].

The potential energy density per year is calculated by multiplying the percentage occurrence in a year, and the results for the three different scenarios are provided in the Tables 7–9 below. From Tables 7 and 8, the hourly resolution data provides annual energy density potential of 105.3 MWh/m, and the daily resolution data provides annual energy density potential of 102 MWh/m. This shows about 3% difference in terms of energy density potential in the same location with different temporal resolution data. This shows that different temporal resolution of the data in same location may not significantly affect the estimation of the total wave energy resource potential present in the location.

Table 7. Annual energy density potential (kWh/(m·yr)) based on hourly resolution data (location-1: Gulf of Mexico).

		Wave Height (m)				
		0–1	**1–2**	**2–3**	**3–4**	**4–5**
Period T(s)	**2–3**	92.76	27.64	14.55	16.47	0.00
	3–4	640.40	188.96	86.45	15.68	0.00
	4–5	1594.86	2206.21	324.17	66.65	0.00
	5–6	2991.75	11,637.70	1664.51	343.43	161.71
	6–7	1754.96	15,724.72	6430.88	719.00	34.30
	7–8	1124.37	15,411.87	14,819.10	3895.31	235.22
	8–9	357.25	3613.04	6323.26	2992.05	363.86
	9–10	184.75	1689.70	3153.47	1921.00	600.31
	10–11	73.31	636.08	897.53	698.62	471.67

Table 8. Annual energy density potential (kWh/(m·yr)) based on daily resolution data (location-1: Gulf of Mexico).

		Wave Height (m)				
		0–1	1–2	2–3	3–4	4–5
Period T(s)	2–3	3.53	0.00	0.00	0.00	0.00
	3–4	371.50	0.00	0.00	0.00	0.00
	4–5	2174.94	728.90	0.00	0.00	0.00
	5–6	3153.07	11,257.95	0.00	0.00	0.00
	6–7	1917.47	24,359.30	4295.80	0.00	0.00
	7–8	545.50	13,581.02	19,637.93	1203.86	0.00
	8–9	222.19	3216.56	7808.62	3555.14	0.00
	9–10	23.52	1034.57	1375.50	1692.93	0.00
	10–11	25.87	103.46	116.39	0.00	0.00

Table 9. Annual energy resource potential (kWh/(m·yr)) based on daily resolution data (location-2: Oregon).

		Wave Height (m)				
		0–1	1–2	2–3	3–4	4–5
Period T(s)	2–3	0.0	0.0	0.0	0.0	0.0
	3–4	17.2	10.1	0.0	0.0	0.0
	4–5	2417.9	156,193.3	0.3	0.0	0.0
	5–6	19,574.2	2,359,410.6	72,159.0	1.2	0.0
	6–7	13,777.7	2,435,869.9	354,993.9	4368.1	0.0
	7–8	23,910.9	5,719,063.8	1,090,173.6	85,867.7	4189.7
	8–9	3031.8	2,811,123.4	747,209.7	88,911.0	19,259.1
	9–10	4047.7	3,452,155.5	1,319,706.9	109,119.2	51,078.2
	10–11	3549.3	3,369,618.8	6,977,068.9	1,443,859.4	219,434.6
	11–12	366.9	196,251.9	531,024.9	252,193.3	38,739.2
	12–13	1917.0	662,058.8	1,760,536.5	1,119,829.5	399,744.1
	13–14	1070.5	110,175.6	213,204.3	209,025.6	92,347.8
	14–15	2283.9	80,617.0	91,425.8	108,845.8	70,770.1
	15–16	0.0	0.0	0.0	0.0	0.0
	16–17	1017.9	28,644.9	27,371.6	49,309.4	27,314.6

Similarly, the resource potential for the location in offshore Oregon is shown in Table 9 above. With a yearly total of about 43,374 MWhr/m of wave energy resource potential, it is confirmed that the location is a high wave energy resource area with resources about 40 times higher than that of the location in the Gulf of Mexico.

4. Power Capture of the WEC

Three scenarios of power capture are examined and compared in this section. After a series of computational fluid dynamics (CFD) diffraction analysis using ANSYS/AQWA suite version 18.1, which is based on potential flow theory, the response amplitude operator (RAO) which gives the motion response of the buoy at different periods is given in Figure 6. The three scenarios and the WEC dimensions are provided in Table 10. Design 1 was tailored to the dominant wave period based on the hourly resolution data while Design 2 dimensions was engineered to capture power optimally at the dominant wave period based on the daily resolution data from location 1 in the Gulf of Mexico. The power capture of design 1 in location 2 in Oregon is also estimated to investigate the impact of different wave conditions on the same WEC design in terms of power capture.

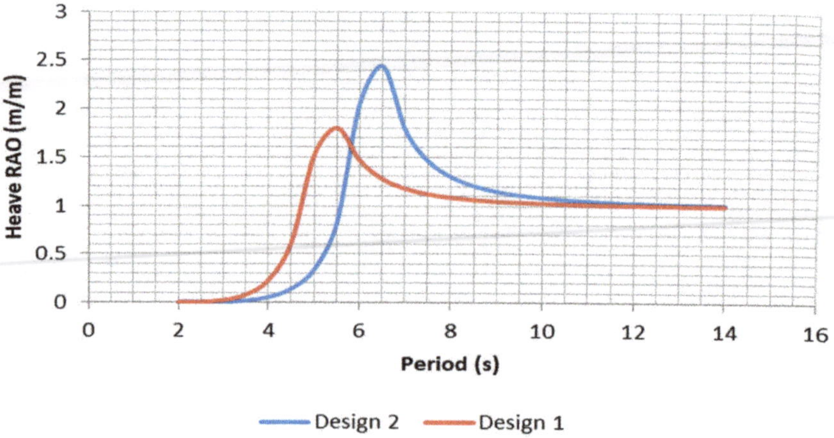

Figure 6. Heave response amplitude for design 1 and design 2.

Table 10. Dimension of the final designs used in three scenarios for estimating power capture. RAO = response amplitude operator.

Scenario	Design	Location	Heave RAO (s)	Depth (m)	Draft (m)	Thickness (m)	Diameter (m)
1	1	GoM	5.5	8	4.5	0.15	8
2	2	GoM	6.5	12.9	7.3	0.15	8
3	1	Oregon	5.5	8	4.5	0.15	8

The power take-off in this study was modeled as pure damper which is assumed to be frequency dependent. The motion equation of the heaving point absorber buoy with the power take off (PTO) can be described by Equation (7) below

$$(M + A)\ddot{x} + B\dot{x} + Cx \;=\; F(t) + \;F_{PTO} \tag{7}$$

where $F_{PTO} = D_{PTO}\dot{x}$, M is mass, A is added mass, x is heave displacements and its derivatives with respect to time; B is damping coefficient and C is hydrostatic force coefficient; $F(t)$ is the external force acting on the buoy while F_{PTO} is the PTO force; and D_{PTO} is the PTO damping coefficient.

The maximum amount of energy captured by the buoy occurs when the PTO damping is equal to the radiation damping of the buoy [39]. Hence the PTO damping will be equal to that of the buoy at resonance. The mean absorbed power by the PTO is given by Equation (8)

$$\frac{1}{2}D_{PTO}\omega^2x^2 \tag{8}$$

where, ω is the angular frequency at resonance.

Using the premises highlighted above, different values of PTO damping were tested on the buoy and the power capture and are shown in Figure 7. The maximum occurred when the damping coefficient was 50 kNm/s for both design 1 and design 2 when the diameter of the buoy was 8 m and 40 kNm/s when the diameter of the buoy was 7 m for both design scenarios.

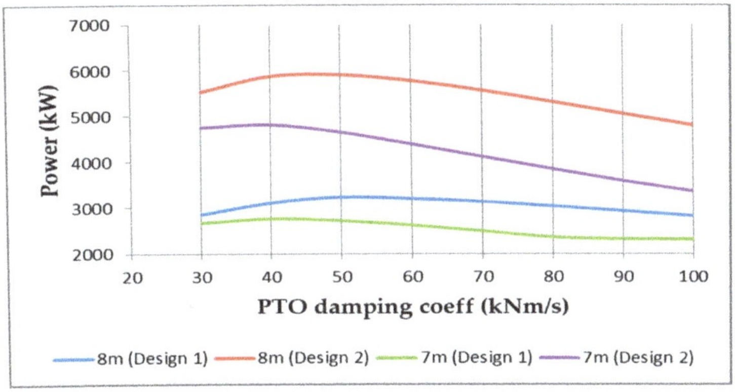

Figure 7. Power take off (PTO) damping vs. power capture.

For both designs, the heaving oscillatory motion of the buoy was converted to power through a power take off (PTO). For the purpose of this design, a PTO damping of 40 kNm/s and 50 kNm/s was used for the WEC devices when the diameter of the buoy was 7 m and 8 m, respectively. Figures 8 and 9 show the change of the power capture with respect to the wave period and wave height, respectively. The figures show, as expected, the power capture peaks when the wave period is at the resonance period of the buoy. However, the power capture increases almost linearly with wave height increase. Hence, while both wave period and wave height affect the power capture, the relationship between the wave period and power capture of a heaving point absorber is more significant due to the resonant behavior pattern.

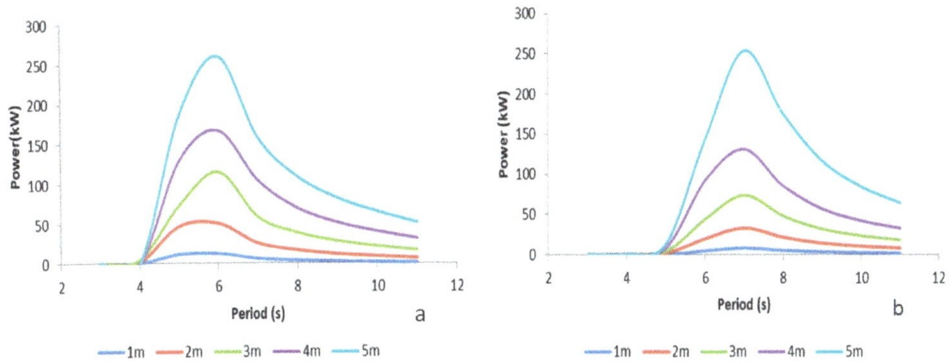

Figure 8. Period vs. power: (**a**) hourly resolution power (**b**) daily resolution power.

The annual energy captured in each bivariate wave height-wave period combinations under different scenarios are given in Tables 11–13. Table 11 shows the yearly energy that can be harvested using design 1 tailored using the hourly data from the GoM, while Table 12 shows the comparative annual energy based on design 2 tailored using daily resolution data. Table 13 shows the annual energy by design 1 when operated in the Oregon location, an offshore region with over 40 times the wave energy potential compared to location 1 in the GoM. When the wave height-wave period combination did not exist due to physics of wave formation, "NA" is input in the tables, while a "0" in the table means the occurrence of that specific combination was zero in the specific time and location.

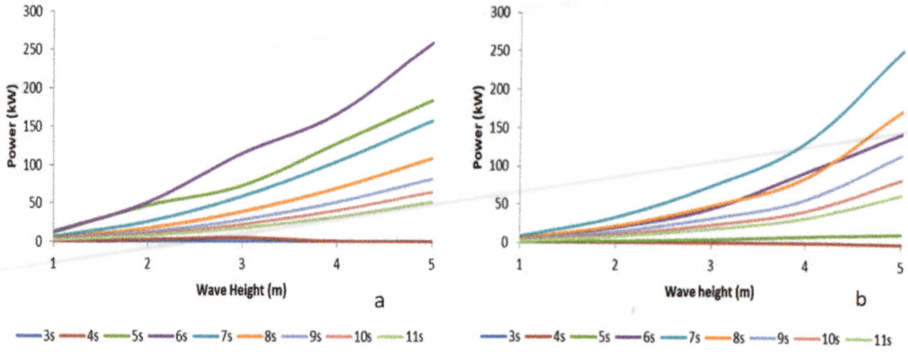

Figure 9. Wave height vs. power: (**a**) hourly resolution power (**b**) daily resolution power.

Table 11. Annual energy matrix (Scenario-1: Gulf of Mexico).

		Wave Height (m)				
		0–1	1–2	2–3	3–4	4–5
Period T (s)	2–3	0.65	0.18	NA	NA	NA
	3–4	214.31	60.68	26.02	NA	NA
	4–5	8482.08	11,789.03	78.05	33.27	0.00
	5–6	14,458.04	56,266.15	8050.77	142.86	78.57
	6–7	3748.43	33,556.00	13,694.89	1526.07	14.28
	7–8	1390.29	19,072.76	18,350.75	4825.71	291.48
	8–9	282.34	2882.82	5104.10	2448.30	302.44
	9–10	101.03	935.14	1776.91	1108.37	356.59
	10–11	28.56	251.38	362.04	289.27	201.52

Table 12. Annual energy matrix (Scenario-2: Oregon).

		Wave Height (m)				
		0–1	1–2	2–3	3–4	4–5
Period T (s)	2–3	0	0	NA	NA	NA
	3–4	4.85	0	0	NA	NA
	4–5	520.58	171.41	0	0	0
	5–6	5948.58	21,176.71	0	0	0
	6–7	5122.98	65,042.87	11,458.13	0	0
	7–8	830.17	20,658.59	29,839.33	1826.05	0
	8–9	198.95	2883.85	7006.98	3191.44	0
	9–10	13.98	616.3	820.62	1011.43	0
	10–11	10.74	43.19	48.9	0	0

The result shows that the total annual energy capture under Scenario 1 was about 212 MWh/yr while Scenario 2 had about 178 MWh/yr. Scenario 3 had about 150 MWh/yr. The results show about 16% difference between Scenarios 1 and 2. It is also interesting to note that the annual energy performance estimation of design 2 was lower than that of that of design 1 even though the dimensions of design 2 were larger. It can be inferred that the temporal resolution of the ocean wave data also significantly affects the overall energy performance of the device in addition to significantly affecting the dimensions of the WEC buoy. When comparing Scenario 1 with Scenario 3, despite the buoys in both scenarios having the same dimensions, the total annual estimated energy capture in Scenario 1 was about 41% greater than that of the Scenario 3 even though the total annual energy theoretical potential in location 2 is over 40 times greater than that of location 1. This reinforces the idea that heaving buoys should be designed and tailored to specific locations. In the next section, the factors involved in the power capture estimation in this paper are further analyzed by conducting detailed statistical analysis.

Table 13. Annual energy matrix (Scenario 3).

		Wave Height (m)				
		0–1	**1–2**	**2–3**	**3–4**	**4–5**
Period T (s)	**2–3**	0	0	NA	NA	NA
	3–4	3.36	5.03	0	NA	NA
	4–5	177.5	2859.96	1.33	0	0
	5–6	527.36	11,582.32	3038.9	4.89	0
	6–7	317.71	8444.96	4830.84	713.31	0
	7–8	340.44	10,534.52	6901.25	2583.02	713.29
	8–9	102.8	6290.16	4892.72	2265.71	1328.49
	9–10	104.14	6119.39	5726.59	2221.71	1927.88
	10–11	86.32	5358.21	11,684.4	7180.4	3554.33
	11–12	24.88	1154.66	2863.26	2648.24	1307.92
	12–13	52.71	1959.88	4799.73	5114.43	3830.67
	13–14	35.39	719.08	1503.07	1988.76	1656.8
	14–15	48.6	578.54	926.02	1350.3	1364.52
	15–16	0	0	0	0	0
	16–17	39.16	293.85	431.14	772.13	718.95
	17–18	11.39	131.39	198.23	284.72	239.69
	18–19	0	0	0	0	0
	19–20	1.42	39.32	98.57	102.09	93.03
	20–21	0	0	0	0	0

5. Design Parameters and Their Effects on WEC's Power Output

Different parameters and their effects on the WEC's power production are tested using statistical hypothesis testing methods such as *t*-tests and factorial analysis. These tests have given more insights into the complexities that exist between some of different parameters that contribute to the energy produced by a WEC. Real ocean data of a 7-day period randomly picked from summer (July, 13–19) and winter (Jan, 1–7) in the GoM. One summer week and one winter week wave data are extracted including both hourly resolution (168 data points) and daily resolution (7 data points). The power output of WECs with different design parameters including design types (design 1& design 2) and diameters (7 m & 8 m) are estimated under the two weeks wave data. These estimated power output values are used as samples for the hypothesis testing. For the two-sample two-tailed *t*-test under hourly and daily data, each sample has 168 and 7 data points, respectively. For the factorial analysis, the estimated power output values are considered as the response/variable, and design type, diameter, and season are considered as factors with each factor having two different levels.

For the hourly resolution data within the week used for design 1, the standard deviations of the wave height and wave period are 0.32 m and 1.03 s for the summer and 0.65 m and 1.05 s for the winter. Similarly, for the daily resolution data used for design 2, the standard deviations of the wave height and wave period are 0.25 m and 0.69 s for the summer and 0.54 m and 0.78 s for the winter. As described in previous sections, design 1 is tailored to capture highest energy between 5–6 s (RAO = 5.5 s), and design 2 is tailored to capture most energy between 6–7 s (RAO = 6 s). So the distribution of wave period within the selected period does influence the results. For the winter season, the wave period falls below 6 s for 95 times out of 168 for the hourly resolution data and 4 times out of 7 times for the daily resolution data. Similarly for the summer season, the wave period falls below 6 s for 23 times for the hourly data and zero time for the daily data. The results of the two-sample two-tailed t-tests are shown in Table 14, where the estimated power outputs are the samples.

Table 14. Results of Series of *t*-tests.

Parameters	*p*-Value
Diameter 7 m, Summer, **Design1** vs. Diameter 7 m, Summer, **Design 2**	0.000
Diameter 7 m, Winter, **Design 1** vs. Diameter 7 m, Winter, **Design 2**	0.045
Diameter 8 m, Summer, **Design 1** vs. Diameter 8 m, Summer, **Design 2**	0.030
Diameter 8 m, Winter, **Design 1** vs. Diameter 8 m, Winter **Design 2**	0.000
Diameter 7 m, Design 1, **Summer** vs. Diameter 7 m, Design 1, **Winter**	0.035
Diameter 8 m, Design 1, **Summer** vs. Diameter 8 m, Design 1, **Winter**	0.040
Diameter 7 m, Design 2, **Summer** vs. Diameter 7 m, Design 2, **Winter**	0.063
Diameter 8 m, Design 2, **Summer** vs. Diameter 8 m, Design 2, **Winter**	0.000
Design 1, Summer, **7 m** vs. Design 1, Summer, **8 m**	0.044
Design 1, Winter, **7 m** vs. Design 1, Winter, **8 m**	0.920
Design 2, Summer, **7 m** vs. Design 2, Summer, **8 m**	0.000
Design 2, Winter, **7 m** vs. Design 2, Winter, **8 m**	0.000

From the results, a p-value less than or equal to 5% indicates that the variables (parameters) significantly influenced the power output. First, different designs (design 1 versus design 2) significantly affected the power output irrespective of the season and diameter. Second, for the seasonal variations, most results were significant except for when the diameter was 7 m in design 2. Last, the size of the diameters in most cases significantly affected the power output of the device, except for design 1 during the winter time.

The factorial analysis was performed on the power output with design, diameter and season as the factors. Each factor had two different levels, so a 2^3 factorial design was formed and tested. The results from the factorial analysis are shown in Table 15. The results show a *p*-value of less than 5% for all the main effects and interaction between design and season, while the remaining two-way and three-way interactions were not significant.

Table 15. Results (*p*-value) of the 2^3 factorial design analysis.

Design	Diameter	Season	Design & Season	Design & Diameter	Diameter & Season	3-Way Interaction
0.020	0.043	0.031	0.012	0.260	0.889	0.745

6. Conclusions

The effects of the spatial and temporal resolution of the ocean wave data on the design of heaving point absorber and its power capture have been analyzed in this study. The effects have been analyzed quantitatively by comparing power capture performance of different designs of heaving point absorber based on different temporal and spatial wave conditions. By applying the normal convention to design the WEC device to resonate, it would capture energy in the most prevalent ocean wave period. The results show that different temporal resolution data lead to different designs, which may capture different amount of energy. However, the difference in temporal resolution of data did not significantly affect the estimation of the ocean wave theoretical power present in a location.

The results confirm the importance of designing a WEC device specific to a location. The WEC device designed to operate in a region in the Gulf of Mexico may generate much less energy in an offshore location in Oregon, even though Oregon location has almost 40 times more ocean wave energy potential than the location in the Gulf of Mexico. Results of these analysis show that the power captured by the same device in the Gulf of Mexico was higher despite having lesser wave energy potential. This is because the WEC was designed to resonate with the prevalent ocean wave condition in the Gulf of Mexico, so its performance was poor in the Oregon location. The influence of the WEC design, diameter and seasonal changes were also examined on the power output of the device. These effects were analyzed using *t*-tests and factorial analysis. The results show that all the parameters to varying degrees have significant effect on the power output. However, only the interaction between design and season shows significant effect. Furthermore, a linear damper of 40 kNm/s and 50 kNm/s was used in this paper to represent the power take off for the 7 m and 8 m buoy. Future studies will perform sensitivity analysis to determine the optimal damper and its nonlinear behavior.

Studying the complexities that exists in the interaction of WEC devices with the ocean waves will increase the knowledge in the design of more effective devices, which will make the penetration of ocean wave energy increase. The scope of this study deals with only the primary capture of mechanical energy from the ocean waves through the hydrodynamic interaction of the WEC device and is most useful at the feasibility stage of the project. This aspect is only a part of the wave energy process, which includes the hydrodynamic conversion, conversion to electrical energy, and transmission. The wave condition data used in this paper was obtained from NOAA directly, which provides hourly and daily resolution. During the feasibility analysis stage of wave energy harvesting and WEC design, it was possible to directly use existing wave condition data with default temporal resolution as many existing studies did [4,6,8,40–45]. Meanwhile, IEC TC114 technical specification recommends 30 min temporal resolution when designing WECs for commercial use. Decision makers and WEC designers should choose the right temporal resolution based on different needs and budget limitations. As the commercialization of WECs draws more attention, it is expected more wave condition data with 30 min temporal resolution may become available, which could reduce the additional cost for collecting wave energy data for WEC designs. In addition, the structural reliability of any WEC devices should also be considered in the design stage, especially their reliability under extreme and hash ocean wave conditions.

Author Contributions: T.A. conducted data collection, product design and data analysis with the suggestions and guidance from H.L. T.A. wrote the initial draft paper under the supervision of H.L. H.L. made major revision on the initial draft paper, and approved the final version to be published. All authors have read and agreed to the published version of the manuscript.

Funding: This research received no external funding.

Acknowledgments: The authors are thankful to the support from Texas A&M University-Kingsville and National Science Foundation (award # EEC-1757812).

Conflicts of Interest: The authors declare no conflict of interest.

References

1. Pecher, A.; Kofoed, J.P. *Handbook of Ocean Wave Energy*; Springer: London, UK, 2017.
2. Aderinto, T.; Li, H. Ocean wave energy converters: Status and challenges. *Energies* **2018**, *11*, 1250. [CrossRef]
3. Lehmann, M.; Karimpour, F.; Goudey, C.A.; Jacobson, P.T.; Alam, M.R. Ocean wave energy in the United States: Current status and future perspectives. *Renew. Sustain. Energy Rev.* **2017**, *74*, 1300–1313. [CrossRef]
4. Reguero, B.G.; Losada, I.J.; Méndez, F.J. A global wave power resource and its seasonal, interannual and long-term variability. *Appl. Energy* **2015**, *148*, 366–380. [CrossRef]
5. Izadparast, A.H.; Niedzwecki, J.M. Estimating the potential of ocean wave power resources. *Ocean Eng.* **2011**, *38*, 177–185. [CrossRef]
6. Jacobson, P.T.; Hagerman, G.; Scott, G. *Mapping and Assessment of the United States Ocean Wave Energy Resource*; Electric Power Research Institute: Palo Alto, CA, USA, 2011.
7. The Crown Estate. *UK Wave and Tidal Key Resource Areas Project—Summary Report*; WWW Document; The Crown Estate: London, UK, 2012.
8. Haces-Fernandez, F.; Li, H.; Jin, K. Investigation into the possibility of extracting wave energy from the Texas coast. *Int. J. Energy Clean Environ.* **2019**, *20*, 23–41. [CrossRef]
9. Fernandez, F.H.; Martinez, A.; Ramirez, D.; Li, H. Characterization of wave energy patterns in Gulf of Mexico. In Proceedings of the Institute of Industrial and Systems Engineers Annual Conference, Pittsburgh, PA, USA, 20–23 May 2017; pp. 1532–1537.
10. Silva, D.; Bento, A.R.; Martinho, P.; Soares, C.G. High resolution local wave energy modelling in the Iberian Peninsula. *Energy* **2015**, *91*, 1099–1112. [CrossRef]
11. Mentaschi, L.; Besio, G.; Cassola, F.; Mazzino, A. Performance evaluation of Wavewatch III in the Mediterranean Sea. *Ocean Model.* **2015**, *90*, 82–94. [CrossRef]

12. Gallagher, S.; Tiron, R.; Whelan, E.; Gleeson, E.; Dias, F.; McGrath, R. The nearshore wind and wave energy potential of Ireland: A high resolution assessment of availability and accessibility. *Renew. Energy* **2016**, *88*, 494–516. [CrossRef]
13. Liang, B.; Shao, Z.; Wu, G.; Shao, M.; Sun, J. New equations of wave energy assessment accounting for the water depth. *Appl. Energy* **2017**, *188*, 130–139. [CrossRef]
14. Chen, X.; Wang, K.; Zhang, Z.; Zeng, Y.; Zhang, Y.; O'Driscoll, K. An assessment of wind and wave climate as potential sources of renewable energy in the nearshore Shenzhen coastal zone of the South China Sea. *Energy* **2017**, *134*, 789–801. [CrossRef]
15. Seemanth, M.; Bhowmick, S.A.; Kumar, R.; Sharma, R. Sensitivity analysis of dissipation parameterizations in a third-generation spectral wave model, WAVEWATCH III for Indian Ocean. *Ocean Eng.* **2016**, *124*, 252–273. [CrossRef]
16. Astariz, S.; Iglesias, G. The economics of wave energy: A review. *Renew. Sustain. Energy Rev.* **2015**, *45*, 397–408. [CrossRef]
17. Witt, M.J.; Sheehan, E.V.; Bearhop, S.; Broderick, A.C.; Conley, D.C.; Cotterell, S.P.; Hosegood, P. Assessing wave energy effects on biodiversity: The Wave Hub experience. *Philos. Trans. R. Soc. A* **2012**, *370*, 502–529. [CrossRef] [PubMed]
18. Elwood, D.; Yim, S.C.; Prudell, J.; Stillinger, C.; Von Jouanne, A.; Brekken, T.; Paasch, R. Design, construction, and ocean testing of a taut-moored dual-body wave energy converter with a linear generator power take-off. *Renew. Energy* **2010**, *35*, 348–354. [CrossRef]
19. Babarit, A. A database of capture width ratio of wave energy converters. *Renew. Energy* **2015**, *80*, 610–628. [CrossRef]
20. Haces-Fernandez, F.; Li, H.; Ramirez, D. Assessment of the potential of energy extracted from waves and wind to supply offshore oil platforms operating in the Gulf of Mexico. *Energies* **2018**, *11*, 1084. [CrossRef]
21. Falnes, J.; Lillebekken, P.M. Budal's latching-controlled-Buoy TypeWave-power plant. In Proceedings of the 5th European Wave Energy Conference, Cork, UK, 17–20 September 2003.
22. Waveroller. Available online: http://aw-energy.com/aboutwaveroller/waveroller-concept (accessed on 30 May 2019).
23. Al Shami, E.; Zhang, R.; Wang, X. Point absorber wave energy harvesters: A review of recent developments. *Energies* **2019**, *12*, 47. [CrossRef]
24. EMEC. Pelamis Wave Power. Available online: http://www.emec.org.uk/about-us/wave-clients/pelamis-wave-power/ (accessed on 30 May 2019).
25. Dallman, A.; Jenne, D.S.; Neary, V.; Driscoll, F.; Thresher, R.; Gunawan, B. Evaluation of performance metrics for the Wave Energy Prize converters tested at 1/20th scale. *Renew. Sustain. Energy Rev.* **2018**, *98*, 79–91. [CrossRef]
26. Thomson, R.C.; Chick, J.P.; Harrison, G.P. An LCA of the Pelamis wave energy converter. *Int. J. Life Cycle Assess.* **2019**, *24*, 51–63. [CrossRef]
27. Evans, D.V. A theory for wave-power absorption by oscillating bodies. *J. Fluid Mech.* **1976**, *77*, 1–25. [CrossRef]
28. Evans, D.V.; Porter, R. Hydrodynamic characteristics of an oscillating water column device. *Appl. Ocean Res.* **1995**, *17*, 155–164. [CrossRef]
29. Mei, C.C. Power extraction from water waves. *J. Ship Res.* **1976**, *20*, 63–66.
30. Shadman, M.; Estefen, S.F.; Rodriguez, C.A.; Nogueira, I.C. A geometrical optimization method applied to a heaving point absorber wave energy converter. *Renew. Energy* **2018**, *115*, 533–546. [CrossRef]
31. Wu, J.; Yao, Y.; Zhou, L.; Göteman, M. Latching and declutching control of the solo duck wave-energy converter with different load types. *Energies* **2017**, *10*, 2070. [CrossRef]
32. Feng, Z.; Kerrigan, E.C. Latching control of wave energy converters using derivative-free optimization. In Proceedings of the 52nd IEEE Conference on Decision and Control, Florence, Italy, 10–13 December 2013; pp. 7474–7479.
33. Li, G.; Belmont, M.R. Model predictive control of sea wave energy converters–Part I: A convex approach for the case of a single device. *Renew. Energy* **2014**, *69*, 453–463. [CrossRef]
34. National Data Buoy Center. Available online: https://www.ndbc.noaa.gov/ (accessed on 30 May 2019).
35. Budar, K.; Falnes, J. A resonant point absorber of ocean-wave power. *Nature* **1975**, *256*, 478. [CrossRef]
36. Hooft, J.P. Oscillatory wave forces on small bodies. *Int. Shipbuild. Prog.* **1970**, *17*, 127–135. [CrossRef]

37. *ANSYS AQWA, version v18.1*; ANSYS Inc.: Canonsburg, PA, USA, 2018.
38. Haces-Fernandez, F.; Li, H.; Ramirez, D. Wave energy characterization and assessment in the US Gulf of Mexico, East and West Coasts with Energy Event concept. *Renew. Energy* **2018**, *123*, 312–322. [CrossRef]
39. Falnes, J.; Perlin, M. Ocean waves and oscillating systems: Linear interactions including wave-energy extraction. *Appl. Mech. Rev.* **2003**, *56*, B3. [CrossRef]
40. Gunn, K.; Stock-Williams, C. Quantifying the global wave power resource. *Renew. Energy* **2012**, *44*, 296–304. [CrossRef]
41. Ahn, S.; Haas, K.A.; Neary, V.S. Wave energy resource classification system for US coastal waters. *Renew. Sustain. Energy Rev.* **2019**, *104*, 54–68. [CrossRef]
42. Robertson, B.; Hiles, C.; Luczko, E.; Buckham, B. Quantifying wave power and wave energy converter array production potential. *Int. J. Mar. Energy* **2016**, *14*, 143–160. [CrossRef]
43. Ferrari, F.; Besio, G.; Cassola, F.; Mazzino, A. Optimized wind and wave energy resource assessment and offshore exploitability in the Mediterranean Sea. *Energy* **2019**, *190*, 116447. [CrossRef]
44. Allahdadi, M.N.; Gunawan, B.; Lai, J.; He, R.; Neary, V.S. Development and validation of a regional-scale high-resolution unstructured model for wave energy resource characterization along the US East Coast. *Renew. Energy* **2019**, *136*, 500–511. [CrossRef]
45. Yang, Z.; Neary, V.S.; Wang, T.; Gunawan, B.; Dallman, A.R.; Wu, W.C. A wave model test bed study for wave energy resource characterization. *Renew. Energy* **2017**, *114*, 132–144. [CrossRef]

sustainability

MDPI

Article

Emergy and Sustainability Ternary Diagrams of Energy Systems: Application to Solar Updraft Tower

Islam Elsayed * and Yoshiki Nishi *

Department of Systems Design for Ocean-Space, Graduate School of Engineering Science, Yokohama National University, 79-5 Tokiwadai, Hodogaya, Yokohama, Kanagawa 240-8501, Japan

* Correspondence: islam-elsayed-bz@ynu.jp (I.E.); nishi-yoshiki-rg@ynu.ac.jp (Y.N.)

Received: 26 November 2020; Accepted: 14 December 2020; Published: 16 December 2020

Abstract: To facilitate sustainable energy development, one has to understand the limited availability of nonrenewable energy resources, and the ability of the earth to renew or recover. Emergy is an instrument that measures environmental loading, ecological economics, and regional sustainable development. In this study, emergy indicators are calculated to investigate the sustainability of solar updraft tower (SUT). SUT produces energy from the hot air, utilizing a combination of a solar collector, central tower, and air turbines. The results demonstrate that the sustainability of SUT grew as the size of the plant increased. Further, emergetic ternary diagrams are drawn to facilitate the comparison between SUT and various technologies. The resources-use efficiency of wind energy and SUT, 200 MW is found to be the lowest among all energy technologies presented in this research. Scenario analysis is performed to explore the future optimization directions. The results demonstrate that the development direction of SUT systems should mainly focus on reducing the materials demanded by the manufacturing and construction of its solar collectors. This study aims to demonstrate the value of emergy as a powerful instrument for drawing long-term sustainable strategies in energy markets for a greener tomorrow.

Keywords: emergy; emergetic ternary diagrams; sustainability; environmental loading; energy systems assessment; solar updraft tower

1. Introduction

The World Commission on Environment and Development in 1987 developed the official definition of the sustainable development concept for the first time, describing it as "the development which meets the needs of current generations without compromising the ability of future generations to meet their own needs" [1,2]. The realization of such development necessitates awareness of the limited availability of nonrenewable resources, and the ability of the earth to renew or recover [3]. The greening of energy emerged as a key element in any region to facilitate sustainable socioeconomic development. Total global carbon emissions from the power sector were about 13.4 billion tons in 2016, about 42% of total carbon emissions from fuel combustion. Coal-fired power plants are one of the largest sources of air pollution. High CO_2 levels accelerated the negative impact on the environment, including, but not limited to, global warming and the rising sea level. An inefficient relationship with the environment will lead to a dramatically negative impact on societies and economies [4]. Therefore, clean energy infrastructure is essential economically, politically, environmentally, and socially for a greener tomorrow.

While the public emphasizes the imperative of green energy infrastructures, engineers and researchers doubt the investment in greener generation technologies like wind and solar unless new evaluation mechanisms are introduced to support renewables and decarbonatization strategies [5]. On the one hand, the de-risking of investments and the idea of reliable sources of energy limit investment strategies in renewables worldwide. On the other hand, some countries around the globe (Denmark, 2019) had nearly half of their total energy consumption powered by renewable resources (wind energy) [4]. Thus, there is a body of academic knowledge that claims that the efficient management of renewable resources and the diversified use of green energy technologies can offer a reliable energy infrastructure. To accelerate and strengthen the greening of the energy industry, one has to understand the complexity of the energy markets and their abundant considerations.

This paper addresses "emergy", proposed by Howard T. Odum [6,7], as a powerful instrument for measuring sustainable development. Emergy is able to facilitate regional sustainable development, ecological management, and environmental conservation for policymakers [8]. Emergy analysis considers that the formation of any system on earth derives from solar energy. The emergy value of any resource is its "solar energy memory" or, in other words, the amount of solar energy directly and/or indirectly used in its formation process. Different types of resources (material, energy, labor, etc.) possess different properties and are expressed in different units. The concept of solar transformity in emergy analysis quantifies the emergy value of any resource in one unified unit. Consequently, the framework of emergy analysis in the environmental accounting of any economic system requires ecological and economic energy transformation chains to determine the total emergy input that drives the system (system empower). As a result, all systems are evaluated on an ecocentrism basis [3,7,9].

M. T. Brown and S. Ulgiati [3] presented emergy in the evaluation of electricity production systems and compared six power plants based on their emergy indicators. Consequently, many researchers reported the evaluation of various power generation technologies based on an emergy analysis. These studies include, but are not limited to, hydropower stations [10,11], geothermal and biomass [12,13], wind energy [14–16], solar photovoltaics and solar thermal collectors [17], and solar tower power plant systems [18]. Furthermore, emergetic ternary diagrams were presented by B. F. Giannetti et al. [19] as graphic tools that offer not only prompt visualization of emergy analysis but also readily recognize system improvements. C. M. V. B. Almeida et al. [20] discussed emergetic ternary diagrams in the application of electricity production systems.

Solar updraft tower (SUT) is one of the green technologies for acquiring renewable energy [21–24]. In nature, the updraft effect causes wind and hurricanes. The geometry of SUT mimics the wind cycle in nature to produce the thermal wind that drives turbines to generate electricity. It generates power using the sun as the only source of energy and using only air as working fluid without phase change; there is no water demand or working mediums, and no cooling mechanisms are needed [25]. While many areas worldwide suffer from severe water shortages, SUT saves gallons of cooling water associated with conventional and solar thermal generation plants [25]. Together with its low-cost thermal energy storage potential, SUT energy can be generated around the clock, eliminating environmental threats associated with the use of batteries in photovoltaic and wind power generation plants. Numerous studies reported cost–benefit models to assess the economic feasibility of SUT power plants [26–28]. Further, other researchers reported the ecological analysis in the lifecycle of SUT power plants [22,29]. I. Elsayed and Y. Nishi examined the sustainability of SUT power generation technology using the Inclusive Impact Index concerning both ecological and economic costs [30]. Despite that, emergy analysis of an SUT power plant has not been reported.

The main contribution of the present study is the evaluation of large-scale SUT power plants using emergy as a benchmark (Figure 1). Emergy accounting and emergy-based indicators are estimated to investigate the relative performance and sustainability of SUT systems. Comparisons of emergy indicators with various conventional and renewable power-generation technologies are then performed. To facilitate the latter comparison, emergetic ternary diagrams are drawn to provide a better interpretation of emergy accounting results and to visualize the interactions between power plants and the environment. Additionally, based on emergetic ternary diagrams, scenario analysis is performed to explore future optimization directions of large-scale SUT power plants. This paper offers a way forward to draw the sustainability pathways of large-scale SUT power systems for optimal socioeconomic decision-making.

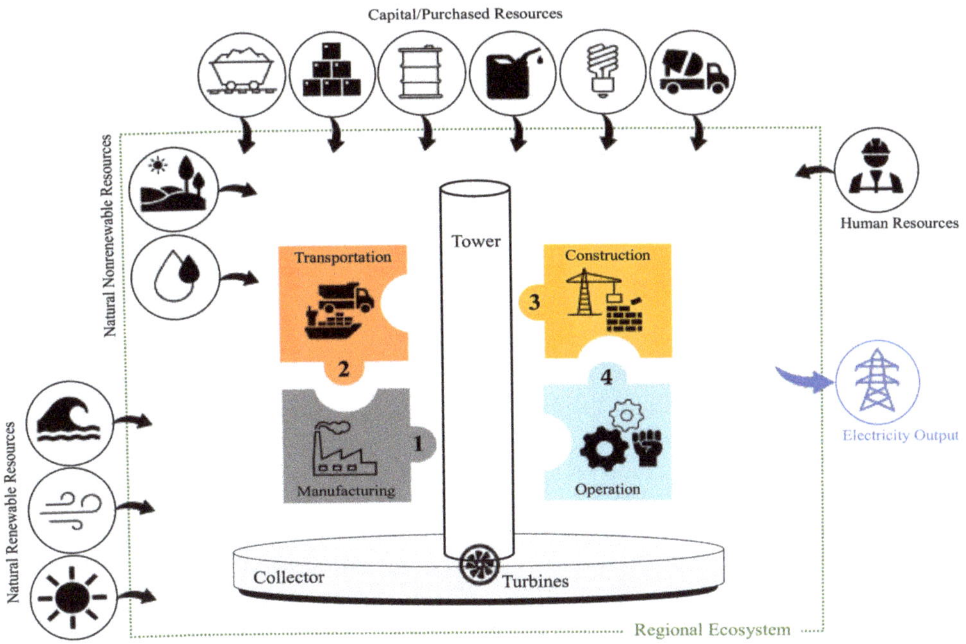

Figure 1. Graphical abstract of emergy analysis for the lifecycle of solar updraft tower power plants.

2. Methodology and Data

2.1. *Solar Updraft Tower: Design Parameters*

SUT produces energy from the hot air, utilizing a combination of a flat plate solar collector, a central tower, and air turbines. The air is heated by the greenhouse effect in the collector; this hot air produced is lighter than the ambient cold air at the top of the tower, and, consequently, the hot air rises up the tower. In other words, the density difference of the air caused by the temperature rise in the collector is converted to a pressure difference, which generates a fluid flow (thermal wind) that drives the turbines to generate electricity [21–25]. In 1983, W. Haaf et al. presented the first large-scale SUT pilot plant, which was in Manzanares, Spain [21]. The plant operated with a tower (steel), height of 194.6 m; a collector (plastic), diameter of 244 m; and a capacity of 50 kW. Following the measurements and investigations from the Manzanares pilot plant, researchers put forward detailed calculations and structural design proposals for large-scale SUT plants. J. Schlaich et al. [22,23] established a design model for commercial large-scale SUT plants of different capacities and calculated the costs and GHG emissions per kWh produced.

The technical constraints of building 1000 m high towers are a challenge; however, they can be built today, and several SUT projects are in development [22]. EnviroMission, an Australian company, began moving forward to build a 200 MW SUT power plant in Arizona, USA [31]. The project proposal includes a concrete tower of 800 m in height, a glass collector of 4800 m in diameter, and 32 horizontal axis turbines (6.25 MW each). Shimizu, a Japanese corporation, is in the process of developing a residential project known as "The Green Float" [32]. With a circular section extending to a diameter of 3000 m, and a central tower section of 1000 m in height, the latter project has the same configurations as a large-scale SUT plant. Shimizu shows that such construction can be done soon and provides insights for the implementation of SUT plants in a desert landscape, as well as nearshore or offshore.

Therefore, it is necessary to perform an impact assessment of SUT power plants and investigate their sustainability. The J. Schlaich et al. [22] model of commercial SUT power plants of different capacities (Table 1) is used in the present study to establish emergy analysis for SUT systems.

Table 1. The design parameters of J. Schlaich et al. [22] for SUT commercial models used in the calculations of the present study.

Item	SUT Power Generation Capacity			
	5 MW	30 MW	100 MW	200 MW
Tower Height (m)	550	750	1000	1000
Tower Diameter (m)	45	90	110	120
Collector Diameter (m)	1250	3750	4300	7000
Collector Height (m)	3	3	3	3
Electricity Output [1] (GWh/y)	14	99	320	680

[1] The SUT power plant site is assumed to have an annual global solar radiation of 2300 kWh/m²/y. Besides, there is no additional thermal storage.

2.2. Materials and Resources in the Lifecycle of SUT Power Plants

Since SUT power plants demand massive structures and materials, questions of technical constraints and energies expended to construct the plant are recurrent [21]. Estimating from the Manzanares pilot, W. Haaf et al. reported the building materials and energies to construct a 100 MW SUT (concrete tower and glass collector) [21]. Subsequently, J. Schlaich and other researchers reported the lifecycle analysis of SUT power plants [22,30]. The turbines are at the periphery of the transitional area between the collector and tower. J. Schlaich et al. reported an approximate number of 24 to 36 turbines for large-scale SUT power plants [22]. Thus, this study considers the use of horizontal-axis turbines with rated capacities of 4.1 and 5 MW. The design theory of the SUT turbines is usually adapted from horizontal-axis wind turbines [22,26,28]. Henceforth, the materials of 4.1 and 5 MW wind turbines are quoted from [33,34].

The construction period of a large-scale SUT power plant is approximately two to three years [22,29,31]. This study assumes that the materials require transportation from a region 1000 km away from the plant's location. The present study adapted the materials, weights, and energy for the construction of the tower and collector from W. Haaf et al. [21]. Table 2 illustrates the materials, component weights, and machinery for construction of the study's SUT power plants.

Table 2. Materials and energy consumption for the construction of commercial SUT power plants of different capacities.

Item		Material	SUT Power Generation Capacity			
			5 MW	30 MW	100 MW	200 MW
Components' manufacturing materials						
Tower		Concrete (ton)	4.54×10^5	4.54×10^5	8.26×10^5	8.26×10^5
Collector		Glass (ton)	1.27×10^4	6.82×10^4	1.60×10^5	3.97×10^5
		Resin and fiberglass (ton)	2.42×10^2	1.21×10^3	5.54×10^3	9.68×10^3
Turbines		Iron (ton)	1.19×10^2	5.94×10^2	2.01×10^3	4.75×10^3
		Steel (ton)	5.93×10^1	2.97×10^2	8.18×10^2	2.37×10^3
		Copper (ton)	9.60×10^0	4.80×10^1	1.33×10^2	3.84×10^2
		Silica (ton)	1.00×10^0	5.00×10^0	1.28×10^1	4.00×10^1
Installation						
Transportation of materials		Diesel fuel (J)	9.43×10^{14}	1.07×10^{15}	1.94×10^{15}	2.49×10^{15}
Construction works	Reinforcement of tower	Steel (ton)	1.65×10^4	2.25×10^4	3.00×10^4	3.00×10^4
	Collector's load-bearing structure	Steel (ton)	6.47×10^4	6.47×10^4	9.60×10^4	1.56×10^5
	Collector's foundations	Concrete (ton)	7.27×10^3	1.69×10^4	2.50×10^4	4.07×10^4
	Machinery	Electricity (MWh)	6.00×10^4	1.10×10^5	1.50×10^5	2.00×10^5

In the present analysis, there is no additional thermal storage, thus no salt or water usage. Additionally, no replacement tower or collector is in place [29]. Therefore, work for the operation and maintenance during the lifetime of the plant is necessary mainly for the turbines. A great deal of data from wind power stations can be used to estimate turbine substitutions' materials in the lifecycle.

The intended service life of commercial SUT power plant ranges from 80 to 120 years [27]; this study considers a lifetime of 30 years.

2.3. Emergy Analysis

The sun is the primary source of energy that fuels the earth's biological and economic growth. Emergy theory, proposed by H. T. Odum, traces the origin of the products and services formation. Considering that solar energy is the main energy input to the Earth, emergy measures how much energy would be needed to form a resource if solar radiation were the only input. Thus, emergy describes the amount of one form of energy invested in the system or, in other words, the "energy memory" of any system. Emergy converts different forms of energy, materials, goods, and human labor and services to equivalents of solar energy in a unified unit, solar embodied joules, abbreviated sej. The emergy required to make one unit of product is called the "solar transformity (sej/unit)" [6,7,9,35].

Emergy assumes all living systems sustain one another by participating in a network of energy flows; even the economy can be incorporated into this energy flow network. Hence, emergy analysis for any technology measures all input and output quantities, considers it as a network of emergy flows, and then determines the total emergy value. This approach helps to analyze the effect of each material and process, and the contribution of the ecosystem (environmental services); thus, it guides the future development of the technology. In emergy community, price does not reflect the real value of the environmental resources. The money is paid only to humans for their services, not to the environmental contributions and resources [35]. The emergy evaluation of any system can be divided into three consecutive steps. The first determines the analysis scale and boundary (analyzing the input and output data of the system). The second computes an emergy diagram and evaluation table. Finally, the third calculates emergy-based indices.

2.3.1. Analysis Scale and System Boundary

Emergy analysis is used in this study to evaluate the sustainability of power generation systems. The research scale is a power generation plant, the solar updraft tower. In the case of power plants, the types of inputs include energy resources (oil, coal, wind, solar radiation, etc.), materials (concrete, steel, glass, etc.), and labor and services necessary to construct, operate, and maintain the plant. The boundary of the analysis covers all phases of the SUT including manufacturing, installation (construction and transportation), and operation and maintenance during its lifetime. The main output of the system is the electricity generated.

2.3.2. Emergy Diagram and Evaluation Table

A typical emergy flow network is shown in Figure 2. The engineering systems comprise renewable and nonrenewable resources powered by both the ecosystem and human work. Therefore, the framework inputs of any engineering system are divided into three types: renewable resources (R), such as solar radiation, wind, rain, etc.; nonrenewable resources (N), such as land area, groundwater, etc. (limited availability); and purchased resources (F), which refers to economic transactions such as electricity, machinery, and human labor, etc. Renewable and nonrenewable resources are considered local resources. The emergy value of an input resource (i) is expressed as follows:

$$\text{Emergy}_i = \text{Quantity}_i \times \text{Solar transformity}_i \quad (1)$$

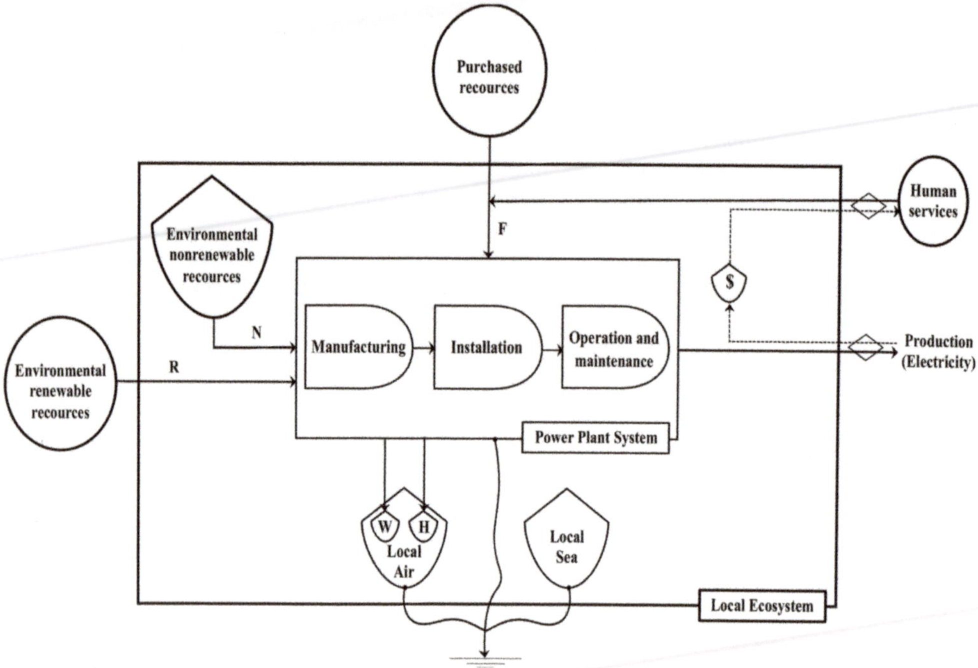

Figure 2. Typical emergy flow diagram of SUT power plant. The diagram shows the environmental renewable (R) and nonrenewable (N) inputs supplied by the local ecosystem, together with purchased inputs (F) from the economic systems. The electricity produced by the plant is the output from the SUT system. Local air storage represents the atmosphere, which receives pollutants (W) and heat released by the power plant (H). Local sea storage represents the marine environment [3,19].

The emergy baseline is a basic parameter for the equation, as it is the base of solar transformities calculations. The emergy transformity data adopted from references should all have the same baseline. The emergy baseline of the solar transformities used in this study is "15.83×10^{24} sej/y". The total emergy input into the system is described as the emergy yield driving the system which equals the sum of emergy inflows ($R + N + F$). The annual electricity production (AEP) of the power plant is converted to joules using the standard conversion of 3.6×10^6 J/kWh.

2.3.3. Emergy indicators

Based on the above characteristics of emergy flows, emergy-based indices were established to investigate the efficiency and sustainability of various systems. In the present work, the most common emergy-based indicators were used to evaluate power generation from SUT power plants [3].

The transformity (Tr) of the system is the emergy invested to generate one unit of electricity. It measures the environmental and economic resources' use efficiency (emergy conversion efficiency) of the system as a whole (Equation (2)) (E_{out} is the total electricity production).

$$Tr = \frac{R + N + F}{E_{out}} \tag{2}$$

Emergy yield ratio (EYR) is the total emergy yield driving the system divided by the outside emergy sources (goods and services) from the economy. It measures the ability of the system to rely on local resources. The higher the value of this index, the larger the contribution to the economy per unit of emergy invested (Equation (3)).

$$EYR = \frac{R + N + F}{F} \quad (3)$$

Environmental loading ratio (*ELR*) is the local nonrenewable resources and total purchased resources from the economy divided by local renewable resources. It measures the stress on the environment and the potential impact of the system on the ecosystem (Equation (4)).

$$ELR = \frac{N + F}{R} \quad (4)$$

Emergy index of sustainability (*EIS*) is the emergy yield ratio per unit of environmental loading. It measures the overall sustainability of the system (Equation (5)). A high resources utilization efficiency (*EYR*) and low stress on the ecosystem means more sustainable technology. A power plant with *EIS* higher than "1" is considered sustainable [3].

$$EIS = \frac{EYR}{ELR} \quad (5)$$

2.4. Emergetic Ternary Diagrams

A ternary diagram is a three-dimensional graph but is illustrated in two dimensions in a triangular coordinate system. This graph type is proposed for the analysis of mixed components. It is commonly used to visualize phase diagrams in the physical sciences [19,20]. Emergetic ternary diagrams were presented by B. F. Giannetti et al. [19] to represent graphically emergy indicators. The authors presented emergetic ternary diagrams as graphic tools that offer not only prompt visualization of emergy data but also readily illustrate system improvements [19].

The emergy driving the system (100%) equals the sum of the percentages of the three types of Emergy resources' inflows (*R*%, *N*%, and *F*%) which are the elements of the emergetic ternary diagram. The diagram is an equilateral triangle where each corner represents an element (resources type). The triangular coordinates are chosen with renewable resources (*R*) on the top apex, nonrenewable resources (*N*) on the left apex, and purchased resources (*F*) on the right apex. The total emergy input into the system is represented by points within the triangle, the percentage of each element being given by the length of the perpendicular line from the given point to the side of the triangle opposite to the appropriate element. Moreover, the emergetic ternary diagram has auxiliary lines to facilitate emergy analysis: resource flow lines, sustainability lines, and sensitivity lines. Resource flow lines define emergy indicators *EYR* and *ELR*. Sustainability lines define the emergy index of sustainability, *EIS*. The sensitivity lines point out the system's behavior with changes in elements (resource fluxes) associated with the apex of each sensitivity line. The latter lines are very useful for identifying areas of improvement to enhance the sustainability of the systems.

In the present work, the emergetic ternary diagrams were depicted to provide a better understanding of emergy flows for SUT systems and to visualize the interactions between the power plant and the environment. Moreover, the graphical analysis facilitates the comparison between SUT systems and various energy technologies presented in this research.

2.5. Data Sources

The studied SUT power plants consist of concrete towers and glass solar collectors. The data in this study are mainly from the feasibility and lifecycle assessment reports of large-scale SUT. The total area of the canopy is considered the plant area, at a site with an annual global solar radiation of 2300 kWh/m^2/year [22]. The total masses and the electricity consumption for the construction of the tower and collector are adapted from the sample estimates of W. Haaf et al. [21]. Turbines' materials are calculated by analogy with similar wind turbines [34]. During the operation and maintenance phase, it is assumed no replacement of the tower or collector is taking place [29]. Meanwhile, one-third of the turbine's materials should be substituted during the average useful life [33,34]. Labor and services costs for installation (engineering, tests, etc.) and operation and maintenance are derived from [22]. In the calculation, considering the energy and materials loss in the production process, and due to the difficulty of getting data, some calculations are still not included [16]. The solar transformity coefficients of each resource are quoted from relevant references.

3. Results: Emergy Accounting

All the input flows during manufacturing, installation, and operation and maintenance of the SUT, 200 MW power generation plant are analyzed and described in Table 3. Energy sources, materials, land loss, and human labor and services expressed in their common units (J, kg, m^3, $, etc.) were converted to emergy flows (sej) by means of appropriate emergy transformation coefficients (called solar transformity, sej/unit). The total emergy of the SUT system accounts for all environmental contributions, energies, and human services used directly and indirectly to produce and maintain the power plant. The fractions of emergy inputs for the SUT, 200 MW plant are depicted in Figure 3. The manufacturing materials of the components are the dominant inputs, constituting 34.9% of total emergy. This is followed by construction works (materials and energy consumption), constituting 29.0% of total emergy. The latter results are the output of the massive structure and materials demanded by a large-scale SUT power plant. Meanwhile, the glass of the canopy makes up the largest proportions of components' manufacturing materials, constituting 20.8% of the total emergy inputs. The emergy invested by local environmental resources like solar radiation and land area contribute small fractions of the total emergy driving the system; this is because of the small solar transformity value of the latter resources. The other generation capacities of the SUT systems are evaluated using emergy tables like Table 3, and the results of emergy flows are listed in Table 4.

Table 3. Emergy evaluation table of SUT, 200 MW power generation plant during lifetime.

Item		Quantity	Unit	Transformity (sej/Unit)	Emergy (sej)
Energy source					
Solar energy (R)		9.56×10^{18}	J	1.00×10^{0} [7]	9.56×10^{18}
Power plant area					
Land use (N)		3.85×10^{7}	m^2/y	8.00×10^{10} [15]	9.24×10^{19}
Components' manufacturing materials					
Tower	Concrete (F)	8.26×10^{5}	ton	5.08×10^{14} [3]	4.19×10^{20}
Canopy	Glass (F)	3.97×10^{5}	ton	1.90×10^{15} [17]	7.55×10^{20}
	Resin and fiberglass (F)	9.68×10^{3}	ton	8.07×10^{15} [14]	7.81×10^{19}
	Iron (F)	4.75×10^{3}	ton	8.60×10^{14} [16]	4.09×10^{18}
Turbines	Steel (F)	2.37×10^{3}	ton	2.77×10^{15} [3]	6.57×10^{18}
	Copper (F)	3.84×10^{2}	ton	2.00×10^{15} [3]	7.68×10^{17}
	Silica (F)	4.00×10^{1}	ton	1.68×10^{15} [16]	6.72×10^{16}
SUBTOTAL					1.26×10^{21}
Installation					
Transportation	Diesel fuel (F)	2.49×10^{15}	J	6.60×10^{4} [7]	1.64×10^{20}
Construction works	Steel reinforcement of tower (F)	3.00×10^{4}	ton	4.82×10^{15} [16]	7.53×10^{20}
	Steel load-bearing structure (F)	1.56×10^{5}	ton	4.82×10^{15} [16]	1.45×10^{20}
	Concrete foundations (F)	4.07×10^{4}	ton	5.08×10^{14} [3]	2.07×10^{19}
	Electricity for machinery (F)	7.20×10^{14}	J	1.85×10^{5} [3]	1.33×10^{20}
SUBTOTAL					1.22×10^{21}
Maintenance resources					
Turbine substitution (33% replacement rate) (F)		5.69×10^{3}	ton		2.96×10^{19}
Human labor and services					
Other costs for installation (26%R and 74%F)		5.04×10^{7}	US$	5.87×10^{12} [36]	2.96×10^{20}
Labor and services for operation (26%R and 74%F)		1.22×10^{8}	US$	5.87×10^{12} [36]	7.18×10^{20}
TOTAL					
Total emergy without human labor and services					2.61×10^{21}
Total emergy with human labor and services					3.63×10^{21}
Production					
Electricity output		2.45×10^{15}	J/y		

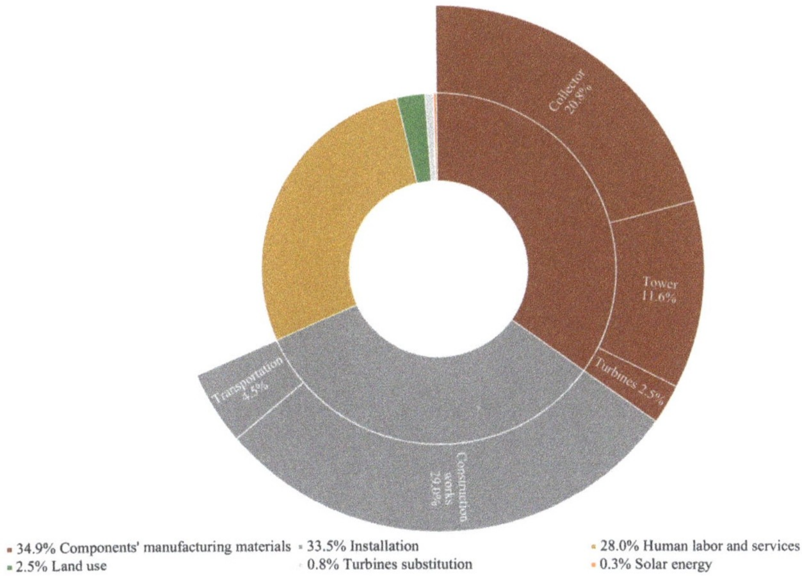

Figure 3. Fractions of emergy input flows for SUT, 200 MW.

Table 4. Emergy input flows and annual energy production of commercial SUT power plants of different capacities.

Emergy Flow		SUT Power Generation Capacity			
		5 MW	30 MW	100 MW	200 MW
R	Environmental renewable resources (sej)	2.05×10^{19}	7.49×10^{19}	1.81×10^{20}	2.73×10^{20}
N	Environmental nonrenewable resources (sej)	2.95×10^{18}	1.59×10^{19}	3.49×10^{19}	9.24×10^{19}
F	Purchased resources (sej)	8.13×10^{20}	1.16×10^{21}	2.14×10^{21}	3.26×10^{21}
Total 1	Total emergy flows, without human labor and services (sej)	7.58×10^{20}	9.65×10^{20}	1.68×10^{21}	2.61×10^{21}
Total 2	Total emergy flows, with human labor and services (sej)	8.36×10^{20}	1.25×10^{21}	2.36×10^{21}	3.63×10^{21}
AEP	Annual energy production (J/y)	5.04×10^{13}	3.56×10^{14}	1.15×10^{15}	2.45×10^{15}

4. Evaluation and Discussion

4.1. *Emergy-Based Indicators*

In order to investigate the relative performance and sustainability of SUT generation systems, their emergy indicators are presented in Table 5, together with comparisons of various conventional and renewable power generation technologies. Although the emergy input flows increased as the capacity of SUT enlarged, the SUT, 100 and 200 MW performed the best with regard to emergy-based indicators. This confirms that the sustainability of SUT grew as the size of the plant increased. Similar conclusions were achieved using a different method to assess SUT power plants [30].

Table 5. Comparison of emergy resources' inflows and emergy indicators for various power generation technologies.

	Power Generation Technology	*R* (sej/y)	*N* (sej/y)	*F* (sej/y)	*Tr* (sej/J)	*EYR*	*ELR*	*EIS*
	Solar Updraft Tower Systems							
#1	5 MW	6.82×10^{17}	9.82×10^{16}	2.71×10^{19}	5.53×10^5	1.03	39.88	0.03
#2	50 MW	2.50×10^{18}	5.28×10^{17}	3.85×10^{19}	1.17×10^5	1.08	15.65	0.07
#3	100 MW	6.04×10^{18}	1.16×10^{18}	7.14×10^{19}	6.82×10^4	1.10	12.01	0.09
#4	200 MW	9.11×10^{18}	3.08×10^{18}	1.09×10^{20}	4.94×10^4	1.11	12.27	0.09
	Renewable Power Generation							
#5	Solar photovoltaics, 18 kW [17]	1.00×10^{14}	3.05×10^{13}	4.88×10^{15}	8.92×10^4	1.03	48.93	0.02
#6	Onshore wind, 50 MW [15]	2.64×10^{18}	1.08×10^{16}	1.54×10^{19}	4.49×10^4	1.17	5.84	0.20
#7	Geothermal, 20 MW [3]	3.36×10^{19}	4.61×10^{18}	1.00×10^{19}	1.47×10^5	4.81	0.44	11.05
#8	Hydro, 85 MW [3]	1.69×10^{19}	4.45×10^{18}	3.21×10^{18}	6.23×10^4	7.65	0.45	16.90
	Conventional Power Generation							
#9	Oil-fired, 1280 MW [3]	3.12×10^{20}	3.32×10^{21}	1.13×10^{21}	2.00×10^5	4.21	14.24	0.30
#10	Coal-fired, 1280 MW [3]	3.68×10^{20}	3.05×10^{21}	7.63×10^{20}	1.71×10^5	5.48	10.37	0.53

Transformity (*Tr*) is an important indicator to measure the overall efficiency of the production system at the biosphere scale. Power plants with greater transformities demand more environmental and economic resources to meet the same electricity demand. Large-scale SUT power plants are excellent in terms of emergy conversion, except for very small generation capacities. The transformity of onshore wind energy and SUT, 200 MW is found to be the lowest among all conventional and renewable generation technologies presented in this research (Figure 4). Further, it is the closest in value to the thermodynamic minimum transformity for electricity production cycles [3]. This is an outcome of the simplicity of this technology as its materials are very conveniently based on environmentally sound production from renewable or recyclable materials.

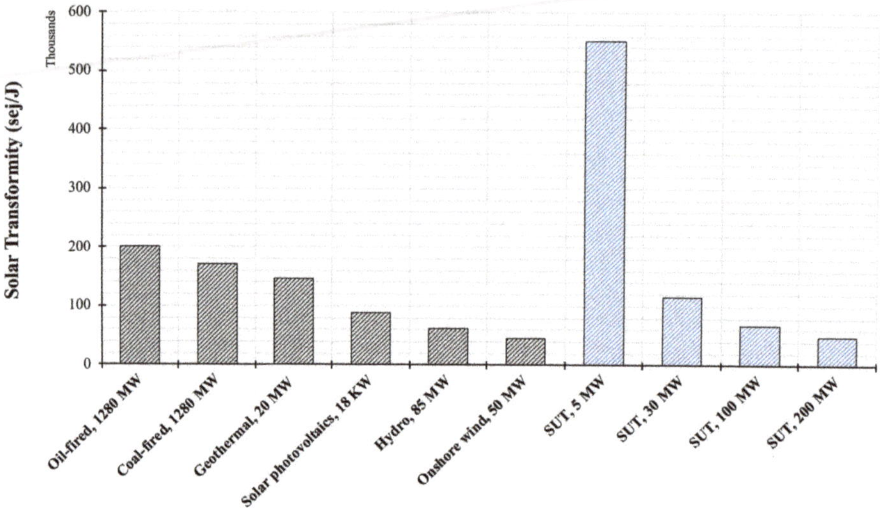

Figure 4. Emergy transformity of various power generation technologies.

The emergy yield ratio (*EYR*) of SUT, 200 MW (1.1) is higher than solar photovoltaics. However, the value is much smaller than those of the other power generation plants. In general, *EYR* evaluates the protentional contribution of the power plant to the economy, and the smaller *EYR* of SUTs indicates a relatively high amount of emergy resources invested by the economy.

The environmental loading ratio (*ELR*) measures the environmental stress of the power plant. The *ELR* of SUT, 200 MW (12.27) is small compared to solar photovoltaics and oil-fired power generation plants. However, the latter value is larger than those of the other renewable power plants. The larger environmental stress is the negative effect of the enormous land area (nonrenewable resources) demanded by the SUT power plant compared to other renewable technologies.

The sustainability of power generation plants is interpreted by the emergy Index of Sustainability (*EIS*), which is the ratio of *EYR* to *ELR*. It should be considered that an *EIS* less than one indicates an unsustainable process from the viewpoint of emergy-based indices [3]. The *EIS* of large-scale SUT (0.09) is higher than the solar photovoltaics. However, the latter value is lower than other power plants and lower than one. This is because of the high value of *ELR*.

4.2. Emergetic Ternary Diagram

The graphical analysis assists the interpretation of emergy accounting results and the discussion. Therefore, emergetic ternary diagrams were drawn to facilitate the comparison between SUT and various energy technologies. In Figure 5, points #1 to #10 represent SUT and various power generation technologies which are presented in this paper (Table 5). The position of any point is plotted based on the percentages of the three emergy resources inflows *R*, *N*, and *F*. To clarify, in Figure 5a, point #10 (coal-fired, 1280 MW) is composed of 9% of *R*, 73% of *N*, and 18% of *F*. The resource flow lines are parallel to the triangle side and indicate a constant value for each resource flow (Figure 5b). They also identify emergy indicators *EYR* (yellow lines) and *ELR* (green lines). The resource flow lines are very useful for comparing the use of resources by each power plant [19,20]. By way of illustration, it can be observed that geothermal (#7) and coal-fired (#10) had almost the same quantity of purchased resources (*F*), but the quantities of renewable and nonrenewable renewable resources (*R* and *N*) are extremely different. Moreover, it can be observed that SUT, solar photovoltaics, and wind-powered plants demanded higher quantities of purchased resources (*F*) than other power generation plants presented in this research.

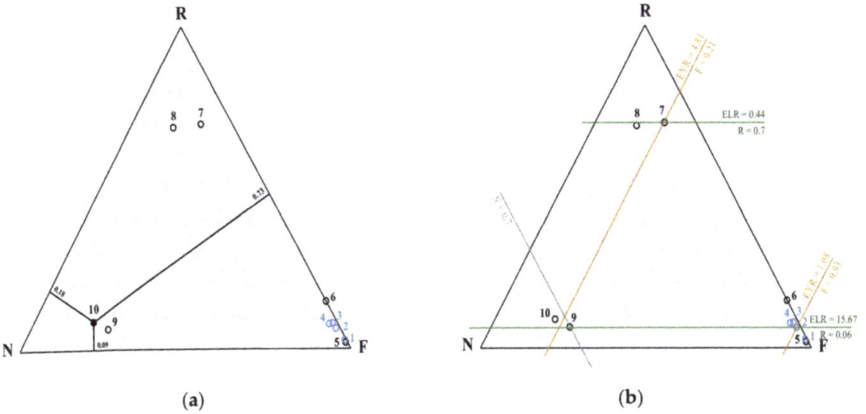

(a) (b)

Figure 5. Emergetic ternary diagrams representing SUT and various power generation systems shown in Table 5. (**a**) represents the composition of point #10 (coal-fired, 1280 MW) and (**b**) the use of resource flow lines (*EYR*: emergy yield ratio; *ELR*: environmental loading ratio).

The emergy index of sustainability *EIS* is presented in emergetic ternary diagrams through the sustainability lines (orange lines) which indicate constant values of *EIS* (Figure 6a). The sustainability lines depart from the *N* apex in the direction of the *RF* side. According to M. T. Brown and S. Ulgiati [3], power plants with an *EIS* higher than one denote sustainable power generation in a long term. Consequently, the sustainability line where *EIS* = 1 divides the triangle into sustainability areas. The presentation of the sensitivity lines (Figure 6b) is very significant for identifying critical elements that may be changed to improve the sustainability of power plant systems. The sensitivity lines are straight lines that depart from an apex towards the opposite side. They can be drawn for any point (system) inside the triangle and trace the impact of changes in a given resource flux (*R*, *N*, or *F*), while the other two fluxes remain constant. By way of illustration, it can be observed that the sustainability of SUT, 200 MW (#4) can be improved by changing the quantities of purchased resources. In the same way, an increase in the renewable services of a coal-fired (#10) plant may enhance its sustainability. As such, the sensitivity lines are a powerful tool to draw the sustainability pathways for decision-makers [19,20].

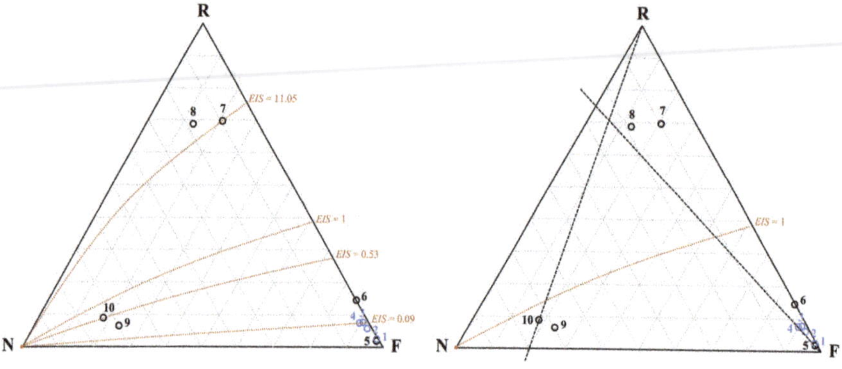

Figure 6. Representation of the auxiliary lines of emergetic ternary diagrams: (**a**) the use of sustainability lines, and (**b**) the use of sensitivity lines for SUT and various power generation systems shown in Table 5 (*EIS*: emergy index of sustainability).

4.3. Scenario Analysis

To further investigate the above results and explore future optimization directions of the SUT power plant, scenario analysis was performed. Based on the emergy index of sustainability, the sustainability of the SUT power generation system is less than coal-fired and oil-fired power generation systems. Further, its value is less than one which means unsustainable generation in the long term. Among the other capacities of SUT, large-scale SUT, 200 MW performs the best with regard to emergy-based indicators. It is worth studying SUT, 200 MW in terms of *EIS* change.

Derived from the sensitivity lines of emergetic ternary diagrams, the sustainability of SUT is strongly relative to purchased resources. The scenario analysis focus on dominant purchased resources includes the materials of the tower, collector, and turbines (manufacturing and construction materials). Additionally, the transportation distance of materials to the plant's location (1000 km in this study) was accounted for. The latter resources are taken as variables and the relationship between *EIS* is shown in Figure 7. The *EIS* of the SUT, 200 MW is mostly affected by the emergy of collector materials, followed by tower materials. On the other side, the turbines' materials and transportation distance have weak impacts. Therefore, the development direction of large-scale SUT power generation systems should mainly focus on reducing materials demanded by the manufacturing and construction of solar collectors.

The service life of a commercial SUT power plant ranges from 80 to 120 years [27]. Figure 8a shows the *ESI* of SUT, 200 MW changes with the length of lifetime (30 years in this study). It can be seen that the length of the SUT lifecycle has a slight impact on its sustainability.

As a final point, the emergy indicators are relative measures and depend on the characteristics of the fractions of emergy inputs. To compare different systems by the emergy index of sustainability, one has to follow the perspective of the evaluation [3]. To clarify further, the renewable fraction of human labor and services changes with the economic level of different regions and time periods, which directly affects *EIS* [3,14]. In fact, numerical variations of emergy indicators can be found in the literature for the same power generation technology. In the present analysis, the renewable fraction of human labor and services is considered as 26% [36]. In Figure 8b, the change of the *EIS* with the percentage of the renewable fraction of human labor and services is observed. Notably, the change in renewable fraction has a relatively high impact on *EIS*. Therefore, in the emergy evaluation, the region and economic level should be unified to facilitate a successful comparison of different power generation systems.

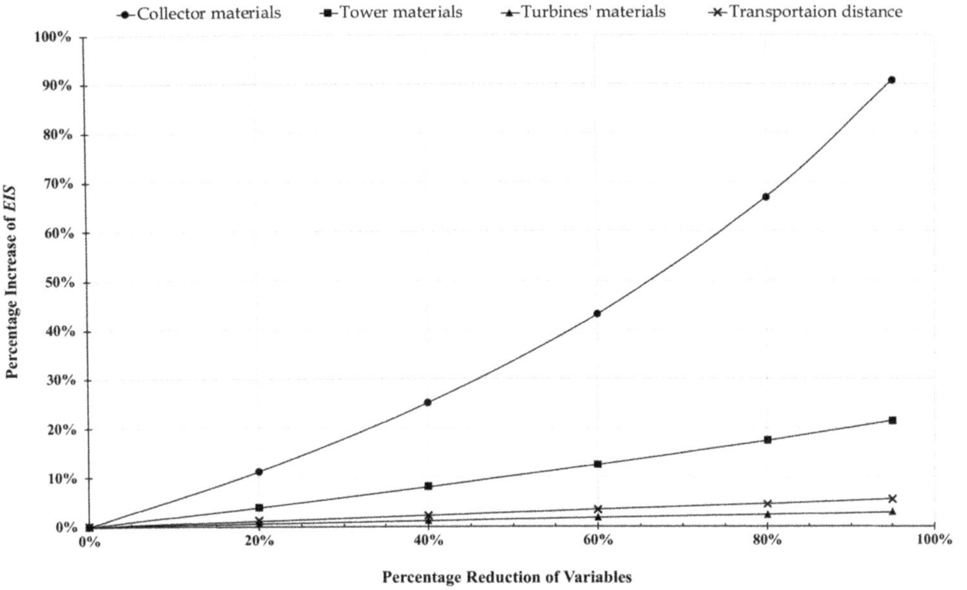

Figure 7. Percentage increase in *EIS* of SUT, 200 MW with the reduction of the tower, collector, and turbines' materials, as well as transportation distance of these materials to the plant's location.

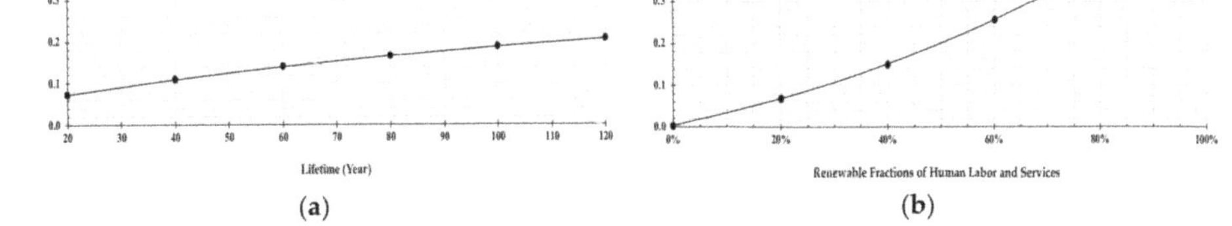

Figure 8. Changes in *EIS* of SUT, 200 MW with: (**a**) lifetime length of the power plant and (**b**) renewable fractions of human labor and services.

5. Conclusions

Clean energy infrastructure is essential economically, politically, environmentally, and socially for a greener tomorrow. To accelerate and strengthen the greening of the energy industry, the sustainable evaluation mechanisms of energy systems should be empowered. This paper addresses "emergy" as a powerful instrument to facilitate efficient energy management for policymakers as they choose alternatives concerning the environment. The framework of emergy analysis for the power plant system involves ecological and economic emergy transformation chains to determine the total emergy input that drives the system. Thereupon, all energy systems are evaluated on an ecocentrism basis. Solar updraft tower (SUT) is one of the green technologies for acquiring renewable energy. SUT produces energy from the hot air, utilizing a combination of a flat plate solar collector, a central tower, and air turbines. Emergy accounting and emergy-based indicators are estimated to investigate the relative performance and sustainability of four power-generation capacities of SUT power plants.

Energy sources, materials, land loss, and human labor and services were converted to emergy flows. The total emergy of the SUT system accounts for all environmental contributions, energies, and human services used directly and indirectly to produce and maintain the power plant. The SUT, 100 and 200 MW performed the best with regard to emergy-based indicators. The results demonstrate that the sustainability of SUT grew as the size of the plant increased. For SUT, 200 MW, the manufacturing materials of the components are the dominant inputs, constituting 34.9% of total emergy. This was followed by the construction works (materials and energy consumption), constituting 29.0% of total emergy. The latter results are the output of the massive structure and materials demanded by a large-scale SUT power plant. Large-scale SUT power plants (100 and 200 MW) are excellent in terms of resources-use efficiency (emergy conversion). The emergy transformity of onshore wind energy and SUT, 200 MW was found to be the lowest among all conventional and renewable generation technologies presented in this research. This is the outcome of the simplicity of this technology as its materials are very conveniently based on environmentally sound production from renewable or recyclable materials.

Emergetic ternary diagrams were drawn to facilitate the comparison between SUT and various energy technologies. The sensitivity lines of emergetic ternary diagrams are very significant in identifying critical elements that may be changed to improve the sustainability of power plant systems. Derived from sensitivity lines, the sustainability of SUT is strongly relative to purchased resources. Scenario analysis was performed to explore future optimization directions of the SUT power plant. The scenario analysis focus on dominant purchased resources included the materials of the tower, collector, and turbines (manufacturing and construction materials). The transportation distance of materials to the plant's location was also taken into account. The *EIS* of the SUT, 200 MW was mostly affected by the emergy of collector materials, followed by that of tower materials. On the other side, the turbines' materials and transportation distance had weak impacts. Therefore, the development direction of large-scale SUT power generation systems should mainly focus on reducing materials demanded by the manufacturing and construction of solar collectors. Moreover, it can be seen from the analysis that the length of the SUT lifecycle had a slight impact on its sustainability. Notably, the change in the percentages of the renewable fraction of human labor and services had a relatively high impact on *EIS*. The renewable fraction changes with the economic level of different regions and time periods. Therefore, in the emergy evaluation, the region and economic level should be unified to facilitate a successful comparison of different power generation systems.

Author Contributions: I.E.: conceptualization, methodology; analysis, investigation, visualization, writing—original draft preparation. Y.N.: methodology, writing—review and editing, supervision, funding acquisition. All authors have read and agreed to the published version of the manuscript.

Funding: This work was supported by Grant-in-Aid for Scientific Research (B) (No. 18H01636) from the Japan Society for the Promotion of Science.

Acknowledgments: The first author, I.E., is indebted to the Japanese Ministry of Education, Culture, Sports, Science and Technology (MEXT) for financial support. Islam is also grateful to Camila Gonzalez Lopez for her support to refine the manuscript.

Conflicts of Interest: The authors declare no conflicts of interest.

References

1. World Commission on Environment and Development. *Our Common Future*; UN Documents, Oxford; Oxford University Press: New York, NY, USA, 1987.
2. Burton, I. Report on Reports: Our Common Future. *Environ. Sci. Policy Sustain. Dev.* **1987**, *29*, 25–29. [CrossRef]
3. Brown, M.T.; Ulgiati, S. Emergy evaluations and environmental loading of electricity production systems. *J. Clean. Prod.* **2002**, *10*, 321–334. [CrossRef]
4. The World Economic Forum. Available online: https://www.weforum.org (accessed on 30 July 2020).
5. Weber, C. Risky being green. *Nat. Energy* **2019**, *4*, 906–907. [CrossRef]
6. Odum, H.T. Self-Organization, Transformity, and Information. *Science* **1988**, *242*, 1132–1139. [CrossRef]
7. Odum, H.T. *Environmental Accounting: Emergy and Environmental Decision Making*; John Wiley: New York, NY, USA, 1996.
8. Wang, C.; Zhang, S.; Yan, W.; Wang, R.; Liu, J.; Wang, Y. Evaluating renewable natural resources flow and net primary productivity with a GIS-Emergy approach: A case study of Hokkaido, Japan. *Sci. Rep.* **2016**, *6*, 37552. [CrossRef]
9. Tilley, D.R.; Howard, T. Odum's contribution to the laws of energy. *Ecol. Model.* **2004**, *178*, 121–125. [CrossRef]
10. Brown, M.; McClanahan, T.R. EMergy analysis perspectives of Thailand and Mekong River dam proposals. *Ecol. Model.* **1996**, *91*, 105–130. [CrossRef]
11. Cui, B.; Hu, B.; Zhai, H. Employing three ratio indices for ecological effect assessment of Manwan Dam construction in the Lancang River, China. *River Res. Appl.* **2011**, *27*, 1000–1022. [CrossRef]
12. Buonocore, E.; Vanoli, L.; Carotenuto, A.; Ulgiati, S. Integrating life cycle assessment and emergy synthesis for the evaluation of a dry steam geothermal power plant in Italy. *Energy* **2015**, *86*, 476–487. [CrossRef]
13. Pang, M.; Zhang, L.; Liang, S.; Liu, G.; Wang, C.; Hao, Y.; Wang, Y.; Xu, M. Trade-off between carbon reduction benefits and ecological costs of biomass-based power plants with carbon capture and storage (CCS) in China. *J. Clean. Prod.* **2017**, *144*, 279–286. [CrossRef]
14. Yang, Q.; Chen, G.; Liao, S.; Zhao, Y.; Peng, H.; Chen, H. Environmental sustainability of wind power: An emergy analysis of a Chinese wind farm. *Renew. Sustain. Energy Rev.* **2013**, *25*, 229–239. [CrossRef]
15. Yang, J.; Chen, B. Emergy-based sustainability evaluation of wind power generation systems. *Appl. Energy* **2016**, *177*, 239–246. [CrossRef]
16. Zhang, X.; Xu, L.; Chen, Y.; Liu, T. Emergy-based ecological footprint analysis of a wind farm in China. *Ecol. Indic.* **2020**, *111*, 106018. [CrossRef]
17. Paoli, C.; Vassallo, P.; Fabiano, M. Solar power: An approach to transformity evaluation. *Ecol. Eng.* **2008**, *34*, 191–206. [CrossRef]
18. Zhang, M.; Wang, Z.; Xu, C.; Jiang, H. Embodied energy and emergy analyses of a concentrating solar power (CSP) system. *Energy Policy* **2012**, *42*, 232–238. [CrossRef]
19. Giannetti, B.F.; Barrella, F.; Almeida, C. A combined tool for environmental scientists and decision makers: Ternary diagrams and emergy accounting. *J. Clean. Prod.* **2006**, *14*, 201–210. [CrossRef]
20. Almeida, C.; Barrella, F.; Giannetti, B.F. Emergetic ternary diagrams: Five examples for application in environmental accounting for decision-making. *J. Clean. Prod.* **2007**, *15*, 63–74. [CrossRef]
21. Haaf, W.; Friedrich, K.; Mayr, G.; Schlaich, J. Solar Chimneys Part I: Principle and Construction of the Pilot Plant in Manzanares. *Int. J. Sol. Energy* **1983**, *2*, 3–20. [CrossRef]
22. Schlaich, J.; Bergermann, R.; Schiel, W.; Weinrebe, G. Sustainable Electricity Generation with Solar Updraft Towers. *Struct. Eng. Int.* **2004**, *14*, 225–229. [CrossRef]
23. Schlaich, J.; Bergermann, R.; Schiel, W.; Weinrebe, G. Design of Commercial Solar Tower Systems: Utilization of Solar Induced Convective Flows for Power Generation. *Sol. Energy* **2003**, *127*, 573–581. [CrossRef]
24. Mullett, L.B. The solar chimney—overall efficiency, design and performance. *Int. J. Ambient. Energy* **1987**, *8*, 35–40. [CrossRef]
25. Ming, T.; Liu, W.; Wu, Y.; Gui, J.; Peng, K.; Pan, T. *Introduction*; Elsevier BV: Amsterdam, The Netherlands, 2016; pp. 1–46.
26. Fluri, T.; Pretorius, J.; Van Dyk, C.; Von Backström, T.; Kröger, D.; Van Zijl, G. Cost analysis of solar chimney power plants. *Sol. Energy* **2009**, *83*, 246–256. [CrossRef]
27. Okoye, C.O.; Solyalı, O.; Taylan, O. A new economic feasibility approach for solar chimney power plant design. *Energy Convers. Manag.* **2016**, *126*, 1013–1027. [CrossRef]

28. Guo, P.; Zhai, Y.; Xu, X.; Li, J. Assessment of levelized cost of electricity for a 10-MW solar chimney power plant in Yinchuan China. *Energy Convers. Manag.* **2017**, *152*, 176–185. [CrossRef]
29. Van Blommestein, B.; Mbohwa, C. Life Cycle Assessment of Solar Chimneys. In *Re-engineering Manufacturing for Sustainability*; Springer Science and Business Media LLC: Berlin/Heidelberg, Germany, 2013; pp. 535–541.
30. Elsayed, I.; Nishi, Y. A Feasibility Study on Power Generation from Solar Thermal Wind Tower: Inclusive Impact Assessment Concerning Environmental and Economic Costs. *Energies* **2018**, *11*, 3181. [CrossRef]
31. EnviroMission Limited. Available online: http://www.enviromission.com.au (accessed on 30 July 2020).
32. Shimizu Corporation. Available online: https://www.shimz.co.jp (accessed on 30 July 2020).
33. Yang, J.; Chang, Y.; Zhang, L.; Hao, Y.; Yan, Q.; Wang, C. The life-cycle energy and environmental emissions of a typical offshore wind farm in China. *J. Clean. Prod.* **2018**, *180*, 316–324. [CrossRef]
34. Nian, V.; Liu, Y.; Zhong, S. Life cycle cost-benefit analysis of offshore wind energy under the climatic conditions in Southeast Asia—Setting the bottom-line for deployment. *Appl. Energy* **2019**, *233–234*, 1003–1014. [CrossRef]
35. Hau, J.L.; Bakshi, B.R. Promise and problems of emergy analysis. *Ecol. Model.* **2004**, *178*, 215–225. [CrossRef]
36. Ren, S.; Feng, X.; Yang, M. Emergy evaluation of power generation systems. *Energy Convers. Manag.* **2020**, *211*, 112749. [CrossRef]

sustainability

MDPI

Article

Evaluation of the Operating Efficiency of a Hybrid Wind–Hydro Powerplant

Francisco Briongos [1], Carlos A. Platero [1], José A. Sánchez-Fernández [2,*] and Christophe Nicolet [3]

1 Department of Automatic Control, Electrical and Electronic Engineering and Industrial Informatics, Universidad Politécnica de Madrid, E-28040 Madrid, Spain; francisco.briongos@gmail.com (F.B.); carlosantonio.platero@upm.es (C.A.P.)

2 Department of Hydraulic, Energy and Environmental Engineering, Universidad Politécnica de Madrid, E-28040 Madrid, Spain

3 Power Vision Engineering Sàrl, CH-1024 Ecublens, Switzerland; christophe.nicolet@powervision-eng.ch

* Correspondence: joseangel.sanchez@upm.es; Tel.: +34-91-067-4332

Received: 17 December 2019; Accepted: 14 January 2020; Published: 16 January 2020

Abstract: This paper analyzes the operating efficiency of a hybrid wind–hydro power plant located in El Hierro Island. This plant combines a wind farm (11.48 MW) and a pumped storage power plant (11.32 MW). It was built with the aim of supplying the island demand from renewable energy instead of using existing diesel units. The paper discusses several operational strategies and proposes an efficiency metric. Using 10 min data, the operation of this power plant has been simulated. From these simulations (more than 50,000 for a year), the operating efficiency and the percentage of demand covered from renewable energy is obtained. The difference between the worst and the best strategy is a twofold increase in efficiency. Moreover, the results of the simulations are compared with the system operational history since June 2015 (when the wind–hydro power plant started operation) until 2018. These comparisons show a reasonable agreement between simulations and operational history.

Keywords: wind farm; pumped storage; isolated systems; power plant efficiency

1. Introduction

Since the beginning of electrification, fossil fuel burning has been the main source of primary energy on which electricity generation is based [1]. However, there is growing evidence that fossil fuel burning for electricity generation is one of the main causes of global warming because of greenhouse gas emissions [2]. For this reason, renewable energy resources have had an important development. Among these renewable energy resources, wind energy has the largest installed capacity in the world [3]. Notwithstanding, wind speed variations and its prediction difficulties create some problems in power system operation. Therefore, when wind penetration increases, power system operators have to take measures to maintain system integrity [4] or consider complementary renewable energy sources [5].

Small islands are of particular interest because, on them, fossil fuel costs are higher [6]. Therefore, renewable energies can be cost competitive. In fact, many islands are installing some renewable energy supplies [7]. To this aim, the support of the local community is an important contribution to success, as the case of Samsø has proven [8]. However, island power systems usually lack the support of a strong interconnected system. So, maintaining system integrity is a harder problem on islands [9]. Notwithstanding, there are several examples of island power systems that have largely increased their share of renewables. To this aim, one successful approach is combining several renewable energy sources [10]. Another option is adding an energy storage system [11]. Furthermore, there are islands that combine both possibilities, such as the Caribbean island of St. Eustatius [12]. In addition, the possible contribution of electric vehicles smart charging to solar and wind integration has been proposed in The Barbados [13].

One of those islands is El Hierro, an island located in the Canary archipelago (Spain). UNESCO declared it a biosphere reserve in 2000. As such, it aims to become entirely free from carbon dioxide emissions [14]. To fulfill this aim, a combined wind–hydro power plant was built. This plant started operation in 2014 but was not fully operational until 2015 [15].

According to the Merriam-Webster dictionary, efficiency is "the ratio of the useful energy delivered by a dynamic system to the energy supplied to it" [16]. Therefore, each energy conversion technology has its own efficiency. However, the evaluation of energy efficiency in power systems has been based on economic metrics [17]. In [18], Maheshwari and Ramakumar described a procedure to design an optimal size microgrid that supplies energy to a rural area, but do not analyze nor define its operating efficiency. In [19], several procedures to calculate the optimum number, size and location of photovoltaic power plants is applied to two different test networks. The optimality condition tries to minimize power line losses and harmonic distortion. So, a technological metric for evaluating power systems efficiency is needed. This paper intends to do so, using an isolated power system that has a wind–hydro hybrid power plant.

This paper is organized as follows: Section 2 describes the El Hierro power system. Section 3 analyzes the operational strategies that were considered. Section 4 presents the simulation model of the power system. Section 5 presents the results of the simulations done and discuss the results comparing them with the operating experience. Finally, Section 6 summarizes the conclusions.

2. Power System Description

The power system considered for the case study is located on El Hierro Island. This Island is placed in the Canary Islands archipelago, Spain. Its electric power supply was based on diesel engine driven generators, which is a common solution in small islands. There are 10 diesel engines, with rated power from 775 kW to 1870 kW, installed in several phases. The generators are connected to the main busbar through transformers. Electric energy is supplied to the island through four 20 kV power lines [20]. Figure 1 shows a simplified one-line diagram of the El Hierro power system.

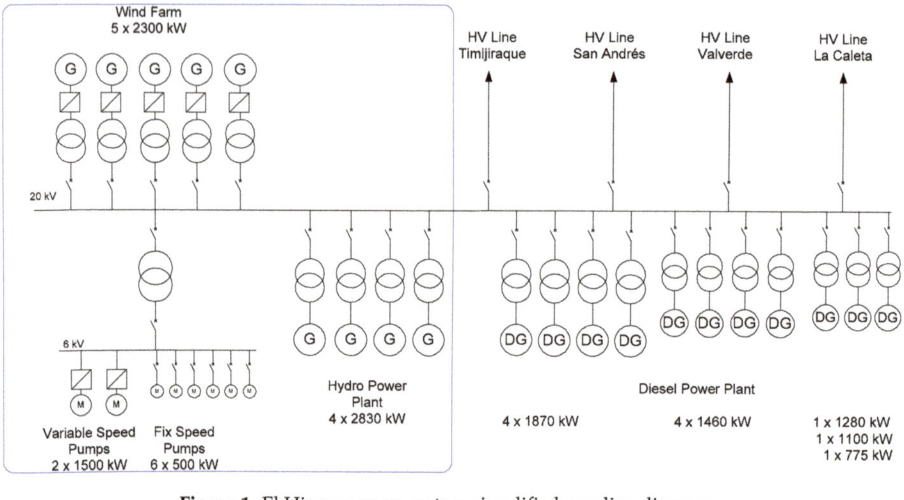

Figure 1. El Hierro power system simplified one-line diagram.

El Hierro Island's orography makes it possible to build an upper and a lower reservoir at a reasonable cost and with a gross height difference larger than 700 m. On the other hand, this island has high wind during long periods. After several studies discussing using fresh or salt water for the hydro power plant, in 2015, a wind farm combined with a pumped storage power plant started commercial operation (using fresh water).

The wind farm comprises five Enercon E-70 wind turbines connected to the 20 kV grid, through separate transformers. The pumped storage power plant has two separate facilities, a hydropower plant and a pumping station. This arrangement is known as 4-machine-type [21]. The Appendix A provides some tables with detailed data on this wind–hydro power plant.

The hydro power plant comprises four Pelton turbines (Table A1), driving synchronous generators (Table A2) with a rated power of 2830 kW, connected to the grid by four transformers (Table A3). The pumping station has eight centrifugal multistage pumps driven by induction motors. There are two variable speed pumps of 1500 kW (Table A6) and six fixed speed pumps of 500 kW (Table A9) [20].

3. Operating efficiency and operational strategies

The efficiency of a system is the ratio of the energy obtained and the energy used. In the case of a power plant is the ratio of energy generated and primary energy used.

The traditional definition of efficiency in case of windmills is the ratio of the wind power converted into electric power and the available wind power. The average efficiency of a windmill is between 32%–43%.

3.1. *Operating Efficiency*

However, in case of the operating efficiency, the definition is slightly different. It is the ratio between the generated energy and the maximum theoretical electric energy available using the best operational strategy.

In the case of a wind farm, the operating efficiency is the relationship of the produced energy ($E_{Wind\ Produced}$) and the maximum possible energy that can be generated in particular wind conditions ($E_{Wind\infty}$). The maximum possible energy corresponds to the operation of the windmills at maximum power mode; so, the pitch should be regulated to get the maximum energy at any wind speed condition and connected to an infinite power grid.

$$\eta = \frac{E_{Wind\ Produced}}{E_{Wind\ \infty}} \tag{1}$$

In order to get a 100% operating efficiency, the produced energy ($E_{Wind\ Produced}$) should be the maximum possible energy ($E_{Wind\infty}$). This is only possible to reach in a large power system, where the power system is able to absorb all the produced energy. In the case of a hybrid wind–hydro power plant installed in a small power system, depending on consumers demand and wind speed, the hydro power plant should store in some periods and generate in others. Therefore, the maximum possible energy cannot always be put on line. Both process, energy storage (pumping mode) and the posterior hydraulic energy generation (turbine mode), have losses. The operating efficiency is defined as the direct wind energy ($E_{Wind\ Grid}$) put on line plus the hydro energy divided by the maximum energy that the windfarm would have produced in a large power system at maximum power mode ($E_{Wind\infty}$).

$$\eta = \frac{E_{Wind\ Grid} + E_{Hydro}}{E_{Wind\ \infty}} \tag{2}$$

The operational strategy of the power system has a large influence on its efficiency. In order to analyze this efficiency, all possible strategies should be considered. They have to take into account the non-dispatchable nature of wind energy. Therefore, these strategies are defined by how they manage wind power generation. In what follows, these operational strategies are discussed.

3.2. *Isolated Wind*

One of the possible operational strategies of this power system consists of splitting the 20 kV bus bar in two. One bar connects the wind farm and the pumping station. The other bar connects the remaining power system.

All the electric energy produced by the wind farm feeds the pumps through the first bus. Through the other bus, the conventional hydro power plant—in combination with the Diesel power plant—feeds the consumers. This is a well-known procedure; the operation is very safe, as the consumers are not exposed to wind power fluctuations. Another important advantage is that the frequency regulation in the windfarm and pumps 20 kV busbar is not critical, as the pumps have no stringent frequency requirements.

The biggest drawback of this operational strategy is the efficiency. This operation mode has a low efficiency, as all renewable energy has been previously pumped and turbined. From the power system point of view, this case is equivalent to a 0% wind power penetration. As wind generators are not electrically connected to the power system, there is no limitation in the wind power. For the above reasons, the operating efficiency of this strategy is around 40%.

3.3. Restricted Wind Power

As wind energy is not fully controllable, the insertion of this type of energy in a small autonomous power system is not an easy task. Close attention should be paid to frequency regulation and wind gusts. A safe strategy is limiting the power injected by the wind farm. In this way, it is possible to get experience of the hybrid wind–hydro power plant operation and, afterwards, slowly increase this limit. When this limit is applied, the wind turbines operate at constant power and, consequently, some wind energy is not harvested.

This was the first strategy followed by El Hierro wind–hydro power plant when its operation started. Figure 2 shows wind generation limited to 4.4 MW.

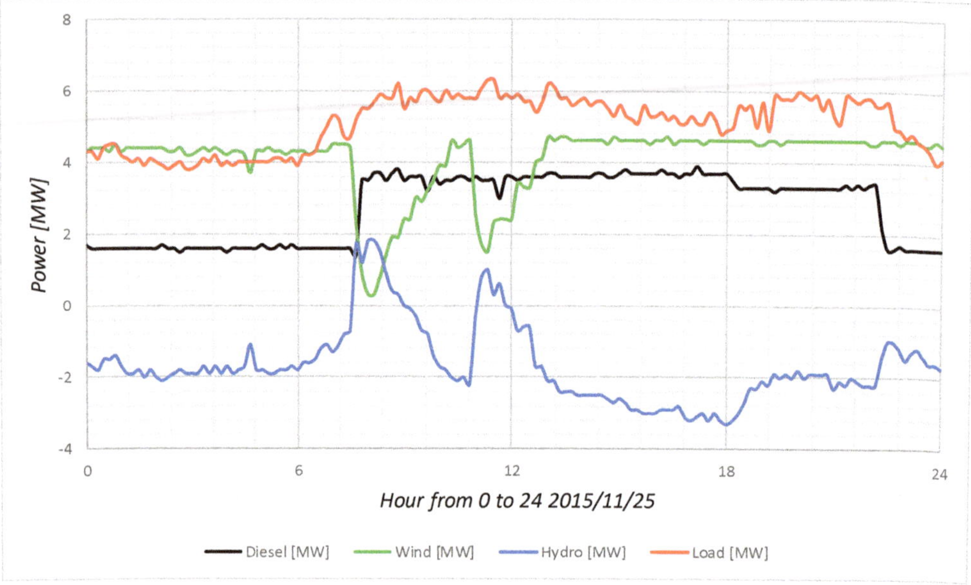

Figure 2. El Hierro power system daily generation and demand 25/11/2015. Source REE data.

3.4. Restricted Wind Penetration

One other safety-minded strategy is limiting the percentage of system demand covered directly by the wind farm. In this way, not all the energy demanded comes from the wind farm. In case of a sudden wind change or a wind generator trip, there are some backup generators in the power system than can supply the missed energy. However, using this strategy, sometimes there is wind energy that is failed to be harvested.

This was the second strategy used by El Hierro wind–hydro power plant. Figure 3 shows wind penetration limited to 70% and wind power limited to 7.2 MW.

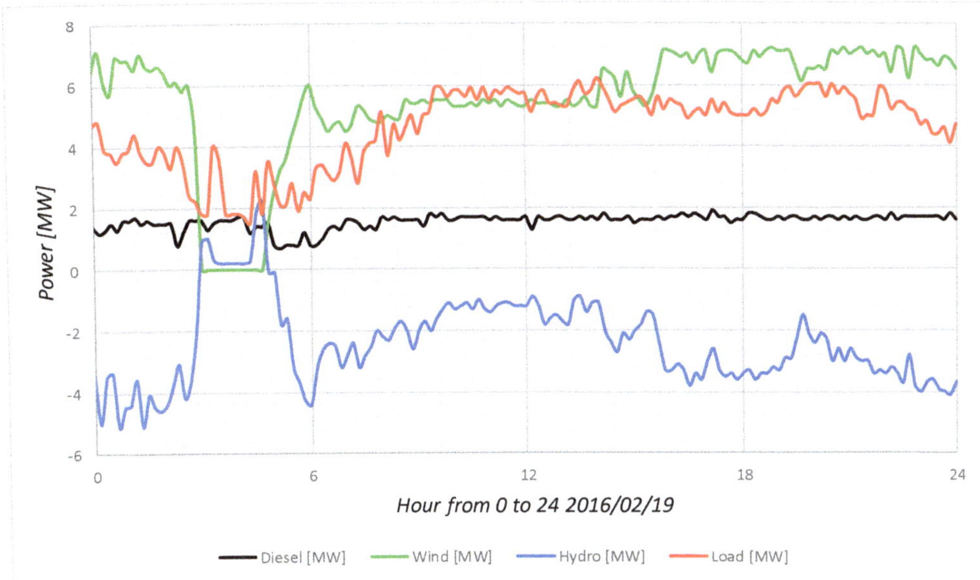

Figure 3. El Hierro power system daily generation and demand 19/02/2016. Source REE data.

3.5. Non-Restricted Wind

On this strategy, wind generators operate as if they were connected to an infinite bus. Therefore, they extract maximum power from the wind, with no limits apart from their own safety. This is only possible if the power system can absorb it. To this aim, the remaining parts of the power plant should adapt their point of operation accordingly. This means increasing or decreasing, in a fast way, the turbines generated, or the pumps demanded, power.

From an efficiency point of view, this is the best strategy. However, integrating all this wind power in El Hierro is a complex problem. The power system should be able to absorb the energy from the wind farm. In case of a wind generator trip, the remaining generators should modify fast enough their generation to maintain frequency. Additionally, in case of wind generation larger than the consumers load, the pumping station should consume the exceeding power. In this way, the frequency could be maintained.

Therefore, there are two different problems in this strategy. The first one is increasing or decreasing, in a fast way, the non-wind generated power in cases of wind generation decreasing below the load [22,23]. The second is increasing or decreasing, also in a fast way, the pumping power demanded [24].

As can be seen in Figure 4, this operational strategy is used nowadays in El Hierro wind–hydro power plant.

4. Simulation Model

The power system has been modelled in MS Excel using Visual Basic for Applications (VBA). The models use data measured every 10 min (wind speed and power system load).

4.1. General Power System Model

In order to analyze how different operational strategies affect the power system, a simulation model has been developed. This model takes into account system restrictions and uses actual wind and load data to obtain the percentage of load covered through renewable energy.

To this aim, all machines of the hybrid wind–hydro powerplant have been modeled. So, the performance of each machine for a given power level can be obtained. Thus, knowing this power for each machine, the overall performance of the system can be calculated.

Each machine will operate at a power level determined by the load share assigned to each generation technology. As the preferred energy source is the wind farm, determining its power output is the first step. Therefore, considering the power available to the wind farm, the hydraulic storage, the diesel power plant and the consumers load, the contribution of each technology can be determined every ten minutes. Adding each of these contributions over a year, the performance of the hybrid wind–hydro power plant and the percentage of renewable energy used in this period can be determined.

When making the power dispatch, first wind farm-available power is compared with load demand. Then, hydraulic storage is checked. Finally, if the load cannot be supplied by wind or hydro, then diesel generators are assigned. As wind and load data are available every ten minutes, this process allows us to evaluate the power assigned to each technology and the variation in hydraulic stored energy every ten minutes.

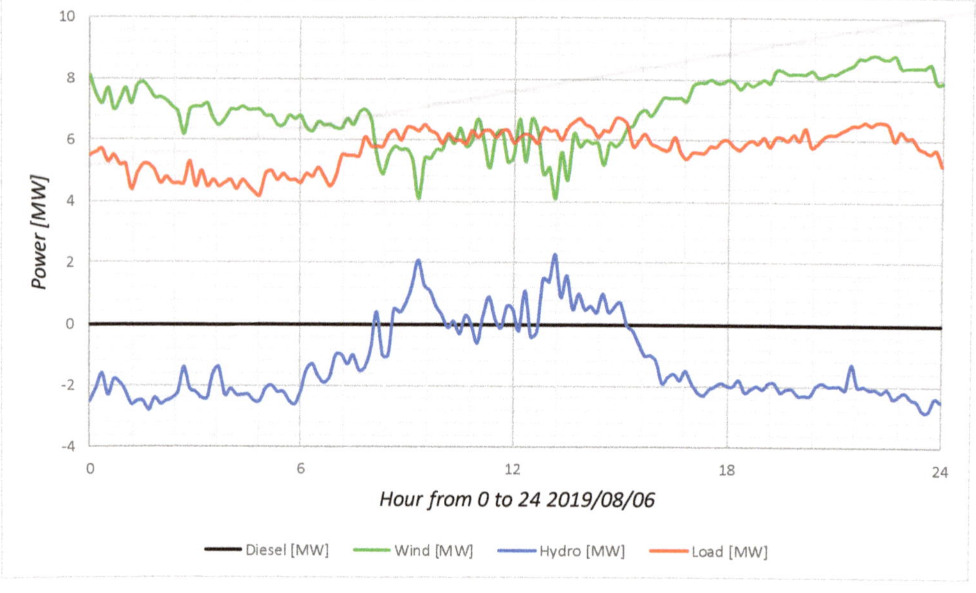

Figure 4. El Hierro power system daily generation and demand 06/08/2019. Source REE data.

Power Assigning Rules

Every 10-min period, the power is assigned according to the following steps:

1) First, curtailments to wind power are checked and, if active, only the allowed power is considered.
2) Then, wind penetration limit is checked (percentage of system demand covered by the wind farm). According to this limit, wind power sent directly to the grid is established.
3) After these checks, if there is a surplus of wind power, pumping conditions will be evaluated and in case of missing power, turbine conditions will be evaluated.

3a) If wind power is larger than load demand, pumping conditions are evaluated:

- Is there enough water in the lower reservoir?
- Is the surplus of wind power lower than the maximum power of the installed pumps?

If there is not enough water in the lower reservoir, pumping will not be carried out and the power supplied by the wind farm will be limited. On the other hand, if the wind farm power surplus is higher than maximum power of the pumps, wind power will be limited to the sum of consumer demand plus the maximum power of the pumps.

3b) If the wind power sent to the grid is not enough to cover consumers demand turbine conditions are evaluated. When the water level of the upper reservoir is not high enough, the turbines will not be activated. So, the diesel power plant will be activated to cover the demand. Additionally, if the wind power plus the turbines power were not enough to cover consumers demand, the diesel plant would also be activated to completely cover the demand.

4.2. Wind Farm Model

For modeling the wind farm in MS Excel, the manufacturer's curve that relates wind speed to the power generated by the wind turbine has been taken as a reference.

Since the known parameter is the wind speed every 10 min, using the manufacturer's curve (shown in Figure 5), the power delivered by each wind turbine is calculated. To make these calculations simpler, a polinomic aproximation of this curve (for wind speeds between 0 and 16 m/s) is used. For wind speeds beteeen 16 and 25 m/s the result is rated power and for wind speeds greater than 25 m/s, zero power is output. These aproximations are programmed in Microsoft Visual Basic for Applications (VBA)

Assuming that all wind turbines in the park receive the same wind speed, the total power produced by the wind farm is obtained multiplying the power generated by one turbine by the number of wind turbines installed in the farm.

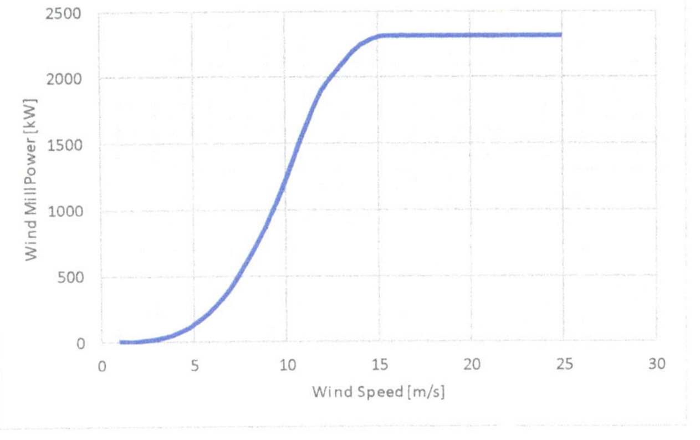

Figure 5. Wind turbine Enercon E-70 power curve.

4.3. Hydropower Plant Model

A similar aproach has been used for making both the turbining and pumping model. The starting data are the power needed and the storage state. From them and the performance of each machine involved, the power generated/consumed by the pumped storage power plant every 10 min can be obtained.

4.3.1. Pelton Turbines Model

For turbining, the performance of a generic Pelton turbine with the same maximum efficiency of the turbines installed in the plant is used (shown in Figure 6). The number of turbines in operation is determined using the criterion of maximum combined performance. Using this criterion and applying the previously mentioned performance curve, a curve for the plant—shown in Figure 7—is obtained.

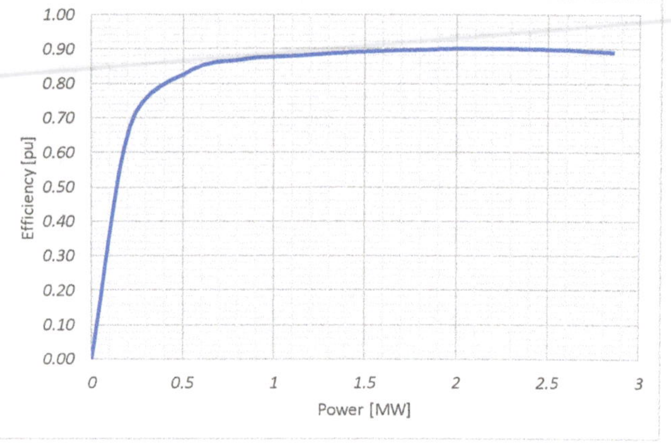

Figure 6. Pelton turbine efficiency curve.

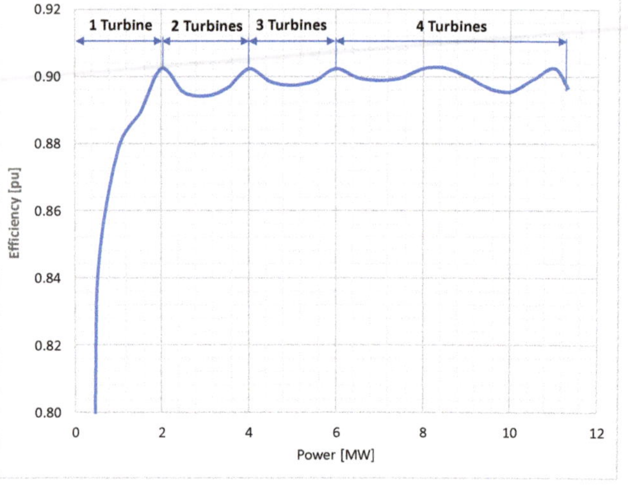

Figure 7. Power plant turbines efficiency curve.

4.3.2. Electric Generators Model

For modeling the generators, generic performance Equations (3)–(5) have been used. The parameters of this Equation are taken from the generators installed in the Gorona plant datasheet (shown in Table A2).

$$P_{Out} = P_{In} - \left[P_{Fix} + P_{Var}\left(\frac{P_{Out}}{FdP\ S_N}\right)^2\right] \tag{3}$$

$$P_{Fix} = P_{Fe} + P_{f\,w} \tag{4}$$

$$P_{Var} = P_{Cu1} + P_{Cu2} + P_{Add} \tag{5}$$

where:

P_{Out}	Output power.
P_{In}	Input power.
P_{Fix}	Synchronous machine fix losses.
P_{Var}	Synchronous machine variable losses.
P_{Cu1}	Stator windings losses.
P_{Cu2}	Rotor windings losses.
P_{Fe}	Iron core losses.
P_{Add}	Additional losses.
P_{fw}	Friction and windage losses.
FdP	Power factor.
S_N	Rated power.

Since each generator is in the same shaft of its turbine, the generators' activating sequence is the same one as of the turbines. Therefore, the overall performance of the generators depend on the turbines operating point.

4.3.3. Main Transformers Model

The last component of the turbining process is the main transformer that connects each generator to the bars of the electrical substation. To calculate the performance of this machine, the well-known Equation (6) is used. Its parameters are in Table A3.

$$P_{Out} = P_{In} - \left[P_0 + P_{Sc}\left(\frac{P_{Out}}{FdP\ S_N}\right)^2\right] \tag{6}$$

where:

P_{Out}	Output power.
P_{In}	Input power.
P_0	No-load losses.
P_{Sc}	Load losses.
FdP	Power factor
S_N	Rated power

The activation sequence of the transformers is the same as the turbines and generators.

4.3.4. Power Plant Model

Finally, the overall performance of turbining, depending on the power required in the substation bars, can be calculated by applying sequentially the performance of all machines involved.

With this overall performance, it is also possible to calculate the water volume from the upper reservoir turbined and transferred to the lower one. In this way, the evolution of the hydraulic storage for the next 10 min can be calculated.

Therefore, in order to implement the whole model of the power plant, a relation between the water volume turbined and the power demanded is applied.

4.4. Pumping Station

4.4.1. Pumps Model

In the Gorona station, there are two types of pumps: fixed speed 500 kW-rated power pumps and variable speed 1500 kW-rated power pumps.

For fixed pumps, as they always work at rated speed and power, their performance is the specified in the manufacturer's data sheet. While for variable speed pumps, taking the operating points from the diagrams provided by the manufacturer, their performance, for each operating point, could be obtained. In this way, taking values from the pump manufacturer's curve, a performance curve for the whole pump set can be calculated (Figure 8).

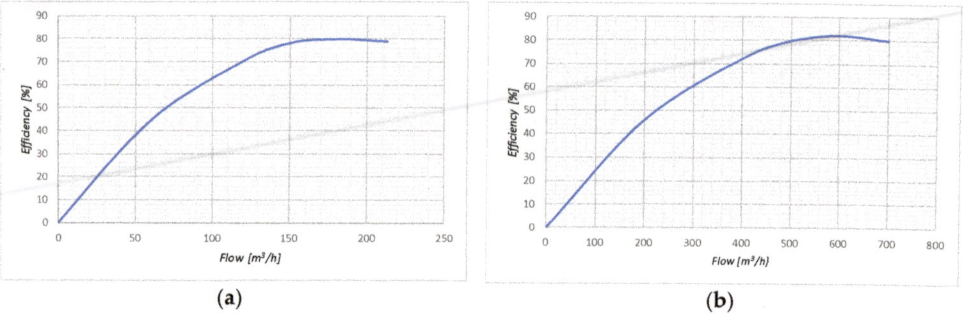

Figure 8. Pumps efficiency curves. (**a**) Fix speed pumps (**b**) Variable speed pumps.

As in the case of turbining, pump operation is only possible if there is enough water stored in the lower reservoir. If this is the case, the activation sequence of the pumps has to take into account that fixed pumps can only be off (zero power consumption) or on (rated power consumption), while variable speed ones can consume any level of power between a minimum and a maximum.

Therefore, if there are enough water for pump operation, the sequence of activation will start with a variable speed pump up to its rated power is reached. Then, a fixed speed pump is started while the variable one reduces its consumption. This process continues until all pumps are in operation at rated power, or the required power consumption is reached. Figure 9 shows the efficiency evolution of the pump station depending on the power required for frequency regulation [24].

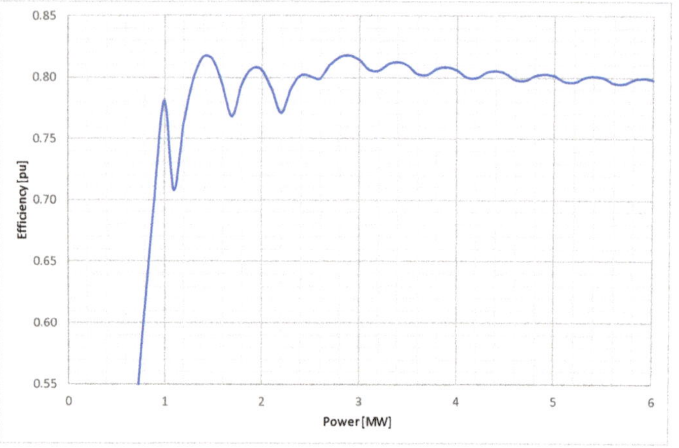

Figure 9. Pumping station efficiency curve.

4.4.2. Electric Motors Model

600 kW motors (Table A7) drive the fixed speed pumps and 1600 kW motors (Table A4) drive the variable speed ones. Both types of motors have a known performance for their rated power according to the manufacturer's specifications. As fixed speed motors will work at constant load, they will exhibit constant efficiency. Regarding variable speed motors, as they are driven by a converter (Table A5) that varies frequency and voltage they will work at constant efficiency too. Therefore, all motors will work at constant efficiency.

As was the case with turbines and generators, each motor is in the same shaft as its pump. Therefore, they have the same activation sequence as that of the pumps.

4.4.3. Main Transformer Model

The pump station transformer is modelled as the hydropower transformers. The transformer data are in Table A8.

4.4.4. Pumping Station Model

As in the turbining process, pump power implies that some water volume will be pumped from the lower reservoir to the upper reservoir for each 10 min period.

The performance of the different pumping machines is applied along the range of operating power to obtain the overall performance of the entire process. In the same way as in turbining, there is a curve that relates pumping power with hydraulic storage variation.

4.5. Diesel Power Plant Model

Due to the aim of this study, the operation of the hybrid wind–hydro power plant, the diesel power plant is taken as a generation source that helps in covering the electrical demand. This is so because the performance of the diesel power plant as a whole is not relevant to evaluate the performance of the hybrid wind–hydro power plant. Only the energy values of the wind farm and the hydroelectric power plant in a specified time period is analyzed. The energy delivered by the diesel power plant is only used to provide the power needed to cover the demand.

5. Results and Discussion

Using the aforementioned simulation models, the performance of the wind–hydro El Hierro power plant has been simulated. To this aim, wind data corresponding to years 2016 and 2017 have been used. As wind data is available every 10 min, and a year has $6 \times 24 \times 365 = 52{,}560$ 10-min periods, simulating an entire year requires running the models 52,560 times. In Figures 10 and 11, the percentage of load demand covered by renewable energy in 2016 and 2017, respectively, is shown. This percentage increases when there is no limit on the power that the wind farm can generate, and also increases when wind penetration is not limited. In both years, the maximum possible renewable load coverage is slightly over 70%.

In those years, the percentage of load demand actually covered by renewable energy has increased. In 2016, it was 40.62%, while it was 46.26% in 2017. Figure 3 shows that in 2016, wind penetration was limited to around 70% and the maximum wind power was around 7.2 MW. Applying this limit to Figure 10 the maximum percentage of demand covered by renewable energy is around 50%, which is in reasonable agreement with the actual value. It should be noted that some factors were not considered in the calculations as penstock losses or evaporation in the upper reservoir. In 2018, this percentage increases again up to 56.62%. Therefore, the power system operation is increasingly approaching theoretical performance.

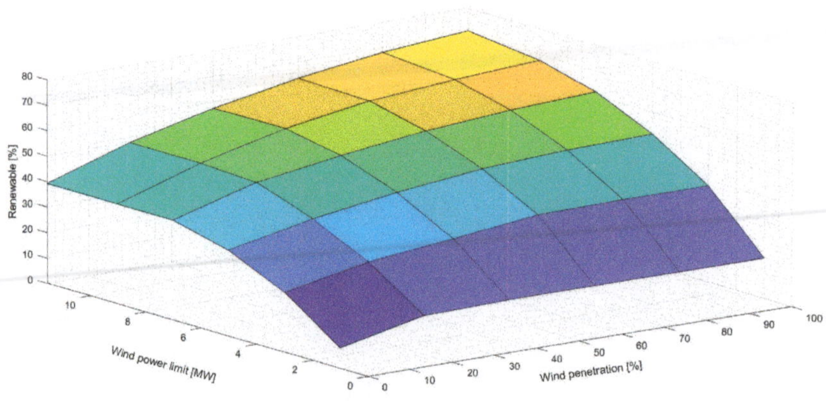

Figure 10. Annual demand covered by renewable energy for 2016.

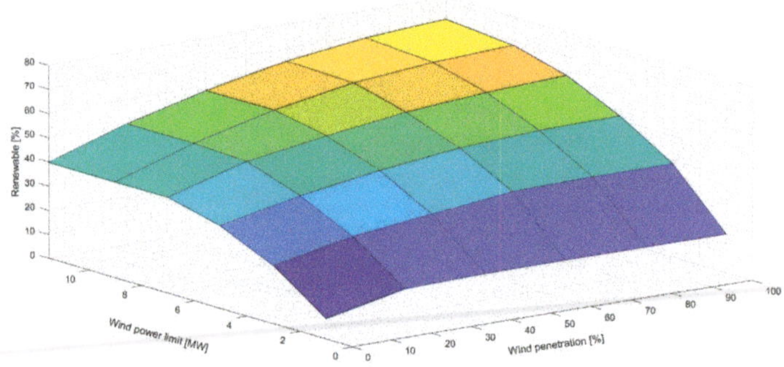

Figure 11. Annual demand covered by renewable energy for 2017.

Regarding the operating efficiency, its value for 2017 wind data is shown in Figure 12. As it was the case in load demand coverage, the efficiency increases when the wind power limit increases and, also, when wind penetration increases. Maximum operating efficiency is achieved when none of these limits apply. In that case, operating efficiency reaches 80%. In Section 3.2, a 40% operating efficiency using the isolated wind strategy was estimated. This means that an unrestricted wind operation achieves a twofold efficiency increase over an isolated wind operation.

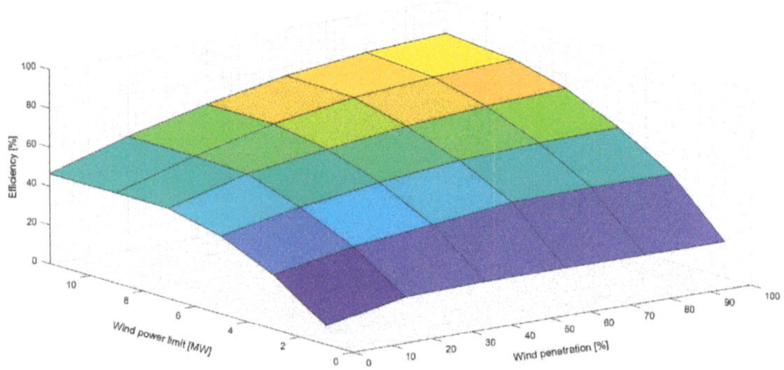

Figure 12. Operational efficiency for 2017.

As shown in Figure 4, nowadays this power system is close to achieving this operating efficiency.

6. Conclusions

This paper presents a model of an isolated power system located in El Hierro Island—a UNESCO biosphere reserve. This model is based on the analysis of energy flows with the aim of evaluating the operating efficiency of a renewable power plant. This power plant is composed of a wind farm and a pumped storage hydropower plant. For every component of the power system, a model that takes into account its energy performance has been presented.

In addition, a definition of operating efficiency adapted to this renewable power plant has been proposed. Operating efficiency is defined through the comparison of the net energy put on the power system by the hybrid power plant with the electrical energy that its wind farm will put in an infinite bus. Therefore, if the wind farm was connected to an infinite bus, its operating efficiency would be 100%. In a small isolated power system, such as El Hierro Island, a storage system is needed and, so, the operating efficiency is always less than 100%.

In order to analyze this operating efficiency, several power system operational strategies have been presented. Several tens of thousands of simulations have been conducted to evaluate the yearly operating efficiency of the renewable power plant depending on the strategy followed. In addition, the percentage of demand covered by the renewable power plant has been also evaluated.

The results of the simulations show that the operating efficiency of the wind–hydro power plant is heavily dependent on the power system operational strategy followed. This efficiency increases when lighter restrictions are imposed on the wind generation. In fact, a twofold increase in efficiency can be achieved between the best (80%) and worst (40%) strategies.

Consequently, the non-restricted wind operational strategy is recommended. Along the operational history of the plant, several of these strategies were used. As the operator's confidence increased, the wind generation restrictions have been reduced. Nowadays, non-restricted wind generation is mostly applied. In fact, there are some periods when the energy supply is drawn 100% from renewable sources.

The percentage of demand covered by this power plant along its operational history has been compared with simulation results showing a reasonable agreement. In addition, as wind generation restrictions have been reduced, the percentage of demand covered by this power plant has increased. The yearly demand coverage by renewable energy was 40.6% in 2016. It has increased to 46.26% in 2017 and to 56.6% in 2018. Nowadays, there are some periods of 100% demand covered by this renewable power plant.

Author Contributions: Conceptualization, C.A.P. and J.A.S.-F.; methodology, F.B.; software, F.B.; validation, C.A.P., J.A.S.-F. and C.N.; formal analysis, F.B.; investigation, C.A.P. and J.A.S.-F.; resources, C.N.; data curation, F.B.; writing—original draft preparation, C.A.P. and J.A.S.-F.; writing—review and editing, C.A.P., J.A.S.-F. and C.N.; visualization, F.B.; supervision, C.N. All authors have read and agreed to the published version of the manuscript.

Funding: This research was funded by Universidad Politécnica de Madrid, grant number RP1604330010.

Conflicts of Interest: The authors declare no conflict of interest

Appendix A

Table A1. Pelton turbines performance data.

Pelton Turbines	**4**	
Rated Power	2854	kW
Rated Flow	0.5	m^3/s
Gross Head	658	m
Net Head	651	m
Rated speed	1000	rpm
Number of jets per runner	1	
Type	Horizontal	
Efficiency	Figure 6	

Table A2. Hydro generator performance data.

Hydro Synchronous Generators	**4**	
Rated apparent power	3300	kVA
Rated Power Factor	0.8	
Rated voltage (±5.0%)	6	kV
Frequency	50	Hz
Rated speed	1000	rpm
Stator windings losses. P_{Cu1}	23.31	kW
Rotor windings losses. P_{Cu2}	10.88	kW
Additional losses. P_{Add}	2.78	kW
Friction and windage losses. P_{fw}	18.85	kW
Iron core losses. P_{Fe}	27.35	kW

Table A3. Hydro generator transformers performance data.

Hydro Generators Transformers	**4**	
Rated apparent power	3300	kVA
Secondary rated voltage	20	kV
Primary rated voltage	6	kV
Frequency	50	Hz
No-load losses	3.5	kW
Load losses	28	kW

Table A4. Variable speed drive motors performance data.

Induction Motor	**2**	
Rated power	1600	kW
Rated Voltage	690	kV
Speed range	0–2979	rpm
Frequency	50	Hz
Efficiency	96.3	%
Power factor	0.88	

Table A5. Variable speed driver performance data.

Frequency Converters	**2**	
Rated power	1750	kW
Rated voltage	690	V
Losses at 1500 kW	<2	%
Losses at 500 kW	<2	%
Input transformer efficiency	>99	%
Global efficiency	≥97	%

Table A6. Variable speed pumps performance data.

Variable Speed Pumps	**2**	
Head	690	m
Rated flow	690	m^3/h
Minimum flow	221	m^3/h
Rated speed	2830	rpm
Efficiency	Figure 8b	

Table A7. Fixed speed motors performance data.

Induction Motor	**6**	
Rated power	600	kW
Rated Voltage	6	kV
Rated speed	2981	rpm
Frequency	50	Hz
Efficiency at		
100%	96.6	%
75%	96.5	%
50%	95.8	%
Power factor at		
100%	0.92	
75%	0.92	
50%	0.89	

Table A8. Pumping station main transformer performance data.

Pumping Station Transformer	**1**	
Rated apparent power	7200	kVA
Primary rated voltage	20	kV
Secondary rated voltage	6.1	kV
Frequency	50	Hz
No-load losses	4.6	kW
Load losses	38	kW

Table A9. Fix speed pumps performance data.

Variable Speed Pumps	**2**	
Head	690	m
Rated flow	210	m^3/h
Rated speed	2965	rpm
Efficiency	Figure 8a	

References

1. Zou, C.; Zhao, Q.; Zhang, G.; Xiong, B. Energy revolution: From a fossil energy era to a new energy era. *Nat. Gas Ind. B* **2016**, *3*, 1–11. [CrossRef]
2. Hardisty, P.E.; Clark, T.S.; Hynes, R.G. Life Cycle Greenhouse Gas Emissions from Electricity Generation: A Comparative Analysis of Australian Energy Sources. *Energies* **2012**, *5*, 872–897. [CrossRef]
3. International Energy Agency Tracking Clean Energy Progress 2017. Available online: https://www.iea.org/publications/freepublications/publication/TrackingCleanEnergyProgress2017.pdf (accessed on 15 February 2019).
4. Kies, A.; Schyska, B.; von Bremen, L. Curtailment in a Highly Renewable Power System and Its Effect on Capacity Factors. *Energies* **2016**, *9*, 955. [CrossRef]
5. de Oliveira Costa Souza Rosa, C.; Alonso Costa, K.; Silva Christo, E.; Braga Bertahome, P. Complementarity of Hydro, Photovoltaic, and Wind Power in Rio de Janeiro State. *Sustainability* **2017**, *9*, 1130. [CrossRef]
6. Ioakimidis, C.S.; Genikomsakis, K.N. Integration of Seawater Pumped-Storage in the Energy System of the Island of São Miguel (Azores). *Sustainability* **2018**, *10*, 3438. [CrossRef]
7. Kuang, Y.; Zhang, Y.; Zhou, B.; Li, C.; Cao, Y.; Li, L.; Zeng, L. A review of renewable energy utilization in islands. *Renew. Sustain. Energy Rev.* **2016**, *59*, 504–513. [CrossRef]
8. Sperling, K. How does a pioneer community energy project succeed in practice? The case of the Samsø Renewable Energy Island. *Renew. Sustain. Energy Rev.* **2017**, *71*, 884–897. [CrossRef]

9. Kumar, S.R.; Gafaro, F.; Daka, A.; Raturi, A. Modelling and analysis of grid integration for high shares of solar PV in small isolated systems—A case of Kiribati. *Renew. Energy* **2017**, *108*, 589–597. [CrossRef]
10. Yoo, K.; Park, E.; Kim, H.; Ohm, J.Y.; Yang, T.; Kim, K.J.; Chang, H.J.; Pobil, A.P. Optimized Renewable and Sustainable Electricity Generation Systems for Ulleungdo Island in South Korea. *Sustainability* **2014**, *6*, 7883–7893. [CrossRef]
11. Ramírez Díaz, A.; Ramos-Real, F.J.; Marrero, G.A.; Perez, Y. Impact of Electric Vehicles as Distributed Energy Storage in Isolated Systems: The Case of Tenerife. *Sustainability* **2015**, *7*, 15152–15178. [CrossRef]
12. Rojas, E.G.; Sadri, H.; Krueger, W. Case study of MW-sized power generation at St. Eustatius island combining photovoltaics, battery storage, and gensets. *Prog. Photovolt.* **2019**. [CrossRef]
13. Taibi, E.; del Valle, C.F.; Howells, M. Strategies for solar and wind integration by leveragingflexibility fromelectric vehicles: The Barbados case study. *Energy* **2018**, *164*, 65–78. [CrossRef]
14. Iglesias, G.; Carballo, R. Wave resource in El Hierro—An island towards energy self-sufficiency. *Renew. Energy* **2011**, *36*, 689–698. [CrossRef]
15. Canary Government. *Anuario Energético de Canarias 2016*; Consejería de Economía, Industria, Comercio y Conocimiento: Santa Cruz de Tenerife, Spain, 2017.
16. Merriam Webster Dictionary. Available online: https://www.merriam-webster.com/dictionary/efficiency (accessed on 27 October 2019).
17. Wang, C.N.; Nguyen, T.D.; Yu, M.C. Energy Use Efficiency Past-to-Future Evaluation: An International Comparison. *Energies* **2019**, *12*, 3804. [CrossRef]
18. Maheshwari, Z.; Ramakumar, R. Smart Integrated Renewable Energy Systems (SIRES): A Novel Approach for Sustainable Development. *Energies* **2017**, *10*, 1145. [CrossRef]
19. Duong, M.Q.; Pham, T.D.; Nguyen, T.T.; Doan, A.T.; Tran, H.V. Determination of Optimal Location and Sizing of Solar Photovoltaic Distribution Generation Units in Radial Distribution Systems. *Energies* **2019**, *12*, 174. [CrossRef]
20. Merino, J.; Veganzones, C.; Sanchez, J.; Martinez, S.; Platero, C. Power System Stability of a Small Sized Isolated Network Supplied by a Combined Wind-Pumped Storage Generation System: A Case Study in the Canary Islands. *Energies* **2012**, *5*, 2351–2369. [CrossRef]
21. Martínez-Lucas, G.; Sarasúa, J.; Sánchez-Fernández, J. Frequency Regulation of a Hybrid Wind-Hydro Power Plant in an Isolated Power System. *Energies* **2018**, *11*, 239. [CrossRef]
22. Platero, C.A.; Nicolet, C.; Sánchez, J.A.; Kawkabani, B. Increasing wind power penetration in autonomous power systems through no-flow operation of Pelton turbines. *Renew. Energy* **2014**, *68*, 515–523. [CrossRef]
23. Platero, C.; Sánchez, J.; Nicolet, C.; Allenbach, P. Hydropower Plants Frequency Regulation Depending on Upper Reservoir Water Level. *Energies* **2019**, *12*, 1637. [CrossRef]
24. Sarasúa, J.; Martínez-Lucas, G.; Platero, C.; Sánchez-Fernández, J. Dual Frequency Regulation in Pumping Mode in a Wind-Hydro Isolated System. *Energies* **2018**, *11*, 2865. [CrossRef]

sustainability

MDPI

Article

Feasibility and Cost Analysis of Photovoltaic-Biomass Hybrid Energy System in Off-Grid Areas of Bangladesh

Nusrat Chowdhury [1], Chowdhury Akram Hossain [2], Michela Longo [3] and Wahiba Yaïci [4,*]

1 Department of Electrical and Electronic Engineering, Daffodil International University, Dhaka 1207, Bangladesh; nusrat.eee@diu.edu.bd
2 Department of Electrical and Electronic Engineering, American International University-Bangladesh, Dhaka 1229, Bangladesh; chowdhury.akram@aiub.edu
3 Department of Energy, Politecnico di Milano, 34–20156 Milano, Italy; michela.longo@polimi.it
4 CanmetENERGY Research Centre, Natural Resources Canada, Ottawa, ON K1A 1M1, Canada
* Correspondence: wahiba.yaici@canada.ca; Tel.: +1-613-996-3734

Received: 27 January 2020; Accepted: 17 February 2020; Published: 19 February 2020

Abstract: In this progressing technological advancement world, hybrid systems for power generation is one of the most promising fields for any researcher. In this context, photovoltaic-biomass hybrid systems with off-grid applications have become extremely popular with both Governments and individual users in rural areas of any part of the world. This system has gained popularity because of low cost, sustainability and very effective outcome with the use of natural resources at the rural areas. In this paper a proposed hybrid system which contains photovoltaics (PV) and biomass along with an additional storage has been considered to find the different aspects from an end user point of view. It also discusses the feasibility of the proposed model for an off-grid power system located in the remote areas of Ashuganj, Bangladesh. In order to analyse the pollutant emissions and calculate the cost parameters of the proposed system, RETScreen simulation software was deployed. This research also carries out a brief financial analysis considering the annual income of the end user and the payback periods for the installed system. It endeavours to provide complete information about different parameters which also includes the environmental impacts involved in establishing the proposed system. The conventional system in the pilot area is a kerosene-based system, hence in this research, a comparison between the proposed and the conventional system has been analysed using simulated results. The simple payback of the project was estimated to be 6.9 years and this model will be able to reduce the CO_2 emissions by approximately 3.81 tonnes per year. The results have significantly supported the proposed system to be more reliable, environmentally-friendly and less costly than the conventional kerosene-based system.

Keywords: photovoltaics (PV); biomass; battery storage; off-grid electrification; feasibility analysis; cost analysis; simple payback period; CO_2 emissions; renewable energy

1. Introduction

Electricity is the major source of energy in most urban systems worldwide. Since the first high-voltage Alternating Current (AC) coal power station was commissioned in London in 1890, electrification of residential and industrial installations has grown exponentially expanding toe 83% of all urban areas by 2010 [1]. Even with a projection of increased electricity use for future energy systems the electrification of rural areas still represents a relevant issue. Such expanded use will likely seriously affect different sectors of developing economies, ranging from industrial to transportation uses [2]. In particular, the augmenting of current electricity distribution grids in rural areas, which are located far from the main national grid, may result in excessive costs in terms of installation, transmission, distribution, and maintenance [3,4]. Indeed, the electrification of rural areas through the extension of grid connections may raise the overall generation costs, which could reach up to seven times the normal price obtainable in urban areas [4]. Some studies have been focussed on operational performance of the Bangladesh rural electrification program and its determinants with a focus on political interference [5–8]. Per [5] for instance, complete electrification through Bangladesh will take many years, and thus the diminishing returns to scale of incremental investment for further rural electrification will be faced in the long run. The authors suggested that both Bangladesh and international donors revisit the original principle of the Rural Electrification Program, eradicate political hindrances from the program, and make sustained efforts to develop more efficient infrastructure of delivering electricity to the rural poor and encouraging local economic development.

Another issue of concern is the unreliability in frequency, blackouts, power losses, and fluctuations of the grid voltage. In such cases, the proper use of Renewable Energy Sources (RESs) in remote areas could represent a viable and economical alternative to the extension of electricity grids [9–11].

Authoritative studies have shown that hybrid stand-alone electricity-generating systems are more economically feasible for off-grid consumers located in distant areas [12,13]. In addition, RES installations can also reduce the amount of CO_2 emissions emanating from electrical energy generation. Studies in Bangladesh have demonstrated, in fact, that 1kWh of electricity generated by solar photovoltaic (PV) systems can reduce the amount of CO_2 emissions by approximately 660 tonnes per year [14]. However, the use of stand-alone PV systems in off-grid applications also presents certain drawbacks, mainly related to the intrinsic intermittency and stochasticity of the solar source. To overcome these drawbacks, the use of electrical Energy Storage System (ESS) solutions is usually implemented [15,16], along with the adoption of hybrid configurations, i.e., by adding traditional controllable power generators (such as diesel-electric motors), to support the intermittency and unreliability of PV systems.

Currently, in off-grid PV systems, the use of large ESS solutions is usually considered economically unviable due to the high investment costs, that is, high costs of storage devices [17,18], whereas the combined adoption of PV with ESS and diesel-electric generators has been widely adopted [19]. Nevertheless, increasing concern about global warming and environmental pollution is propelling the replacement of generators that were traditionally powered by fossil fuels, while there is more interest in adopting greener solutions. To this end, the use of biomass generators could indeed represent an interesting alternative, due both to its low carbon impact and its lower investment and developmental costs.

As indicated in the literature, numerous feasibility and techno-economic studies were performed on micro-grid projects in different countries, specifically on stand-alone hybrid energy systems for applications in remote areas [20–45]. The inference drawn is that it appears that no feasibility study or other similar work has been conducted on such a system in the remote communities of Bangladesh; thus, the present study is an original research initiative and a firm contribution to knowledge.

Considering the environmental and cost concerns described earlier, the purpose of this paper is a feasibility study of the potentials and likely impact of a hybrid PV-biomass system as a possible option for the provision of power in a rural area of Bangladesh. The primary objective of the study is to present a preliminary design of a hybrid PV-biomass system that is capable of satisfying the energy needs of the selected off-grid application, and to compare its economic performance in respect of existing solutions, by making use of discounted cash flow and payback period analyses. The paper examines the technical, economic and environmental feasibility, both the integration and sizing of a hybrid PV-biomass system, and the energy storage of microgrid for remote electrification within Ashuganaj, Bangladesh. In addition, the impact of related CO_2 emissions is also analysed, and compared to traditional solutions. The study is essentially performed employing the simulation tool referred to as RETScreen [46].

The paper is organised as follows: Section 2 gives a full description of the proposed project location as well as the simulation methods; also provides the load profiles and climate data related to the project location and Section 3 presents details on the proposed PV-biomass hybrid system. Simulation results are discussed in Section 4, including the sizing of the system, the cost and emissions analyses, while conclusions are drawn in Section 5.

2. Project Location and Simulation Methods

2.1. Simulation Method

Many studies have analysed renewable energy systems using RETScreen [47–51]. RETScreen software was employed in this design to achieve the least energy costs that people will find affordable [52]. The RET Screen Clean Energy Management Software (usually shortened to RET Screen) is a clean energy software package developed by Ministry of Natural Resources Canada (Government of Canada) for evaluating both financial and environmental costs and benefits of different renewable energy technologies for any location in the world. This software uses visual basic and C language as working platform. RETScreen PV model also covers off-grid PV applications and include stand-alone, hybrid and water pumping systems also. It has a global climate data database of more than 6000 ground stations (month wise solar irradiation and temperature data for the year), energy resource maps (i.e., wind maps), hydrology data, product data like solar photovoltaic panel details and wind turbine power curves. It also provides link to NASA climate database. It enables comprehensive identification, assessment and optimization of the technical and financial viability of potential renewable energy and energy-efficient projects. It also allows measurement and verification of the actual performance of facilities and identification of energy savings/production opportunities. The software tool can determine the technical and financial viability of renewable energy, energy efficiency and cogeneration projects. The area number of worksheets for performing detailed project analysis includes Energy Modelling, Cost Analysis, Emissions Analysis, Financial Analysis and Sensitivity and Risk Analyses sheets. The analysis of different types of energy efficient and renewable technologies (RETS) covers mainly energy production, life-cycle costs and greenhouse gas emission reduction. In summary, RETScreen Plus is a Windows-based energy management software tool to study the energy performance [51,52].

This simulation process empowers experts and decision-makers to identify and evaluate the technical and financial viability of potential clean energy projects and also to measure and verify the actual and ongoing energy performance of energy-efficiency projects. The process also enables the evaluation of energy production, life-cycle costs and greenhouse gas emission reductions for the proposed Hybrid system [46]. The flowchart used in the simulation tool is shown in Figure 1.

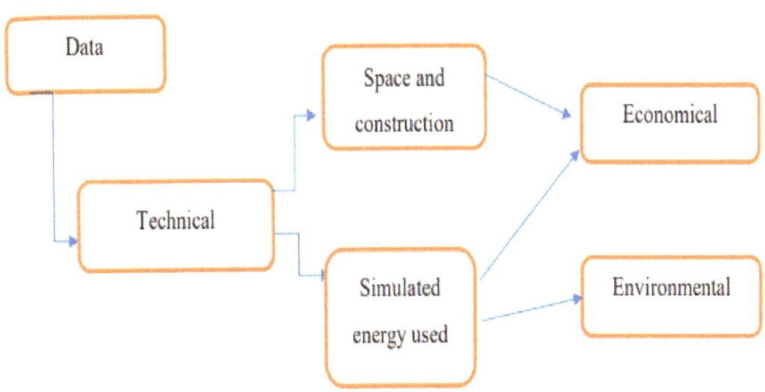

Figure 1. RETScreen technical evaluation structure flowchart [53].

Ashuganj city located in the Brahmanbaria District of Chittagong Division in Bangladesh, was selected as the reference site for the application and evaluation of the proposed hybrid PV-biomass system. The city was selected because of its residents' pervasive dependence on fossil fuels to meet their energy needs, due to the absence of a power grid extension. Residents situated close to the city depend mainly on horticulture and animal husbandry for their livelihood. Therefore, in rural areas, it is economical to use hybrid systems comprising solar and biomass, provided the biomass supply is consistently available. In most of the areas, manure, crop wastes and cooking wastes are accessible free of charge.

2.2. Load Profile and Climate Data of Project Location

The load demand of the proposed system is estimated by reference to a household load (4 LED light, 2 DC fan and 1 DC TV) for a typical lower middle-class family. The study evaluated fifty houses for provision of electricity for the proposed hybrid system (Table 1). The actual daily average energy demand for fifty houses is calculated as 45.6 kWh.

Table 1. Load profile.

Daily Load Profile						
Appliance	**Type**	**Unit**	**Watt**	**Hour of Operation (h)**	**Total Power (W)**	**Total Energy Demand (Wh)**
LED	DC	4	6	8	24	192
Fan	DC	2	20	14	20	560
TV	DC	1	40	4	40	160
	Daily Demand for each house				84	912
	Daily Demand for 50 houses				4.2 kW	45.6 kWh

The load characteristics are calculated on the basis of three seasons: summer (March–June), spring (July–October) and winter (November–February). The RETScreen software calculates the monthly and yearly energy consumptions for the load based on the following indices: appliance used, season, weather condition based on location, number of people per house and number of total houses. Considering the system loss, the daily average energy demand is estimated as 47.2 kWh which is higher than the actual demand and annual energy demand is 14.161 MWh for both the base case and the proposed case. The detailed data of monthly and yearly energy consumption is shown in Table 2. In winter season the load demand is comparatively lower than the other two seasons. Therefore, it is considered 54%. In spring season considering the weather condition, the percentage of energy use is 93%. Due to the extreme hot weather in summer the percentage of energy is considered 99%.

As mentioned earlier, this paper discusses the feasibility study of the proposed hybrid system with RETScreen. The financial analysis has been carried out for systematic progression towards a PV-Biomass-based hybrid system. We need to determine the latitude and longitude of the study area and analysis is done to derive the climate data. The RETScreen software gives complete weather details based on climate location. Table 3 indicates the climate data location and the project location which has been collected from NASA by RETScreen. The daily solar radiation, air temperature, humidity and earth temperature data are also collected for the reference location to check the feasibility of solar project implementation. The detailed data are shown in Table 4.

Table 2. Load characteristics of base and proposed cases.

Load Characteristics		
Electricity-DC	**Base Case**	**Proposed Case**
Daily	47.2 kWh	47.2 kWh
Annual	14.161 MWh	14.161 MWh
Peak Load-Annual	4.7 kW	
Percentage of Month Used		
Month	**Base Case**	**Proposed Case**
January	54%	54%
February	54%	54%
March	93%	93%
April	93%	93%
May	93%	93%
June	93%	93%
July	99%	99%
August	99%	99%
September	99%	99%
October	99%	99%
November	54%	54%
December	54%	54%

Table 3. Ashuganj site reference data.

Parameter	Unit	Climate Data Location	Project Location
Latitude	°N	24.1	24.1
Longitude	°E	91.9	91.9
Elevation	m	14	14
Heating design temperature	°C	13	
Cooling design temperature	°C	30.9	
Earth temperature amplitude	°C	13.5	

Table 4. Site reference conditions.

Month	Ambient Air Temperature °C	Relative Humidity %	Daily Solar Radiation $kWh/m^2/d$	Earth Temperature °C
January	20.4	54.7	4.42	21.6
February	22.7	55.3	4.98	23.9
March	25.2	61.7	5.44	27
April	26.3	73.1	5.51	28.1
May	27.1	79.1	5.11	28.8
June	27.5	84.7	4.16	28.4
July	27.3	85.9	4.04	27.9
August	27.1	85.5	4.18	27.8
September	26.7	84.1	4.02	27.6
October	26	77.9	4.28	26.8
November	23.8	69.4	4.25	24.4
December	21.3	60.1	4.28	22.1
Annual	25.1	72.7	4.55	26.2

3. The Proposed PV-Biomass Hybrid System

The proposed system consists of electric DC loads, solar PV, biomass generator, battery, and converter. Figure 2 demonstrates the block diagram of the proposed system. The system is fed by PV arrays and a biomass generator. There is no grid connection between the systems. Biomass is an abundant source of energy around the world, which is composed of organic matter including agricultural residues, and wood, animal and human wastes. Use of biomass for the purpose of power generation has become very popular, especially since it is an easily obtainable source of energy in the rural parts of Bangladesh. Additionally, it is a cleaner source of energy than fossils throughout the world. Its relative abundance makes it a viable option for use as a potential source of energy for electricity-generation in the country where it comprises animal manure that can either be converted through the absorption process or its residues extracted through the combustion process. In this system hybrid solar and biomass system was chosen as biomass is accessible effectively in the form of manure throughout the year. This hybrid system using biomass and solar with a battery as a storage system for electricity generation is more economical, because it can generate electricity during cloudy days also.

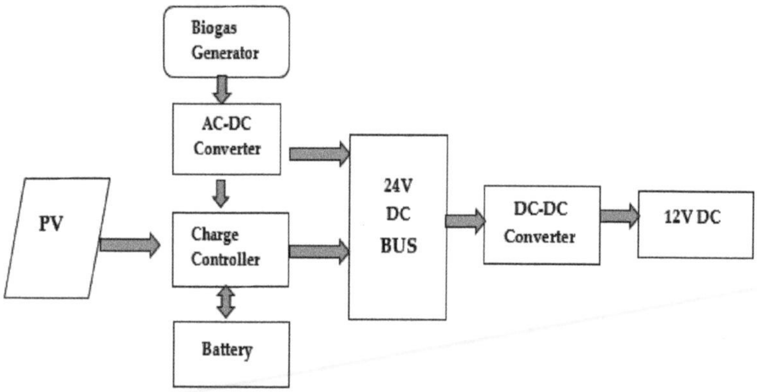

Figure 2. Block diagram of proposed photovoltaics (PV)-biomass hybrid energy system.

In the proposed system, the AC generator is used. Here, the output of the generator is converted into 24 V DC using the AC to DC converter and then connected to 24 V DC bus bar. Finally, DC to DC converter is used in every house to convert 24 V DC to 12 V DC to operate the DC house load.

The total capital cost of 1 kW biogas fuel-based generator considered as BDT (Bangladeshi Taka) 60,000/kW (USD $714.29/kW) and the lifetime of the generator is specified in hours of operation. The lifetime of the generator is considered as 15,000 hr. The efficiency of the generator is considered as 80%. We estimate the cost per tonne of biogas at BDT 70 (USD $0.833) based on biomass resources being obtainable almost free of charge.

4. Results and Discussion

RETScreen software has been used to analyse the different parameters of the proposed case and finally, the proposed case was compared with the base case. After the comparison, based on the financial viability, annual savings and evaluation of GHG emissions it can be easily deduced that the proposed case is more beneficial than the existing one. Analysis types and methods selected in RETScreen are mini-grid and method two respectively.

4.1. Base Case Power Study

Having chosen an off-grid area, the kerosene lamp was evaluated as a power source in the base case. Estimated total power capacity for base case is 4.70 kW. We estimated the cost of a litre of kerosene to be BDT 65 (USD $0.77) and so the total cost of electricity is calculated as BDT 474,702 (USD $5651.21) for the existing kerosene-based system by RETScreen. Table 5 indicates the unit cost and total electricity cost for the base case.

Table 5. Base case power system.

Parameter	Value
Grid Type	Off-grid
Fuel type	Kerosene-L
Fuel rate	65 BDT/L
Capacity	4.7 kW
Heat rate	8 kJ/kWh
Annual O& M cost	BDT 474,500
Electricity rate-base case	33.512 BDT/kWh
Total electricity cost	**BDT 474,702**

From the simulation, it was evident that the unit cost of electricity for the existing system is very high and also harmful to the environment. A new system has been proposed for this reason.

4.2. Proposed Case Power Study

In the proposed case analysis, mono-crystalline silicon PV solar cell with a power capacity of 12.9 kWp, and an efficiency of 13.1% was used. In this model maximum power point tracker is used as a control method and miscellaneous losses are considered as 5%. To fulfil the peak time energy demand biomass generator is used where the biomass rate is BDT 70/tonne (USD $0.83/tonne) and capacity of the generator is considered as 1 kW. As a storage system and an emergency backup, battery has been used. In this system, a total of 9 batteries are considered where each battery has a capacity of 24 V, 200 Ah. A one-day autonomy has been estimated for reliable power supply. The total capacity of the battery bank of 1800 Ah and 43 kWh is considered. Figure 3 presents the battery, PV and biomass generator specification that is given as an input in RETScreen and also shows that for the given combination approximately 81.2% of total energy comes from PV while the remainder of the energy comes from biomass generator. Thus, the total annual energy delivered to the load from PV and biomass generator are 13.98 MWh and 3.2 MWh respectively. Therefore, from the proposed system yearly 17.185 MWh energy can be produced which can easily fulfil the required load demand of 14.161 MWh.

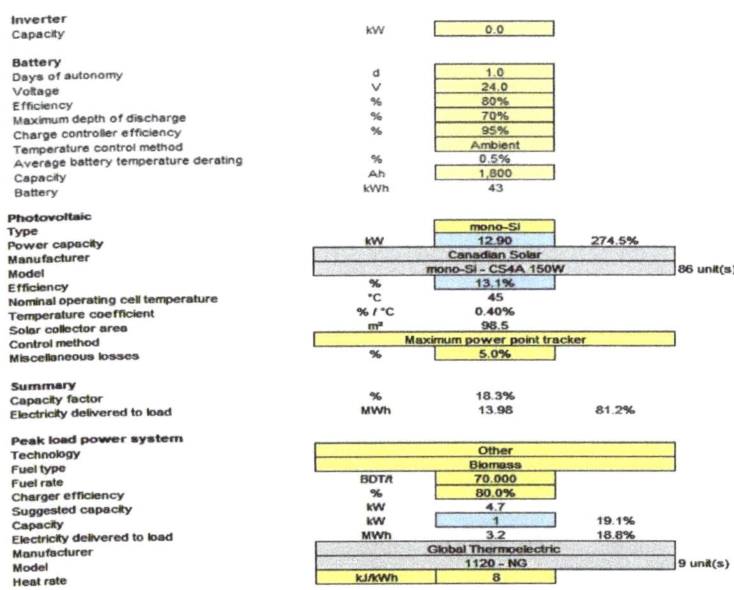

Figure 3. The specification of different components and energy produced by each source. Notation—Mono-Si: Monocrystalline silicon; CS4A: Multi-contact connector type 4.

In the proposed system, since most of the energy comes from PV, it produces clean energy and reduces CO_2 emissions compared to the existing system.

4.3. Cost Analysis

In order to analyse the costs of the proposed system, the initial, annual, and periodic costs as well as credits for any base case costs that are avoided in the proposed system are analysed. Before implementing the project in Ashuganj it is necessary to test the feasibility of the project to determine its suitability for the selected area. To check the climate feasibility, some testing is necessary which comes at a cost. For a "Feasibility analysis," more detailed and more accurate information is usually required. The calculations performed by the RETScreen Software for this step are straightforward and relatively simple (addition and multiplication). Figure 4 shows the detailed cost calculation for the project.

Initial costs (credits)	Unit	Quantity	Unit cost	Amount
Feasibility study				
Feasibility study	cost	2	BDT 3,000	BDT 6,000
Subtotal:				BDT 6,000
Development				
Development	cost	2	BDT 3,000	BDT 6,000
Subtotal:				BDT 6,000
Engineering				
Engineering	cost	2	BDT 4,000	BDT 8,000
Subtotal:				BDT 8,000
Power system				
Base load - Photovoltaic	kW	12.90	BDT 28,000	BDT 361,200
Peak load - Other	kW	0.90	BDT 60,000	BDT 54,000
Road construction	km	0	BDT -	BDT -
Transmission line	km	1	BDT 1,200,000	BDT 1,200,000
Substation	project	1	BDT 100,000	BDT 100,000
Energy efficiency measures	project	1	BDT 20,000	BDT 20,000
Battery	cost	9	BDT 24,000	BDT 216,000
				BDT -
Subtotal:				BDT 1,951,200
Balance of system & miscellaneous				
Spare parts	%	0.0%	BDT -	BDT -
Transportation	project	15	BDT 400	BDT 6,000
Training & commissioning	p-d	8	BDT 1,000	BDT 8,000
User-defined	cost		BDT -	BDT -
Contingencies	%	10.0%	BDT 1,985,200	BDT 198,520
Interest during construction	7.00%	1 month(s)	BDT 2,183,720	BDT 6,369
Subtotal:				BDT 218,889
Total initial costs				BDT 2,190,089

Figure 4. Initial cost of proposed system. Notation—BDT: Bangladeshi Taka.

In cost analysis, it has been observed that the total initial cost is BDT 2,190,089 (USD $26,072.49) where 89.1% cost comes from power system sources such as PV, battery, biomass generator while the remaining cost components are from feasibility study and system miscellaneous. In the proposed system the lifetime of the PV, battery and converter have been based on 25 years, 5 years and 10 years respectively. Therefore, at full project life in 25 years the battery will require replacement 4 times, while the converter will be replaced twice. For the proposed system the total annual cost is BDT 167,696 (USD $1996.38). In periodic cost, it is seen that the battery replacement cost in full project life is BDT 403,200 (USD $4800), and converter cost is BDT 89,600 (USD $1066.67). After all the expenses the annual savings from the project is BDT 474,702 (USD $5,651.21) which is provided in Table 6.

Table 6. Annual saving of proposed system.

Project Costs and Saving/Income Summary			
Initial Cost			
Feasibility study	0.3%	BDT	6000
Development	0.3%	BDT	6000
Engineering	0.4%	BDT	8000
Power system	89.1%	BDT	1,951,200
Balance of system & misc.	10.0%	BDT	218,889
Total initial costs	100%	BDT	2,190,089
Annual Costs & Debt Payments			
O & M		BDT	159,375
Fuel cost-proposed case		BDT	0
Debt payments-25 yrs		BDT	8321
Total Annual costs		BDT	167,696
Periodic Costs (credits)			
Battery-5 yrs		BDT	403,200
Converter-10 yrs		BDT	89,600
End of project life-cost		BDT	376,832
Annual Savings and Income			
Fuel cost-base case		BDT	474,702
Total annual savings and income		BDT	474,702

4.4. *Financial Viability & Cumulative Cash Flow Analysis*

The RETScreen Software enables a user to input various forms of financial data such as discount rates, etc., which it automatically calculates to produce key financial feasibility indicators such as simple payback, equity payback, and net present value. Based on the data entered by the user, financial indicators for the project being analysed are provided, thus deriving vital information which facilitates the project evaluation process for planners and decision-makers [54].

The simple payback SP is the number of years it takes for the cash flow (excluding debt payments) to equal the total investment (which is equal to the sum of debt and equity):

$$SP = \frac{C - IG}{\left(C_{energy} + C_{capacity} + C_{RE} + C_{GHG}\right) - \left(C_{O\&M} + C_{fuel}\right)} \tag{1}$$

where, C is the total initial cost of the project and IG is the value of incentives and grants. C_{energy}, $C_{capacity}$, C_{RE}, and C_{GHG} are annual energy saving or income, annual capacity saving or income, annual renewable energy production credit income and greenhouse gas reduction income respectively. $C_{O\&M}$, C_{fuel} represent the yearly operation and maintenance cost and yearly cost of fuel or electricity respectively.

Similarly, the year-to-positive cash flow (also equity payback), N_{PCF} is the first year that the cumulative cash flows for the project are positive. It is calculated by solving the following equation for N_{PCF}:

$$0 = \sum_{n=0}^{N_{PCF}} \overset{>}{C}_n \tag{2}$$

where, $\overset{>}{C}_n$ is the after tax cash flow in year n.

The net present value NPV of a project is calculated by discounting all cash flows as given in the following equation:

$$NPV = \sum_{n=0}^{N} \frac{\overset{>}{C}_n}{(1+r)^N} \tag{3}$$

where, N is the project life in years and r is the discount rate.

Discounted payback period, DPBP can be calculated that can be calculated that the discounted cash flow method discounts each inflow considering the time value of money until NPV equals zero at the certain year n of the system operation as indicated in Equation (4) [55,56].

$$\sum_{n=0}^{DPBT} \frac{CIn - COn}{(1+c)^n} = 0 \tag{4}$$

where, CI: cash inflow; CO: cash outflow; C: cost opportunity of capital; n: time period.

The annual life cycle savings ALCS is calculated using the following formula:

$$ALCS = \frac{NPV}{1 - \frac{1}{(1+r)^N}} \tag{5}$$

From the cash flow diagram depicted in Figure 5 using RETScreen, it can be estimated that it takes 7.2 years for cash flow to become positive and that the simple payback period will be 6.9 years. From the financial viability analysis, we get a Net Present Value (NPV) of BDT 855,428 (USD $10,183.67) and annual life cycle saving is BDT 73,405 (USD $873.87) and equity payback period is 7.2 years. In terms of the project's economics, we can say that the proposed hybrid system is the most economical one because after 7 years the project will start to generate profit and reduce the system's overall costs.

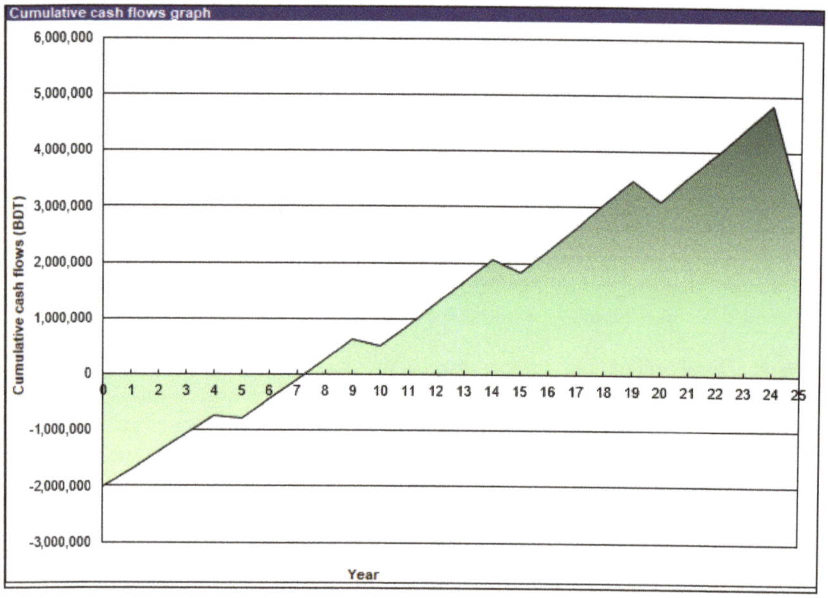

Figure 5. Cumulative cash flows.

4.5. Emissions Analysis

The emissions' analysis estimates the greenhouse gas emission-reduction (mitigation) potential of the proposed case. RETScreen estimates the annual GHG emission reduction, ΔGHG of electricity by utilising the following Equation:

$$\Delta_{GHG} = \left(e_{base} - e_{proposed}\right)E_{proposed}\left(1 - \lambda_{proposed}\right)\left(1 - e_{credit}\right) \quad (6)$$

where e_{base}: base case GHG emission factor; $e_{proposed}$: proposed case GHG emission factor; $E_{proposed}$: proposed case annual electricity produced; $\lambda_{proposed}$: the fraction of electricity lost in transmission and distribution (T&D) for the proposed case. For off grid system consider the value of $\lambda_{proposed}$ is zero; e_{credit}: the GHG emission reduction credit transaction fee.

For a single fuel type or source, the following formula is used to calculate the base case electricity system GHG emission factor, e_{base}:

$$e_{base} = \left(e_{CO2}GWP_{CO2} + e_{CH4}GWP_{CH4} + e_{N2O}GWP_{N2O}\right)\frac{1}{\eta}\frac{1}{1-\lambda} \quad (7)$$

where e_{CO2}, e_{CH4}, and e_{N2O} are respectively the CO_2, CH_4 and N_2O emission factors for the fuel/source considered, GWP_{CO2}, GWP_{CH4}, and GWP_{N2O} are the global warming potentials for CO_2, CH_4 and N_2O, η is the fuel conversion efficiency, and λ is the fraction of electricity lost in transmission and distribution. For standard analysis consider GWP_{CO2}, GWP_{CH4}, and GWP_{N2O} as 1, 21, and 310.

In cases for which there are a number of fuel types or sources, the GHG emission factor $e_{proposed}$ for the electricity mix is calculated as the weighted sum of emission factors calculated for each individual fuel source:

$$e_{proposed} = \sum_{i=1}^{n} f_i e_{proposed,i} \quad (8)$$

where n is the number of fuels/sources in the mix, f_i is the fraction of end-use electricity coming from fuel/source i, and $e_{proposed,i}$, is the emission factor for fuel i, calculated through a formula similar to Equation (8):

$$e_{proposed,i} = \left(e_{CO2,i}GWP_{CO2} + e_{CH4,i}GWP_{CH4} + e_{N2O,i}GWP_{N2O}\right)\frac{1}{\eta_i}\frac{1}{1-\lambda_i} \quad (9)$$

where, $e_{CO2,i}$, $e_{CH4,i}$, and $e_{N2O,i}$ are respectively the CO_2, CH_4 and N_2O emission factors for fuel/source i, η i is the fuel conversion efficiency for fuel i, and λ_i is fraction of electricity lost in transmission and distribution for fuel i. Consider all λ_i are zero in case of mix of duel/sources [54].

Table 7 reports the estimated GHG emissions reduction in the proposed system. From the result, it is evident that the proposed system reduces the CO_2 emissions by 0.269 tonne/MWh compared to the base case.

Thus, using Equation (6), we can calculate the yearly CO_2 reduction in the proposed system for 14.161MWh annual electricity production which is 3.81 tonnes. Therefore, we can say that the proposed system is cost effective and also environmentally-friendly.

$$\Delta_{GHG} = (0.269 - 0)X\ 14.161 = 3.81\ tonnes \quad (10)$$

Table 7. Emissions analysis.

Base Case System GHG Summary (Baseline)						
Fuel type	**Fuel Mix %**	**CO_2 emission factor Kg/GJ**	**CH_4 emission factor Kg/GJ**	**N_2O emission factor Kg/GJ**	**Fuel Consumption MWh**	**GHG emission factor tCO_2/MWh**
Kerosene	100%	73.9	0.0070	0.0020	0	0.269
Total	100%	73.9	0.0070	0.0020	0	0.269
Proposed Case System GHG Summary (Power Proposed Project)						
Fuel type	**Fuel Mix %**	**CO_2 emission factor Kg/GJ**	**CH_4 emission factor Kg/GJ**	**N_2O emission factor Kg/GJ**	**Fuel Consumption MWh**	**GHG emission factor tCO_2/MWh**
Biomass	0.1	0	0.032	0.004	0	0.007
Solar	99.9	0	0	0	14	0
Total	100%	0	0	0	14	0

5. Conclusions

This paper highlighted the benefits of using a hybrid energy system consisting of both solar energy and biomass energy to reduce energy costs and CO_2 emissions. The design was compared with data from the RETScreen data and also with the existing kerosene-based system. Due to the lack of the regional power grid, and available local resources, photovoltaic panel, biogas generator along with battery storage bank are the best solution for providing electricity in future. The paper demonstrated that the hybrid mini-grid system is the most economical and reliable for rural areas. Another useful part of employing a hybrid system is the minimal use of biomass generator which ultimately reduces the greenhouse gas emissions. The only drawback in this system is the battery cost. So, to make this proposed system a reliable one the government should take step to reduce the battery cost. In this hybrid energy model, simulation results showed that 81.2% of total energy is produced by PV and the rest of the energy comes from biomass generator. The simple payback of the project was estimated to be 6.9 years and this project will be able to reduce the CO_2 emissions by approximately 3.81 tonnes per year. The study proved that the proposed system is more reliable and cost-effective and also more environmentally friendly when compared with the kerosene-based system.

Author Contributions: N.C. and C.A.H. proposed the core idea, developed the models. N.C. performed the simulations, exported the results and analysed the data. N.C., C.A.H., M.L. and W.Y. contributed to the design of the models and the writing of this manuscript. M.L. and W.Y. revised the paper. All authors have read and agreed to the published version of the manuscript.

Funding: This research received no external funding.

Conflicts of Interest: The authors declare no conflict of interest.

References

1. Pasetti, M.; Rinaldi, S.; Manerba, D. A Virtual Power Plant Architecture for the Demand-Side Management of Smart Prosumers. *Appl. Sci.* **2018**, *8*, 432. [CrossRef]
2. Rinaldi, S.; Pasetti, M.; Sisinni, E.; Bonafini, F.; Ferrari, P.; Rizzi, M.; Flammini, A. On the Mobile Communication Requirements for the Demand-Side Management of Electric Vehicles. *Energies* **2018**, *11*, 1220. [CrossRef]
3. Zomers, A. The challenge of rural electrification. *Energy Sustain. Dev.* **2003**, *7*, 69–76. [CrossRef]
4. Alavi, S.M. Techno-Economic Pre-Feasibility Study of Wind and Solar Electricity Generating Systems for Households in Central Finland. Master's Thesis, University of Jyväskylä, Jyväskylä, Finland, 2014.
5. Taniguchi, M.; Kaneko, S. Operational performance of the Bangladesh rural electrification program and its determinants with a focus on political interference. *Energy Policy* **2009**, *37*, 2433–2439. [CrossRef]

6. Islam, A.; Chan, E.S.; Taufiq-Yap, Y.H.; Mondal, M.A.H.; Moniruzzaman, M.; Mridha, M. Energy security in Bangladesh perspective—An assessment and implication. *Renew. Sustain. Energy Rev.* **2014**, *32*, 154–171. [CrossRef]
7. Mandelli, S.; Barbieri, J.; Mereu, R.; Colombo, E. Off-grid systems for rural electrification in developing countries: Definitions, classification and a comprehensive literature review. *Renew. Sustain. Energy Rev.* **2016**, *58*, 1621–1646. [CrossRef]
8. Bhattacharyya, S.C.; Palit, D. Mini-grid based off-gridelectrification to enhance electricity access in developing countries: What policies may be required? *Energy Policy* **2016**, *94*, 166–178. [CrossRef]
9. Kobos, P.H.; Erickson, J.D.; Drennen, T.E. Technological learning and renewable energy costs: Implications for US renewable energy policy. *Energy Policy* **2006**, *34*, 1645–1658. [CrossRef]
10. Nguyen, K.Q. Alternatives to grid extension for rural electrification: Decentralized renewable energy technologies in Vietnam. *Energy Policy* **2007**, *35*, 2579–2589. [CrossRef]
11. Longo, M.; Hossain, C.A.; Roscia, M. Smart Mobility for Green University Campus. In Proceedings of the Asia-Pacific Power and Energy Engineering Conference (APPEEC), Kowloon, China, 8–11 December 2013; pp. 1–6.
12. Bernal-Agustín, J.L.; Dufo-López, R. Simulation and optimization of stand-alone hybrid renewable energy systems. *Renew. Sustain. Energy Rev.* **2009**, *13*, 2111–2118. [CrossRef]
13. Ho, W.S.; Hashim, H.; Hassim, M.H.; Muis, Z.A.; Shamsuddin, N.L.M. Design of distributed energy system through Electric System Cascade Analysis (ESCA). *Appl. Energy* **2012**, *99*, 309–315. [CrossRef]
14. Chowdhury, N.; Hossain, C.A.; Longo, M.; Yaïci, W. Optimization of solar energy system for the electric vehicle at university campus in Dhaka, Bangladesh. *Energies* **2018**, *11*, 2433. [CrossRef]
15. Dedé, A.; Della Giustina, D.; Massa, G.; Pasetti, M.; Rinaldi, S. A Smart PV Module with Integrated Electrical Storage for Smart Grid Applications. In Proceedings of the IEEE International Symposium on Power Electronics, Electrical Drives, Automation and Motion (SPEEDAM), Anacapri, Italy, 22–24 June 2016; pp. 895–900.
16. Marchi, B.; Pasetti, M.; Zanoni, S. Life Cycle Cost Analysis for BESS Optimal Sizing. In Proceedings of the Energy Procedia, Special issue of the 2016 International Scientific Conference on Environmental and Climate Technologies (CONECT), Riga, Latvia, 12–14 October 2016; Volume 113, pp. 127–134.
17. Marchi, B.; Zanoni, S.; Pasetti, M. A Techno-Economic Analysis of Li-ion Battery Energy Storage Systems in Support of PV Distributed Generation. In Proceedings of the 21st Summer School F. Turco of Industrial Systems Engineering, Naples, Italy, 13–15 September 2016; pp. 145–149.
18. Marchi, B.; Pasetti, M.; Zanoni, S.; Zavanella, L.E. The Italian Reform of Electricity Tariffs for Non Household Customers: The Impact on Distributed Generation and Energy Storage. In Proceedings of the 22nd Summer School F. Turco of Industrial Systems Engineering, Palermo, Italy, 13–15 September 2017; pp. 103–109.
19. Hossain, C.A.; Chowdhury, N.; Longo, M.; Yaïci, W. System and Cost Analysis of Stand-Alone Solar Home System Applied to a Developing Country. *Sustainability* **2019**, *11*, 1403. [CrossRef]
20. Khan, M.J.; Iqbal, M.T. Pre-feasibility study of stand-alone hybrid energy systems for applications in Newfoundland. *Renew. Energy* **2005**, *30*, 835–854. [CrossRef]
21. Akella, A.K.; Sharma, M.P.; Saini, R.P. Optimum utilization of renewable energy sources in a remote area. *Renew. Sustain. Energy Rev.* **2007**, *11*, 894–908. [CrossRef]
22. Shaahid, S.M.; Elhadidy, M.A. Technical and economic assessment of grid-independent hybrid photovoltaic–diesel–battery power systems for commercial loads in desert environments. *Renew. Sustain. Energy Rev.* **2007**, *11*, 1794–1810. [CrossRef]
23. Kenfack, J.; Neirac, F.P.; Tatietse, T.T.; Mayer, D.; Fogue, M.D.; Lejeune, A. Micro hydro–PV–hybrid system: Sizing a small hydro–PV–hybrid system for rural electrification in developing countries, technical note. *Renew. Energy* **2009**, *34*, 2259–2263. [CrossRef]
24. Silva, S.B.; de Oliveira, M.A.G.; Severino, M.M. Economic evaluation and optimization of a photovoltaic–fuel cell–batteries hybrid system for use in the Brazilian Amazon. *Energy Policy* **2010**, *38*, 6713–6723. [CrossRef]
25. Ma, T.; Yang, H.; Lu, L. A feasibility study of a stand-alone hybrid solar–wind–battery system for a remote island. *Appl. Energy* **2014**, *121*, 149–158. [CrossRef]
26. Okedu, K.E.; Al-Hashmi, M. Assessment of the cost of various renewable energy systems to provide power for a small community: Case of Bukha, Oman. *Int. J. Smart Grid* **2018**, *2*, 3.

27. Sepulveda, T.T.; Martinez, L. Optimization of a hybrid energy system for an isolated community in Brazil. *Int. J. Renew. Energy Res.* **2016**, *6*, 1476–1481.
28. Bhattarai, P.R.; Thompson, S. Optimizing an off-grid electrical system in Brochet, Manitoba, Canada. *Renew. Sustain. Energy Rev.* **2016**, *53*, 709–719. [CrossRef]
29. Lipu, M.S.H.; Hafiz, M.G.; Ullah, M.S.; Hossain, A.; Munia, F.Y. Design Optimization and Sensitivity Analysis of Hybrid Renewable Energy Systems: A case of Saint Martin Island in Bangladesh. *Int. J. Renew. Energy Res.* **2017**, *7*, 2.
30. Brenna, M.; Longo, M.; Yaici, W.; Abegaz, T.D. Simulation and Optimization of Integration of Hybrid Renewable Energy Sources and Storages for Remote Communities Electrification. In Proceedings of the IEEE PES Innovative Smart Grid Technologies Conference Europe (ISGT-Europe), Torino, Italy, 26–29 September 2017.
31. Longo, M.; Yaïci, W.; Foidelli, F. Hybrid Renewable Energy System with Storage for Electrification—Case Study of Remote Northern Community in Canada. *Int. J. Smart Grid* **2019**, *3*, 63–71.
32. Ahmed, S.; Islam, M.T.; Karim, M.A.; Karim, N.M. Exploitation of renewable energy for sustainable development and overcoming power crisis in Bangladesh. *Renew. Energy* **2014**, *72*, 223–235. [CrossRef]
33. Islam, M.T.; Shahir, S.A.; Uddin, T.M.I.; Saifullah, A.Z.A. Current energy scenario and future prospect of renewable energy in Bangladesh. *Renew. Sustain. Energy Rev.* **2014**, *39*, 1074–1088. [CrossRef]
34. Kanagawa, M.; Nakata, T. Assessment of access to electricity and the socio-economic impacts in rural areas of developing countries. *Energy Policy* **2008**, *36*, 2016–2029. [CrossRef]
35. Siddaiah, R.; Saini, R.P. A review on planning, configurations, modeling and optimization techniques of hybrid renewable energy systems for off grid applications. *Renew. Sustain. Energy Rev.* **2016**, *58*, 376–396. [CrossRef]
36. Guo, S.; Liu, Q.; Sun, J.; Jin, H. A review on the utilization of hybrid renewable energy. *Renew. Sustain. Energy Rev.* **2018**, *91*, 1121–1147. [CrossRef]
37. Islam, M.S.; Akhter, R.; Rahman, M.A. A thorough investigation on hybrid application of biomass gasifier and PV resources to meet energy needs for a northern rural off-grid region of Bangladesh: A potential solution to replicate in rural off-grid areas or not? *Energy* **2018**, *145*, 338–355. [CrossRef]
38. Murugaperumala, K.; Vimal Raj, P.A.D. Feasibility design and techno-economic analysis of hybrid renewable energy system for rural electrification. *Sol. Energy* **2019**, *188*, 1068–1083. [CrossRef]
39. Rad, M.A.V.; Ghasempour, R.; Rahdan, P.; Mousavi, S.; Arastounia, M. Techno-economic analysis of a hybrid power system based on the cost-effective hydrogen production method for rural electrification, a case study in Iran. *Energy* **2020**, *190*, 116421. [CrossRef]
40. Tanim, M.M.; Chowdhury, N.A.; Rahman, M.M.; Ferdous, J. Design of a Photovoltaic-Biogas Hybrid Power Generation System for Bangladeshi Remote area Using HOMER Software. In Proceedings of the 2014 3rd International Conference on the Developments in Renewable Energy Technology ICDRET, Dhak, Bangladesh, 29–31 May 2014.
41. Habib, M.A.; Chungpaibulpatana, S. Utilization of Solar and Biomass for Rural Electrification in Bangladesh. In Proceedings of the 2014 International Conference and Utility Exhibition on Green Energy for Sustainable Development ICUE, Pattaya City, Thailand, 19–21 March 2014.
42. Ahsan-Uz-Zaman, K.M.; Wahed, A.; Sayam, A.S.M.; Faruk, O.; Sarker, B.C. Solar-Biomass Hybrid System, an Approach for Rural Electrification in Bangladesh. In Proceedings of the 4th International Conference on Electrical Engineering and Information and Communication Technology iCEEiCT, Dhaka, Bangladesh, 13–15 September 2018; pp. 187–192.
43. Das, B.K.; Hoque, N.; Mandal, S.; Pal, T.K.; Raihan, M.A. A techno-economic feasibility of a stand-alone hybrid power generation for remote area application in Bangladesh. *Energy* **2017**, *134*, 775–788. [CrossRef]
44. Khan, E.U.; Mainali, B.; Martin, A.; Silveira, S. Techno-economic analysis of small scale biogas based polygeneration systems: Bangladesh case study. *Sustain. Energy Technol. Assess.* **2014**, *7*, 68–74. [CrossRef]
45. Mandal, S.; Das, B.K.; Hoque, N. Optimum sizing of a stand-alone hybrid energy system for rural electrification in Bangladesh. *J. Clean Prod.* **2018**, *200*, 12–27. [CrossRef]
46. Natural Resources Canada, RETScreen Data Analysis Software and Modelling Tool. Available online: https://www.nrcan.gc.ca/energy/software-tools/7465 (accessed on 15 January 2020).
47. Afzal, A. Performance Analysis of Integrated Wind, Photovoltaic and Biomass Energy Systems. In Proceedings of the World Renewable Energy Congress 2011, Linköping, Sweden, 8–13 May 2011; pp. 818–825.

48. Liqun, L.; Chunxia, L. Feasibility analyses of hybrid wind-PV-battery power system in Dongwangsha, Shanghai. *Przegląd Elektrotechniczny* **2013**, *89*, 239–242.
49. Kalinchyk, I.; Pfeiffer, C.F.; Inshekov, E. RETScreen Modeling for Combined Energy Systems Fertilizers Plant Case. In Proceedings of the 55th Conference on Simulation and Modelling, Modelling, Simulation and Optimization (SIMS 55) 2014, Aalborg, Denmark, 21–22 October 2014; pp. 7–15.
50. Hasan, M.M.; Chowdhury, N.; Hossain, C.A.; Longo, M. State of Art on Possibility & Optimization of Solar PV-Wind Hybrid System. In Proceedings of the 2019 International Conference on Robotics, Electrical and Signal Processing Techniques (ICREST), Dhaka, Bangladesh, 10–12 January 2019; pp. 598–601.
51. Sinha, S.; Chandel, S.S. Review of software tools for hybrid renewable energy systems. *Renew. Sustain. Energy Rev.* **2014**, 32, 192–205. [CrossRef]
52. Dwivedy, D.; Singh, S.K.; Choudhury, M.K.; Pradhan, S.R. Study of Cost Analysis and Emission Analysis for Grid Connected PV Systems using RETSCREEN 4 Simulation Software. *Int. J. Eng. Res. Technol.* **2015**, *4*, 203–207.
53. Wattblock; Witheridge, S.; Stegen, S. Comparison of Solar PV vs Solar Thermal Hot Water Systems to Provide Energy Solutions for Strata Buildings. Master's Thesis, Griffith School of Engineering Griffith University, Mount Gravatt, Australia, 22 May 2017.
54. Clean Energy Project Analysis: RETScreen Engineering & Cases Textbook. Available online: http://msessd.ioe.edu.np/wp-content/uploads/2017/04/Textbook-clean-energy-project-analysis.pdf (accessed on 15 January 2020).
55. D'Adamo, I. The Profitability of Residential Photovoltaic Systems. A New Scheme of Subsidies Based on the Price of CO2 in a Developed PV Market. *Soc. Sci.* **2018**, *7*, 148. [CrossRef]
56. Nam, H.; Mukai, K.; Konishi, S.; Nam, K. Biomass gasification with high temperature heat and economic assessment of fusion-biomass hybrid system. *Fusion Eng. Des.* **2019**, *146*, 1838–1842. [CrossRef]

sustainability

MDPI

Article

From NEDC to WLTP: Effect on the Energy Consumption, NEV Credits, and Subsidies Policies of PHEV in the Chinese Market

Xinglong Liu [1,2], Fuquan Zhao [1,2], Han Hao [1,2,3], Kangda Chen [1,2], Zongwei Liu [1,2,4,*], Hassan Babiker [5] and Amer Ahmad Amer [5]

1 State Key Laboratory of Automotive Safety and Energy, Tsinghua University, Beijing 100084, China; lxl19@mails.tsinghua.edu.cn (X.L.); zhaofuquan@tsinghua.edu.cn (F.Z.); hao@tsinghua.edu.cn (H.H.); ckd16@mails.tsinghua.edu.cn (K.C.)

2 Tsinghua Automotive Strategy Research Institute, Tsinghua University, Beijing 100084, China

3 China Automotive Energy Research Center, Tsinghua University, Beijing 100084, China

4 Sloan Automotive Laboratory, Massachusetts Institute of Technology, Cambridge, MA 02139, USA

5 Research and Development Center, Saudi Aramco, Dhahran 31311, Saudi Arabia; hassan.babiker@aramco.com (H.B.); amer.amer.4@aramco.com (A.A.A.)

* Correspondence: liuzongwei@tsinghua.edu.cn

Received: 30 June 2020; Accepted: 15 July 2020; Published: 17 July 2020

Abstract: The switching from new European driving cycle (NEDC) to worldwide harmonized light vehicles test procedure (WLTP) will affect the energy consumption of plug-in hybrid electric vehicle (PHEV), and then affect the new energy vehicle (NEV) credit regulation and subsidy policy for PHEVs. This paper reveals the impact on energy consumption, NEV credit regulation, and subsidy policy for PHEV in the Chinese market of the switching from NEDC to WLTP based on qualitative analysis and quantitative calculation. The results show that the WLTP procedure is stricter than NEDC in the determination of road load, test mass, driving resistance forces, and tire selection. Firstly, the electricity consumption (EC) of PHEV in charge-depleting mode (CD) under the WLTP procedure is 26% higher than NEDC on average, which makes the all-electric range (AER) significantly lower under WLTP. The weight EC tested in the WLTP procedure is higher than NEDC. Secondly, the fuel consumption (FC) of PHEV in CD mode is related to the adjustment of the engine management system (EMS) and the size of battery energy under the WLTP procedure. For the FC in the charge-sustaining (CS) mode of PHEV under the WLTP procedure is 20% higher than NEDC on average. However, the weight fuel consumption of PHEVs under WLTP with a long AER may be lower than that of NEDC due to the characteristics of utility factor in the WLTP procedure. Thirdly, most PHEVs fail to meet the requirements of 50 km AER due to the switching of the test procedures. However, the Chinese government reduced the technical specification of PHEV's AER under the WLTP procedure to 43 km to support the development of PHEV technology. It ensures that the switching of test procedures does not change the treatment that they could obtain, the NEV credits, and subsidy as a NEV in China. However, the increasing of the EC in CD mode and the FC in CS mode under the WLTP procedure makes the PHEV obtain lower credit and subsidy multiple compared with NEDC procedure.

Keywords: PHEV; NEDC; WLTP; energy consumption; NEV credit regulation; subsidy policy

1. Introduction

With the development of the automobile industry, China is facing the problems of energy security and environmental pollution [1–3]. The relevant national agencies have made a series of regulatory constraints on the energy consumption and emissions of vehicles to solve these problems [4,5]. According to the technical roadmap of energy-saving and new energy vehicles, the average fuel consumption (FC) of new passenger cars in 2020 and 2025 will reach 5 L/100 km and 4 L/100 km, respectively [6]. Because of the characteristics of the traditional internal combustion engine (ICE), it has gradually been unable to meet the stringent regulatory requirements [7]. Therefore, electrical upgrading based on the traditional internal combustion engine has become an effective way to solve the problem of energy consumption and emissions [8]. However, considering the battery costs and charging problems, there are still many obstacles in large-scale promotion of fully electrified battery electric vehicles (BEV). Therefore, the plug-in hybrid electric vehicle (PHEV), combining the ideal characteristics of BEVs with the range of traditional internal combustion engine vehicles (ICEV), has become the preferred technology choice for automobile manufacturers to meet increasingly stringent regulations [9–11].

Furthermore, it is an advantage of PHEVs that they can operate in two different modes due to its complicate powertrain configuration [12]. First, in charge-depleting mode (CD), the motor is responsible for propulsion and the ICE is switched off. Secondly, in charge-sustaining (CS) mode, the ICE provides power to drive the wheels and keep the SOC within a certain range [13]. However, due to the complexity of PHEV operation mode, it is difficult to accurately evaluate its energy consumption in tests [14]. In the past, the NEDC has been used to evaluate the fuel economy of Europe and China. However, the European Union has stipulated the gradual implementation of WLTP instead of NEDC to assess the fuel economy of light vehicles since September 2017. Furthermore, WLTP will be adopted as the FC test regulation of light vehicles in China from 2021 to 2025.

However, the switching of test procedures will change the energy consumption and emissions of the whole vehicle, which will have a certain impact on the development strategy of automobile manufacturers [15]. For example, several major European automakers have changed their development strategies for engine turbocharging and discontinued some PHEVs due to the switching from NEDC to WLTP. Thus, the switching of test procedure would affect the energy consumption of PHEV. Moreover, PHEV is one of the new energy vehicles vigorously promoted by the Chinese government. The change of test procedures will further affect the NEV credits and subsidy availability of PHEV on the basis of energy consumption. Therefore, it is necessary to study the effect of NEDC switching to WLTP in advance. On the one hand, it can help the government to accurately evaluate the energy-saving and emission reduction effect brought by the promotion of PHEV. On the other hand, it could help automotive manufacturers to develop new energy vehicle technology development routes.

However, there are few studies on the energy consumption of PHEV under different test procedures. Pavlovic et al. compared the differences from the energy consumption of PHEVs between NEDC and WLTP through experimental tests [16]. They found that the all-electric range (AER) determined by the WLTP procedure was significantly lower than NEDC. However, the FC tested from WLTP were often lower than the corresponding NEDC with the increase of battery energy. Tsiakmakis et al. studied the driving range and FC of PHEV under WLTP and NEDC based on the simulation method [17,18]. They found that the ratio of carbon dioxide emissions of PHEVs at WLTP to NEDC was largely dependent on battery energy. With the increase of battery energy, the proportion decreases rapidly. They also concluded that the ratio of energy consumption of PHEV between WLTP and NEDC is 1 when the battery energy is 25 kWh. Soulouk et al. investigated the main changes that the WLTP test procedure implies to a mid-size sedan electrified vehicle design (series, parallel P2, and power split) and quantifies their impact on the vehicles fuel economy [19]. They found that across different electrified vehicle architectures, the vehicles' fuel economy under the WLTP procedure in CS mode substantially decreases compared to the NEDC. Moreover, the battery needs to deliver more energy in the WLTP cycle compared to the NEDC to meet the AER requirement. The above-mentioned literatures analyze the change of energy consumption due to the switching of test procedures, but it does not systematically compare the differences between the NEDC and WLTP procedures for PHEVs. Moreover, there is a gap in the impact of the switching of test procedures on the policies for PHEVs, such as NEVs credit regulation and subsidy policy.

Therefore, this paper aims to reveal the impact of the switching from NEDC to WLTP on PHEV energy consumption and its external policies. This paper is organized as follows: In the first section, the significance of the research is introduced. Next, the differences between the NEDC and WLTP procedures are analyzed using qualitative analysis method. Then, the impact of test procedures switching on the energy consumption, NEVs credits regulation, and subsidy policy for PHEV is studied in the way of quantitative calculation method. Following that, the paper puts forward relevant policy suggestions for the problems caused by test procedure switching. Finally, the last section summarizes all findings from this paper.

2. Test Procedure Differences between the NEDC and WLTP Procedure for PHEV

2.1. Differences in Road Load Determination between NEDC and WLTP

The first difference is the determination of test mass between NEDC and WLTP. The mass of the test vehicle used to determine the road load is equal to the curb mass plus 100 kg in the NEDC procedure [20]. While for the WLTP procedure, the test mass is equal to the reference mass plus the mass of the fitted equipment of specific vehicles and the representative load mass [21]. It can be seen that test mass determination from WLTP will be significantly higher than NEDC. Therefore, the driving resistance of the whole vehicle will increase during the energy consumption tests, which will improve the energy consumption of PHEV under the WLTP procedure.

The second difference is the tire selection between NEDC and WLTP. The rolling resistance coefficient of a tire is the main contributor to the total rolling resistance. It is well-known that the influencing factors of the rolling resistance coefficient of the tire are mainly the width and circumference of the tire, the inflation pressure, and the tread depth [20]. In tire selection, NEDC requires that the widest tire must be selected for testing, while WLTP selects the tire according to the rolling resistance level of the tire according to the tested vehicles. Although NEDC is stricter than WLTP in terms of requirements, the rolling resistance coefficient of the widest tire specified in NEDC is not the largest. Therefore, the rolling resistance coefficient of the tire selected in WLTP is larger than that in NEDC to a certain extent. Generally speaking, the greater the inflation pressure, the lower the rolling resistance coefficient. There is no regulation on tire pressure in NEDC, so it is usually done to inflate the tire to the maximum allowable pressure. However, the road load is determined when the tire pressure is set to the minimum value as specified in WLTP. In terms of tire pressure, NEDC has more advantages than WLTP. Finally, it is known that the greater the tread depth, the greater the rolling resistance coefficient. The WLTP procedure for the minimum tire tread depth is more stringent (80%) than the NEDC requirement of 50% [21]. In a word, WLTP is stricter than NEDC in terms of tire selection, which leads to the higher energy consumption of PHEV when testing in WLTP.

The third difference is the determination of the coefficient of resistance force between NEDC and WLTP [22]. In the process of the coast down tests, paired runs in alternate directions must be performed due to the practical impossibility to have a perfectly flat test track. In the calculation method, NEDC averages the up and down test time. Unlike this, WLTP averages the resistance force not time in both directions. Due to the difference of test time in two different directions, the average time of the test track is relatively long compared to the real test time. Therefore, this method of average time leads to some errors and results in the final road load coefficient is lower than the coefficient calculated utilizing average force. Therefore, WLTP is stricter than NEDC in the way of calculating driving resistance. It will make PHEV require more energy in tests. Also, NEDC ignores the moment of the inertia effect of components in the process of determining the resistance force. The resistance force determined by the WLTP procedure is about 3% higher than NEDC due to the effect of the moment of inertia [20].

2.2. Differences in Test Protocol and Driving Cycles between NEDC and WLTP

The differences between the driving cycles of NEDC and WLTP are shown in Figure 1. The test cycle corresponding to the WLTP procedure is Worldwide harmonized light vehicles test cycle (WLTC). It can be seen from the figure that the WLTC cycle has a longer test time and distance compared with the NEDC cycle. Furthermore, the WLTC cycle has a significantly higher average maximum speed and acceleration, which will significantly improve the testing energy consumption of PHEV [23]. Moreover, the large decrease in the idle ratio in the WLTC cycle will weaken the fuel-saving effect of the start/stop and hybrid power technology [24].

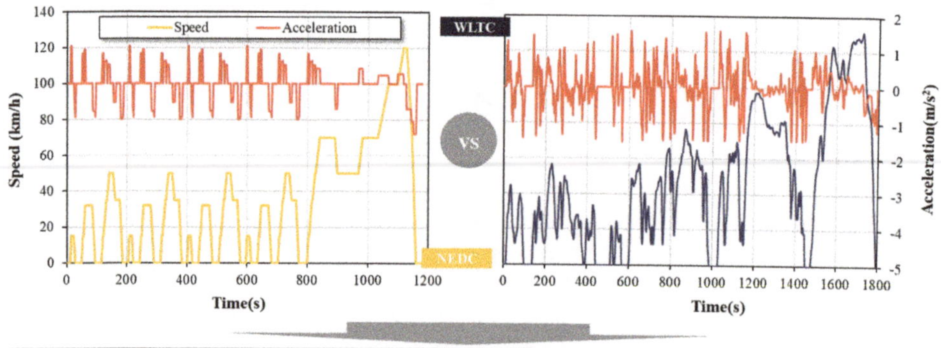

Parameters	Duration (s)	Distance (km)	Maximum speed (km/h)	Average speed (km/h)	Stop duration (%)	Constant driving (%)	Acceleration (%)	Deceleration (%)	Maximum acceleration (m/s²)	Minimum acceleration (m/s²)
NEDC	1180	11.03	33.6	120.0	23.7	40.3	20.9	43.8	1.04	-1.39
WLTC	1800	23.27	46.5	131.3	12.6	3.7	15.1	39.9	1.67	-1.50

Figure 1. Comparison of new European driving cycle (NEDC) and worldwide harmonized light vehicles test cycle (WLTC) driving cycles.

However, the higher engine loads experienced by vehicles under the WLTC cycle can make the engine work in the high-efficiency range, which might compensate for the FC caused by the higher dynamic effect to a certain extent [25]. Furthermore, the proportion of cold start for vehicles under the WLTC cycle is lower than the NEDC cycle, which will reduce the effect of cold start and further reduce FC. Moreover, the WLTP procedure introduced the specific gearshift strategy with a manual transmission by calculated the engine speed and vehicle characteristics, while in the NEDC procedure, the same fixed gear positions are used for all vehicles [21]. Thus, this new WLTP gearshift strategy results in engine speeds ranging in their lower end, which consequently provide better engine efficiency and lower FC.

In a word, the FC for the ICEVs will change limited (far less than expected) by simply analyzing the change from NEDC to the WLTC cycle assuming that other conditions remain unchanged. The amplitude of the FC change is related to the characteristics of the vehicle and the engine. This is also confirmed by other scholars [26–29]. They found that the CO_2 emission ratio of WLTC and NEDC cycle is between 0.89 and 1.16 when only considering the difference of the test cycle. However, the EC for BEVs will be significantly improved under the worse WLTC cycle not having the compensating effect from the engine.

Furthermore, the NEDC procedure only tests one cycle in CD mode for PHEVs, while WLTP extends to the whole CD mode until the battery reaches the SOC level in CS mode [16]. Therefore, the real EC of PHEV in CD mode cannot be accurately measured according to the measurement of NEDC if the EC of PHEV in CD mode is non-linear. Moreover, with the increase of the discharge depth, the energy consumption of the battery gradually increases, considering the discharge characteristics of the power battery [18]. It will result in that the EC in CD mode under WLTP will be significantly higher than NEDC. Meanwhile, there exist some differences when calculating the FC for PHEVs in CS mode between WLTP and NEDC. Under the WLTP procedure, PHEVs should be corrected the FC in CS mode for the difference of the SOC of the battery between the start and end of the CS test. If the change of SOC is greater than 0.5% and SOC is reduced (corresponding to battery discharge) in WLTP,

FC correction will be enforced [16]. This is not considered under the NEDC procedure. From this point of view, WLTP will further increase FC in the tests.

2.3. *Differences in Post-Processing of the Energy Consumption Data between NEDC and WLTP*

For the calculation of energy consumption under the NEDC procedure, the final FC and EC are calculated by the following formulas [30]:

$$C_{weight}^{NEDC} = \frac{D_e * C_{CD}^{NEDC} + D_{av} * C_{CS}^{NEDC}}{D_e + D_{av}} = UF^{NEDC} * C_{CD}^{NEDC} + \left(1 - UF^{NEDC}\right) * C_{CS}^{NEDC} \tag{1}$$

$$E_{weight}^{NEDC} = \frac{D_e * E_{CD}^{NEDC}}{D_e + D_{av}} = UF^{NEDC} * E_{CD}^{NEDC} \tag{2}$$

where $C_{weight}^{NEDC}\left(E_{weight}^{NEDC}\right)$ represent the FC (EC) for PHEV; $C_{CD}^{NEDC}\left(E_{CD}^{NEDC}\right)$ indicate the FC (EC) in CD mode; C_{CS}^{NEDC} indicates the FC in CS mode. D_e is the electric range of the PHEV and D_{av} is the average distance has driven in CS mode, 25 km. To facilitate comparison with the method to calculate the weight energy consumption of the WLTP procedure, UF^{NEDC} is assumed the equivalent utilization factor. It assumes that the EC in CS mode is 0 under the NEDC procedure, which is the same as the WLTP procedure. It makes the comparison simplified between the EC under different test procedures in this paper.

As abovementioned, NEDC only tests one cycle for PHEV in CD mode, so the FC of PHEVs whose electric range is over 11.03 km (one NEDC cycle) in CD mode is 0. Considering that the current AER of PHEV is more than 50 km, the formula adopted to calculate the weight FC of PHEV under NEDC procedure can be simplified as follows:

$$C_{weight}^{NEDC} = \frac{D_{av} * C_{CS}^{NEDC}}{D_e + D_{av}} = \left(1 - UF^{NEDC}\right) * C_{CS}^{NEDC} \tag{3}$$

However, the favorable testing assumptions under the NEDC procedure in CD mode will be eliminated with the improvement of WLTP. Thus, the FC in the CD mode test cannot be ignored, which will make the FC of PHEV in CD mode higher than NEDC. In WLTP, the FC in CD mode and final weight FC is calculated according to the different weights of each phase in CD mode. The formula is as follows [21]:

$$\begin{cases} C_{weight}^{WLTP} = \left(\sum\limits_{j=1}^{k} UF_j\right) * C_{CD}^{WLTP} + \left(1 - \sum\limits_{j=1}^{k} UF_j^{WLTP}\right) * C_{CS}^{WLTP} \\ C_{CD}^{WLTP} = \frac{\sum_{j=1}^{k}\left(UF_j^{WLTP} * C_{CD,j}\right)}{\sum_{j=1}^{k} UF_j} \end{cases} \tag{4}$$

where C_{weight}^{WLTP} is the weight FC under the WLTP procedure for PHEVs, L/100 km, UF_j^{WLTP} is the utility factor of the CD phase j under the WLTP procedure; C_{CD}^{WLTP} is the FC under the WLTP procedure in CD mode and C_{CS}^{WLTP} is the FC under the WLTP procedure in CD mode, L/100 km; k represents the number of velocity segments tested from the CD test to the transition cycle; $C_{CD,j}$ is the FC of phase j in CD mode, L/100 km.

When calculating the weight EC under the WLTP procedure, the EC in CS mode is not considered, so the calculation formula is as follows:

$$\begin{cases} E_{weight}^{WLTP} = \left(\sum\limits_{j=1}^{k} UF_j\right) * E_{CD}^{WLTP} \\ E_{CD}^{WLTP} = \frac{\sum_{j=1}^{k}\left(UF_j^{WLTP} * E_{CD,j}\right)}{\sum_{j=1}^{k} UF_j} \end{cases} \tag{5}$$

where E_{weight}^{WLTP} is the weight EC under the WLTP procedure for PHEVs, kWh/100 km, E_{CD}^{WLTP} is the EC under the WLTP procedure in CD mode, kWh/100 km; $E_{CD,j}$ is the EC of phase j in CD mode, kWh/100 km.

By comparing the calculation formula of weight energy consumption between NEDC and WLTP, the results of a PHEV are strongly affected by the utility factor related to the electric range [31,32]. Thus, it is not accurate for the NEDC procedure to set 25 km as the average distance of CS mode, which does not conform to the current travel characteristics [32]. It depends on the AER of the specific vehicle. If the vehicle has a high AER, it is likely to drive mainly in CD mode. If the AER of the vehicle is lower, the probability of driving in CS mode is higher. The WLTP procedure introduces the utility factor based on the driver's travel characteristics to more accurately describe the driving probability in CD and CS modes, as shown in Figure 2. It should be noted that since China's current test regulations refer to Europe, its utility factor curve also comes from Europe. However, considering the difference in travel characteristics between China and Europe, China should develop the utility factor that adapts to the travel characteristics of China [33].

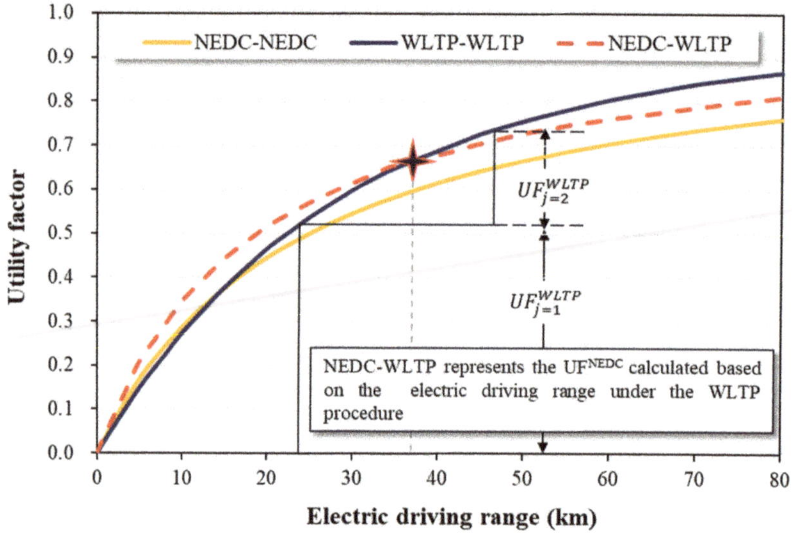

Figure 2. Comparison of utility factor between NEDC and worldwide harmonized light vehicles test procedure (WLTP).

It can be seen from Figure 2 that the utility factor calculated based on NEDC-NEDC and WLTP-WLTP is approximately equal in the range of short driving range. However, the energy consumption of CD mode under the WLTP procedure is calculated by weighting the utility factor of each phase of cycles, so it is more consistent with the travel behaviors. Also, considering that the actual electric range under the two test procedures is not the same, to better evaluate the difference between the two utility factors, the equivalent utility factor of NEDC is calculated by using AER under the WLTP procedure (making the electric range in NEDC transferred to the electric range in WLTP), which is the NEDC-WLTP utility factor curve as shown in Figure 2. It can be found that the equivalent utility factor of the NEDC procedure is higher than that of WLTP when the AER is short. Therefore, this will reduce the proportion of FC of PHEV in CS mode when calculating FC. It results in the FC in CS mode measured by WLTP being relatively high compared to NEDC, derived from Formula (4). Meanwhile, the utility factor of WLTP is higher than the equivalent utility factor of NEDC when the AER of PHEV under WLTP is greater than 38 km (intersection point in Figure 2). Therefore, when calculating the weight FC, the FC of PHEV with a high driving range under the WLTP procedure will below. On the contrary, with the continuous increase of the AER, the utility factor of WLTP is significantly higher than the equivalent utility factor of NEDC. Hence, the weight value of NEDC may be lower than that of WLTP, when calculating the weight FC.

3. Effect on the Energy Consumption of PHEVs from NEDC to WLTP

In essence, the energy consumption test of PHEVs is to test the energy consumption in CD mode and CS mode, respectively. Then, calculate the weight energy consumption by different weighting methods according to the NEDC and WLTP procedures. To analyze the impact of switching from NEDC to WLTP on the energy consumption of PHEV accurately, this part makes quantitative analysis from three parts: Energy consumption in CD mode, energy consumption in CS mode, and final weight energy consumption. It should be noted that the energy consumption data are mainly collected from the literature [16,18,19], tested by the China Automotive Technology and Research Center (CATRC) [34], and provided by relevant automobile manufacturers in this paper.

3.1. Analysis of the Energy Consumption in CD Mode (C_{CD} and E_{CD})

As abovementioned, the NEDC procedure only tests one cycle for PHEV in CD mode, the FC will be 0 if the AER of the testing vehicle over 11.03 km. However, the WLTP procedure tests the complete CD phase. The engine of PHEV will start in the condition that the power demand is higher than the power provided by the battery or the SOC of the battery is relatively lower [16]. To understand the test procedure and the determination of energy consumption for PHEVs under WLTP, the testing results of two PHEVs tested in the Joint Research Centre (JRC) of the European Commission Laboratories are cited in this study [16], as shown in Figure 3. The results showed the driving cycles, SOC of battery, and engine revolutions per minute (RPM).

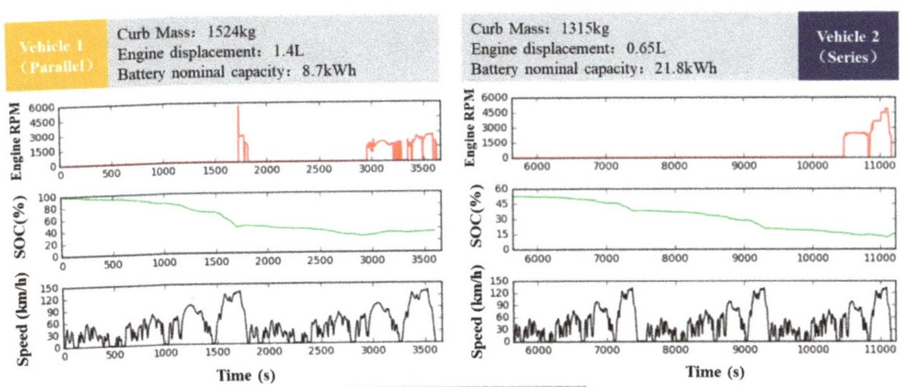

Testing Vehicles	Vehicle 1	Vehicle 2
FC in CD mode (L/100km)	1.366	0.067

Figure 3. WLTP charge depleting test sequence for two plug-in hybrid electric vehicles (PHEVs) from the study of Joint Research Centre (JRC).

For vehicle 1 in Figure 3, although the SOC was not already at its minimum value, the ICE started during the extra-high speed part of the first WLTC. It may be the consequence that the engine management system (EMS) for the specific vehicle was tuned by the OEMs to behave in a certain way on the NEDC cycle, thus bringing this unexpected behavior in the WLTP testing [16]. As abovementioned, the maximum speed of WLTC is much higher than that of NEDC, and that most likely was the cause of the behavior that ICE started early.

For parallel or hybrid PHEVs, if the EMS has not been tuned based on WLTC or the motor power is lower than the required power at the maximum speed, the engine will start even if the SOC is not at its minimum value. Meanwhile, the engine must start in the last WLTC due to the low SOC. Therefore, to reduce the FC of PHEV in CD mode, the vehicle manufacturer needs to adjust the EMS to avoid starting the engine of PHEV frequently in high SOC state. However, the engine decouples from the wheels or the series PHEV. The engine only started in the last WLTC in CD mode. Therefore, the FC mainly comes from the last cycle of CD mode if the EMS for the engine for PHEV is tuned based on WLTC, as shown in Figure 3.

At the same time, it can be found that the FC of Vehicle 1 in CD mode is significantly lower than that of Vehicle 2. This is mainly because the Vehicle 2 has a large power battery capacity so that the utility factor in the last cycle is significantly smaller than that of the Vehicle 2 (0.2% < 22.2%). According to Formula (4), the utility factor in the last cycle is lower, its FC in CD mode is lower. Therefore, the more WLTC cycles the PHEV experiences, the lower the FC in CD mode. In a word, the contribution of FC in CD mode to the real FC gets lower with the increase of battery capacity.

The EC in CD mode under the NEDC procedure is obtained by the meaning of dividing the electricity consumed in a test cycle by the range of an NEDC cycle (11.03 km). The EC in the CD mode under the WLTP procedure is to consider the weight value of each speed phase in the whole CD mode, as shown in Formula (5). Through collecting the EC of PHEVs from the CATRC in China and other research institutions in the world, it is the EC of 10 vehicles in the CD mode under the WLTP and NEDC, as shown in Figure 4.

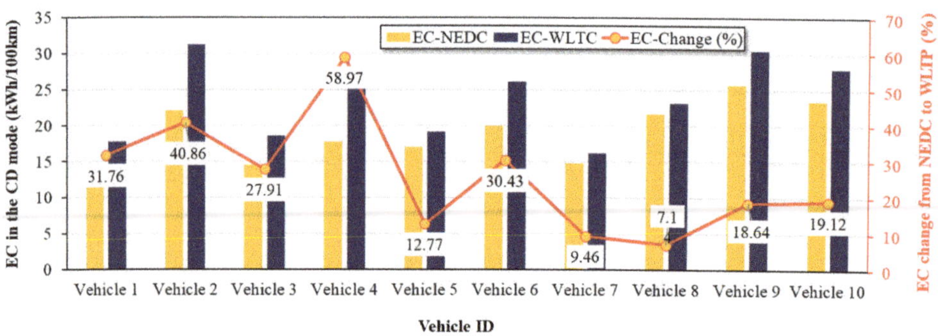

Figure 4. Electricity consumption (EC) of different PHEVs in charge-depleting (CD) mode under the NEDC and WLTP procedures.

It can be found that the EC in CD mode under the WLTP procedure is significantly higher than that under the NEDC procedure, with an average of about 26%. The results are consistent with those of JRC and ICCT reports [18,32]. It can be explained that the calculation of curb weight, tire selection, and driving resistance calculation of the WLTP procedure are stricter than those under the NEDC procedure. Meanwhile, The EC test of PHEV in CD mode is similar to that of battery electric vehicles, which is more sensitive to driving cycles as abovementioned. So, severe WLTP will cause higher EC. Furthermore, the JRC report points out that the EC ratio of the WLTP and NEDC procedure is related to the PHEV types [18]. It concluded that the heavier the vehicle type is, the larger the ratio is. This is mainly resulting from that the determination of road load is related to the curb weight. The calculation of curb weight under the WLTP procedure is stricter than that under the NEDC procedure. Because there are few vehicles selected in this study, there is no comparison on vehicle types. It could investigate the factor of vehicle type using the simulation method in the future work.

According to Formula (5), the EC of PHEV in CD mode could represent the AER of the corresponding vehicle. Therefore, it can be concluded that the AER of PHEV decreases by 26% on average with the switching of the testing procedure. Therefore, some PHEVs less than 63 km (NEDC) no longer meet the present regulatory requirements with the 50 km AER. So, the automotive manufacturers need to further increase the energy of the battery to meet the regulatory requirement for PHEV when designing after the switching of test procedures.

3.2. *Analysis of the FC in CS Mode (C_{CS})*

It can be seen from the calculation formula of weight energy consumption of PHEV that the EC in the CS mode is generally not considered. Therefore, this paper mainly analyzes the FC of PHEV in the CS mode. Considering that PHEV has the same FC characteristics as HEVs with a higher mixing degree when tested in the CS mode, the FC data of the HEVs are taken into account to analyze the FC in the CS mode for PHEV. The FC data are collected from the literature [16,18,19] and provided by an automotive manufacturer as shown in Figure 5.

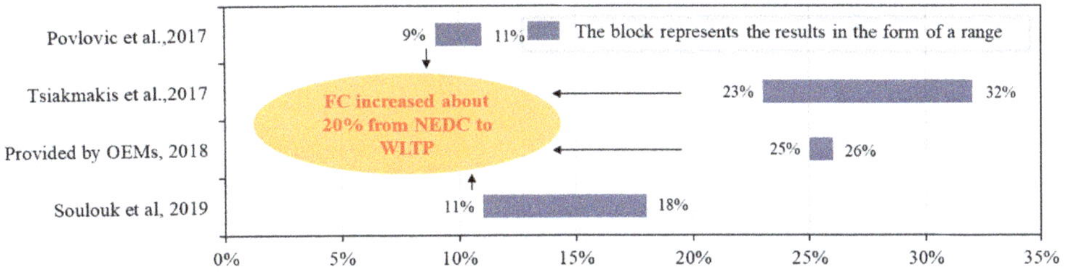

Figure 5. Fuel consumption (FC) change in the charge-sustaining (CS) mode of PHEVs from NEDC to WLTP.

As can be seen from Figure 5, FC in CS mode for PHEV under the WLTP procedure is significantly higher than that under the NEDC procedure, with an average of about 20%. This is explained in three aspects. First, the WLTP procedure is stricter than the NEDC procedure in the determination of test mass, tire selection, and driving resistance. Second, the correction of SOC under the WLTP procedure will also increase the FC of PHEV. Finally, the driving cycle of WLTC is stricter than the NEDC cycle for PHEV. On the one hand, the load can greatly improve FC due to the characteristics of the WLTC cycle. On the other hand, a small proportion of idle speed will reduce the fuel-saving capacity of the hybrid power system, which will also increase the FC of PHEV under the WLTP procedure.

3.3. *Analysis of the Weight Energy Consumption for PHEV*

The weight energy consumption of PHEV is calculated by weighting the energy consumption in the CD mode and CS mode. Table 1 shows the weight energy consumption of 10 different PHEVs, among which vehicle 1 and vehicle 2 are measured by JRC through experiments, while other data of eight PHEVs are from CATRC. It can be found that the weight EC of various PHEVs is lower than that in the CD mode, which is mainly due to the advantage of the weighted formula for EC. Furthermore, the weight EC of 10 PHEVs averagely increases by about 38% when switching from NEDC to WLTP, which is significantly higher than that in CD mode (26%). This shows that WLTP is stricter than NEDC in the calculation of weight EC of PHEV.

The weight FC of PHEV under the WLTP and NEDC procedure is shown in Figure 6. It can be seen that the weight energy consumption of PHEVs has increased significantly except for vehicle 1. The decrease of weight FC of vehicle 1 shows that the weight FC under the WLTP procedure might be lower than that under the NEDC procedure. It gives the credit to the difference of the calculation method of weight FC between NEDC and WLTP. In other words, the calculation method of WLTP is in favor of the PHEVs with a longer AER. Moreover, this phenomenon will be more significant with the increase in battery capacity. Thus, it can be said that the WLTP procedure is more friendly to the weight FC of PHEV with a larger battery.

Table 1. Weight energy consumption of 10 PHEVs.

Vehicle ID	Curb Mass (kg)	Engine Displacement (L)	Battery Capacity (kWh)	AER (km)		Weight FC (L/100 km)		Weight EC (kWh/100 km)	
				NEDC	WLTP	NEDC	WLTP	NEDC	WLTP
Vehicle 1	1315	0.65	21.8	161.8	122.8	0.68	0.40	11.67	16.71
Vehicle 2	1524	1.40	8.7	39.3	27.9	2.23	2.77	13.53	22.83
Vehicle 3	1505	1.80	8.0	55.0	43.0	1.30	2.70	10.00	13.58
Vehicle 4	1548	1.50	11.0	62.0	39.0	1.30	4.00	12.64	20.65
Vehicle 5	1665	1.50	9.0	53.0	47.0	1.40	3.30	11.54	14.02
Vehicle 6	1730	1.50	12.0	60.0	46.0	1.50	3.30	14.12	19.04
Vehicle 7	1955	1.50	12.0	81.0	74.0	1.30	2.30	11.32	13.56
Vehicle 8	2037	1.50	13.0	60.0	56.0	1.60	3.30	15.29	19.41
Vehicle 9	2281	2.00	18	70.0	59.0	1.60	3.80	18.95	25.51
Vehicle10	2390	2.00	19	81.0	68.0	1.80	2.90	17.92	23.36

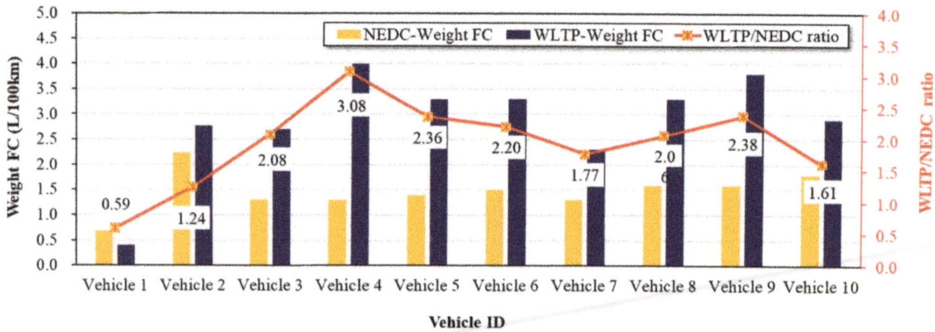

Figure 6. Weight FC of PHEVs under the WLTP and NEDC procedures.

It can be seen from the figure that the WLTP/NEDC ratio of weight FC is relatively large, with an average of about 1.94. This is quite different from the results (0.6 ~ 1.2) in the JRC report [18]. It needs to be explained that the weight FC data in this paper are all from the vehicle model currently on sale. The EMS of the engine of PHEV developed in China is tuned based on the NEDC cycle now, so the engine will start frequently under the WLTP procedure with the stricter driving cycle. Therefore, the engine will start frequently even if the SOC is high and to generate high FC in the CD mode. With the implementation of the WLTP procedure in China in 2021, automobile manufacturers will tune the EMS according to the WLTC driving cycle. This will make the engine not start frequently in the CD mode under the WLTP procedure. Then, the WLTP/NEDC ratio of weight FC will be further reduced in the future.

4. Effect on the Policies of PHEV from NEDC to WLTP

To promote the rapid development of the new energy automobile industry, the Chinese government proposed corresponding mandatory regulations and policy incentives, such as the NEV credit policy and subsidy policy [35–37]. In particular, the Chinese government delayed the decline of subsidies for two years in order to maintain the development of NEVs due to the COVID-19 outbreak in China. In order to ensure the development of NEVs driven by technology, the threshold value to meet the NEV credits regulations and subsidy policies is directly related to the energy-saving technology of vehicles.

The energy consumption of NEVs will change with switching from NEDC to WLTP, which will affect NEV credits and subsidy policies. The newly issued NEV credits regulation [38] and subsidy policy [39] in China, the following analysis is made on the impact of testing procedure switching on the policies for PHEVs. The NEV credit policy and subsidy policy for PHEVs are evaluated by the EC index in CD mode and the FC index in CS mode [38,39]. Table 2 presents the EC in CD mode and FC in the CS mode of the above 10 PHEVs under the NEDC and WLTP procedures. Among them, the FC in the CS mode of PHEV (Vehicle 3-Vehicle 8) in the China market is not given specific data. So, this paper assumes that the FC in CS mode under the WLTP procedure is 1.2 times of that in the NEDC procedure based on the above research.

Table 2. EC in CD mode and FC in CS mode of different PHEVs under the NEDC and WLTP procedures.

Energy Consumption	Vehicle ID									
	1	2	3	4	5	6	7	8	9	10
NEDC-CD EC (kWh/100 km)	13.47	22.14	14.55	17.74	16.98	20.00	14.81	21.67	25.71	23.46
NEDC-CS FC (L/100 km)	6.12	5.21	4.30	4.60	4.30	5.20	5.50	5.50	6.10	7.50
WLTP-CD EC (kWh/100 km)	17.75	31.18	18.60	28.21	19.15	26.09	16.22	23.21	30.51	27.91
WLTP-CS FC (L/100 km)	6.66	5.78	5.16	5.52	5.16	6.24	6.60	6.60	7.32	9.00

4.1. Effect on the NEV Credit Regulation of PHEV from NEDC to WLTP

The NEVs credits obtained by PHEV is equal to the basic credit multiplied by the credit multiple. According to the latest NEV credit regulation, the basic credit of PHEV is 1.6, while the credit multiple is determined by the values of the EC in CD mode and the FC in CS mode, which are 1 or 0.5. If the EC in CD mode for PHEV is greater than 135% of the EC targets of the same vehicle type BEV, or the FC in CS mode is greater than 70% of the FC limits for passenger car for the corresponding vehicle type, the credit multiple is 0.5, while the other is 1. The specific calculation is shown in Formula (6):

$$CR = \begin{cases} 0.8, \left(if\ E_{CD} \geq 135\%\ EC_{target}\ or\ C_{cs} \geq 70\%\ FC_{limit}\right) \\ 1.6,\ other \end{cases} \quad (6)$$

where CR is the credits that could be obtained by PHEV; EC_{target} indicates the EC targets for the same vehicle type BEV, which is the function of curb mass, kwh/100 km; FC_{limit} is the FC limits for the corresponding vehicle type, which is the function of curb mass, L/100 km.

The NEVs credits distribution for 10 PHEVs under the NEDC procedure is as shown in Figure 7. It should be noted that only when the EC in CD mode and the FC in CS mode are lower than the target and limit curves at the same time can one credit multiplier be obtained. It can be seen from the figure that 9 PHEVs can obtain NEVs credits except for vehicle 2 whose AER is less than 50 km. Among them, vehicle 3 and vehicle 9 can only get 0.5 times credit multiplier due to high EC, and the other seven PHEVs can get 1 credit multiplier.

The NEVs credits distribution for 10 PHEVs under the WLTP procedure is as shown in Figure 8. It can be seen from the figure that only two PHEVs can obtain one credit multiplier and three PHEVs can obtain a 0.5 credit multiplier after the test procedure is switched from NEDC to WLTP. The remaining five PHEVs could not obtain the NEVs credits. This shows that the switching of the test procedure has a great influence on the NEVs credit acquisition of PHEVs. The reason that five PHEVs cannot obtain NEVs credit is that their AER is less than 50 km under the WLTP procedure. It makes the PHEVs not meet the origin regulatory requirements and not enjoy policy privilege. Therefore, to obtain the NEVs credit of PHEV, the automobile manufacturers need to further increase the total energy of the battery and improve the AER of PHEV in the future. Furthermore, for the PHEV obtaining a 0.5 credit multiplier, the main reason is that the EC is too high, but the FC is still within the technological threshold of the NEV credit policy. It indicates that the influence of switching test procedures on the NEV credit policy for PHEV is higher in CD mode than in CS mode. Thus, the automobile manufacturers need to further consider the energy-saving technology of EC in the future vehicle design, such as the improvement of battery and motor efficiency, the optimization of the battery management system, and the further implementation of vehicle lightweight technology.

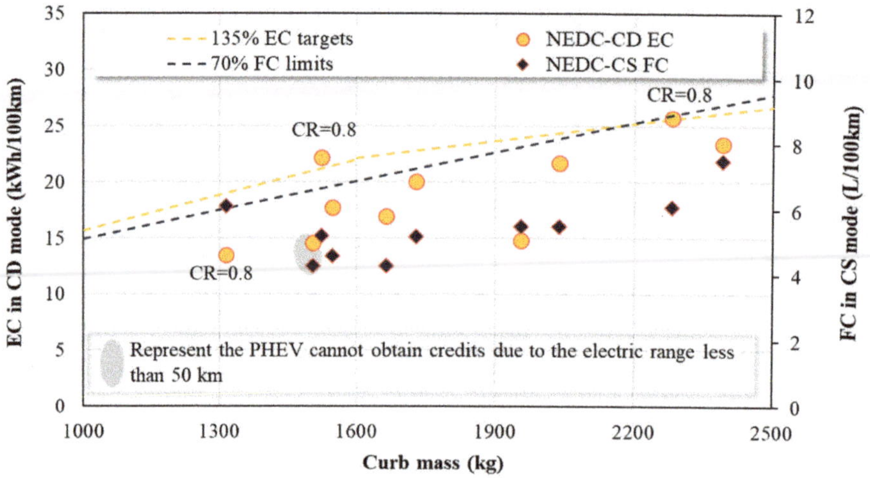

Figure 7. The new energy vehicles (NEVs) credits distribution for 10 PHEVs under the NEDC procedure.

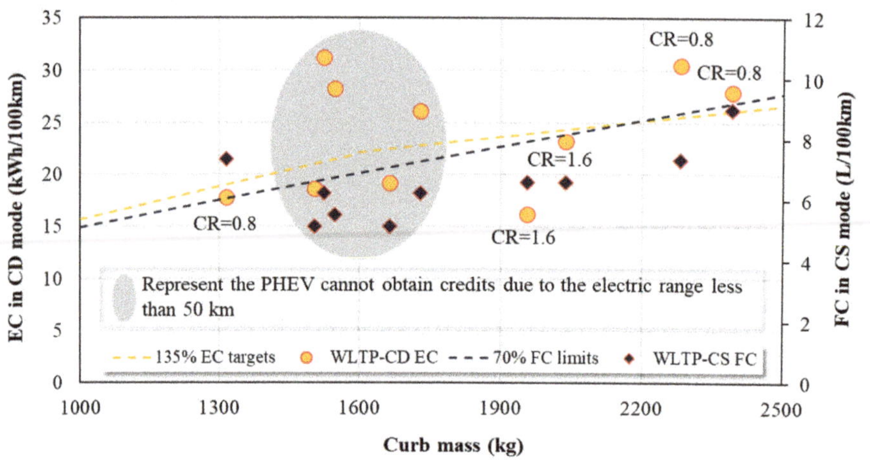

Figure 8. The NEVs credits distribution for 10 PHEVs under the WLTP procedure.

4.2. *Effect on the Subsidy Policy of PHEV from NEDC to WLTP*

According to the latest NEVs subsidy policy, the basic subsidy of PHEV is 8500 ¥ (RMB). The subsidy multiplier is related to the AER of PHEV and the FC in CS mode. According to the technological thresholds of subsidy policy, the FC in CS mode should be less than 60% compared with the FC limit of the same vehicle type for PHEVs with an AER of less than 80 km under the NEDC procedure. When the ratio is between 55% and 60%, the subsidy multiplier of the subsidy for PHEVs is 0.5. When the ration is less than 55%, the subsidy multiplier of the subsidy for PHEVs is 1. For PHEVs with AER under the NEDC procedure greater than or equal to 80 km, the EC in CD mode shall meet the requirements of the technological threshold for battery electric passenger vehicles in 2019. The specific calculation is shown in Formula (7):

$$SU = \begin{cases} 8500 * 0.5, \ (if \ \ 55\% FC_{limit} \le C_{CS} < 60\% \ FC_{limit}) \\ 8500 * 1.0, (if \ C_{CS} < 55\% \ FC_{limit} \ \ or \ AER \ge 80 \text{ km}) \end{cases} \tag{7}$$

where SU is the subsidy that PHEV could obtain, ¥ (RMB); AER is the all-electric range of PHEV, km.

The subsidy distribution for 10 PHEVs under the NEDC and WLTP procedure is shown in Figure 9. It can be seen from the figure that all vehicles could obtain the one multiplier of subsidy except Vehicle 1 under the NEDC procedure. It should be noted that although the FC in CS mode of Vehicle 10 is within the range of 0.5 multiplier subsidy, it can also obtain 1 multiplier subsidy due to its AER over 80 km (81 km). However, only three PHEVs could obtain 0.5 multiplier subsidies, but no one could obtain one times subsidies when switching from NEDC to WLTP procedure. Five PHEVs could not obtain subsidies because their AER is less than 50 km, which does not meet range requirements for PHEVs in China. Two PHEVs cannot obtain subsidies due to their high FC in CS mode, which is over the FC limits. It can be concluded that the switching of the test procedure has a great impact on the subsidy policy of PHEV. Thus, the automobile manufacturers need to further increase the total energy of the battery and expand the AER to fit the switching from NEDC to WLTP, which is also the strategy commonly used by European automobile manufacturers. Therefore, to obtain more subsidies for PHEV, it is necessary to further develop energy-saving technology based on increasing battery capacity, to reduce the FC in CS mode to the scope of subsidies.

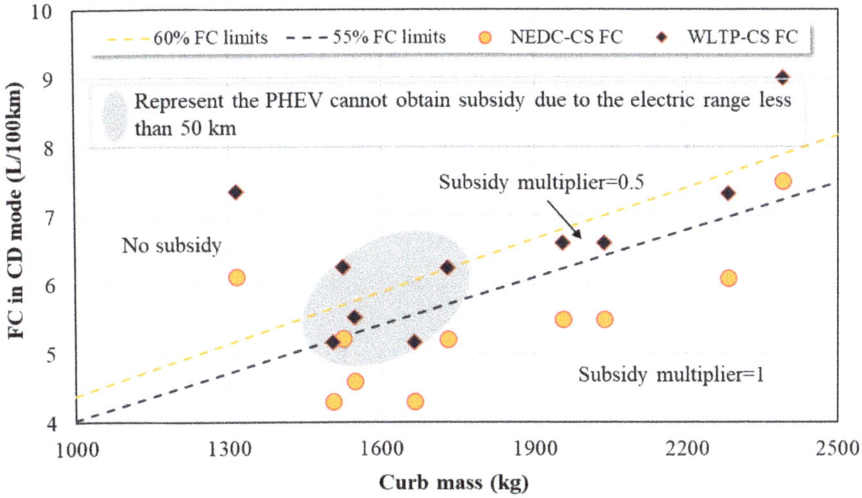

Figure 9. The subsidy distribution for 10 PHEVs under the NEDC and WLTP procedure.

In a word, the switch from NEDC to WLTP makes PHEV in a disadvantageous situation in the NEVs policy and subsidy policy. The main influence comes from two aspects: First, the switching of working conditions makes the AER of PHEV greatly reduced, which does not meet the requirements for the technical specifications of PHEV of the Chinese government, and cannot continue to enjoy the NEV credit and subsidy policies privilege; on the other hand, the switching of test procedures makes the energy consumption of PHEV in CD mode and CS mode increase, so that the PHEV can obtain the corresponding multipliers of the NEV credit and subsidy reduced.

However, the Chinese government has always adhered to the development route of PHEV. Although the test procedures are switched, it is not intended to make it difficult for automotive manufacturers to develop PHEV. In order to ensure that PHEV can still meet the requirements of technical specifications and obtain corresponding NEV credits and subsidies after the test procedure switching, the Chinese government has reduced the AER of PHEV to 43 km in the latest technical requirements.

Figure 10 shows the distribution of NEV credits and subsidies obtained under WLTP after reducing the AER technological specification of PHEVs. It can be seen from Figure 10a that three PHEVs can obtain one multiplier NEV credits after the AER of technological requirement is reduced. Moreover, five PHEVs can obtain 0.5 multiplier NEV credits while only two PHEVs cannot obtain NEV credits due to the AER is less than 43 km. Compared to the Figure 8, the decrease of AER technological requirement could improve the accessibility of NEV credits. Compared with Figure 7, the switching of test procedures could not alter the accessibility of NEV credits when the technological requirement of AER reducing to 43 km under the WLTP procedure. However, the increasing of EC and FC in the CD and CS mode under the WLTP procedure could decrease the multiplier of NEV credits compared to NEDC.

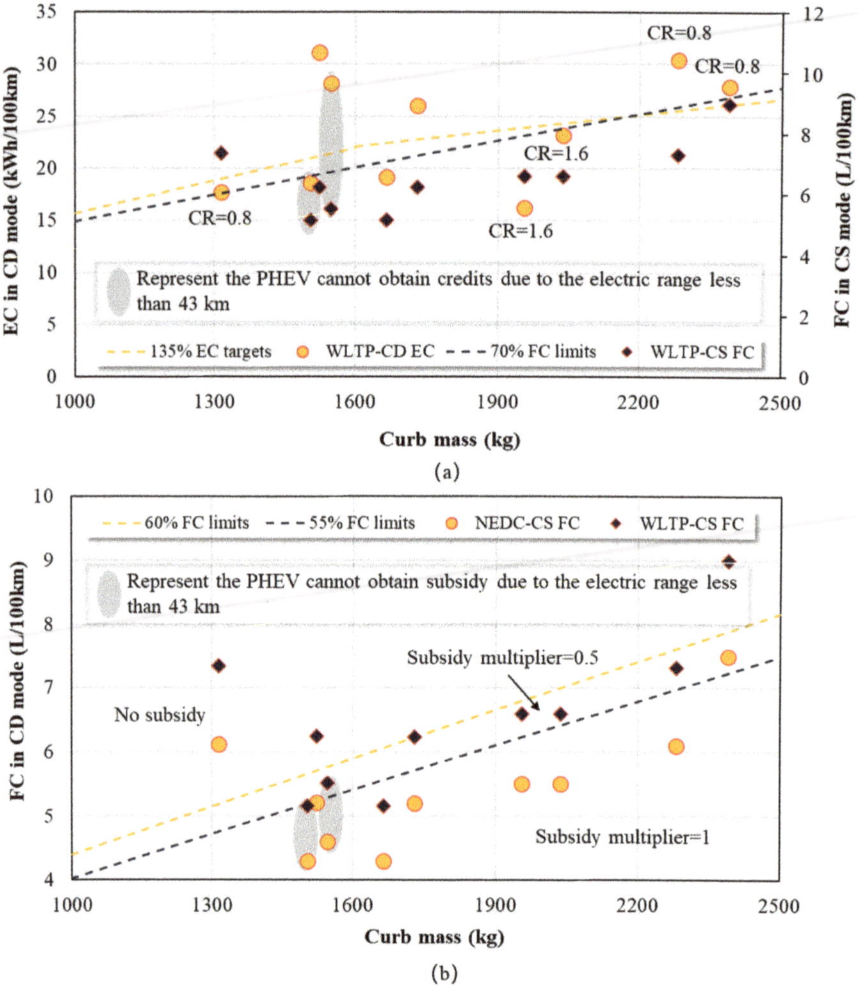

Figure 10. The NEV credits and subsidy distribution for 10 PHEVs under WLTP procedure after adjusting the all-electric range (AER) to 43 km in Chinese market: (**a**) NEV credits distribution; (**b**) subsidy distribution.

Meanwhile, it can be seen from Figure 10b that six PHEVs can get 0.5 multiplier subsidy under the WLTP procedure after the AER is reduced to 43 km, mainly because the FC in CS mode is too high to meet the technical thresholds of 1 multiplier subsidy. The remaining four PHEVs are not subsidized, of which two are due to the AER less than 43 km, and the other two are due to the FC in CS mode exceeding the threshold value of subsidy, so they are not subsidized. Compared to Figure 9, the decrease of AER technological requirement could improve the accessibility of subsidy while the higher FC in CS mode could lead to the lower multiplier subsidy.

This shows that the switching of test procedures has a limited impact on PHEV, which does not affect the availability of NEV credit and subsidies for PHEV with the support of the Chinese government. The switching of test procedure decreases the multipliers of NEV credit and subsidy due to the higher energy consumption. Thus, in order to further obtain the NEV credits and subsidies of PHEV, automobile manufacturers need to carry out the research and development of vehicle energy-saving technology, so as to reduce the EC in CD mode and the FC of CS mode. Therefore, it can be concluded that the switching of test procedure mainly puts forward higher requirements for the energy-saving technology of PHEV, but does not alter the availability of NEV credits and subsidies [40].

5. Policy Suggestion

The switching from NEDC to WLTP makes great changes in the energy consumption of PHEV in CD and CS modes to the extent that weight energy consumption, which affects the acquisition of NEV credit and subsidy of PHEVs to a extend. Therefore, to promote the reasonable development of PHEV in China after the test procedure change, the following suggestions are proposed:

First, the switching from NEDC to WLTP procedure increases the energy consumption of PHEV and makes most PHEVs unable to meet the requirements of 50 km AER. The Chinese government has changed the AER of PHEV to 43 km, indicating that the government has not tightened the requirements of PHEV, and the automotive manufacturer does not need to increase the battery to improve AER. However, the EC in CD mode and the FC of CS mode under the WLTP procedure are increased due to the condition switching. If the automotive manufacturers want to reach the corresponding fuel regulations, they need to further develop the energy-saving technology of the whole vehicle.

Secondly, the Chinese government's relaxation of PHEV's AER has little impact on the availability of PHEV's NEV credits and subsidies, so automotive manufacturers should not worry too much. However, with the increase of EC in CD mode and FC in CS mode, PHEV will obtain NEV credit and subsidy multiples. Thus, in order to obtain higher NEV credit and subsidy multiples, automotive manufactures need to research and develop energy-saving technologies of PHEV.

Thirdly, China automotive test cycle (CATC) will be used instead of WLTP to test the energy consumption of PHEVs after 2025. The Chinese government needs to evaluate the effect of the switching from WLTP to CATC in advance to make a smooth transition of PHEV's technical route. At the same time, automotive manufactures should prepare in advance, and develop and design PHEV according to the difference between CATC and WLTP.

Fourthly, China only considers the FC in the energy consumption test of PHEVs at present. In order to reasonably evaluate the energy consumption of PHEV, it is necessary to take the power consumption into account for comprehensive evaluation in the future. Then, the Chinese government should do a good job in policy planning, and automotive manufacturers should be prepared in advance.

6. Conclusions and Prospect

This paper studies the impact of switching from NEDC to WLTP on energy consumption, NEV credits, and subsidy policies of PHEVs through qualitative analysis and quantitative calculation methods. Firstly, the differences between the NEDC and WLTP procedures for testing energy consumption of PHEVs are qualitatively compared. Secondly, the effect of switching of test procedures on the energy consumption of PHEV is quantitative analysis. Finally, the NEV credit and subsidy changes of PHEVs are analyzed based on the energy consumption analysis after test procedure switching. According to the analysis results, the following conclusions are obtained:

(1) For the testing of PHEVs, the WLTP procedure is stricter in the determination of road load and test mass than those in the NEDC procedure. The weight calculation method for weight EC in the WLTP procedure is stricter than that in the NEDC procedure. The weight calculation method for weight FC in the WLTP procedure is stricter than that in the NEDC procedure for the PHEVs with a short AER. However, with the increase of the AER, the weight energy consumption under

the WLTP procedure may be lower than NEDC due to the weight calculation for weight FC in the WLTP procedure friendly to PHEVs with longer AER.

(2) The EC of PHEV in CD mode under the procedure is about 26% higher than that of NEDC, which greatly reduces the AER in WLTP. The FC in CD mode is related to the adjustment of the engine and the size of battery capacity. The higher the battery capacity, the lower the FC in CD mode. Furthermore, the FC in CS mode under the WLTP procedure is about 20% higher than that of NEDC. Because the EMS of PHEV on sale at present in China is tuned based on the NEDC cycle, not the WLTC cycle, the weight FC ratio of WLTP/NEDC is about 1.94.

(3) Most PHEVs do not meet the requirement of 50 km driving range due to the switching of test procedures. However, the relaxation of the requirement of 43 km AER under the WLTP procedure by the Chinese government has little impact on the availability of PHEV NEV credits and subsidies. However, the increase of the EC in the CD mode and the FC in CS mode have a great impact on the NEV credit multiple and subsidy multiple that PHEV can obtain. Therefore, automobile manufacturers need to further increase the research and development of vehicle energy-saving technology to obtain higher NEV credits and subsidies in the future.

(4) The Chinese government has reduced the technical specifications to 43 km of PHEV's AER under the WLTP procedure mainly for improving PHEV in term of the energy-saving technology after test procedure switching.

This paper only analyzes the differences between NEDC and WLTP procedures for testing energy consumption and to explore the effect of switching of test procedures on the NEV credits and subsidies for PHEV. However, it fails to give how to choose the power assembly of engine, battery, and motor to reduce the energy consumption of PHEV under the WLTP procedure to obtain more NEVs credit and subsidies. Moreover, the design value of the AER of PHEV under the WLTP procedure as well as the specific improvement measures of NEV credits and subsidy policies are not studied. Therefore, the battery and engine sizes of PHEV will be studied based on the WLTP procedure, and then evaluate the reasonability of NEV credit and subsidy policies based on the abovementioned study in the future work.

Author Contributions: Conceptualization, X.L., F.Z., and Z.L.; methodology, X.L. and Z.L.; software, X.L.; validation, X.L. and K.C.; formal analysis, K.C.; investigation, X.L.; resources, X.L.; data curation, X.L.; writing—original draft preparation, X.L., Z.L., and K.C.; writing—review and editing, Z.L. and H.H.; visualization, X.L.; supervision, A.A.A. and H.B.; project administration, F.Z. and Z.L.; funding acquisition, Z.L.; A.A.A., and H.B.. All authors have read and agreed to the published version of the manuscript.

Funding: This study is supported by the National Natural Science Foundation of China (U1764265, 71774100, 71690241) and the Chinese Academy of Engineering (2019-XZ-55-01-01).

Acknowledgments: The authors would like to thank Saudi Aramco for generous financial support, and Xing He for discussions. The authors would like to thank the anonymous reviewers for their reviews and comments.

Conflicts of Interest: The authors declare no conflict of interest.

Abbreviations

NEDC	New European Driving Cycle
WLTP	Worldwide Harmonized Light Vehicles Test Procedure
WLTC	Worldwide Harmonized Light Vehicles Test Cycle
CATC	China Automotive Test Cycle
PHEV	Plug-in Hybrid Electric Vehicles
AER	All-electric Range
EAER	Equivalent All-electric Range
CD mode	Charge-Depleting mode
CS mode	Charge-Sustaining mode

NEV	New Energy Vehicle
BEV	Battery Electric Vehicle
ICEV	Internal Combustion Engine Vehicles
SOC	State of Charge
EMS	Engine Management System
RPM	Revolutions Per Minute
EC	Electricity consumption
FC	Fuel Consumption
JRC	Joint Research Centre
CATRC	China Automotive Technology & Research Center

References

1. Chen, K.; Zhao, F.; Hao, H.; Liu, Z. Synergistic Impacts of China's Subsidy Policy and New Energy Vehicle Credit Regulation on the Technological Development of Battery Electric Vehicles. *Energies* **2018**, *11*, 3193. [CrossRef]
2. Hao, H.; Cheng, X.; Liu, Z.; Zhao, F. Electric vehicles for greenhouse gas reduction in China: A cost-effectiveness analysis. *Transp. Res. Part D Transp. Environ.* **2017**, *56*, 68–84. [CrossRef]
3. Hao, H.; Wang, S.; Liu, Z.; Zhao, F. Impact of China's CAFC standards on light-weighting strategy of automotive manufacturers. In Proceedings of the 2015 China Society of Automotive Engineers, Shanghai, China, 27 October 2015.
4. Ministry of Industry and Information Technology of People's Republic of China. Regulations and Policies of the Energy-Efficient and New Energy Vehicle. Available online: http://www.miit.gov.cn/n1146285/n1146352/n3054355/n3057585/n3057592/index.html (accessed on 25 December 2016).
5. Wang, S.; Chen, K.; Zhao, F.; Hao, H. Technology pathways for complying with Corporate Average Fuel Consumption regulations up to 2030: A case study of China. *Appl. Energy* **2019**, *241*, 257–277. [CrossRef]
6. SAE-China. *Technology Roadmap for Energy-Saving and New Energy Vehicles*; Chinese Society of Automotive Engineers: Beijing, China, 2016.
7. Liu, Z.; Hao, H.; Cheng, X.; Zhao, F. Critical issues of energy efficient and new energy vehicles development in China. *Energy Policy* **2018**, *115*, 92–97. [CrossRef]
8. Zhao, F.; Hao, H.; Liu, Z. Technology strategy to meet China's 5 L/100 km fuel consumption target for passenger vehicles in 2020. *Clean Technol. Environ. Policy* **2015**, *18*, 7–15. [CrossRef]
9. Weiss, M.; Zerfass, A.; Helmers, E. Fully electric and plug-in hybrid cars–An analysis of learning rates, user costs, and costs for mitigating CO2 and air pollutant emissions. *J. Clean Prod.* **2019**, *212*, 1478–1489. [CrossRef] [PubMed]
10. Badin, F. Energy efficiency evaluation of a Plug-in hybrid vehicle under European procedure, world harmonized procedure and actual use. *World Electr. Veh. J.* **2015**, *7*, 475–488. [CrossRef]
11. Ahmadian, A. A Review on Plug-in Electric Vehicles: Introduction, Current Status, and Load Modeling Techniques. *J. Mod. Power Syst. Clean Energy* **2020**, *8*, 412–425. [CrossRef]
12. Singh, K.V.; Bansal, H.O.; Singh, D. A comprehensive review on hybrid electric vehicles: Architectures and components. *J. Mod. Transp.* **2019**, *27*, 77–107. [CrossRef]
13. Shams-Zahraei, M. A Study on Plug-in Hybrid Electic Vehicles. In Proceedings of the 2009 IEEE Region 10 Conference, Singapore, 23–26 November 2009; pp. 1–5.
14. Li, M.; Peng, W.; Yan, F. An investigation into the comprehensive evaluation method of the energy consumption of PHEV. *Automot. Eng.* **2014**, *8*, 919–922.
15. Cubito, C.; Millo, F.; Boccardo, G.; Di Pierro, G.; Ciuffo, B.; Fontaras, G.; Serra, S.; Garcia, M.O.; Trentadue, G. Impact of Different Driving Cycles and Operating Conditions on CO_2 Emissions and Energy Management Strategies of a Euro-6 Hybrid Electric Vehicle. *Energies* **2017**, *10*, 1590. [CrossRef]
16. Pavlovic, J.; Tansini, A.; Fontaras, G.; Ciuffo, B.; Otura, M.G.; Trentadue, G.; Suarez-Bertoa, R.; Millo, F. *The Impact of WLTP on the Official Fuel Consumption and Electric Range of Plug-in Hybrid Electric Vehicles in Europe*; SAE: Warrendale, PA, USA, 2017.
17. Tsiakmakis, S.; Fontaras, G.; Ciuffo, B.; Samaras, Z. A simulation-based methodology for quantifying European passenger car fleet CO_2 emissions. *Appl. Energy* **2017**, *199*, 447–465. [CrossRef]

18. Tsiakmakis, S.; Fontaras, G.; Cubito, C.; Pavlovic, J.; Anagnostopoulos, K.; Ciuffo, B. *From NEDC to WLTP: Effect on the Type-Approval CO_2 Emissions of Light-Duty Vehicles*; Publications Office of the European Union: Brussels, Belgium, 2017; p. 50.
19. Solouk, A.; Kapadia, J.; Masterson, B.; Shakiba-herfeh, M. *Impacts of WLTP Test Procedure on Fuel Consumption Estimation of Common Electrified Powertrains*; SAE: Warrendale, PA, USA, 2019.
20. Pavlovic, J.; Ciuffo, B.; Fontaras, G.; Valverde, V.; Marotta, A. How much difference in type-approval CO2 emissions from passenger cars in Europe can be expected from changing to the new test procedure (NEDC vs. WLTP)? *Transp. Res. Part A Policy Pract.* **2018**, *111*, 136–147. [CrossRef]
21. GB/T 19753-2019. Test Methods for Energy Consumption of Light-Duty Hybrid Electric Vehicles. 2019. Available online: http://www.catarc.org.cn/StandardRevision/detail/1472.html (accessed on 23 September 2019).
22. Zhang, P.; Wang, X.; Tang, H. Analysis and comparison of Vehicle Sliding Resistance Based on CHINA 5 and CHINA 6. *Automible Applied Technol.* **2018**, *17*, 152–153.
23. Kim, N.; Moawad, A.; Shidore, N.; Rousseau, A. Fuel Consumption and Cost Potential of Different Plug-In Hybrid Vehicle Architectures. *SAE Int. J. Altern. Powertrains* **2015**, *4*, 88–99. [CrossRef]
24. Marotta, A.; Pavlovic, J.; Ciuffo, B.; Serra, S.; Fontaras, G. Gaseous Emissions from Light-Duty Vehicles: Moving from NEDC to the New WLTP Test Procedure. *Environ. Sci. Technol.* **2015**, *49*, 8315–8322. [CrossRef]
25. Guo, Q.; Zhao, D.; Chen, P. Comparison of WLTC & NEDC and Preliminary Study of Their Impact on Automobile Fuel Consumption. *Chin. J. Automot. Eng.* **2017**, *7*, 196–204.
26. Jon, A.; John, M.; Cecile, F.; Dirk, B.; Simon, D.V.; Matthew, H.; Matthew, K.; Jonathon, M. On-Road and Chassis Dynamometer Evaluations of Emissions from Two Euro 6 Diesel Vehicles. *SAE Int. J. Fuels Lubr.* **2014**, *7*, 919–934.
27. Bielaczyc, P.; Woodburn, J.; Szczotka, A. In Exhaust Emissions of Gaseous and Solid Pollutants Measured over the NEDC, FTP-75 and WLTC Chassis Dynamometer Driving Cycles. In Proceedings of the SAE 2016 World Congress and Exhibition, Detroit, MI, USA, 12–14 April 2016.
28. Bielaczyc, P.; Woodburn, J.; Szczotka, A. *A Comparison of Carbon Dioxide Exhaust Emissions and Fuel Consumption for Vehicles Tested over the NEDC, FTP-75 and WLTC Chassis Dynamometer Test Cycles*; SAE: Warrendale, PA, USA, 2015.
29. Pavlovic, J.; Marotta, A.; Ciuffo, B. CO_2 emissions and energy demands of vehicles tested under the NEDC and the new WLTP type approval test procedures. *Appl. Energy* **2016**, *177*, 661–670. [CrossRef]
30. GB/T 19753-2013. Test Methods for Energy Consumption of Light-Duty Hybrid Electric Vehicles. 2013. Available online: http://std.samr.gov.cn/gb/search/gbDetailed?id=71F772D7EDD0D3A7E05397BE0A0AB82A (accessed on 23 December 2013).
31. Paffumi, E.; De Gennaro, M.; Martini, G. Alternative utility factor versus the SAE J2841 standard method for PHEV and BEV applications. *Transp. Policy* **2018**, *68*, 80–97. [CrossRef]
32. ICCT (The International Council On Clean Transportation). *How to Measure Fuel Consumption and CO2 Emission of Plug-in Hybrid Vehicles, Today and in the Future*; ICCT: San Francisco, CA, USA, 2017.
33. Goebel, D.; Plötz, P. Machine learning estimates of plug-in hybrid electric vehicle utility factors. *Transp. Res. Part D Transp. Environ.* **2019**, *72*, 36–46. [CrossRef]
34. National Technical Committee of Auto Standardization. Draft for Consultation on Technical Conditions of Plug-in Hybrid Electric Passenger Vehicles, 2019. Available online: http://www.catarc.org.cn/StandardRevision/detail/1471.html (accessed on 23 September 2019).
35. Wang, S.; Zhao, F.; Liu, Z.; Hao, H. Impacts of a super credit policy on electric vehicle penetration and compliance with China's Corporate Average Fuel Consumption regulation. *Energy* **2018**, *155*, 746–762. [CrossRef]
36. Zhao, F.; Chen, K.; Hao, H.; Wang, S.; Liu, Z. Technology development for electric vehicles under new energy vehicle credit regulation in China: Scenarios through 2030. *Clean Technol. Environ. Policy* **2018**, *21*, 275–289. [CrossRef]
37. Hao, H.; Ou, X.; Du, J.; Wang, H.; Ouyang, M. China's electric vehicle subsidy scheme: Rationale and impacts. *Energy Policy* **2014**, *73*, 722–732. [CrossRef]
38. Ministry of Industry and Information Technology of People's Republic of China. Amendments to the Measures for Parallel Management of Average Fuel Consumption of Passenger Vehicle Enterprises and Points of New Energy Vehicles (Draft for Comments), 2019. Available online: http://www.miit.gov.cn/n1146285/n1146352/n3054355/n3057585/n3057592/c7027601/content.html (accessed on 9 June 2019).

39. The State Council. Notice on Improving Fiscal Subsidy Policies for the Promotion and Application of New Energy Vehicles, 2020. Available online: http://www.gov.cn/zhengce/zhengceku/2020-04/23/content_5505502.htm (accessed on 23 April 2020).
40. Zhou, B.; Zhang, S.; Wu, Y.; Ke, W.; He, X.; Hao, J. Energy-saving benefits from plug-in hybrid electric vehicles: Perspectives based on real-world measurements. *Mitig. Adapt. Strateg. Global Chang.* **2017**, *23*, 735–756. [CrossRef]

sustainability

MDPI

Article

Hybrid Wind–PV Frequency Control Strategy under Variable Weather Conditions in Isolated Power Systems

Ana Fernández-Guillamón [1,*,†], Guillermo Martínez-Lucas [2,†], Ángel Molina-García [1,†] and Jose-Ignacio Sarasua [2,†]

1 Department of Automatics, Electrical Engineering and Electronic Technology, Universidad Politécnica de Cartagena, 30202 Cartagena, Spain; angel.molina@upct.es

2 Department of Hydraulic, Energy and Environmental Engineering, Universidad Politécnica de Madrid, 28040 Madrid, Spain; guillermo.martinez@upm.es (G.M.-L.); joseignacio.sarasua@upm.es (J.-I.S.)

* Correspondence: ana.fernandez@upct.es; Tel.: +34-968-325-357

† These authors contributed equally to this work.

Received: 7 August 2020; Accepted: 15 September 2020; Published: 19 September 2020

Abstract: Over the last two decades, variable renewable energy technologies (i.e., variable-speed wind turbines (VSWTs) and photovoltaic (PV) power plants) have gradually replaced conventional generation units. However, these renewable generators are connected to the grid through power converters decoupled from the grid and do not provide any rotational inertia, subsequently decreasing the overall power system's inertia. Moreover, the variable and stochastic nature of wind speed and solar irradiation may lead to large frequency deviations, especially in isolated power systems. This paper proposes a hybrid wind–PV frequency control strategy for isolated power systems with high renewable energy source integration under variable weather conditions. A new PV controller monitoring the VSWTs' rotational speed deviation is presented in order to modify the PV-generated power accordingly and improve the rotational speed deviations of VSWTs. The power systems modeled include thermal, hydro-power, VSWT, and PV power plants, with generation mixes in line with future European scenarios. The hybrid wind–PV strategy is compared to three other frequency strategies already presented in the specific literature, and gets better results in terms of frequency deviation (reducing the mean squared error between 20% and 95%). Additionally, the rotational speed deviation of VSWTs is also reduced with the proposed approach, providing the same mean squared error as the case in which VSWTs do not participate in frequency control. However, this hybrid strategy requires up to a 30% reduction in the PV-generated energy. Extensive detailing of results and discussion can be also found in the paper.

Keywords: frequency control; power system stability; variable renewable energy sources; wind power plants; photovoltaic power plants

1. Introduction

Over the last two decades, power systems' generation has slowly been changing, replacing conventional generation units (mainly based on the fossil and nuclear fuels) with variable renewable energy sources (vRESs) [1]. This transition has been supported by several aspects, including environmental concern (especially greenhouse gas emissions) and the aim of decreasing the energy dependence of third countries [2–4]. vRES refers to variable-speed wind turbines (VSWTs) and photovoltaic (PV) power plants, which have a stochastic behavior due to their dependence on weather conditions [5]. Together with this point, vRESs are connected to the grid through power inverters decoupled from the grid and, hence, do not inherently provide any inertial or frequency response under power imbalances [6]. Consequently, the high penetration levels of vRESs will substantially challenge the supply security and reliability of future power systems, thus requiring more flexibility to maintain the balance between generation and demand [7–9]. Moreover, this real-time power balance is required for a stable grid frequency [10].

As an attempt to improve the frequency response of power systems with high vRES integration, several frequency control techniques have been proposed for such generation units in the specific literature [11,12]. In fact, they are usually classified as depicted in Figure 1. Together with these strategies, energy storage systems (i.e., batteries, flywheels, and supercapacitors) have also been considered as suitable options to maintain grid stability [13,14]. With regard to PV power plants, as they are static elements without any rotating parts, they do not have any stored kinetic energy [15]. Therefore, their synchronous inertia constant is nearly zero ($H_{PV} \approx 0$) [16]. Consequently, they can only provide frequency response based on de-loading strategies. The de-loading technique implies the reduction of the supplied active power under normal operation conditions, going towards the right part of the P–V curve (refer to Figure 2). In this way, when grid frequency is lower than its nominal value, the PV installations can slightly decrease their voltage (from V_{del} to V_{MPP}, see Figure 2) and subsequently increase the corresponding generated power [17]. The ΔV decrease is usually estimated based on the grid frequency deviation Δf through a proportional or proportional–integral controller. Similarly to PV power plants, VSWTs can also provide frequency response using the de-loading technique. In this case, it is performed by pitching the blades or over-speeding the rotor [18]. De-loading strategies imply a reduction of the electrical power generated (both for PV and VSWTs) and, subsequently, a considerable decrease in the benefits for their owners [19]. As VSWTs have some stored kinetic energy due to their blades, drive train, and the electrical generator, some authors affirm that VSWTs' inertia constant H_{WT} is in line with those of conventional power plants, but it is hidden from the power system point of view due to the power converter [20,21]. Consequently, together with the de-loading approach, VSWTs can provide such rotational inertia to the grid through different inertial response strategies. These strategies require at least one supplementary control loop to be included in the power controller [22]. Under frequency deviation conditions, the frequency control loop(s) will provide an additional input to the VSWTs, aiming to temporarily increase their active power generation. With this type of controller, VSWTs can always generate their maximum active power according to each wind speed value, in contrast to the de-loading technique, in which they slightly curtailed certain amounts of power. As a result, inertial response approaches are preferable to the de-loading technique, since the wind resource is better used [23]. However, inertial response approaches are limited by the VSWTs' speed governor, which prevents the rotational speed of the VSWTs from decreasing below the minimum allowed value [24]. Tables 1 and 2 summarize different vRES frequency control strategies proposed in the specific literature over the last years, including the type of control, PV/VSWT integration level, and the power imbalance (ΔP) that causes the frequency deviation. As can be seen in Tables 1 and 2, these frequency control strategies are usually analyzed under severe power imbalances (up to 50%). However, in isolated power systems with high wind–PV integration, the variability of wind speed and solar irradiation commonly causes large deviations in the system frequency [25]. Indeed, these fluctuations pose stress on the power system operation, as transmission and distribution system operators (TSOs/DSOs) deal with not only unmanageable demand, but also uncontrollable generation [26]. Consequently, there is an acute need to improve the frequency response of isolated power systems with high vRES integration, especially under typical variable meteorological changes. In fact, over the last decade, some studies have already focused on this topic. However, most of the works proposed are based on including energy storage systems [27–34], demand response [35], and electric vehicles [36], or using independent frequency controllers for VSWT and PV power plants [37–39].

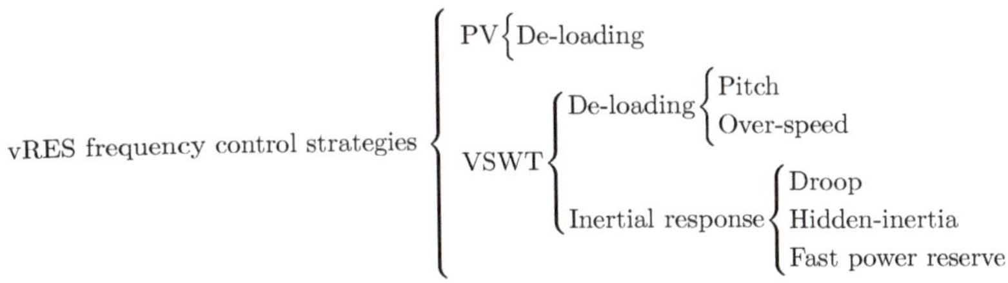

Figure 1. Frequency control strategies for variable renewable energy sources (vRESs).

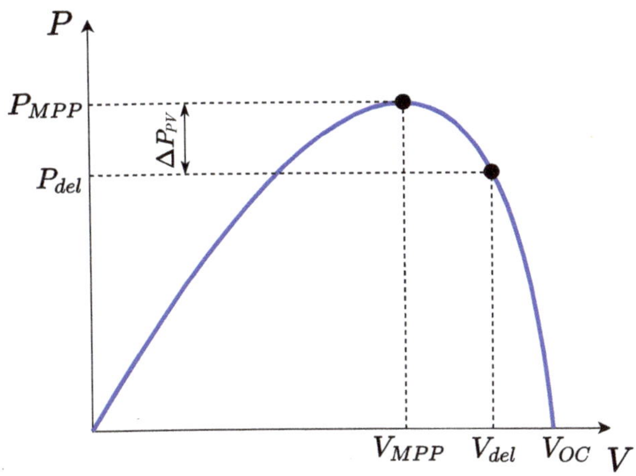

Figure 2. De-loading technique of photovoltaic (PV) power plants.

Table 1. PV frequency control proposals.

Ref.	Type of Control	PV Integration (%)	ΔP (%)	Year
[40]	De-loading	9	5	2012
[41]	De-loading	9	5	2012
[42]	De-loading	16	–	2013
[43]	De-loading	22	8.1	2014
[44]	De-loading	23–48	50	2017
[45]	De-loading	10–20–30	10	2019
[46]	De-loading	10	10	2019

Table 2. Variable-speed wind turbine (VSWT) frequency control proposals.

Ref.	Type of Control	VSWT Integration (%)	ΔP (%)	Year
[47]	De-loading (pitch)	—	—	2016
[48]	Droop	11, 29.5	15	2013
[49]	Hidden-inertia emulation	20	8.3	2015
[50]	Hidden-inertia emulation	—	10	2016
[51]	Hidden-inertia emulation	—	10	2019
[52]	Fast power reserve	20	10	2015
[53]	Fast power reserve	16.7–33	16.7	2016
[54]	Fast power reserve	5–45	2.5–10	2018

In contrast to the previous studies, this paper proposes a hybrid wind–PV frequency control strategy in which the VSWTs' rotational speed deviation is the proportional–integral (PI) controller input of the PV frequency response strategy. Similar approaches have been previously proposed by the authors, but by linking the rotational speed deviation of VSWTs to hydro-power plants instead of PV power plants [55]. The main contributions of this paper can be summarized as:

- A new hybrid wind–PV frequency control strategy is proposed. VSWTs include the hidden-inertia emulation technique, whereas PV power plants use the de-loading approach. The novelty of the hybrid control is that the PV frequency controller receives the VSWTs' rotational speed deviation as an input instead of the grid frequency deviation.
- The proposed controller is tested on an isolated power system consisting of thermal, hydro-power, VSWT, and PV power plants under six different scenarios. Frequency deviations are the result of the variability of both wind speed and solar irradiation, synthetically estimated (wind speed) and based on real measured values (solar irradiation).
- The frequency response is compared to three different frequency strategies: (*i*) conventional power plants; (*ii*) conventional power plants and wind power plants; and (*iii*) conventional power plants, wind power plants, and PV power plants with frequency deviation as input. Minor frequency oscillations were obtained with the hybrid wind–PV frequency strategy in terms of minimum and maximum frequency deviations and mean squared error (MSE) of frequency, as well as in terms of minimum and maximum rotational speed of the VSWTs and MSE of their rotational speed deviation.

The rest of the paper is organized as follows: The models used for the simulations are described in Section 2; the methodology followed to carry out this study is presented in Section 3; in Section 4, the results of the study are shown and analyzed; finally, Section 5 presents the main conclusions of the paper.

2. Proposed System Modeling

A mathematical simulation model is used to analyze the proposed hybrid wind–PV frequency controller. This power system and, consequently, the model implemented in Matlab/Simulink includes conventional generators (hydro-power and thermal power plants), PV power plants, and VSWTs as well as the power demand, as shown in Figure 3. The frequency control is in line with current European requirements (primary and secondary controls), as will be presented in Section 3. In the following subsections, the main components of the proposed model are described in detail.

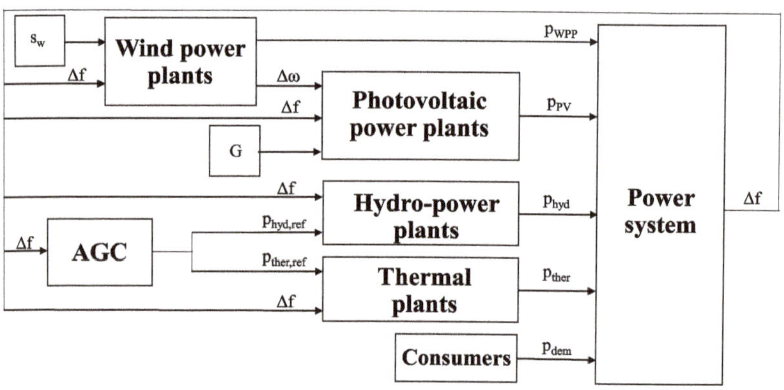

Figure 3. Block diagram of the model.

2.1. Power System and System Inertia

The power system is modeled considering an aggregated inertial model as proposed in [56]. This formulation has been previously used to model the Irish isolated power system [57] and El Hierro isolated power system [58], among others. Grid frequency variations are the result of the imbalance between the power generation and the corresponding demand (see Equation (1)):

$$f\,\frac{df}{dt} = \frac{1}{2\,H_{eq}}\left(p_w + p_{PV} + p_{hyd} + p_{ther} - p_{dem} - D_{net}\,\Delta f\right), \tag{1}$$

where p_w, p_{PV}, p_{hyd}, and p_{ther} represent the power supplied by VSWT, PV, hydro-power, and thermal power plants, respectively; p_{dem} is the total power demand; and D_{net} is the consumer load sensitivity factor to frequency variations. System inertia (H_{eq}) is estimated as [59]:

$$H_{eq} = H_{hyd} + H_{ther}, \tag{2}$$

where H_{hyd} and H_{ther} are the hydro-power and thermal power plants' inertia constants, respectively. The VSWT and PV power plants are connected to the grid through power converters and, therefore, do not inherently provide synchronous inertia to the grid, as already discussed in Section 1.

2.2. Conventional Power Plants

The conventional power plants (CPPs) considered in this paper are reheat thermal and hydro-power plants. Both of them are modeled following the transfer functions proposed in [60]. Thermal power plants' transfer function, shown in Figure 4a, provides the power variations of these power plants from the frequency deviation (Δf) and the automatic generation control (AGC) power reference ($p_{ther,ref}$). Figure 4b shows the transfer function to model the hydro-power plant, including the conduits' dynamics. The generated power p_{hyd} also depends on the frequency deviation Δf and the power reference signal provided by the AGC ($p_{hydr,ref}$). The parameters for both models in Figure 4 are listed in Table 3.

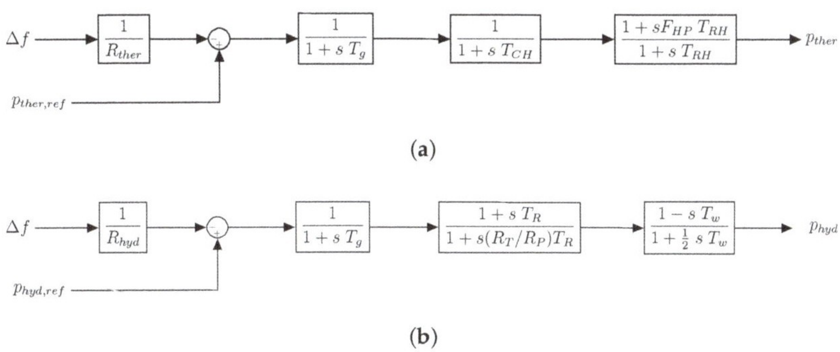

Figure 4. Block diagram of conventional power plants: (**a**) thermal power plants' transfer function and (**b**) hydro-power plants' transfer function.

2.3. Wind Power Plants

Four different wind power plants (WPPs) are considered by modeling each WPP with an aggregated wind turbine. Therefore, one equivalent wind turbine is then simulated by multiplying the corresponding generated power by the number of VSWTs of such a WPP [61]. The VSWT model includes the wind power model, which determines the power extracted from the wind speed; the blade pitch control to regulate the wind input torque; the torque maximum power point (MPP) tracking control, which restores the optimal rotor speed after a rotational speed deviation; and the frequency controller, which is discussed in Section 3.2. Further information with regard to the VSWT model can be found in [62]. The one-mass rotor mechanical model is used for simulations, and is acceptable when the voltage is assumed to be constant [63,64]:

$$\omega_{WT} = \frac{p_{mt} - p_W}{2\,H_{WT}\,s}\,. \tag{3}$$

The reference rotational speed ω_{ref} is determined from the measured power p_{ef}, which is the active generated power p_W after a delay T_f:

$$\omega_{ref} = -0.67 \times p_{ef}^2 + 1.42 \times p_{ef} + 0.51\,. \tag{4}$$

The VSWT model is represented in Figure 5. The values of the different parameters are listed in Table 3.

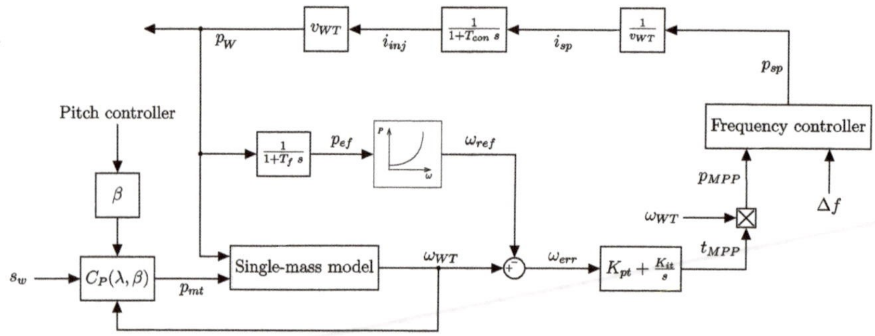

Figure 5. Block diagram of a VSWT.

2.4. PV Power Plant

Four different PV power plants are also considered. They are modeled following the diode equation, where the output current I_{PV} is given by [65,66]:

$$I_{PV} = N_P \times I_{ph} - N_P \times I_{rs} \times \left[\exp\left(\frac{q}{k_B \times T_C \times A} \times \frac{V_{PV}}{N_s}\right) - 1\right]. \tag{5}$$

N_s is the number of cells in series, N_P is the number of strings in parallel, I_{ph} is the photo-current, q is the electron charge, k_B is Boltzmann's constant, T_C is the temperature of the cell, A is the diode ideality factor, and V_{PV} is the PV voltage. The photo-current I_{ph} is calculated by:

$$I_{ph} = [I_{sc} + k_I(T_C - T_{STC})] \times G, \tag{6}$$

where I_{sc} is the PV cell's short-circuit current, k_I is the short-circuit current temperature coefficient of the PV cell, T_{STC} is the temperature under Standard Test Conditions (STC), and G is the solar irradiation. The PV cell's reverse saturation current I_{rs} follows Equation (7), where $I_{r,ref}$ is the reverse saturation current at T_{STC}, and E_G is the band-gap energy of the PV cell's material:

$$I_{rs} = I_{r,ref} \times \left(\frac{T_C}{T_{STC}}\right)^3 \times \exp\left[\frac{q \times E_G}{k_B \times A} \times \left(\frac{1}{T_{STC}} - \frac{1}{T_C}\right)\right]. \tag{7}$$

The voltage at the MPP (V_{MPP}) for a given irradiation G and cell temperature T_C is estimated depending on its value at STC and a correction factor α_V:

$$V_{MPP} = V_{MPP,STC} \times \left(\frac{\ln(G)}{\ln(1000)}\right) \times (1 + \alpha_V \times (T_C - T_{STC})). \tag{8}$$

By multiplying the PV voltage V_{PV} by the output PV current I_{PV}, the active power of the PV generator is then estimated. In the case that the PV voltage corresponds to the MPP voltage V_{MPP}, the output power of the PV power plants would be the maximum available active power under such G and T_C conditions. The values of the different parameters considered are listed in Table 3.

Table 3. Power plant parameters [60,62,66].

Thermal		Hydro-Power		VSWTs		PV	
Parameter	**Value**	**Parameter**	**Value**	**Parameter**	**Value**	**Parameter**	**Value**
R_{ther}	0.05	R_{hyd}	0.05	T_f	5	k_I	0.0017
T_g	0.2	T_g	0.2	T_{con}	0.02	$I_{r,ref}$	1.2×10^{-7}
T_{CH}	0.3	T_R	5	v_{WT}	1	α_v	-5.53×10^{-3}
F_{HP}	0.3	R_T	0.38	K_{pt}	1	q	1.602×10^{-19}
T_{RH}	7	R_P	0.05	K_{it}	0.2	k_B	1.38×10^{-23}
H_{ther}	5	T_w	1			A	1.92
		H_{hydro}	3.3			I_{sc}	8.03
						E_g	1.12

3. Methodology

Frequency control strategies focus on minimizing grid frequency variations due to generation–demand mismatches [67]. With this aim, generation units must increase/decrease their generation to equal the total power demand (plus the power system losses) [68]. In the following subsections, frequency controls for CPPs, VSWTs, and PV power plants are described.

3.1. Frequency Control in Conventional Power Plants

CPPs based on synchronous generators inherently release or absorb kinetic energy as a natural inertial response to frequency deviations. In Europe, frequency control with CPPs has a hierarchical structure, usually organized in primary control (frequency containment reserves), secondary control (frequency restoration reserves), and tertiary control (replacement reserves) [69]. Primary frequency control (PFC) is automatically activated by the generator units some seconds after the power imbalance. The CPP power response is proportional to the frequency deviation Δf, following Equation (9), in which R is the droop characteristic [70,71]:

$$\Delta p = -\frac{\Delta f}{R}. \tag{9}$$

According to the European network of transmission system operators for electricity (ENTSO-E), R should range between 2% and 12%, with a dead-band between 10 and 30 mHz in which the PFC is not activated [72]. For this work, R is considered as 5% for both thermal and hydro-power plants, and the dead-band is established at 30 mHz. Subsequent to the PFC action, there is still some remaining frequency deviation due to the power imbalance. The secondary frequency control is then required to completely remove this frequency deviation [73,74]. In fact, this secondary frequency control is in charge of modifying the power generation set-point accordingly [75]. The AGC coordinates the effort's dispatch among the different CPPs of the secondary frequency control. In this paper, the equivalent total secondary regulation effort (ΔRR) is determined by:

$$\Delta RR = -\Delta f \times K_f, \tag{10}$$

where K_f is estimated following the ENTSO-E's recommendations [76]. This ΔRR is distributed between the thermal and hydro-power plants depending on their participation factors ($K_{u,i}$), determined according to the droop of each power plant [77]:

$$\Delta p_{i,ref} = \frac{1}{T_{u,i}} \int \Delta RR \times K_{u,i}\, dt = \frac{-1}{T_{u,i}} \times K_{u,i} \times K_f \times \int \Delta f\, dt, \tag{11}$$

where i represents *hyd* or *ther*, and $K_{u,hyd} + K_{u,ther} = 1$.

3.2. VSWT Frequency Control Strategy

Among the three different inertial response strategies for VSWTs (refer to Figure 1), the hidden-inertia emulation technique is selected for this work. The hidden-inertia emulation technique is based on a proportional–derivative (PD) control loop, with Δf as input. This PD controller provides an additional power proportional to the frequency deviation and its derivative value:

$$\Delta p_{FC} = K_d \times \frac{d\Delta f}{dt} + K_p \times \Delta f\,. \tag{12}$$

The derivative part represents the inertial control loop (emulating the hidden-inertia of the VSWTs). These inertial (derivative) and proportional control loops allow that some of the kinetic energy stored in the rotating masses of the VSWTs (i.e., rotor, drive train, and electrical generator) is released to provide a fast frequency response from the power converter's capability. This Δp_{FC} signal is added to the power reference output depending on the wind speed.

3.3. PV Frequency Control Strategy

The de-loading technique is used for PV frequency control. This strategy curtails a certain amount of active power, working the PV power plant on a de-loaded voltage value (V_{del}) on the right side of the MPP voltage (V_{MPP}), as shown in Figure 2. Note that $V_{del} > V_{MPP}$. In the following subsections, the traditional de-loading strategy and the proposed hybrid wind–PV frequency control approach are analyzed.

3.3.1. Conventional De-Loaded PV Frequency Control Strategy

To provide frequency response, the PV power plant usually works at the de-loaded voltage V_{del}. In the case of frequency reduction, the PV voltage should be also reduced, moving towards V_{MPP} to increase its generated active power. The strategies previously proposed in the specific literature include a proportional (P) or proportional–integral (PI) controller, with the frequency deviation (Δf) as an input of such a controller [40–43]. Frequency deviation (Δf) passes through the corresponding control loop (P or PI), giving an additional voltage ΔV (refer to Figure 6). The modified PV voltage is then determined as:

$$V_{PV} = V_{del} - \Delta V\,. \tag{13}$$

This V_{PV} is used in Equation (5) to determine the PV current, I_{PV}, which is the corresponding PV active power:

$$P_{PV} = I_{PV} \times V_{PV}\,. \tag{14}$$

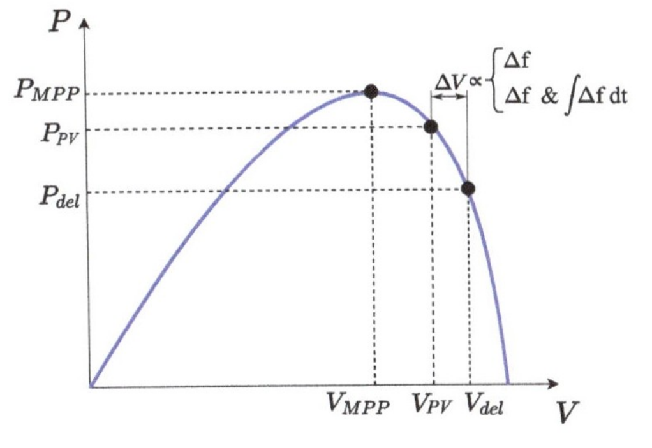

Figure 6. Conventional de-loaded PV frequency control.

3.3.2. Hybrid Wind–PV Frequency Control

In this work, and in contrast to previous studies, the authors propose the use of the VSWT rotational speed deviation $\Delta\omega$ as an input for the de-loaded PV frequency controller. Due to the VSWTs' hidden-inertia frequency controller, the rotational speed ω_j of each WPP deviates from the reference value $\omega_{ref,j}$, following Equation (15):

$$\Delta\omega_j = \omega_j - \omega_{ref,j}\,, \tag{15}$$

where $\omega_{ref,j}$ is calculated from Equation (4). The global $\Delta\omega$ sent to the PV power plants is the sum of the $\Delta\omega_j$ of each WPP:

$$\Delta\omega = \sum_{j=1}^{4} \Delta\omega_j. \tag{16}$$

This global $\Delta\omega$ will pass through the PI controller, obtaining the value of ΔV of Equation (13). An overview of the hybrid wind–PV frequency control approach is shown in Figure 7. By using $\Delta\omega$ as an input, PV power plants modify their generated power according to this rotational speed deviation, improving the VSWTs' rotational speed control and subsequently increasing the inertial control performance of the VSWTs.

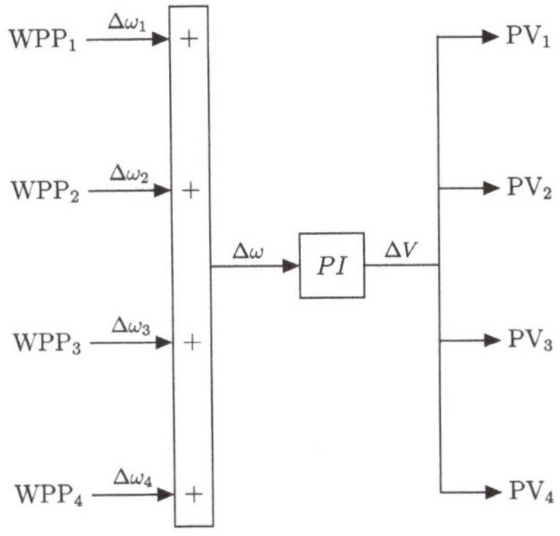

Figure 7. Hybrid wind–PV frequency control approach.

4. Results

4.1. Scenarios under Consideration

The Ten-Year Network Development Plan (TYNDP) 2020 Scenario Report is used to propose the generation mix under consideration [78]. National trend scenarios are taken into account by the authors, which account for both the supply and demand data collected from the European electricity TSOs. Figure 8 shows the generation mix considered in this paper following [78] for the years 2025 and 2040.

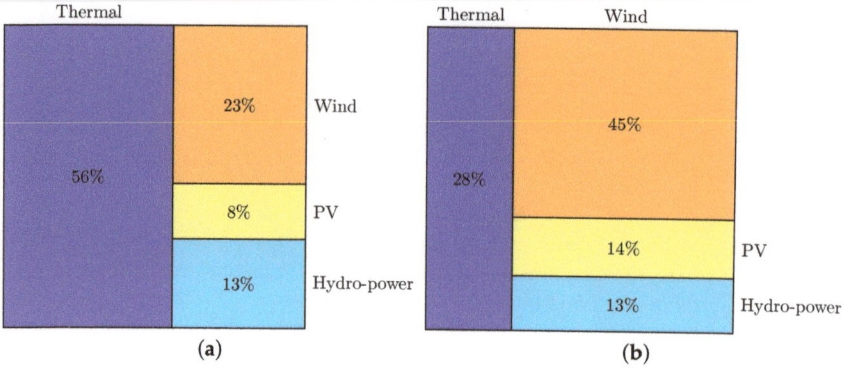

Figure 8. Generation mix following the national trends of TYNDP in (**a**) year 2025 and (**b**) year 2040.

Three different electricity demand values (i.e., valley, mean, and peak periods) are also studied based on the Gran Canaria isolated power system (Spain). This power system has been extensively described in [79]. Consequently, by combining the two generation mixes and the three different demand values, six scenarios are under analysis. Power demand is assumed as constant during the simulations, as the timeframe under analysis is 5 min (300 s). As presented in Table 4, this assumption is in line with most TSOs.

Table 4. Time interval where demand is considered as constant for different transmission system operators (TSOs).

TSO	Location	Time (min)	Website
ENMAX	Canada	15	[80]
ERGON	Australia	15	[81]
RTE	France	15	[82]
REE	Spain	10	[83]
IESO	Canada	5	[84]
CAISO	California	5	[85]
TEPCO	Japan	5	[86]
TRANSPOWER	New Zealand	5	[87]

Wind speed and solar irradiation present some oscillations during the simulation time intervals. These variations cause certain frequency deviations from the nominal value (50 Hz). Real measured values of solar irradiation from a PV power plant in Albacete (Spain) are used for simulation purposes. In addition, synthetic wind speeds estimated from the methodology detailed in [88] are also included in the scenarios. Such synthetic wind speeds have been statistically compared to real measured wind speed series, obtaining similar values.

As mentioned in Section 2, four different WPPs and PV power plants are considered:

$$p_{WPP} = \sum_{j=1}^{4} p_{WPP,j}, \tag{17}$$

$$p_{PV} = \sum_{k=1}^{4} p_{PV,k}. \tag{18}$$

A heterogeneous distribution of wind- and PV-generated power is estimated to give more realistic simulations, considering that WPP_1 and PV_1 account for 50% of the total p_{WPP} and p_{PV}; WPP_2 and PV_2 are 25% of the total p_{WPP} and p_{PV}; and WPP_3 =WPP_4 and PV_3 =PV_4 are 12.5% of the total p_{WPP} and p_{PV}, respectively. The wind speed values s_w and solar irradiation G for each WPP and PV power plant are depicted in Figures 9 and 10.

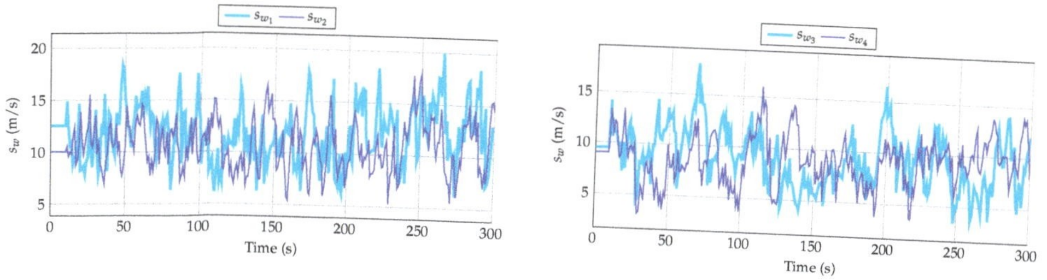

(**a**) (**b**)

Figure 9. Wind speeds considered for the wind power plants (WPPs): (**a**) WPP_1 and WPP_2 as well as (**b**) WPP_3 and WPP_4.

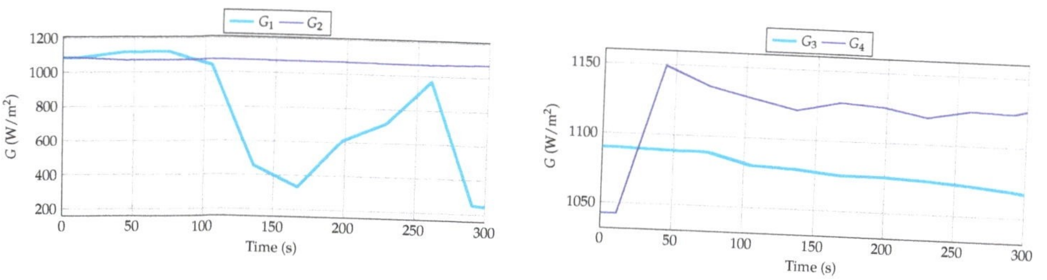

(**a**) (**b**)

Figure 10. Solar irradiations considered for (**a**) PV_1 and PV_2 as well as (**b**) PV_3 and PV_4.

4.2. Simulation Results

The six scenarios presented in Section 4.1 were simulated in a Matlab/Simulink environment. A fixed step with the ode3 (Bogcki-Shampine) solver and a step size of 10^{-2} s was used. The Matlab version was r2016a. Four different frequency control strategies were compared:

1. Frequency control is only provided by conventional power plants (referred to as *CPP*).
2. Frequency control is provided by conventional power plants and WPPs with a hidden-inertia emulation technique (referred to as *WPP*).
3. Frequency control is provided by conventional power plants, WPPs with a hidden-inertia emulation technique, and PV power plants with 10% de-loading and a P controller with Δf as input (referred to as *PV* (f)).

4. Frequency control is provided by conventional power plants, WPPs with a hidden-inertia emulation technique, and PV power plants with 10% de-loading and a *PI* controller with $\Delta\omega$ of the VSWTs as input (referred to as *PV* (ω)), which is the hybrid wind–PV frequency strategy proposed in this paper.

In each simulation, the authors considered different criteria to compare the frequency control strategies under consideration. Specifically, the parameters taken into account are the following: the minimum and maximum frequency values, the MSE of the grid frequency, the PV-generated energy, the MSE of the thermal and hydro-power plants with respect to their initial assigned power, the minimum and maximum values of the rotational speed of each WPP, and the MSE of the rotational speed of each WPP with respect to their reference value (ω_{ref}) for each wind speed. Tables 5–9 show the results of the parameters for the six scenarios simulated under the different frequency control strategies. The PV-generated energy is determined by multiplying the active power P_{PV} by the time interval under analysis (5 min).

Table 5. Simulation results: frequency, PV electrical energy, and the conventional power plants' mean square error (MSE).

Scenario	Load (MW)	250		400		550	
	Year	2025	2040	2025	2040	2025	2040
f_{min} (Hz)	CPP	49.28	44.89	49.31	46.24	49.34	45.74
	WPP	49.54	48.49	49.55	48.59	49.57	48.63
	PV (f)	49.58	48.88	49.61	48.85	49.60	49.00
	PV (ω)	49.58	48.94	49.61	48.87	49.61	49.20
f_{max} (Hz)	CPP	50.85	54.30	50.84	54.16	50.78	54.08
	WPP	50.47	50.74	50.47	50.71	50.45	50.72
	PV (f)	50.26	50.41	50.29	50.48	50.32	50.48
	PV (ω)	50.23	50.25	50.18	50.23	50.17	50.54
MSE_f (Hz^2)	CPP	0.072	1.481	0.069	1.123	0.063	1.154
	WPP	0.035	0.243	0.034	0.216	0.032	0.203
	PV (f)	0.019	0.108	0.021	0.111	0.023	0.104
	PV (ω)	0.015	0.087	0.014	0.082	0.015	0.085
E_{PV} (MWh)	CPP	1.536	2.878	2.636	4.645	3.424	5.955
	WPP	1.536	2.878	2.636	4.645	3.424	5.955
	PV (f)	1.269	2.391	2.251	3.930	3.096	5.010
	PV (ω)	1.072	2.186	1.962	3.610	2.740	4.091
$MSE_{P_{ther}}$ (MW^2)	CPP	105.4	198.7	258.1	515.6	442.0	932.0
	WPP	83.40	171.8	206.3	442.2	370.9	795.8
	PV (f)	56.98	128.2	143.9	320.5	290.7	555.3
	PV (ω)	73.62	150.3	170.5	359.4	328.7	773.7
$MSE_{P_{hyd}}$ (MW^2)	CPP	2.931	41.51	7.272	97.38	12.51	177.1
	WPP	2.579	31.51	6.426	74.60	11.69	130.8
	PV (f)	2.029	20.28	4.939	49.68	9.798	81.92
	PV (ω)	3.021	26.07	7.021	61.54	13.19	127.3

With respect to the frequency deviations, the proposed hybrid wind–PV technique results in the maximum f_{min} and the minimum f_{max} values. This means that smaller frequency deviations are obtained, which is also shown in the MSE. In fact:

- A reduction of the MSE between 75% and 95% is obtained when the proposed hybrid wind–PV frequency strategy is used, in contrast to the CPP approach.
- A reduction of the MSE between 50% and 65% is obtained when the proposed hybrid wind–PV frequency strategy is used, in contrast to the WPP approach.
- A reduction of the MSE between 20% and 35% is obtained when the proposed hybrid wind–PV frequency strategy is used, in contrast to the PV(f) approach.

Moreover, the authors would like to highlight that similar results are obtained for the three different power demand scenarios in terms of minimum and maximum frequency deviations, together with the MSE of frequency. Consequently, it can be affirmed that this study is scalable to other isolated power systems, and similar results will be obtained independently of the demand as long as the generation mixes are similar to those considered here, which are in line with future European renewable energy integration roadmaps. It is important to remark that, for the 2025 scenario, frequency deviations are nearly within the acceptable range proposed by the ENTSO-E, i.e., ±800 mHz [89], even if the vRESs do not participate in frequency control. However, as vRESs massively replace CPPs, frequency response will be substantially deteriorated. In fact, minimum and maximum frequency values under 45 Hz and over 54 Hz, respectively, are obtained if only CPPs are considered for frequency control response for year 2040. This means that, in the medium term, it is a crucial need that vRESs participate in frequency control to avoid such negative effects due to important (and negative) frequency oscillations.

Focusing on the PV-generated energy, the hybrid wind–PV technique gets the minimum values, mainly due to the initial 10% power de-loading and the action based on the $\Delta\omega$ that comes from VSWTs. Specifically:

- Considering the CPP and WPP strategies, the PV power plants work on their MPP and, subsequently, their generated energy is the maximum among the four strategies (and the same for both cases).
- Considering the PV(f) strategy, the PV power plants are de-loaded by 10%. A reduction between 10% and 20% of the PV-generated energy is then obtained in comparison to the CPP and WPP strategies.
- A reduction between 20% and 30% of the PV-generated energy is obtained by using the proposed hybrid wind–PV control compared to the CPP and WPP strategies.

Therefore, the hybrid wind–PV strategy implies an additional PV energy reduction of up to 10% in comparison to a conventional PV frequency control strategy. This aspect should be subsequently evaluated by the PV installation operators. Indeed, their benefits should be partially reduced if TSOs/DSOs do not reward them for providing frequency control services.

With regard to the conventional power plants, including any frequency strategy of vRESs reduces the contribution of thermal and hydro-power to the frequency response. The MSE between the power generated by them and their initial assigned value is thus reduced:

- Comparing the CPP and WPP strategies, the use of VSWTs for frequency control reduces the MSE of thermal power plants by between 14%–20%, with a reduction between 12%–24% for hydro-power plants.
- Comparing the CPP and PV(f) strategies, including a conventional de-loading frequency control strategy for PV power plants reduces the MSE of thermal power plants by between 35%–46%, with a reduction between 31%–51% for hydro-power plants.
- Comparing the CPP and PV(ω) strategies, using the hybrid wind–PV control approach reduces the MSE of thermal power plants between 24%–30%, with a reduction between 5%–35% for hydro-power plants. Moreover, there are some cases in which the MSE of the hydro-power plant is slightly increased.

The PV(f) strategy is then the best technique from the point of view of the MSE of conventional power plants.

Finally, focusing on the rotational speed of WPPs (Tables 6–9):

- The CPP strategy has the smallest variations of rotational speeds. This is due to the fact that such rotational speed variations are only the result of wind speed changes.
- The WPP strategy has the largest variations of rotational speed values. In fact, both the minimum/maximum values of ω_j are obtained with this technique, even though they are small

variations of around 5%–10%. Consequently, the maximum MSE is obtained with the WPP strategy. In some cases, the MSE result is three times higher than the value obtained with the CPP approach. Naturally, the speed deviations with this strategy are the result of both the wind speed changes and the hidden-inertia frequency control approach.

- The PV(f) technique slightly improves the minimum/maximum rotational speed values and the MSE (if comparing to the WPP strategy). However, these values are still worse than with the CPP approach.
- The PV(ω) strategy reduces the minimum and maximum values of the rotational speed even more, and, consequently, reduces the MSE (if comparing to the WPP and PV(f) strategies). In fact, there are some cases in which the MSE is quite similar for both the CPP and PV(ω) techniques.

It can be affirmed that the PV(ω) technique is the best overall frequency control strategy, as frequency deviations are reduced and the rotational speed variations of the WPPs are closer to their reference values. Nevertheless, this frequency control solution implies certain reductions in PV-generated energy that should be analyzed by the PV installation operators to compensate for possible decreases in benefits. In Figure 11, the frequency evolution, active power of the four generation units, and $\Delta\omega$ input for the PI controller of PV power plants for the hybrid wind–PV strategy are presented for the 550 MW load and 2025 year scenario. As can be seen, when CPPs are only considered for frequency control, there are severe frequency oscillations, which are reduced by including vRESs into the frequency control. As was discussed, among the three different vRES frequency control strategies, the proposed hybrid wind–PV technique reduces frequency oscillations, together with lower fluctuations in both thermal and hydro-power generation units. Even though CPPs oscillate less than with the other approaches, they are working further than their initially assigned values, as demonstrated with the MSE of Table 5. Similar output wind power values are determined for the three vRES strategies, as the same PD constants are assumed. This response is also less oscillatory than the case in which only CPPs are responding under imbalance conditions. Finally, the PV output power is equal if only the CPP or WPP strategies are considered, as PV active power only changes due to the variations in the solar irradiation G. When the de-loading technique is implemented, the PV power plants include a curtailed 10% active power (from 50 to 45 MW, approximately). Moreover, note that considering the $\Delta\omega$ as input significantly reduces the PV-generated power and, consequently, the generated energy (refer to Table 5), but the frequency deviations are also reduced in comparison to the PV(f) strategy. In Figures 12 and 13, the active power of the four WPP and PV power plants are shown, respectively.

Table 6. Simulation results: rotational speed of WPP_1.

Scenario	Load (MW)	250		400		550	
	Year	**2025**	**2040**	**2025**	**2040**	**2025**	**2040**
$\omega_{1,min}$ (pu)	CPP	1.067	1.067	1.067	1.067	1.067	1.067
	WPP	1.000	0.971	1.003	0.983	1.005	0.976
	PV (f)	1.024	1.000	1.019	1.001	1.016	0.995
	PV (ω)	1.030	1.000	1.033	1.026	1.030	0.995
$\omega_{1,max}$ (pu)	CPP	1.359	1.359	1.359	1.359	1.359	1.359
	WPP	1.394	1.405	1.391	1.409	1.389	1.408
	PV (f)	1.381	1.399	1.385	1.405	1.385	1.403
	PV (ω)	1.372	1.370	1.377	1.376	1.379	1.403
MSE_{ω_1}, $\times10^{-3}$ (pu^2)	CPP	3.631	2.649	2.649	2.649	2.479	2.649
	WPP	4.406	6.485	4.081	5.582	4.098	5.966
	PV (f)	3.738	4.956	3.613	4.517	3.770	4.892
	PV (ω)	3.631	4.551	3.497	3.985	3.589	5.075

Table 7. Simulation results: rotational speed of WPP_2.

Scenario	Load (MW)	250		400		550	
	Year	2025	2040	2025	2040	2025	2040
$\omega_{2,min}$ (pu)	CPP	1.052	1.052	1.052	1.052	1.038	1.052
	WPP	1.011	0.885	1.014	0.945	1.015	0.919
	PV (f)	1.037	0.906	1.032	0.966	1.027	0.938
	PV (ω)	1.043	0.891	1.041	0.945	1.040	0.926
$\omega_{2,max}$ (pu)	CPP	1.327	1.327	1.327	1.327	1.327	1.327
	WPP	1.407	1.385	1.406	1.389	1.404	1.389
	PV (f)	1.393	1.393	1.396	1.390	1.399	1.391
	PV (ω)	1.348	1.373	1.355	1.376	1.360	1.403
MSE_{ω_2}, $\times10^{-3}$ (pu^2)	CPP	3.037	3.037	3.037	3.037	3.517	3.037
	WPP	5.455	9.638	5.470	8.692	5.268	8.902
	PV (f)	4.485	7.363	4.663	7.191	4.718	7.317
	PV (ω)	3.765	6.432	3.948	6.138	4.031	7.117

Table 8. Simulation results: rotational speed of WPP_3.

Scenario	Load (MW)	250		400		550	
	Year	2025	2040	2025	2040	2025	2040
$\omega_{3,min}$ (pu)	CPP	0.917	0.917	0.917	0.917	0.766	0.917
	WPP	0.915	0.885	0.912	0.894	0.914	0.891
	PV (f)	0.917	0.902	0.914	0.917	0.915	0.912
	PV (ω)	0.922	0.917	0.919	0.925	0.919	0.916
$\omega_{3,max}$ (pu)	CPP	1.345	1.345	1.345	1.345	1.366	1.345
	WPP	1.410	1.410	1.408	1.413	1.406	1.412
	PV (f)	1.386	1.414	1.390	1.410	1.395	1.414
	PV (ω)	1.352	1.383	1.367	1.380	1.369	1.401
MSE_{ω_3}, $\times10^{-3}$ (pu^2)	CPP	11.778	11.778	11.778	11.778	24.351	11.778
	WPP	13.636	15.890	13.653	15.304	13.572	15.314
	PV (f)	12.912	14.462	13.120	14.287	13.230	14.492
	PV (ω)	11.394	12.575	11.650	12.774	11.803	13.708

Table 9. Simulation results: rotational speed of WPP_4.

Scenario	Load (MW)	250		400		550	
	Year	2025	2040	2025	2040	2025	2040
$\omega_{4,min}$ (pu)	CPP	0.925	0.925	0.925	0.925	0.925	0.925
	WPP	0.946	0.817	0.947	0.865	0.945	0.848
	PV (f)	0.933	0.847	0.936	0.871	0.938	0.874
	PV (ω)	0.924	0.866	0.925	0.879	0.926	0.895
$\omega_{4,max}$ (pu)	CPP	1.332	1.332	1.332	1.332	1.323	1.332
	WPP	1.376	1.455	1.377	1.431	1.373	1.443
	PV (f)	1.366	1.438	1.370	1.437	1.368	1.433
	PV (ω)	1.356	1.440	1.361	1.436	1.361	1.427
MSE_{ω_4}, $\times10^{-3}$ (pu^2)	CPP	8.601	8.601	8.601	8.601	11.235	8.601
	WPP	9.380	13.266	9.593	12.791	9.485	12.584
	PV (f)	9.113	11.317	9.300	11.347	9.276	11.096
	PV (ω)	8.820	10.696	8.846	10.446	8.854	11.237

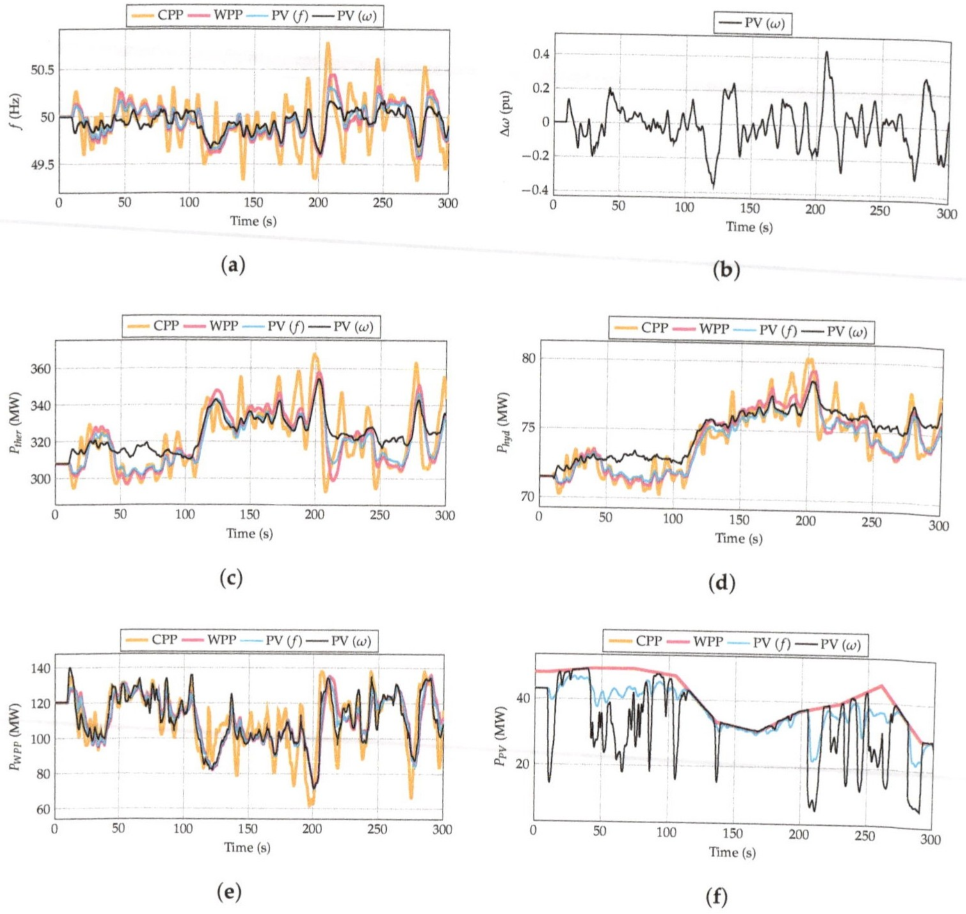

Figure 11. Simulation results for 550 MW demand and year 2025: (**a**) frequency evolution, (**b**) rotational speed deviation, (**c**) thermal power, (**d**) hydro-power, (**e**) VSWT power, and (**f**) PV power.

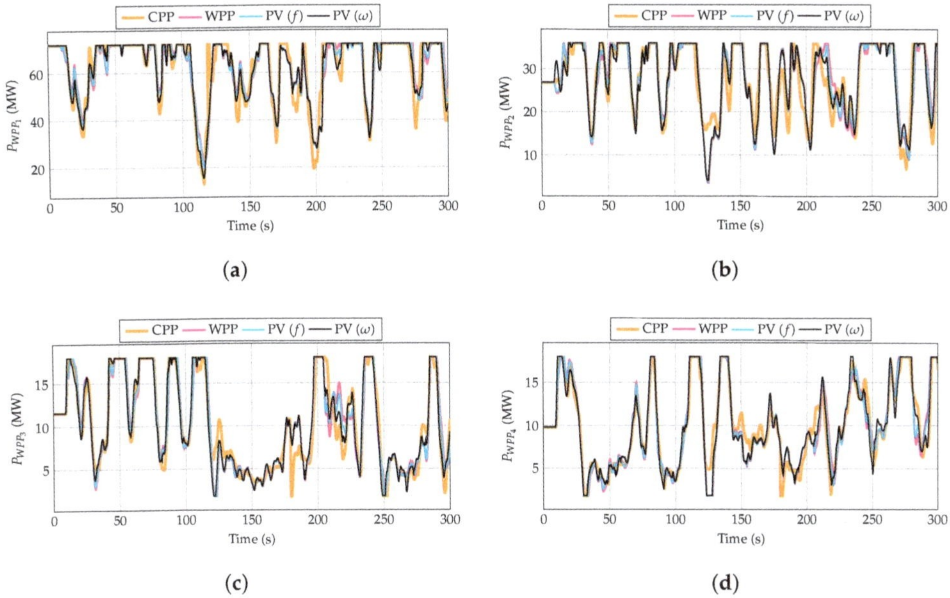

Figure 12. WPP power for 550 MW demand and year 2025: (**a**) WPP_1, (**b**) WPP_2, (**c**) WPP_3, and (**d**) WPP_4.

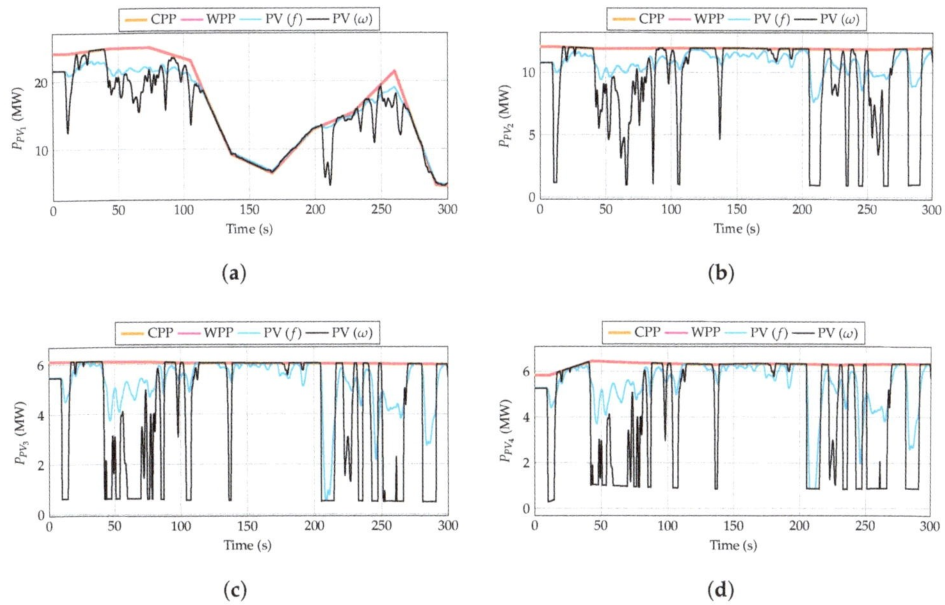

Figure 13. PV power for 550 MW demand and year 2025: (**a**) PV_1, (**b**) PV_2, (**c**) PV_3, and (**d**) PV_4.

Figure 14 depicts the frequency evolution, active power of the four generation units, and $\Delta\omega$ input for the PI controller of PV power plants for the hybrid wind–PV strategy, corresponding to the 250 MW demand and year 2040 scenario. These results are similar to the previous simulations depicted in Figure 11. In this case, when only CPPs participate in frequency control, the frequency oscillations are higher due to the reduced synchronous inertia of the power system. Grid frequency oscillations are drastically reduced by including vRESs into the frequency response, especially with the proposed hybrid wind–PV solution. Similar active wind power is obtained if they also participate in frequency regulation, providing lower oscillations than the case in which only CPPs are considered. When PV power plants include the de-loading technique, a 10% power reduction is required (from 38 to 34 MW, approximately). When the PV installations receive the $\Delta\omega$ as an input, their active power is reduced accordingly, in line with the generated energy shown in Table 5.

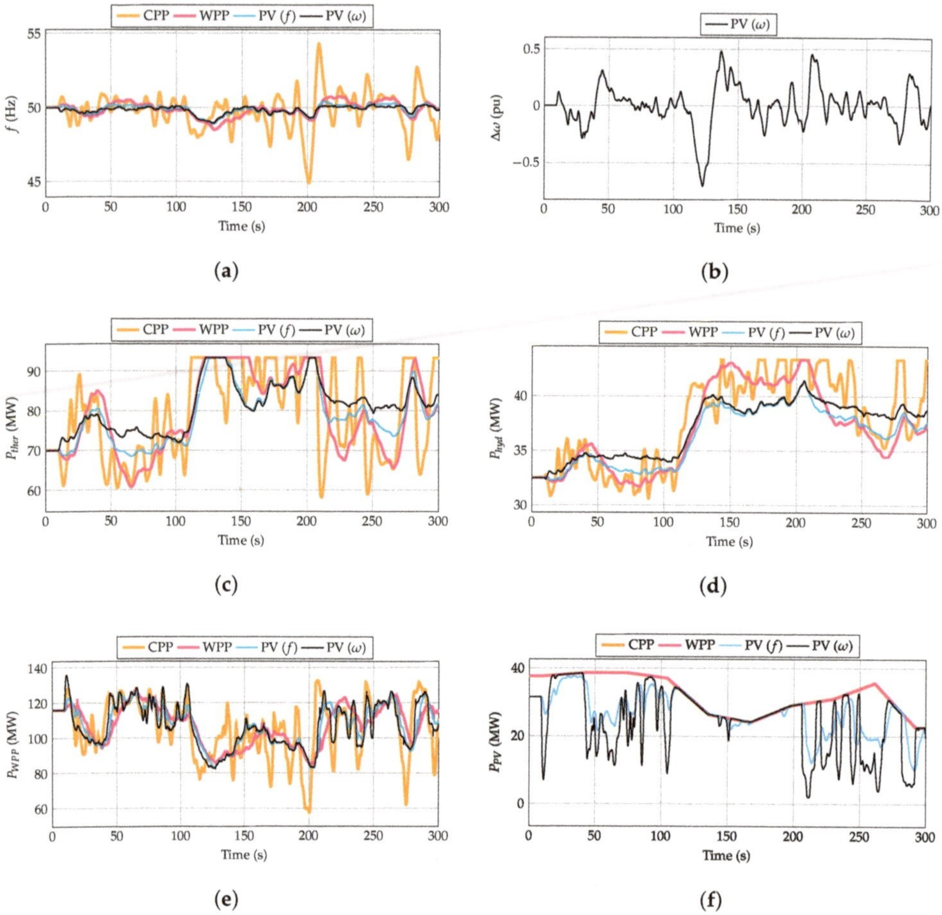

Figure 14. Simulation results for 250 MW demand and year 2040: (**a**) frequency evolution, (**b**) rotational speed deviation, (**c**) thermal power, (**d**) hydro-power, (**e**) VSWT power, and (**f**) PV power.

4.3. Limitations and Further Work

This paper tests the benefits of introducing a hybrid wind–PV frequency controller that is able to monitor not only frequency deviations, but also VSWTs' rotational speed deviations from their optimal reference values. This generalization allows conclusions to be drawn in a high casuistry of scenarios, but it makes it necessary to neglect or simplify some aspects without losing accuracy. These simplifications are:

- Thermal units are supposed to work at the same operating point, considering a single equivalent turbine. In addition, only one thermal power plant technology is assumed (reheat thermal). Hydro-power plants are modeled analogously (including only one kind of hydro turbine). Finally, each wind power plant is modeled as one equivalent VSWT.
- The initial assigned power (generation programming) of each one of the generation units was not obtained with technical–economic criteria (unit commitment), nor were their PFC reserves and secondary control action. Once each generation unit is individually modeled, it is reasonable to assign an initial power to each unit according to both technical and economic criteria.
- To obtain each vRES penetration level, the ENTSO-E recommendations for interconnected power systems were followed by the authors. However, isolated power systems can have different vRES integration levels. However, this hypothesis is assumed to give generality to the present study.
- As described in Section 2, power line dynamics are neglected, as well as the communication lines between the wind and PV power plants for the proposed hybrid wind–PV controller set-up.

As the results and conclusions of this study can be considered as positive, the next step proposed by the authors is to analyze the impact of establishment in some specific isolated power systems, such as one of the existing ones in the Canary archipelago, in the Aegean islands, or in the Azores archipelago. Since those power systems are well known, it is reasonable to individually introduce each generation unit into the dynamic model. Additionally, current and future renewable energy penetration as well as real wind speed and irradiation data can be used for future simulations. Once the locations of both wind and PV power plants are known, the influence of the communication time on the control actions can be also analyzed. The corresponding time delays with the measurement of errors and the transmission of control signals between wind and PV power plants would then be included in the model. Likewise, this proposal will be able to optimize the power system's operation by associating the control of some specific wind power plants to certain geographically neighboring PV power plants.

5. Conclusions

This paper proposes a hybrid wind–PV frequency control approach for isolated power systems with high vRES integration under variable weather conditions. The proposed controller consists of VSWTs that include a hidden-inertia emulation technique and PV power plants working with a de-load power of 10%. The input of the PV frequency control is the rotational speed deviation of the VSWTs, in contrast to previous studies where the frequency deviation is the input of the de-loaded PV controller. In this way, the PV power plants modify their generated power by following the rotational speed deviation of the VSWTs, reducing the deviation of such rotational speeds, and, consequently, minimizing the frequency deviations under power imbalances. The proposed hybrid frequency approach is compared to three different frequency strategies: (*i*) conventional power plants, (*ii*) conventional power plants and wind power plants, and (*iii*) conventional power plants, wind power plants, and PV power plants with grid frequency as input. The results show that frequency oscillations are drastically reduced by including the proposed hybrid controller. In fact, the mean squared error of frequency variations is reduced by up to 95% in comparison to the case in which only conventional power plants are controlling the frequency variations. Severe reductions are also found when comparing the hybrid wind–PV strategy to the other two strategies in which vRESs participate in frequency control (between 20% and 65%). The rotational speed deviation of the VSWTs also decreases with the hybrid wind–PV frequency approach, even getting the same mean squared error as when VSWTs do not participate in frequency control. However, the energy generated by the PV power plants decreases by between 20% and 30% when using the proposed approach, and should be subsequently analyzed by the transmission/distribution operators to guarantee some additional benefits for the owners of such generation units. Based on the results, the authors also wish to remark on the importance of vRESs participating in frequency control in the near future in order to avoid significant frequency deviations that will occur if these sources keep replacing conventional power plants without providing any frequency response. Consequently, hybrid frequency control strategies in line with this work should be proposed and analyzed to minimize frequency variations linked to a massive vRES integration.

Author Contributions: Conceptualization, A.F.-G. and J.-I.S.; methodology, A.F.-G.; software, A.F.-G. and G.M.-L.; validation, G.M.-L. and J.-I.S.; formal analysis, J.-I.S.; investigation, Á.M.-G. and J.-I.S.; resources, A.F.-G.; data curation, A.F.-G. and J.-I.S.; writing—original draft preparation, A.F.-G. and G.M.-L.; writing—review and editing, J.-I.S. and Á.M.-G.; visualization, Á.M.-G.; supervision, J.-I.S.; project administration, A.F.-G.; funding acquisition, A.F.-G., Á.M.G, J.-I.S., and G.M.-L. All authors have read and agreed to the published version of the manuscript.

Funding: This work was partially supported by 'Ministerio de Educación, Cultura y Deporte' of Spain (ref. FPU16/04282) and by 'Ministerio de Economía y Competitividad' under the project 'Value of pumped-hydro energy storage in isolated power systems with high wind power penetration' of the National Plan for Scientific and Technical Research and Innovation 2013-2016, grant number ENE2016-77951-R.

Conflicts of Interest: The authors declare no conflict of interest.

Abbreviations

The following abbreviations and nomenclature are used in this manuscript:

α_V	Correction factor depending on the cell's temperature
$\Delta\omega$	Rotational speed deviation of VSWT
Δf	Frequency deviation
ΔP	Power imbalance
ΔRR	Total secondary regulation effort
ΔV	Additional voltage for PV frequency control
f	Grid frequency
k_B	Boltzmann's constant
k_I	Short-circuit current temperature coefficient
p	Active power (pu)
q	Electron charge
s_w	Wind speed
A	Diode ideality factor
D_{net}	Consumer loads' sensitivity to frequency deviations
E_G	Band-gap energy
G	Sun irradiation
H	Inertia constant
I	Current
I_{ph}	Photo-current
$I_{r,ref}$	Reverse saturation current at T_{STC}
I_{rs}	Reverse saturation current
I_{sc}	Short-circuit current
K_u	Participation factor on AGC
N_P	Number of PV strings in parallel
N_s	Number of PV cells in series
P	Active power (MW)
R	Droop characteristic
S_B	Base power
T	Temperature
T_C	Temperature of the PV cell
V	Voltage

del	De-load (subscript)
dem	Power demand (subscript)
hyd	Hydro-power (subscript)
PV	Photovoltaic (subscript)
ther	Thermal (subscript)
MPP	Maximum power point (subscript)
STC	Standard test conditions (subscript)
WPP	Wind (subscript)
vRES	Variable renewable energy source
AGC	Automatic generation control
CPP	Conventional power plant
DSO	Distribution system operator
ENTSO-E	European Network of Transmission System Operators for Electricity
MPP	Maximum power point
MSE	Mean squared error
PFC	Primary frequency control
PV	Photovoltaic
STC	Standard test conditions
TSO	Transmission system operator
VSWT	Variable-speed wind turbine
WPP	Wind power plant

References

1. Aquila, G.; de Oliveira Pamplona, E.; de Queiroz, A.R.; Junior, P.R.; Fonseca, M.N. An overview of incentive policies for the expansion of renewable energy generation in electricity power systems and the Brazilian experience. *Renew. Sustain. Energy Rev.* **2017**, *70*, 1090–1098.
2. Bjelic, I.B.; Ciric, R.M. Optimal distributed generation planning at a local level—A review of Serbian renewable energy development. *Renew. Sustain. Energy Rev.* **2014**, *39*, 79–86.
3. Zappa, W.; Van Den Broek, M. Analysing the potential of integrating wind and solar power in Europe using spatial optimisation under various scenarios. *Renew. Sustain. Energy Rev.* **2018**, *94*, 1192–1216.
4. Fernández-Guillamón, A.; Das, K.; Cutululis, N.A.; Molina-García, Á. Offshore wind power integration into future power systems: Overview and trends. *J. Mar. Sci. Eng.* **2019**, *7*, 399.
5. Beaudin, M.; Zareipour, H.; Schellenberglabe, A.; Rosehart, W. Energy storage for mitigating the variability of renewable electricity sources: An updated review. *Energy Sustain. Dev.* **2010**, *14*, 302–314.
6. Ulbig, A.; Borsche, T.S.; Andersson, G. Impact of low rotational inertia on power system stability and operation. *IFAC Proc. Vol.* **2014**, *47*, 7290–7297.
7. Bouffard, F.; Ortega-Vazquez, M. The value of operational flexibility in power systems with significant wind power generation. In Proceedings of the 2011 IEEE Power and Energy Society General Meeting, Detroit, MI, USA, 24–28 July 2011; pp. 1–5.
8. Schaber, K.; Steinke, F.; Mühlich, P.; Hamacher, T. Parametric study of variable renewable energy integration in Europe: Advantages and costs of transmission grid extensions. *Energy Policy* **2012**, *42*, 498–508.
9. Osorio, S.; van Ackere, A. From nuclear phase-out to renewable energies in the Swiss electricity market. *Energy Policy* **2016**, *93*, 8–22.
10. Van Stiphout, A.; Poncelet, K.; De Vos, K.; Deconinck, G. The impact of operating reserves in generation expansion planning with high shares of renewable energy sources. In Proceedings of the IAEE European Energy Conference, Sustainable Energy Policy and Strategies for Europe, Rome, Italy, 28–31 October 2014; pp. 1–15.
11. Rakhshani, E.; Rodriguez, P. Active power and frequency control considering large-scale RES. In *Large Scale Renewable Power Generation*; Springer: 2014; pp. 233–271.
12. Fernández-Guillamón, A.; Gómez-Lázaro, E.; Muljadi, E.; Molina-Garcia, Á. A Review of Virtual Inertia Techniques for Renewable Energy-Based Generators. In *Power Systems*; IntechOpen: 2020.
13. Serban, I.; Teodorescu, R.; Marinescu, C. Energy storage systems impact on the short-term frequency stability of distributed autonomous microgrids, an analysis using aggregate models. *IET Renew. Power Gener.* **2013**, *7*, 531–539.

14. Adrees, A.; Milanovic, J.V. Study of frequency response in power system with renewable generation and energy storage. In Proceedings of the 2016 Power Systems Computation Conference (PSCC), Genoa, Italy, 20–24 June 2016; pp. 1–7.
15. Hosseinipour, A.; Hojabri, H. Virtual inertia control of PV systems for dynamic performance and damping enhancement of DC microgrids with constant power loads. *IET Renew. Power Gener.* **2017**, *12*, 430–438.
16. Fernández-Guillamón, A.; Gómez-Lázaro, E.; Muljadi, E.; Molina-García, Á. Power systems with high renewable energy sources: A review of inertia and frequency control strategies over time. *Renew. Sustain. Energy Rev.* **2019**, *115*, 109369.
17. Alatrash, H.; Mensah, A.; Mark, E.; Haddad, G.; Enslin, J. Generator emulation controls for photovoltaic inverters. *IEEE Trans. Smart Grid* **2012**, *3*, 996–1011.
18. Zhang, X.; Chen, Y.; Wang, Y.; Zha, X.; Yue, S.; Cheng, X.; Gao, L. Deloading power coordinated distribution method for frequency regulation by wind farms considering wind speed differences. *IEEE Access* **2019**, *7*, 122573–122582.
19. Fang, X.; Krishnan, V.; Hodge, B.M. Strategic offering for wind power producers considering energy and flexible ramping products. *Energies* **2018**, *11*, 1239.
20. Yingcheng, X.; Nengling, T. Review of contribution to frequency control through variable speed wind turbine. *Renew. Energy* **2011**, *36*, 1671–1677.
21. Fernández-Guillamón, A.; Vigueras-Rodríguez, A.; Molina-García, Á. Analysis of power system inertia estimation in high wind power plant integration scenarios. *IET Renew. Power Gener.* **2019**, *13*, 2807–2816.
22. Martínez-Lucas, G.; Sarasúa, J.I.; Sánchez-Fernández, J.Á. Eigen analysis of wind–hydro joint frequency regulation in an isolated power system. *Int. J. Electr. Power Energy Syst.* **2018**, *103*, 511–524.
23. Martínez-Lucas, G.; Sarasúa, J.I.; Sánchez-Fernández, J.Á. Frequency regulation of a hybrid wind–hydro power plant in an isolated power system. *Energies* **2018**, *11*, 239.
24. Ullah, N.R.; Thiringer, T.; Karlsson, D. Temporary primary frequency control support by variable speed wind turbines—Potential and applications. *IEEE Trans. Power Syst.* **2008**, *23*, 601–612.
25. Pandey, S.K.; Mohanty, S.R.; Kishor, N. A literature survey on load–frequency control for conventional and distribution generation power systems. *Renew. Sustain. Energy Rev.* **2013**, *25*, 318–334.
26. Fernández-Guillamón, A.; Molina-García, A.; Vigueras-Rodríguez, A.; GÓmez-Lázaro, E. Frequency Response and Inertia Analysis in Power Systems with High Wind Energy Integration. In Proceedings of the 2019 International Conference on Clean Electrical Power (ICCEP), Otranto, Italy, 2–4 July 2019; pp. 388–393.
27. Li, X.; Li, Y.; Han, X.; Hui, D. Application of fuzzy wavelet transform to smooth wind/PV hybrid power system output with battery energy storage system. *Energy Procedia* **2011**, *12*, 994–1001.
28. Nayeripour, M.; Hoseintabar, M.; Niknam, T. Frequency deviation control by coordination control of FC and double-layer capacitor in an autonomous hybrid renewable energy power generation system. *Renew. Energy* **2011**, *36*, 1741–1746.
29. Li, X.; Hui, D.; Lai, X. Battery energy storage station (BESS)-based smoothing control of photovoltaic (PV) and wind power generation fluctuations. *IEEE Trans. Sustain. Energy* **2013**, *4*, 464–473.
30. Ma, Y.; Yang, P.; Wang, Y.; Zhou, S.; He, P. Frequency control of islanded microgrid based on wind-PV-diesel-battery hybrid energy sources. In Proceedings of the 2014 17th International Conference on Electrical Machines and Systems (ICEMS), Hangzhou, China, 22–25 October 2014; pp. 290–294.
31. Gatta, F.M.; Geri, A.; Lamedica, R.; Lauria, S.; Maccioni, M.; Palone, F.; Rebolini, M.; Ruvio, A. Application of a LiFePO4 battery energy storage system to primary frequency control: Simulations and experimental results. *Energies* **2016**, *9*, 887.
32. Krishnan, M.S.; Ramkumar, M.S.; Amudha, A. Frequency Deviation Control In Hybrid Renewable Energy System Using Fc-Uc. *Int. J. Control Theory Appl.* **2017**, *10*, 333–344.
33. Salama, H.S.; Aly, M.M.; Vokony, I. Voltage/Frequency control of isolated unbalanced radial distribution system fed from intermittent wind/PV power using fuzzy logic controlled-SMES. In Proceedings of the 2019 International Conference on Innovative Trends in Computer Engineering (ITCE), Hangzhou, China, 22–25 October 2019; pp. 414–419.
34. Das, S.; Akella, A. A fuzzy logic-based frequency control scheme for an isolated AC coupled PV-wind-battery hybrid system. *Int. J. Model. Simul.* **2020**, *40*, 308–320.

35. Marchese, K.; Pourmousavi, S.; Nehrir, M. The application of demand response for frequency regulation in an islanded microgrid with high penetration of renewable generation. In Proceedings of the 2013 North American Power Symposium (NAPS), Manhattan, KS, USA, 22–24 September 2013; pp. 1–6.
36. Vahedipour-Dahraie, M.; Rashidizaheh-Kermani, H.; Najafi, H.R.; Anvari-Moghaddam, A.; Guerrero, J.M. Coordination of EVs participation for load frequency control in isolated microgrids. *Appl. Sci.* **2017**, *7*, 539.
37. Almi, M.; Arrouf, M.; Belmili, H.; Boulouma, S.; Bendib, B. Energy management of wind/PV and battery hybrid system. *Int. J. New Comput. Archit. Appl. (IJNCAA)* **2014**, *4*, 30–38.
38. Liu, Y.; You, S.; Liu, Y. Study of wind and PV frequency control in US power grids—EI and TI case studies. *IEEE Power Energy Technol. Syst. J.* **2017**, *4*, 65–73.
39. Vattigunta, R.R.; Rather, Z.H.; Gokaraju, R. Fast frequency support from hybrid solar PV and wind power plant. In Proceedings of the 2018 IEEE International Conference on Power Electronics, Drives and Energy Systems (PEDES), Chennai, India, 18–21 December 2018; pp. 1–6.
40. Zarina, P.; Mishra, S.; Sekhar, P. Deriving inertial response from a non-inertial PV system for frequency regulation. In Proceedings of the 2012 IEEE International Conference on Power Electronics, Drives and Energy Systems (PEDES), Bengaluru, India, 16–19 December 2012; pp. 1–5.
41. Zarina, P.; Mishra, S.; Sekhar, P. Photovoltaic system based transient mitigation and frequency regulation. In Proceedings of the 2012 Annual IEEE India Conference (INDICON), Kochi, India, 7–9 December 2012; pp. 1245–1249.
42. Mishra, S.; Zarina, P.; Sekhar, P. A novel controller for frequency regulation in a hybrid system with high PV penetration. In Proceedings of the 2013 IEEE Power & Energy Society General Meeting, Vancouver, BC, Canada, 21–25 July 2013; pp. 1–5.
43. Rahmann, C.; Castillo, A. Fast frequency response capability of photovoltaic power plants: The necessity of new grid requirements and definitions. *Energies* **2014**, *7*, 6306–6322.
44. Jietan, Z.; Linan, Q.; Pestana, R.; Fengkui, L.; Libin, Y. Dynamic frequency support by photovoltaic generation with "synthetic" inertia and frequency droop control. In Proceedings of the 2017 IEEE Conference on Energy Internet and Energy System Integration (EI2), Beijing, China, 26–28 November 2017; pp. 1–6.
45. Jibji-Bukar, F.; Anaya-Lara, O. Frequency support from photovoltaic power plants using offline maximum power point tracking and variable droop control. *IET Renew. Power Gener.* **2019**, *13*, 2278–2286.
46. Li, Q.; Baran, M.E. A Novel Frequency Support Control Method for PV Plants using Tracking LQR. *IEEE Trans. Sustain. Energy* **2019**, doi:10.1109/TSTE.2019.2953684.
47. Wilches-Bernal, F.; Chow, J.H.; Sanchez-Gasca, J.J. A fundamental study of applying wind turbines for power system frequency control. *IEEE Trans. Power Syst.* **2016**, *31*, 1496–1505.
48. Wang, Y.; Delille, G.; Bayem, H.; Guillaud, X.; Francois, B. High wind power penetration in isolated power systems—Assessment of wind inertial and primary frequency responses. *IEEE Trans. Power Syst.* **2013**, *28*, 2412–2420.
49. You, R.; Barahona, B.; Chai, J.; Cutululis, N.A. Frequency support capability of variable speed wind turbine based on electromagnetic coupler. *Renew. Energy* **2015**, *74*, 681–688.
50. Hafiz, F.; Abdennour, A. An adaptive neuro-fuzzy inertia controller for variable-speed wind turbines. *Renew. Energy* **2016**, *92*, 136–146.
51. Magnus, D.; Pfitscher, L.; Scharlau, C. A Synergy Analysis of Synthetic Inertia and Speed Controllers on Variable Speed Wind Turbines. In Proceedings of the 2019 IEEE PES Innovative Smart Grid Technologies Conference-Latin America (ISGT Latin America), Gramado, Brazil, 15–18 September 2019; pp. 1–6.
52. Hafiz, F.; Abdennour, A. Optimal use of kinetic energy for the inertial support from variable speed wind turbines. *Renew. Energy* **2015**, *80*, 629–643.
53. Kang, M.; Kim, K.; Muljadi, E.; Park, J.; Kang, Y.C. Frequency Control Support of a Doubly-Fed Induction Generator Based on the Torque Limit. *IEEE Trans. Power Syst.* **2016**, *31*, 4575–4583, doi:10.1109/TPWRS.2015.2514240.
54. Fernández-Guillamón, A.; Villena-Lapaz, J.; Vigueras-Rodríguez, A.; García-Sánchez, T.; Molina-García, Á. An Adaptive Frequency Strategy for Variable Speed Wind Turbines: Application to High Wind Integration Into Power Systems. *Energies* **2018**, *11*, 1436.
55. Fernández Guillamón, A. Análisis y Simulación de Estrategias Agregadas de Control de Frecuencia Entre Grandes Parques Eólicos y Aprovechamientos Hidroeléctricos. 2017. Available online: https://repositorio.upct.es/handle/10317/6222 (accessed on 10 July 2020).

56. Mansoor, S.; Jones, D.; Bradley, D.A.; Aris, F.; Jones, G. Reproducing oscillatory behaviour of a hydroelectric power station by computer simulation. *Control Eng. Pract.* **2000**, *8*, 1261–1272.
57. O'Sullivan, J.; Rogers, A.; Flynn, D.; Smith, P.; Mullane, A.; O'Malley, M. Studying the maximum instantaneous non-synchronous generation in an island system—Frequency stability challenges in Ireland. *IEEE Trans. Power Syst.* **2014**, *29*, 2943–2951.
58. Sarasúa, J.I.; Martínez-Lucas, G.; Lafoz, M. Analysis of alternative frequency control schemes for increasing renewable energy penetration in El Hierro Island power system. *Int. J. Electr. Power Energy Syst.* **2019**, *113*, 807–823.
59. Fernández-Guillamón, A.; Vigueras-Rodríguez, A.; Gómez-Lázaro, E.; Molina-García, Á. Fast power reserve emulation strategy for VSWT supporting frequency control in multi-area power systems. *Energies* **2018**, *11*, 2775.
60. Kundur, P.; Balu, N.J.; Lauby, M.G. *Power System Stability and Control*; McGraw-Hill: New York, NY, USA, 1994; Volume 7,
61. Pöller, M.; Achilles, S. Aggregated wind park models for analyzing power system dynamics. In Proceedings of the 4th International Workshop on Large-Scale Integration of Wind Power and Transmission Networks for Offshore Wind Farms, Billund, Denmark, Billund, Denmark, 20–21 October 2003.
62. Clark, K.; Miller, N.W.; Sanchez-Gasca, J.J. Modeling of GE wind turbine-generators for grid studies. *GE Energy* **2010**, *4*, 0885–8950.
63. Zhao, S.; Nair, N.K. Assessment of wind farm models from a transmission system operator perspective using field measurements. *IET Renew. Power Gener.* **2011**, *5*, 455–464.
64. Fortmann, J. *Modeling of Wind Turbines With Doubly Fed Generator System*; Springer: 2014.
65. Zarina, P.; Mishra, S.; Sekhar, P. Exploring frequency control capability of a PV system in a hybrid PV-rotating machine-without storage system. *Int. J. Electr. Power Energy Syst.* **2014**, *60*, 258–267.
66. Sekhar, P.; Mishra, S. Storage free smart energy management for frequency control in a diesel-PV-fuel cell-based hybrid AC microgrid. *IEEE Trans. Neural Netw. Learn. Syst.* **2015**, *27*, 1657–1671.
67. Martínez-Lucas, G.; Sarasúa, J.I.; Sánchez-Fernández, J.Á.; Wilhelmi, J.R. Power-frequency control of hydropower plants with long penstocks in isolated systems with wind generation. *Renew. Energy* **2015**, *83*, 245–255.
68. Mancarella, P.; Chicco, G.; Capuder, T. Arbitrage opportunities for distributed multi-energy systems in providing power system ancillary services. *Energy* **2018**, *161*, 381–395.
69. ENTSO-E. Electricity Balancing in Europe. Available online https://docstore.entsoe.eu/ (accessed on 15 July 2020).
70. Díaz-González, F.; Hau, M.; Sumper, A.; Gomis-Bellmunt, O. Participation of wind power plants in system frequency control: Review of grid code requirements and control methods. *Renew. Sustain. Energy Rev.* **2014**, *34*, 551–564.
71. Ersdal, A.M.; Imsland, L.; Uhlen, K. Model predictive load-frequency control. *IEEE Trans. Power Syst.* **2015**, *31*, 777–785.
72. ENTSO-E. ENTSO-E Network Code for Requirements for Grid Connection Applicable to all Generators. June 2011. Available online: https://consultations.entsoe.eu/ (accessed on 20 July 2020)
73. Guo, F.; Wen, C.; Mao, J.; Song, Y.D. Distributed secondary voltage and frequency restoration control of droop-controlled inverter-based microgrids. *IEEE Trans. Ind. Electron.* **2014**, *62*, 4355–4364.
74. Zhao, C.; Mallada, E.; Low, S.H. Distributed generator and load-side secondary frequency control in power networks. In Proceedings of the 2015 49th Annual Conference on Information Sciences and Systems (CISS), Baltimore, MD, USA, 18–20 March 2015; pp. 1–6.
75. Zhao, C.; Mallada, E.; Low, S.; Bialek, J. A unified framework for frequency control and congestion management. In Proceedings of the 2016 Power Systems Computation Conference (PSCC), Genoa, Italy, 20–24 June 2016; pp. 1–7.
76. UCTE (Union for the Co-ordination of Transmission of Electricity). Operation Handbook. 2004. Available online: https://www.entsoe.eu/publications/system-operations-reports/operation-handbook (accessed on 17 July 2020)
77. Wood, A.J.; Wollenberg, B.F.; Sheblé, G.B. *Power Generation, Operation, and Control*; John Wiley & Sons: Hoboken, NJ, USA, 2013.

78. ENTSO-E. TYNDP 2020—Scenario Report; 2020. Available online: https://www.entsos-tyndp2020-scenarios.eu/wp-content/uploads/2020/06/TYNDP_2020_Joint_ScenarioReport_final.pdf (accessed on 1 July 2020).
79. Fernández-Guillamón, A.; Sarasúa, J.I.; Chazarra, M.; Vigueras-Rodríguez, A.; Fernández-Muñoz, D.; Molina-García, A. Frequency control analysis based on unit commitment schemes with high wind power integration: A Spanish isolated power system case study. *Int. J. Electr. Power Energy Syst.* **2020**, *121*, 106044.
80. ENMAX. Available online: https://www.enmax.com/generation-wires/real-time-system-demand (accessed on 6 July 2020).
81. ERGON. Available online: www.ergon.com.au/network/manage-your-energy/home-energy-tips/peak-demand-at-home/network-demand (accessed on 6 July 2020).
82. RTE. Available online: https://www.rte-france.com/en/eco2mix/eco2mix-consommation-en (accessed on 6 July 2020).
83. REE. Available online: https://demanda.ree.es/visiona/peninsula/demanda/tablas/2020-07-04/1 (accessed on 6 July 2020).
84. IESO. Available online: http://www.ieso.ca/power-data (accessed on 6 July 2020).
85. CAISO. Available online: http://www.caiso.com/TodaysOutlook/Pages/default.aspx (accessed on 6 July 2020).
86. TEPCO. Available online: https://www4.tepco.co.jp/en/forecast/html/index-e.html (accessed on 6 July 2020).
87. TRANSPOWER. Available online: https://www.transpower.co.nz/power-system-live-data (accessed on 6 July 2020).
88. Martínez-Lucas, G.; Sarasúa, J.I.; Pérez-Díaz, J.I.; Martínez, S.; Ochoa, D. Analysis of the Implementation of the Primary and/or Inertial Frequency Control in Variable Speed Wind Turbines in an Isolated Power System with High Renewable Penetration. Case Study: El Hierro Power System. *Electronics* **2020**, *9*, 901.
89. ENTSO-E. *Frequency Stability Evaluation Criteria for the Synchronous Zone of Continental Europe*; ENTSO-E: Brussels, Belgium, 2016.

sustainability

Article

On the Design of an Integrated System for Wave Energy Conversion Purpose with the Reaction Mass on Board

Jinming Wu * and Zhonghua Ni

School of Mechanical Engineering, Southeast University, Nanjing 211189, China; nzh2003@seu.edu.cn

* Correspondence: jinmingwu@seu.edu.cn;

Received: 12 March 2020; Accepted: 2 April 2020; Published: 3 April 2020

Abstract: In this paper, we investigate the design of an integrated system consisting of two non-rigidly connected bodies: A floating buoy and an emerged offshore structure. When waves excite the buoy to oscillate, the relative motion between the two bodies are converted to useful energy through a spring damper system, resulting in wave energy being absorbed. The parameter to design includes the mass and underwater shape of the buoy. The spring stiffness of the power take-off (PTO) system is constrained to be non-negative with the concerns of complexity in implementation and system stability. Results suggest that a larger mass of the buoy is advantageous due to smaller optimal spring stiffness and damping coefficient of the PTO system, more absorbed wave power, and less motion amplitude of the two bodies. A favorable underwater shape of the buoy is characterized by large diameter to draft ratio, with the section profile preferring a circle or square rather than an equilateral triangle. Investigations on the designed buoy in irregular waves show that the integrated system presents its peak power absorption within the common range of energy period, and the motion amplitude of the offshore structure is larger than the wave amplitude in a certain range of sea states.

Keywords: integrated system; floating buoy; offshore structure; wave energy converter; non-negative spring stiffness

---

1. Introduction

Due to high energy density and excellent predictability, ocean wave energy stands up as a promise substitute of fossil energies, which give rise to environmental issues. Although the first wave energy converter (WEC) had been patented over two centuries ago [1], wave energy conversion technology is still immature and at a pre-commercialized stage. The levelized cost of wave energy is estimated to be \$1.0~1.5/kWh, and not ready for the present power economic request when compared to \$0.07/kWh of onshore wind energy [2]. A possible solution to minimize this discrepancy is to combine WECs with offshore structures, such as navigation aids [3,4], offshore solar [5] and wind [6] electricity generation systems, and even offshore aquaculture devices [2], so that the substructure, mooring system, and electrical grid connections, which account for respectively 20%, 10%, and 6% of the construction cost of an individual WEC [7], can be shared.

In view of wave energy absorption, the aforementioned integrated systems belong to the two-body type, which uses the relative motion between two bodies to harvest energy. A two-body system usually consists of two parts: A floating buoy that moves along with the incoming wave and a reaction mass, which can be submerged in deep water [8,9] or on board [10]. A two-body system with the reaction mass on board has the advantage of lacking submerged moving bodies and actuators, and may be easier for the installation and operation process [10]. Studies on the two-body system with reaction mass on board can be traced back to Parks [11]. French et al. [12] applied phase control to a heaving buoy reacting against an internal mass. Korde [10] revealed that the on board system shows a power capture width ratio comparable with that of the submerged system over a wide range of wave frequency, demonstrating its potential in wave power absorption. Additionally, Korde [13] proposed a motion compensation system to achieve stationarity of the reaction mass to further increase the absorbed power for the on board system.

As concerns how to integrate WECs with offshore structures to form a two-body system, one option is that the offshore structure be on board a floating buoy, which is demonstrated in [14] that a multi-purpose platform is on board floating buoyancy columns with actuators connecting them to absorb wave energy. In order to achieve the best wave power absorption performance, previous studies on two-body systems with reaction mass on board tried to optimize the parameter of the reaction mass [10–13,15]. However, in the integrated system of the offshore structure and the floating buoy, the reaction mass (the offshore structure) is normally provided as fixed parameters ahead of time, hence only the parameter of the floating buoy can be optimized.

The objective of this paper is to design the parameter of the floating buoy with an offshore structure on board to achieve satisfied performance of the integrated system. Here, the parameter of the floating buoy consists of its mass and underwater shape, while the performance of the integrated system includes absorbed wave power and motion amplitude of the two bodies. A 5 MW offshore wind turbine is chosen as the offshore structure for a case study. When WECs are resonant with waves, the absorbed wave power reaches the maximum. However, in order to achieve resonance, the optimal spring stiffness of the power take-off (PTO) system may sometimes be negative, which not only is difficult to implement in reality, but may also shift WECs to unsteady states. Therefore, the spring stiffness of the PTO system is constrained to be non-negative in this paper. The designing process is performed in regular waves, and the designed integrated system is then investigated in irregular waves.

2. Methods

Figure 1 shows the integrated two-body system consisting of the offshore structure and the floating buoy, and with the offshore structure on board the floating buoy. The PTO system makes use of the relative motion between the two bodies to produce useful power, and is simplified as a pair of spring and damper connected in parallel. The stiffness of the spring and the damping coefficient of the damper are denoted as k_p and c_p, respectively. The mass and displacement of the offshore structure are respectively denoted as m_s and z_s, and that of the buoy as m_b and z_b. For simplicity, the offshore structure and the buoy are subjected to heave motion only.

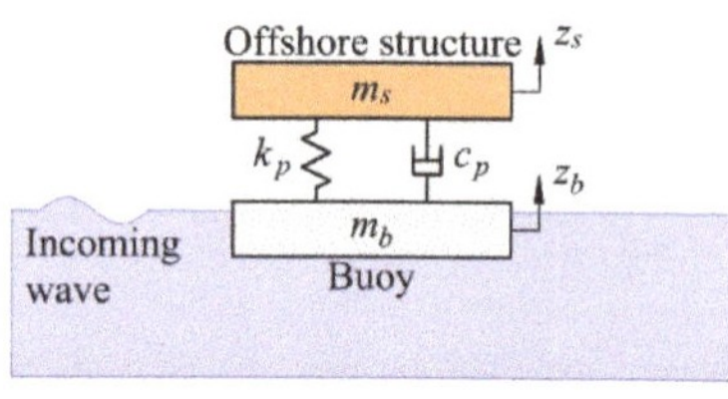

Figure 1. Schematic diagram of the integrated system.

In regular waves, according to the fundamental investigation on two-body systems by Falnes [16], the equation of motion of the offshore structure and the buoy can be written as

$$\begin{aligned} m_s\ddot{z}_s + c_p\left(\dot{z}_s - \dot{z}_b\right) + k_p(z_s - z_b) &= 0; \\ [m_b + m_{bh}(\omega)]\ddot{z}_b + c_{bh}(\omega)\dot{z}_b + k_{bh}z_b - c_p\left(\dot{z}_s - \dot{z}_b\right) - k_p(z_s - z_b) &= f_d; \end{aligned} \tag{1}$$

where m_{bh}, c_{bh}, and k_{bh} denote the added mass, radiation damping coefficient, and hydrostatic stiffness of the buoy, ω denotes the wave frequency, and f_d denotes the excitation force the incoming wave acts on the buoy. The hydrodynamic coefficients and excitation force are calculated numerically by the ANSYS AQWA software [17]. The hydrostatic stiffness of the buoy in heave is calculated as

$$k_{bh} = \rho g S, \tag{2}$$

where ρ is the water density, g is the gravity acceleration, and S is the area of the waterplane. By applying the form of complex amplitude [18,19], which is denoted by a symbol '^' on top of the variable, Equation (1) can be written as

$$\begin{aligned} -\omega^2 m_s \hat{z}_s + \left(k_p + \mathrm{i}\omega c_p\right)(\hat{z}_s - \hat{z}_b) = 0; \\ -\omega^2[m_b + m_{bh}(\omega)]\hat{z}_b + [k_{bh} + \mathrm{i}\omega c_{bh}(\omega)]\hat{z}_b - \left(k_p + \mathrm{i}\omega c_p\right)(\hat{z}_s - \hat{z}_b) = \hat{f}_d; \end{aligned} \tag{3}$$

Equation (3) can be transferred to the matrix form as

$$\begin{bmatrix} -\omega^2 m_s + k_p + \mathrm{i}\omega c_p & -\left(k_p + \mathrm{i}\omega c_p\right) \\ -\left(k_p + \mathrm{i}\omega c_p\right) & -\omega^2[m_b + m_{bh}(\omega)] + k_{bh} + \mathrm{i}\omega c_{bh}(\omega) + k_p + \mathrm{i}\omega c_p \end{bmatrix} \begin{bmatrix} \hat{z}_s \\ \hat{z}_b \end{bmatrix} = \begin{bmatrix} 0 \\ \hat{f}_d \end{bmatrix}. \tag{4}$$

Here, we denote the impedance of the offshore structure, the buoy, and the PTO system as Z_s, Z_b, and Z_p, and their real and imaginary parts are denoted by Z_{sr}, $\mathrm{i}Z_{si}$, Z_{br}, $\mathrm{i}Z_{bi}$, Z_{pr}, and $\mathrm{i}Z_{pi}$, respectively. The impedance is defined as

$$\begin{aligned} Z_{sr} &= -\omega^2 m_s; \\ Z_{br} &= -\omega^2[m_b + m_{bh}(\omega)] + k_{bh}; \\ Z_{bi} &= \omega c_{bh}(\omega); \\ Z_{pr} &= k_p; \\ Z_{pi} &= \omega c_p. \end{aligned} \tag{5}$$

Since the offshore structure is out of water, Z_{si} = 0. Then, Equation (4) can be written as

$$\begin{bmatrix} Z_s + Z_p & -Z_p \\ -Z_p & Z_b + Z_p \end{bmatrix} \begin{bmatrix} \hat{z}_s \\ \hat{z}_b \end{bmatrix} = \begin{bmatrix} 0 \\ \hat{f}_d \end{bmatrix}. \tag{6}$$

By finding the inverse of the matrix in the left term of Equation (6) and matrix multiplication, it gives

$$\begin{bmatrix} \hat{z}_s \\ \hat{z}_b \end{bmatrix} = \frac{1}{Z_s Z_b + (Z_s + Z_b) Z_p} \begin{bmatrix} Z_p \hat{f}_d \\ \left(Z_s + Z_p\right)\hat{f}_d \end{bmatrix} \tag{7}$$

For a two-body system, the time-averaged absorbed power can be calculated by [20]

$$P = \frac{\omega^2 c_p}{2} |\hat{z}_s - \hat{z}_b|^2 \tag{8}$$

It can be obtained from Equation (7) that

$$\hat{z}_s - \hat{z}_b = \frac{-\hat{f}_d}{Z_b + \left(1 + \frac{Z_b}{Z_s}\right) Z_p}. \tag{9}$$

By denoting

$$G = 1 + \frac{Z_b}{Z_s}, \tag{10}$$

Equation (9) can be written as

$$\hat{z}_s - \hat{z}_b = \frac{-\hat{f}_d}{Z_b + G Z_p}. \tag{11}$$

Here, we denote the real and imaginary part of G as G_r and iG_i, respectively. Then, we have

$$\begin{aligned} G_r &= 1 + \tfrac{Z_{br}}{Z_{sr}}; \\ G_i &= \tfrac{Z_{bi}}{Z_{sr}}; \end{aligned} \tag{12}$$

Inserting Equation (11) to Equation (8) gives

$$P = \frac{\omega^2 c_p}{2} \frac{\hat{f}_d \hat{f}_d^*}{\left(Z_b + GZ_p\right)\left(Z_b^* + G^* Z_p^*\right)}. \tag{13}$$

Decomposition of the complex parameters to real and imaginary parts makes Equation (13) be written as

$$P = \frac{\omega}{2} \frac{\left|\hat{f}_d\right|^2}{|G|^2 Z_{pi} + 2(G_r Z_{bi} - G_i Z_{br}) + \frac{|G|^2 Z_{pr}^2 + 2(G_r Z_{br} + G_i Z_{bi}) Z_{pr} + |Z_b|^2}{Z_{pi}}} \tag{14}$$

When there is no constraint on the parameters of the PTO system, it can be seen that when

$$Z_{pr} = Z_{pro} = -\frac{G_r Z_{br} + G_i Z_{bi}}{|G|^2} \tag{15}$$

And

$$Z_{pi} = Z_{pio} = \frac{G_r Z_{bi} - G_i Z_{br}}{|G|^2}, \tag{16}$$

the absorbed power reaches the maximum

$$P_{\max} = \frac{\omega}{8} \frac{\left|\hat{f}_d\right|^2}{G_r Z_{bi} - G_i Z_{br}}. \tag{17}$$

Equation (5) and Equation (12) gives

$$G_r Z_{bi} - G_i Z_{br} = \left(1 + \frac{Z_{br}}{Z_{sr}}\right) Z_{bi} - \frac{Z_{bi}}{Z_{sr}} Z_{br} = Z_{bi}. \tag{18}$$

From Equation (15) to Equation (18) and the relation between Z_{pr}, Z_{pi}, and k_p, c_p as shown in Equation (5), we can obtain the optimal values of k_p, c_p, and maximum absorbed power as

$$k_{po} = -\frac{G_r Z_{br} + G_i Z_{bi}}{|G|^2} \tag{19}$$

$$c_{po} = \frac{c_{bh}}{|G|^2}; \tag{20}$$

$$P_{\max} = \frac{\left|\hat{f}_d\right|^2}{8 c_{bh}}. \tag{21}$$

It agrees with the finding of Korde [10] that the maximum absorbed power only depends on the excitation force and radiation damping coefficient of the buoy for an on board system. It is interesting to find that although the integrated system oscillates in two degrees of freedom, it behaves as a single-degree-of-freedom system in terms of power absorption.

However, in some cases, Equation (19) gives the optimal spring stiffness of the PTO system a negative value, which is not only difficult to implement in reality, but also may offset the system to unsteadiness. In this paper, we constrain the spring stiffness of the PTO system to be non-negative.

Taking into consideration this constraint, we update the optimal spring stiffness and damping coefficient of the PTO system and the maximum absorbed power of the integrated system as

$$k_{po} = \max\left\{-\frac{G_r Z_{br} + G_i Z_{bi}}{|G|^2}, 0\right\}, \tag{22}$$

$$c_{po} = \begin{cases} \frac{c_{bh}}{|G|^2}, & \text{if } k_{po} > 0 \\ \left|\frac{Z_b}{\omega G}\right|, & \text{if } k_{po} = 0 \end{cases}, \tag{23}$$

And

$$P_{\max} = \begin{cases} \frac{\left|\hat{f}_d\right|^2}{8c_{bh}}, & \text{if } k_{po} > 0 \\ \frac{\omega}{4}\frac{|\hat{f}_d|^2}{|G||Z_b| + Z_{bi}}, & \text{if } k_{po} = 0 \end{cases} \tag{24}$$

In this paper, a 5 MW offshore wind turbine, which is used in the WindFloat concept [21] proposed by the Principle Power Inc. for offshore wind farm application, is chosen as the offshore structure for a case study. The offshore structure is restricted to oscillate in heave mode. The total mass of the offshore structure is found to be 854 mt [21]. The water depth is considered in offshore scenario to be h = 100 m, and the wave amplitude, which is denoted by A, is fixed at 1 m for regular waves.

3. Effect of the Constraint on The Spring Stiffness of the PTO System

Since the spring stiffness of the PTO system is constrained to be non-negative, which is not considered in previous literature for the on board two-body system, we investigate the effect of the constraint first. From Equation (19), we find that it is the numerator of the right term that causes the unconstrained optimal spring stiffness of the PTO system to be negative since the denominator is absolutely non-negative. Here we denote

$$H = -(G_r Z_{br} + G_i Z_{bi}). \tag{25}$$

Inserting Equation (12) to Equation (25) gives

$$H = -\frac{1}{Z_{sr}}\left(Z_{br}^2 + Z_{sr} Z_{br} + Z_{bi}^2\right). \tag{26}$$

According to the characteristic of parabolic functions, we find that H will be negative only when

$$Z_{sr}^2 - 4Z_{bi}^2 \geq 0 \tag{27}$$

i.e.,

$$\frac{Z_{bi}}{|Z_{sr}|} \leq \frac{1}{2}, \tag{28}$$

and

$$\frac{Z_{br}}{|Z_{sr}|} \to \frac{1}{2}, \frac{Z_{br}}{|Z_{sr}|} \in \left[\frac{1}{2} - \sqrt{\frac{1}{4} - \left(\frac{Z_{bi}}{|Z_{sr}|}\right)^2}, \frac{1}{2} + \sqrt{\frac{1}{4} - \left(\frac{Z_{bi}}{|Z_{sr}|}\right)^2}\right] \tag{29}$$

Here, $Z_{br}/|Z_{sr}|$ and $Z_{bi}/|Z_{sr}|$ can be recognized as the real and imaginary part of the non-dimensional buoy impedance, respectively. It can be seen that a negative unconstrained optimal spring stiffness of the PTO system results from the imaginary part of the non-dimensional buoy impedance being smaller than 0.5 and the real part approaching 0.5, and the extent of approaching is determined from the imaginary part. To avoid negative values, the spring stiffness of the PTO system should be constrained.

We selected a truncated circular cylinder as the reference underwater shape of the floating buoy as shown in Figure 2, where d denotes the draft and D for the diameter. The mass of the buoy is set as

$m_b = \mu m_s$, in which μ is defined as the mass ratio. As for a simple case study here, we set the diameter to draft ratio $\gamma = D/d = 2$, and $\mu = 0.5$. Since the displacement of the buoy should be equal to the mass of both the offshore structure and the buoy, the dimension of the underwater part of the buoy can be obtained based on the mass of the offshore structure as

$$d = \sqrt[3]{\frac{4(1+\mu)m_s}{\pi\rho\gamma^2}} \tag{30}$$

$$D = d\gamma \tag{31}$$

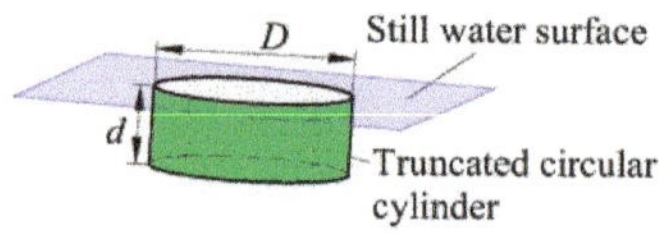

Figure 2. Diagram of the truncated circular cylinder.

Figure 3 shows the hydrodynamic coefficients and non-dimensional impedance of the buoy as a function of wave frequency. The hydrodynamic coefficients are calculated by the boundary element method (BEM) using the ANSYS AQWA software. Based on the hydrodynamic coefficients as shown in Figure 3a, the non-dimensional impedance of the buoy can be easily obtained as shown in Figure 3b. We find that, within the investigated range of wave frequency, the imaginary part of the non-dimensional buoy impedance is smaller than 0.5, which implies that the unconstrained optimal spring stiffness of the PTO system is possible to be negative. Further, the real part of the non-dimensional buoy impedance decreases monotonously with wave frequency, and crosses 0.5 at $\omega = 1.06$ rad/s. Therefore, within the wave frequency range determined from Equation (29), the unconstrained optimal spring stiffness of the PTO system will be negative.

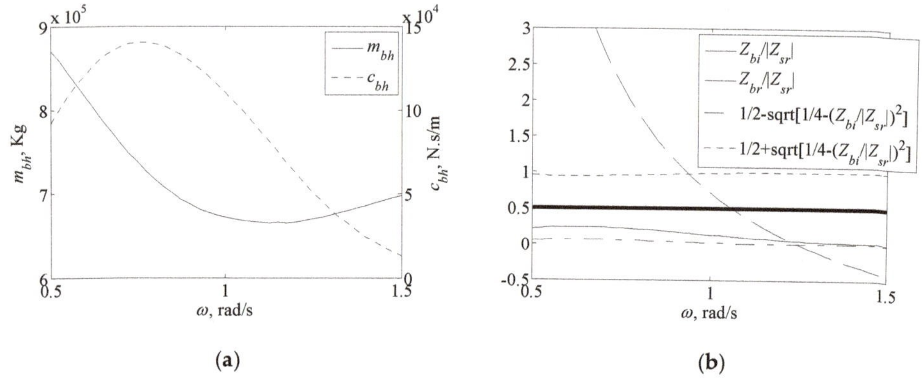

Figure 3. Hydrodynamic coefficients and non-dimensional impedance of the buoy as a function of wave frequency: (**a**) hydrodynamic coefficients; (**b**) non-dimensional impedance.

Figure 4 shows the optimal parameters of the PTO system and maximum absorbed power of the integrated system as a function of wave frequency. It can be clearly seen that, within the wave frequency range from 0.94 rad/s to 1.22 rad/s, the constrained optimal spring stiffness of the PTO system is compulsorily tuned to 0 since the unconstrained optimal value is negative, as inferred from Figure 3. Another obvious finding in Figure 4a is that, around $\omega = 0.93$ rad/s, both the optimal spring stiffness and damping coefficient of the PTO system are impractically large, which can be attributed to the norm of G. The definition of G can be found in Equation (10) and Equation (12). The imaginary part of G is the negative of the imaginary part of the non-dimensional buoy impedance, while the real part of G is the negative of the real part of the non-dimensional buoy impedance plus one. From Figure 3, we find that the imaginary part of the non-dimensional buoy impedance is much smaller in magnitude than the real part, indicating that the norm of G is dominated by the real part. Around the wave frequency of 0.93 rad/s, the real part of G is crossing the zero point, which will result in impractically large values of k_{po} and c_{po} since the norm of G is in the denominator of both. It is observed in Figure 4b that the maximum absorbed power is less than the unconstrained case, which is denoted as $P_{maxunconstrained}$, within the frequency range when the constrained optimal spring stiffness of the PTO system is tuned to zero. This is as expected since the global extreme point of the absorbed power cannot be reached due to the constraint. This suggests that, to absorb the most wave power, the integrated two-body system should be carefully designed so that the wave frequency range where the constrained optimal spring stiffness of the PTO system is tuned to zero stays far away from the peak frequency of a spectrum, around which wave energy is concentrated.

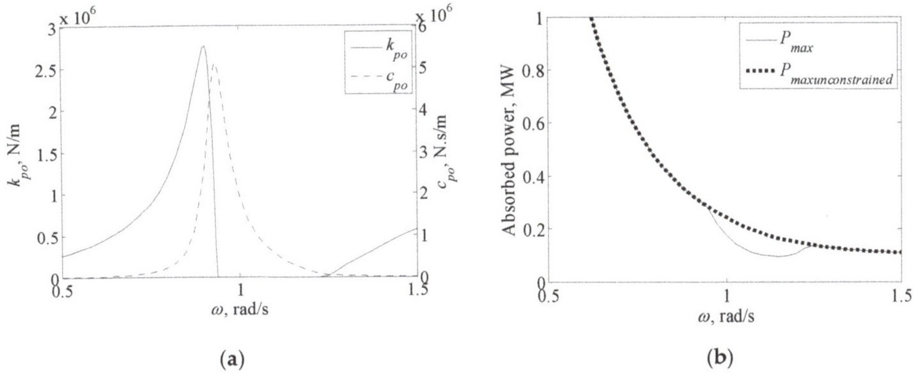

Figure 4. Optimal parameters of the PTO system and maximum absorbed power of the integrated system as a function of wave frequency: (**a**) Optimal parameters of the PTO system; (**b**) maximum absorbed power.

4. Design of the Integrated System

For a given offshore structure to achieve satisfied performance of the integrated system, parameters of the floating buoy should be properly designed. In this paper, parameters of the buoy designed include the mass and the underwater shape, while performance of the integrated system is focused on the absorbed wave power and motion amplitude of the two bodies.

4.1. Design of the Mass of the Buoy

In order to investigate the influence of the mass of the buoy on the performance of the integrated system, we fixed the underwater shape of the buoy while varying the mass. However, when varying the mass of the buoy, to support the varying total mass of the integrated system, the underwater shape will be changed accordingly to provide adequate displacement. Therefore, we take an alternative way to set the key feature of the underwater shape of the buoy to be fixed, and the diameter to draft ratio is chosen. Here, the underwater shape of the buoy employs the truncated circular cylinder, as shown in Figure 2, and the diameter to draft ratio is set to 2. It is assumed that the density of the buoy can be arbitrarily adjusted to fulfill the relation between the volume and mass of the buoy.

Figure 5 shows the optimal parameters of the PTO system as a function of wave frequency at different mass ratios. A general finding is that the maximum values of k_{po} and c_{po} decrease with the mass ratio, and the range of wave frequency when k_{po} is tuned to zero becomes narrower as well. Both suggest that the mass of the buoy should be larger to achieve better design of the PTO system, since smaller spring stiffness and damping coefficient seem easier to implement in engineering.

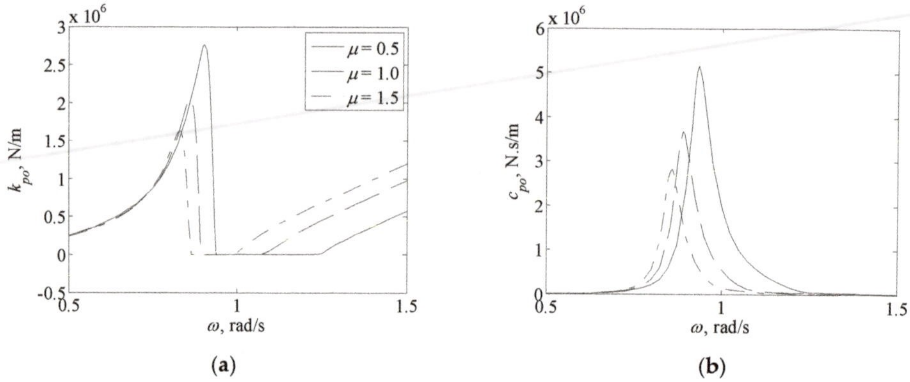

Figure 5. Optimal parameters of the PTO system as a function of wave frequency at different mass ratios: (**a**) Optimal spring stiffness; (**b**) Optimal damping coefficient.

Figure 6 shows the maximum absorbed power of the integrated system as a function of wave frequency at different mass ratios. A narrower range of wave frequency, where power absorption ability degrades due to the constraint on the spring stiffness of the PTO system, is observed at a larger mass ratio. Additionally, the absorbed power increases slightly with the mass ratio when the wave frequency is large. The above findings indicate that an integrated system with a larger mass ratio shows better power absorption performance.

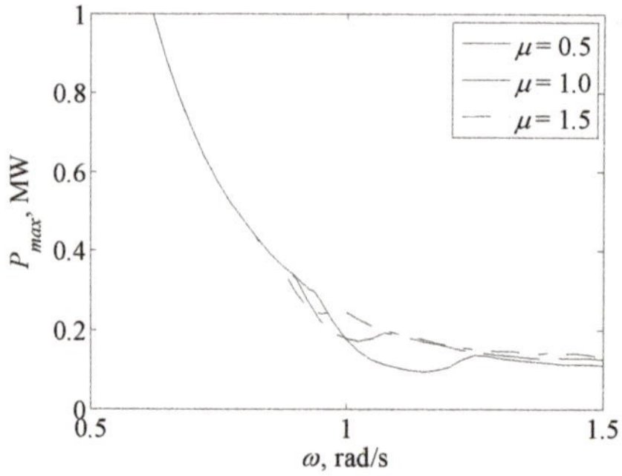

Figure 6. Maximum absorbed power of the integrated system as a function of wave frequency at different mass ratios.

Equation (7) gives the response of the offshore structure and the floating buoy respectively as

$$\hat{z}_s = \frac{Z_p \hat{f}_d}{Z_s Z_b + (Z_s + Z_b) Z_p} \tag{32}$$

$$\hat{z}_b = \frac{\left(Z_s + Z_p\right) \hat{f}_d}{Z_s Z_b + (Z_s + Z_b) Z_p}. \tag{33}$$

Figure 7 shows the non-dimensional motion amplitude of the integrated system, which is defined by the motion amplitude divided by the wave amplitude imitating the definition of RAO as in [22], i.e., $z_{sn} = |\hat{z}_s|/A$ and $z_{bn} = |\hat{z}_b|/A$, as a function of wave frequency at different mass ratios. The non-dimensional motion amplitude of both the offshore structure and the floating buoy generally shows a tendency of fall first and then rise with the wave frequency. However, the motion amplitude of the buoy seems to be more sensitive to the constraint on the spring stiffness of the PTO system shown as the obvious reduction of the motion amplitude at corresponding wave frequencies, while the motion amplitude of the offshore structure seems to be only slightly influenced. A larger mass ratio generally shifts both curves of zsn and zbn to the left. Within the common range of wave period in the ocean, i.e., from 6 s to 10 s (wave frequency from 1.05 rad/s to 0.63 rad/s), a larger mass ratio results in smaller motion amplitude of both the offshore structure and the floating buoy, which is preferred since smaller motion amplitude will not only reduce the influence on the function of the offshore structure due to the motion of itself, but also reduce viscous energy loss due to the motion of the buoy.

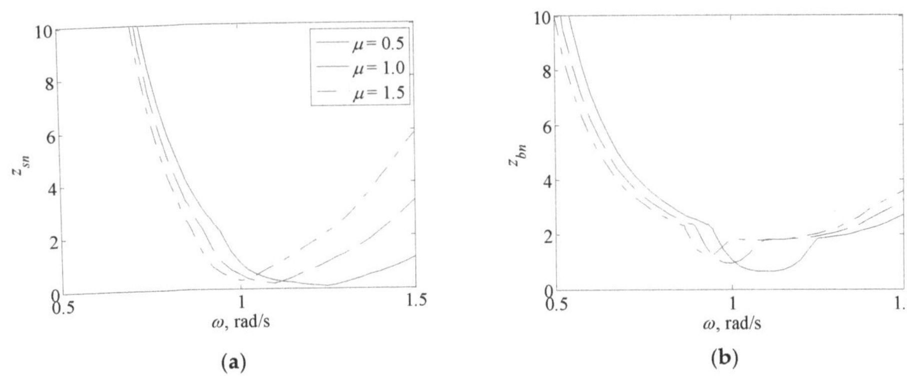

Figure 7. Non-dimensional motion amplitude of the integrated system as a function of wave frequency at different mass ratios: (**a**) For the offshore structure; (**b**) for the floating buoy.

Combined with the findings in above paragraphs, we conclude that a larger mass ratio is preferred for the integrated system due to smaller optimal parameters of the PTO system, more absorbed wave power, and less motion amplitude of the two bodies. However, in practical applications, the mass ratio should not be too large since it will introduce additional structural costs. Therefore, a balance between the performance of the integrated system and the structural cost should be made so that the overall profit reaches the maximum. In the following section, we set the mass ratio to be one for the designing process of the underwater shape of the buoy.

4.2. Design of the Underwater Shape of the Buoy

First of all, we focus on the influence of the diameter to draft ratio γ of the underwater shape of the buoy. Figure 8 shows the maximum absorbed power and non-dimensional motion amplitude of the integrated system as a function of wave frequency at different diameter to draft ratios. At $\gamma = 1$, the absorbed power is strongly influenced by the constraint on the spring stiffness of the PTO system, while at $\gamma = 4$, the influence almost vanishes. Although smaller motion amplitude of the offshore structure from 0.76 rad/s to 1 rad/s is observed at $\gamma = 1$, it is at the expense of the maximum absorbed power being significantly decreased due to the constraint. An overall comparison suggests that a larger diameter to draft ratio results in both more absorbed power and reasonable small motion amplitude of the two bodies. Therefore, we set the diameter to draft ratio to be 4 for further studies in the following.

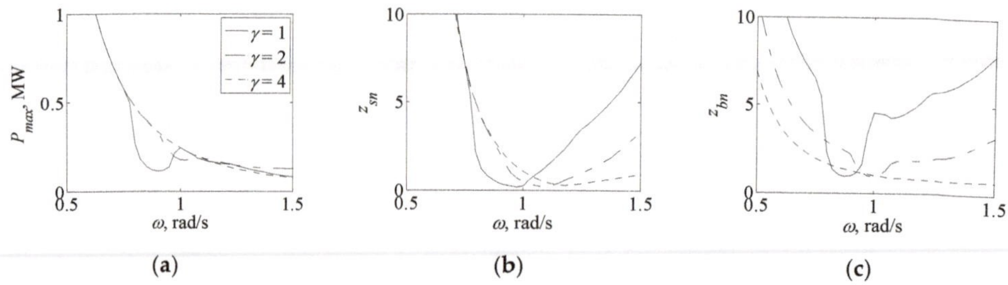

Figure 8. Maximum absorbed power and non-dimensional motion amplitude of the integrated system as a function of wave frequency at different diameter to draft ratios: (**a**) Maximum absorbed power; (**b**) non-dimensional motion amplitude of the offshore structure; (**c**) non-dimensional motion amplitude of the floating buoy.

Secondly, we focus on the influence of non-axisymmetry of the section of the underwater shape. Two representative non-axisymmetric section profiles, i.e., the equilateral triangle and the square as shown in Figure 9, are analyzed as opposed to the circle with axisymmetric section. In Figure 9, D represents the side length of the triangle and the square. Due to non-axisymmetry, the performance of the integrated system varies at different wave directions, which are defined by the wave angle β in Figure 9. Here, we take two typical wave directions for both non-axisymmetric buoys into consideration, i.e., 0° and 60° for the triangle, and 0° and 45° for the square. Figure 10 shows the maximum absorbed power and non-dimensional motion amplitude of the integrated system as a function of wave frequency at different section profiles. An interesting finding is that non-axisymmetry of the section does not bring about significant influence on the performance of the integrated system, and only a slight increase of the motion amplitude of the buoy is observed for the triangular section. Here, the circular or square sections are recognized as the better choice for the integrated system.

Figure 9. Two representative non-axisymmetric section profiles: (**a**) Equilateral triangular section; (**b**) square section.

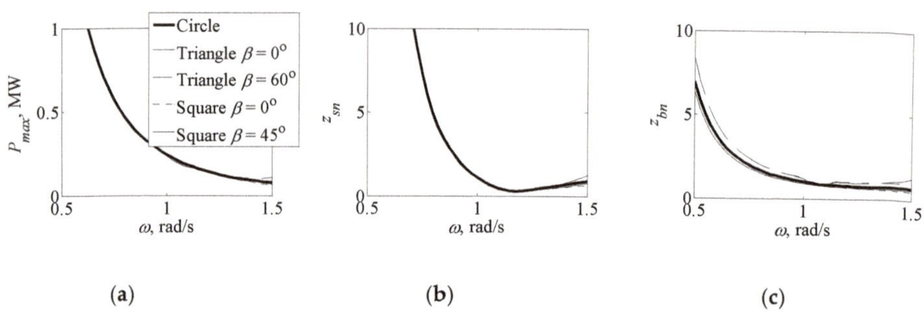

Figure 10. Maximum absorbed power and non-dimensional motion amplitude of the integrated system as a function of wave frequency at different section profiles: (**a**) Maximum absorbed power; (**b**) non-dimensional motion amplitude of the offshore structure; (**c**) non-dimensional motion amplitude of the floating buoy.

5. Performance of the Designed Integrated System in Irregular Waves

The designing process in the previous section suggests that a larger mass of the buoy is better, and a favorable underwater shape of the buoy is characterized by large diameter to draft ratio and constant section size, with the section profile preferring a circle or square rather than an equilateral triangle. In this section, we progress the investigation into irregular waves to study the performance of the designed integrated system in real ocean sea. The mass ratio is set to 1, and the truncated circular cylinder with a diameter to draft ratio of 4 is chosen as the underwater shape of the buoy.

An irregular wave can be recognized as the superposition of regular wave components [23] with the wave amplitude weighted by the energy distribution function, which in this paper employs the Pierson-Moskowitz spectrum [24], defined as

$$S(\omega) = 262.9\frac{H_s^2}{\omega^5 T_e^4}\exp\left(-\frac{1054}{\omega^4 T_e^4}\right), \tag{34}$$

where H_s is the significant wave height and T_e is the energy period. For the n-th regular wave component with frequency ω_n, the wave amplitude is

$$A_n = \sqrt{2S(\omega_n)\Delta\omega}. \tag{35}$$

The response of the offshore structure and the floating buoy is given by

$$z_s = \sum_n \mathrm{Re}\left\{\frac{Z_p \hat{f}_d(\omega_n)}{Z_s(\omega_n)Z_b(\omega_n) + [Z_s(\omega_n) + Z_b(\omega_n)]Z_p}\exp[\mathrm{i}(\omega_n t + \varphi_n)]\right\}; \tag{36}$$

$$z_b = \sum_n \mathrm{Re}\left\{\frac{\left[Z_s(\omega_n) + Z_p\right]\hat{f}_d(\omega_n)}{Z_s(\omega_n)Z_b(\omega_n) + [Z_s(\omega_n) + Z_b(\omega_n)]Z_p}\exp[\mathrm{i}(\omega_n t + \varphi_n)]\right\} \tag{37}$$

where t is the time, and φ_n is the phase of the n-th regular wave designated randomly. The absorbed power of the integrated system is given as

$$P = \frac{c_p}{t_2 - t_1}\int_{t_1}^{t_2}(z_s - z_b)^2 \mathrm{d}t. \tag{38}$$

Due to orthogonality, influence of the coupled terms between different wave frequencies when inserting Equation (36) and Equation (37) into Equation (38) will vanish if the time interval between t_1 and t_2 is long enough [25]. Then, the absorbed power of the integrated system reduces to

$$P = \sum_n \frac{\omega_n}{2}\frac{\left|\hat{f}_d(\omega_n)\right|^2}{\left|G(\omega_n)\right|^2 Z_{pi} + 2Z_{bi}(\omega_n) + \frac{\left|G(\omega_n)\right|^2 Z_{pr}^2 + 2[G_r(\omega_n)Z_{br}(\omega_n) + G_i(\omega_n)Z_{bi}(\omega_n)]Z_{pr} + \left|Z_b(\omega_n)\right|^2}{Z_{pi}}} \tag{39}$$

Figure 11 shows the contour of the absorbed power as a function of the spring stiffness and damping coefficient of the PTO system at H_s = 2 m and T_e = 8 s. It is found that the absorbed power is a unimodal function of the spring stiffness and damping coefficient of the PTO system. As with previous sections, the spring stiffness of the PTO system should be constrained to be non-negative. Therefore, finding the maximum absorbed power of the integrated system is a constrained optimization problem, which is solved here by employing the *fmincon* function using the active-set method within the MATLAB environment.

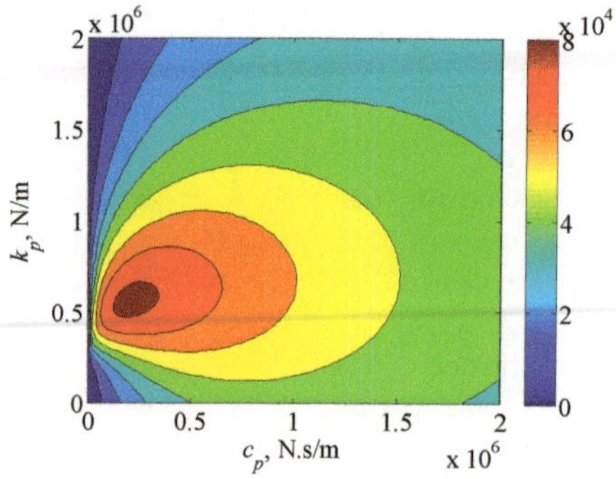

Figure 11. Contour of the absorbed power as a function of the spring stiffness and damping coefficient of the PTO system at $H_s = 2$ m and $T_e = 8$ s.

Figure 12 shows the optimal parameters of the PTO system and maximum absorbed power of the integrated system as a function of energy period at $H_s = 2$ m. It is found that the optimal spring stiffness of the PTO system automatically fulfills the constraint within the investigated range of energy period without being compulsorily tuned to zero. Both the optimal spring stiffness and damping coefficient experience rather milder variation with the energy period than in regular wave conditions, as shown in Figure 4; Figure 5. This can be explained by the fact that irregular waves are composed of wave components of different wave periods, and the behavior of the integrated system in irregular waves is the average of that in regular waves weighted by the energy distribution function. As a result, the violent variation tendency is filtered out by the averaging process. The maximum absorbed power peaks at the energy period of 7.3 s, which is within the common range of wave period in the ocean, i.e., from 6 s to 10 s, indicating that the power absorption ability of the integrated system is at the favorable state. When wave periods are larger than 10 s, it is found that the curves in both subfigures of Figure 12 keep the current tendency and monotonically decrease with the wave period. Therefore, to focus on the WEC behavior in the common range of wave period, the data out of the range is not shown in the figure.

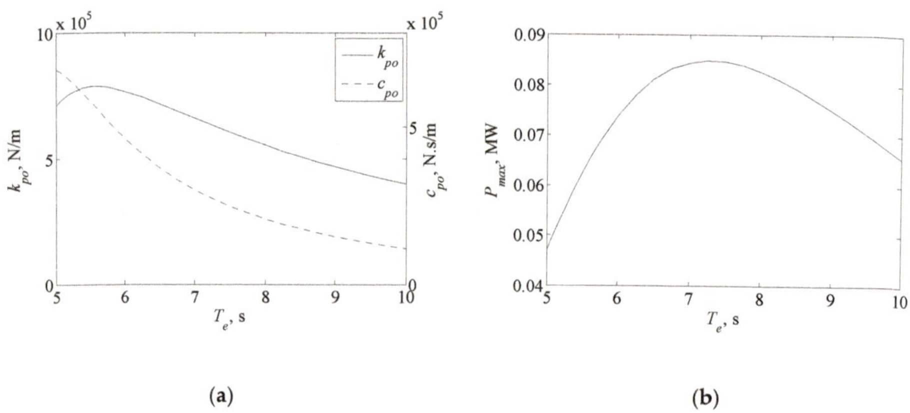

Figure 12. Optimal parameters of the PTO system and maximum absorbed power of the integrated system as a function of energy period at $H_s = 2$ m: (**a**) Optimal parameters of the PTO system; (**b**) maximum absorbed power of the integrated system.

Inserting the optimal parameters of the PTO system into Equation (36) and Equation (37) gives the motion response of the offshore structure and the floating buoy when the absorbed wave power is maximized. Figure 13 shows an example of the time history of the motion response of the integrated system at $T_e = 6$ s and $H_s = 2$ m when the absorbed wave power is maximized. It can be observed that the offshore structure always follows the motion of the buoy, but with a different magnitude and lag phase.

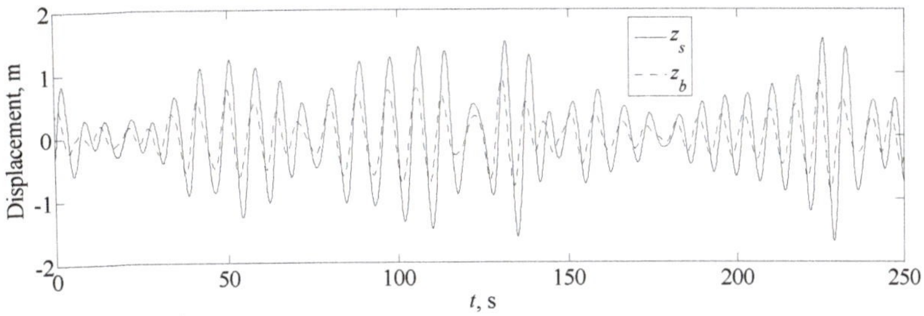

Figure 13. An example of the time history of the motion response of the integrated system at $T_e = 6$ s and $H_s = 2$ m when the absorbed wave power is maximized.

To quantitatively estimate the motion amplitude of the integrated system in irregular waves, we imitate the definition of the significant wave height, which is the mean of the highest one-third wave height in a wave record [26], to define the significant motion amplitude as the mean of largest one-third motion amplitude in a time history record. The significant motion amplitude of the offshore structure and the buoy are denoted by z_{ss} and z_{bs}, respectively, and they probably represent the visual estimate of the motion amplitude of integrated system. As with previous representation of the motion amplitude of the integrated system, the non-dimensional form of z_{ss} and z_{bs} are respectively defined as

$$z_{ssn} = \frac{z_{ss}}{H_s/2}; \quad (40)$$

$$z_{bsn} = \frac{z_{bs}}{H_s/2}. \quad (41)$$

Figure 14 shows the non-dimensional significant motion amplitude of the integrated system as a function of energy period when the absorbed wave power is maximized. The calculated result is the mean of 10 different cases of random phases for the regular wave components. Within the investigated range of energy period, the motion amplitude of the offshore structure is larger than the buoy, and both increase with energy period. One may note that, in the designing process, we have tried our best to choose the design with smaller motion amplitude of the offshore structure, which however exceeds the wave amplitude for a certain range of energy period. This suggests that the offshore structure should be sophisticatedly designed to function insensitively to its heave motion. When offshore structures are not integrated with WECs, they are normally designed to suppress oscillation with waves. However, when they are integrated with WECs, their motion amplitude is even larger than waves. Therefore, to employ offshore structures to absorb wave power, functional devices in the offshore structure should work normally even when the structural support is under large oscillation. In this condition, measures may be taken to suppress oscillation of functional devices, such as motion compensation systems, vibration attenuation systems, etc. On the other hand, this is the accompanying price when the offshore structure is employed to absorb wave power.

Levelized costs of energy (LCOE) are defined as the ratio of total lifetime costs versus total expected outputs, and are used to measure economical competitiveness of a renewable energy technology. Total lifetime costs usually include investment costs (i.e., construction costs), operations, and maintenance costs. Since wave energy is still at an infant stage, lifetime costs of WECs vary significantly depending on many factors, such as the initial capital cost, the capture ratio, the discount rate, and the installation size. Some papers have done preliminary work on estimating levelized costs of wave energy. However, due to immaturity of WEC technology, the estimation process is usually based on many assumptions, and the estimated results show high level of uncertainty. For the integrated system in this paper, levelized costs of wave energy are also quite hard to estimate precisely. However, we can employ some basic data in the literature to provide a rough estimation to find the improvement of levelized costs of wave energy when WECs are integrated to offshore structures. It can be seen from [7] that levelized costs of wave energy are about 190 £/MWh, in which the construction component occupies 155 £/MWh. When integrating WECs to offshore structures, the substructure, mooring system, and electrical grid connections can be shared. Therefore, when calculating levelized costs of wave energy of the integrated system, the above three parts that account for 36% of the construction cost [7] can be subtracted. As a result, the total construction costs reduce by 55.8 £/MWh. If we assume other costs beside construction to be consistent, levelized costs reduce by 29.4%. Therefore, when integrating WECs with offshore structures, levelized costs of wave energy will be significantly reduced. However, we should note that the above calculation is quite simple and under a high level of uncertainty. Precise improvement of levelized costs of wave energy should take into consideration more factors and will be the work of next phase.

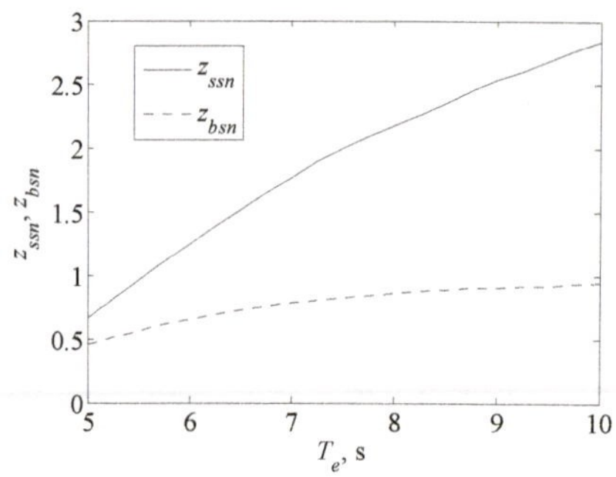

Figure 14. Non-dimensional significant motion amplitude of the integrated system as a function of energy period when the absorbed wave power is maximized.

6. Conclusions

In this work, our main interest is focused on the design of an integrated system, which consists of a floating buoy and an offshore structure on board and employs the relative motion between the two bodies to absorb wave energy, so that satisfied performance in the power absorption process can be achieved. The parameter to design is the mass and underwater shape of the buoy, and the performance concerned includes the absorbed power and motion amplitude of the two bodies. The spring stiffness of the power take-off (PTO) system is constrained to be non-negative with the concerns of complexity in implementation and system stability.

When PTO parameters are not constrained, the optimal spring stiffness will be negative when imaginary and real parts of the non-dimensional buoy impedance are smaller than and close to 0.5, respectively. The constraint on the spring stiffness of the PTO system will reduce the maximum absorbed power when the unconstrained optimal value constrained is negative.

A larger mass of the buoy is advantageous due to smaller optimal spring stiffness and damping coefficient of the PTO system, more absorbed wave power, and less motion amplitude of the two bodies. The best underwater shape of the buoy is characterized by large diameter to draft ratio with the section profile prefers a circle or square rather than an equilateral triangle. Investigations on the designed buoy in irregular waves show that the integrated system presents its peak power absorption ability within the common range of energy period, and the motion amplitude of the offshore structure is larger than the wave amplitude in a certain range of sea states.

Author Contributions: Conceptualization, Z.N.; formal analysis, Z.N.; methodology, J.W.; writing—original draft, J.W.; writing—review & editing, J.W. All authors have read and agreed to the published version of the manuscript.

Funding: This research was funded by the NSFC of China, grant number 51905096, and the Basic Research Plan of Jiangsu Province, grant number BK20190373, and was also supported by "the Fundamental Research Funds for the Central Universities".

Conflicts of Interest: The authors declare no conflict of interest.

References

1. Buccino, M.; Stagonas, D.; Vicinanza, D.; Muller, G. Development of a composite sea wall wave energy converter system. *Renew. Energy* **2015**, *81*, 509–522. [CrossRef]
2. Foteinis, S.; Tsoutsos, T. Strategies to improve sustainability and offset the initial high capital expenditure of wave energy converters (WECs). *Renew. Sustain. Energy Rev.* **2017**, *70*, 775–785. [CrossRef]
3. Heath, T.V. A review of oscillating water columns. *Philos. Trans. R. Soc. Lond. Ser. A* **2012**, *370*, 235–245. [CrossRef] [PubMed]
4. Oh, J.S.; Han, S.H. Inlet geometry effect of wave energy conversion system. *J. Mech. Sci. Technol.* **2012**, *26*, 2793–2798. [CrossRef]
5. Rebhi, R.; Ibala, A.; Masmoudi, A. On the modeling of a doubly-excited brushless claw pole alternator: Application to a solar-wind-wave hybrid energy conversion system. In Proceedings of the 8th International Conference and Exhibition on Ecological Vehicles and Renewable Energies (EVER), Monte Carlo, Monaco, 27–30 March 2013.
6. Astariz, S.; Iglesias, G. Output power smoothing and reduced downtime period by combined wind and wave energy farms. *Energy* **2016**, *97*, 69–81. [CrossRef]
7. Allan, G.; Gilmartin, M.; McGregor, P.; Swales, K. Levelised costs of Wave and Tidal energy in the UK: Cost competitiveness and the importance of "banded" Renewables Obligation Certificates. *Energy Policy* **2011**, *39*, 23–39. [CrossRef]
8. Liang, C.; Zuo, L. On the dynamics and design of a two-body wave energy converter. *Renew. Energy* **2017**, *101*, 265–274. [CrossRef]
9. Engström, J.; Kurupath, V.; Isberg, J.; Leijon, M. A resonant two body system for a point absorbing wave energy converter with direct-driven linear generator. *J. Appl. Phys.* **2011**, *110*, 124904. [CrossRef]
10. Korde, U.A. Systems of reactively loaded coupled oscillating bodies in wave energy conversion. *Appl. Ocean Res.* **2003**, *25*, 79–91. [CrossRef]
11. Parks, P.C. Wedges, plates, and waves: Some simple mathematical models for wave power machines. In *Power from Sea Waves*; Count, B.M., Ed.; Academic Press: London, UK, 1980.
12. French, M.J.; Bracewell, R.H. Heaving point absorbers reacting against an internal mass. In *Hydrodynamics of Ocean-Wave Energy Utilization*; Evans, D.V., Falcao, A.F., Eds.; Springer: Berlin, German, 1986.
13. Korde, U.A. On providing a reaction for efficient wave-energy absorption by floating bodies. *Appl. Ocean Res.* **1999**, *21*, 587–599. [CrossRef]
14. Zhang, H.; Xu, D.; Zhao, H.; Xia, S.; Wu, Y. Energy extraction of wave energy converters embedded in a very large modularized floating platform. *Energy* **2018**, *158*, 317–329. [CrossRef]
15. Chaplin, R.V.; French, M.J.; Bracewell, R.H. An improved version of PS Frog. In Proceedings of the 2nd European Wave Power Conference, Lisbon, Portugal, 8–10 November 1995.
16. Falnes, J. Wave-energy conversion through relative motion between two single-mode oscillating bodies. *J. Offshore Mech. Arct. Eng.* **1999**, *121*, 32–38. [CrossRef]
17. Ansys, Inc. *AQWA Reference Manual*; Version 16.0; Ansys, Inc.: Canonsburg, PA, USA, 2015.
18. Falnes, J. *Ocean Waves and Oscillating Systems: Linear Interaction Including Wave-Energy Extraction*; Cambridge University Press: Cambridge, UK, 2002.
19. Wu, J.; Yao, Y.; Zhou, L.; Chen, N.; Yu, H.; Li, W.; Göteman, M. Performance analysis of solo Duck wave energy converter arrays under motion constraints. *Energy* **2017**, *139*, 155–169. [CrossRef]
20. Cândido, J.J.; Justino, P.A.P.S. Modelling, control and Pontryagin Maximum Principle for a two-body wave energy device. *Renew. Energy* **2011**, *36*, 1545–1557. [CrossRef]
21. Roddier, D.; Cermelli, C.; Aubault, A.; Weinstein, A. WindFloat: A floating foundation for offshore wind turbines. *J. Renew. Sustain. Energy* **2010**, *2*, 033104. [CrossRef]

22. López, M.; Taveira-Pinto, F.; Rosa-Santos, P. Influence of the power take-off characteristics on the performance of CECO wave energy converter. *Energy* **2017**, *120*, 686–697. [CrossRef]
23. Babarit, A.; Duclos, G.; Clément, A.H. Comparison of latching control strategies for a heaving wave energy device in random sea. *Appl. Ocean Res.* **2004**, *26*, 227–238. [CrossRef]
24. Vicente, P.C.; Falcão, A.F.; Gato, L.M.C.; Justino, P.A.P. Dynamics of arrays of floating point-absorber wave energy converters with inter-body and bottom slack-mooring connections. *Appl. Ocean Res.* **2009**, *31*, 267–281. [CrossRef]
25. Wu, J.; Yao, Y.; Zhou, L.; Göteman, M. Real-time latching control strategies for the solo Duck wave energy converter in irregular waves. *Appl. Energy* **2018**, *222*, 717–728. [CrossRef]
26. Holthuijsen, L.H. *Waves in Oceanic and Coastal Waters*; Cambridge University Press: Cambridge, UK, 2007.

sustainability

Article

Optimizing Wave Overtopping Energy Converters by ANN Modelling: Evaluating the Overtopping Rate Forecasting as the First Step

José Manuel Oliver [1,2,*], Maria Dolores Esteban [2], José-Santos López-Gutiérrez [2], Vicente Negro [2] and Maria Graça Neves [3]

1 CIOPU SL, 12004 Castelló de la Plana, Spain
2 Grupo de Investigación de Medio Marino, Costero y Portuario, y Otras Áreas Sensibles, Universidad Politécnica de Madrid, 28040 Madrid, Spain; mariadolores.esteban@upm.es (M.D.E.); josesantos.lopez@upm.es (J.-S.L.-G.); vicente.negro@upm.es (V.N.)
3 Laboratorio Nacional de Engenharia Civil, 1700-066 Lisbon, Portugal; gneves@lnec.pt
* Correspondence: josemanuel.oliver@alumnos.upm.es; Tel.: +34-679-993-244

Abstract: Artificial neural networks (ANN) are extremely powerful analytical, parallel processing elements that can successfully approximate any complex non-linear process, and which form a key piece in Artificial Intelligence models. Its field of application, being very wide, is especially suitable for the field of prediction. In this article, its application for the prediction of the overtopping rate is presented, as part of a strategy for the sustainable optimization of coastal or harbor defense structures and their conversion into Waves Energy Converters (WEC). This would allow, among others benefits, reducing their initial high capital expenditure. For the construction of the predictive model, classical multivariate statistical techniques such as Principal Component Analysis (PCA), or unsupervised clustering methods like Self Organized Maps (SOM), are used, demonstrating that this close alliance is always methodologically beneficial. The specific application carried out, based on the data provided by the CLASH and EurOtop 2018 databases, involves the creation of a useful application to predict overtopping rates in both sloping breakwaters and seawalls, with good results both in terms of prediction error, such as correlation of the estimated variable.

Keywords: artificial neural network; principal component analysis; wave energy converters; wave overtopping rate

check for updates

Citation: Oliver, J.M.; Esteban, M.D.; López-Gutiérrez, J.-S.; Negro, V.; Neves, M.G. Optimizing Wave Overtopping Energy Converters by ANN Modelling: Evaluating the Overtopping Rate Forecasting as the First Step. *Sustainability* **2021**, *13*, 1483. https://doi.org/10.3390/su13031483

Academic Editor: Gregorio Iglesias Rodriguez

Received: 7 November 2020
Accepted: 21 January 2021
Published: 1 February 2021

1. Introduction

During the last decade, the power generation sector has experienced a huge rise in what are known as renewable energy. The sea as one of the most powerful energy sources on earth, over 120,000 TWh/year capacity [1], is one of the key pieces in this sustainability strategy, such that it has even been included in the Sustainable Development Goals (SDGs) [2]. Among its potentials as a generator of clean energy [3] can be cited those derived from the use of marine currents [4], tides [5], thermal gradients [6], salt flats [7], and finally the use of waves to generate energy [8]. These uses have experienced one of the highest growth rates among renewable energy technology in recent years [9], and this may mean a change in trend in the production of sustainable energy, although it is not without major drawbacks at this early stage of development, one of which is precisely associated with the disparity in technologies [10].

This change in trend has undoubtedly been favored by technological advances in all the sectors involved. This is especially so in relation to hydrodynamic systems, with the improvement of equipment and control systems, as well as the utilization of new, more durable materials in such a strongly aggressive environment as the marine one [11]. But these are not the only motivations that direct the focus of interest towards this sector. There is no denying the change in trend in energy production strategies, linked it with a change in social sensitivity. A sensitivity which advocates the search for new sources of energy not linked to the consumptive use of finite natural resources, and where the search for more sustainable harvesting strategies is necessary. Especially if these coexist with a catalytic impulse advocated by different administrations [12].

Among the quoted wide range of uses, the one attributed to the inexhaustible energy of the waves stands out for its potential, especially when it is estimated that its associated energy power ranges between 8000–80,000 TWh/year [1]. Although this is a figure currently under intense debate, since different and detailed approximations will cause variation when specific dependent factors are introduced in the analysis [13]. Although faced with the common drawbacks that characterize developing marine technologies, advantages are presented that make structures for harnessing wave energy strategically advantageous alternatives compared to other marine energy converting technologies. For example, the clear correlation between areas of high energy demand such as densely populated coastal regions and production areas, or the greater stationarity of the generating capacity of the waves compared to more established renewable energy sources, such as wind energy [14].

As a crucial part of this search we focused on those that profit from energy generated when a singular structure is overtopped by incoming waves. Due to the fact that wave height decreases as waves travel from offshore to onshore, more powerful energy generation will commonly be associated with floating structures placed offshore [15], like popular Wave Dragon [16]. There also exists the possibility of combining power generation with other infrastructural needs. It can be combined with building defense structures, or take advantage of existing ones, which will result in a significant reduction of installation costs [17,18], and maintenance costs will be lower because of the accessibility of sites compared to those associated with offshore structures [15]. The sharing of construction techniques commonly used in coastal engineering structures is well documented [18,19], and may be extended to taking advantage of those areas associated with wave propagation, such as refraction, making its implementation even more interesting [20]. These reasons justify that one part of the research effort should be directed to the use of this type of energy converter in existing breakwater structures.

One of the most attractive challenges related to the wave energy generation field of application, is the application of Artificial Intelligence (AI) techniques. This can be applied to both the characterization of the problems associated with the technologies of the converters, and for the evaluation of the energy resources themselves [21]. As proof of its high potential and versatility, AI techniques are commonly applied in the majority of scientific fields, especially in those applied sciences such as engineering, sometimes with significant success [22].

The present work particularly explores the capacities of one of the most popular AI techniques: Artificial Neural Networks, applied in the field of coastal engineering, and specifically for the forecasting of the overtopping rate as an essential part of the design of wave energy converter structures of the overtopping type [11,19].

The potential power of the wave energy converter during its life time T can be defined by the following expression:

$$Energy = \int_0^T f P_h \tag{1}$$

where P_h is the hydrodynamic power, and f is a factor that comprises several efficiency related factors, as electrical, mechanical, or relative to the electrical energy transmission.

There is a direct relationship between the hydrodynamic power energy in the incoming waves with the power stored in the reservoirs of the overtopping breakwater for energy conversion from potential energy to useful energy, but it is necessary to adequately determine certain parameters as the height of the crest or the slope characteristics, to determine the overtopping rate associated with it. The accuracy of such information together with the stochastic nature of the variation both in height and in period that characterize the wave field make it a complex problem and difficult to solve, which is not always adequately solved in the related literature [11]. It is in this multivariate, nonlinear, uncertain situation that ANN can outperform classical approaches [23] and is what directs this research.

The research focuses on building an ANN model where its predictive properties can be applied to a wide range of structures. Data from the CLASH project and EurOtop 2018 are used to develop a new methodological approach that can be differentiated from those proposed in contemporary works [24–26]. The clear objective being the incorporation of this knowledge into the design of wave energy converters for breakwaters and coastal structures, maintaining their defense purpose.

As previously mentioned, other more specific works have developed ANN applications [27], some introduce the prediction of new parameters related to the transmission coefficient [28,29], or the reflection coefficient [26,30,31], or, in one exceptional case, used an ANN as an indirect means for the proposal of a new empirical expression for the calculation of the overtopping [32]. All of them validate the application of artificial intelligence techniques in this specific field of maritime engineering. The application of which has already been related to the more classic parameters of the waves or coastal engineering [33], and even in fields as specific as the scour depth around of marine structures [34–37], or newer ones, such as interaction with coastal biocenosis [38].

2. Materials and Methods

The proposed methodological approach includes a first phase of descriptive analysis of the available variables in the database, where the data are scaled to a common scale, and then proceeds to the detection and elimination of outliers, using univariate and multivariate techniques. As the next step, the identification of the most significant predictor variables is carried out. This is followed by moving to a process of dimensionality reduction of the input data space, using various techniques: Principal Component Analysis, and artificial neural networks with clustering ability (Kohonen networks). Once these pre-processing phases have been completed, the predictive Artificial Neural Network modelling is carried out, and the results of the prediction are analyzed.

2.1. The Data Base and Its Component Parameters

The overtopping is defined as the physical phenomenon that causes a certain flow over the top of a structure when the height of top is less than the rise of the waves (run-up) of the successive wave trains that affect that structure. Its quantification is carried out mainly by means of the variable called overtopping rate (q), which is the flow that exceeds a length of structure per unit of time, when a certain number of waves impact on it.

In the case during time interval t_0 there were N_0 waves falling upon the structure, of heights and periods (H_i, T_i), where each wave produces a certain volume of overflow V_i (H_i, T_i), the overtopping rate can be defined as:

$$q = \frac{1}{t_0} \sum_{i=1}^{N_0} V_i(H_{i,}, T_i) \quad (2)$$

where is the overtopping rate (m^3/s/m or m^2/s); N_0 is the number of total waves; $H_{i,}$, T_i are the height and period of every i wave that fall upon the structure (m; s); $V_i(H_{i,}, T_i)$ is the overflow volume produced by each wave of the wave series, per unit length; and $t_0 = \sum_{i=1}^{N_0} T_i$ is the duration of the record of waves in a storm (s).

Quantification is usually carried out by both prototype test and reduced scale tests, but the latter is mainly used due to its greater economy and simplicity [37]. Both methodological approaches have been used to compile the database that concerns this investigation: CLASH-EurOtop.

The improvement of the databases that have led to their application, has recently borne fruit in a new edition of the EurOtop manual [25]. There are several parameters that must be taken into account when building any overtopping rate prediction model, most of which are collected in this database, as are those used in this study. Among them, the main ones will be geometric and hydraulic parameters [39]. Hydraulic parameters such as the significant wave height (*Hs*), the mean and peak wave period (*T*), the direction of incidence of the waves on the structure (β). Geometric, or related to the typology of the structure itself, such as the freeboard of the structure (R_c). Other factors that also influence, although to a lesser extent, would be the bottom slope, the depth at the foot of the structure, the wind (direction and intensity), the wave grouping, the run-up interference, and so on. So, any database related to the quantification of that phenomenon will necessarily be enriched by data of this nature.

There are many empirical formulations in the specialized literature [40], most of them limited to a certain field or range of application, set either by the nature or design of the structure, or by other conditions specific to the environment (wind, wave conditions, angle of incidence). The desire and need to unify all these existing formulations, at least as far as the data from trials in which they had their genesis, converged in the CLASH Project and its subsequent modifications at EurOtop 2018 [25,41,42]. Therefore, they are the sources of data input used in the modelling of this work.

The parameters that describe each test, with an initial total number of 34, including the measurement of the overtopping rate, were grouped into three categories: general, hydraulic and structural parameters. The general parameters refer to the reliability that is given to each test carried out (*RF*), and his degree of complexity (*CF*). The hydraulic parameters refer to those characteristics related to the waves. While the structural parameters, a total of 17, are proposed to geometrically define the structures under analysis, as well as their boundary conditions relative to depths of water (See Figure 1).

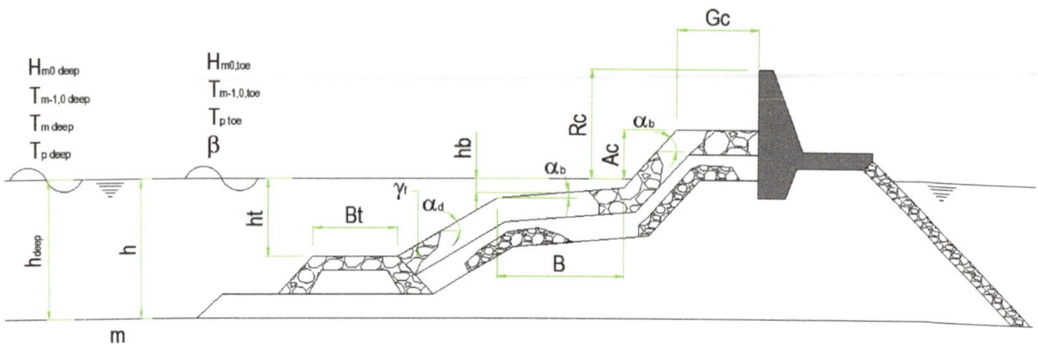

Figure 1. Schematic of the initial descriptive parameters of the model, representing a generic model of a coastal structure.

The hydraulic parameters related to the wave field proposed are: the spectral significant wave height offshore and at the toe of the structure ($H_{m0\ deep}$; H_{m0}), the mean spectral wave period offshore and at the toe ($T_{m-1,0\ deep}$; $T_{m-1,0}$), the average and peak period offshore ($T_{m,deep}$; $T_{p,deep}$), the peak period at toe of the structure ($T_{p,0}$), and the wave incidence angle (β). The structural parameters, considered to geometrically define the structure: the water depth offshore (h_{deep}), the water depth in front of the structure (h), the water depth at the toe of the structure (h_t), the width of the toe berm (B_t), the width of the berm (B), the berm submergence (h_b), the slope of the structure downward of the berm ($cot\alpha_d$), the slope of the structure upward of the berm ($cot\alpha_u$), the average co-tangent, considering the contribution or not of the berm and the slope ($cot\alpha_{incl}$; $cot\alpha_{excl}$), the slope of the berm ($tan\alpha_b$), the crest freeboard of the structure (R_c), the armour crest freeboard of the structure (A_c) and the crest width of the structure (G_c), and finally among the structural parameters, those related with the armour elements characterization, like the permeability/roughness factor of the armour layer (γ_f) or the size of the structure elements along the slope (D).

2.2. *Artificial Neural Network Models*

ANNs are data-driven, parallel processing structures that offer highly nonlinear problem solving applications that are very difficult to solve using traditional techniques. Among the great variety of existing processing paradigms, the use of two of the most common is proposed in this study: Multilayer Perceptrons (MLP) [43], and the Kohonen Neural Networks (KNN) which is also named as Self Organizing Maps (SOM) [44].

2.2.1. Multilayer Perceptron

MLP networks are a supervised training type of ANN, where neurons are strongly interconnected with the previous layer, from where they receive information, and with the posterior layer, towards where they transmit it. In the present case, the output layer is composed of a single neuron, corresponding to the overtopping rate, whose output response can be represented mathematically such that:

$$y = \sum_{k=1}^{n} v_k \varphi\left(\sum_{j=1}^{m} w_{kj} \cdot x_j + b_k\right) \tag{3}$$

with input data represented by the matrix $X \in R^m$, output data $y \in R$, the weight matrix being $V \in R^{1xn}$, $W \in R^{nxm}$, and where the vector of bias terms $b \in R^n$, where m is the dimension of the input space, n is the number of neurons in the hidden layer, and R is the set of real numbers. More detailed information for the MLP networks can be found in Haykin (1991) [43].

2.2.2. Kohonen Neural Network

The KNN is an ANN model, with the ability to classify data according to their similarity, in a preserving topological way.

Kohonen networks are basically made up of two layers, the input layer and the competitive or exit layer (or self-organized map), both layers are fully interconnected, and unlike MLPs they respond to unsupervised training.

In a sequence of phases of competition, comparison, cooperation and adaptation, a process is structured in which a neuron of the output layer will be activated by a process of comparison or measurement of similarity between the input pattern and that neuron, the candidate for winning. This measure of similarity is usually the Euclidean distance d_j between the input vector X and the vector of synaptic weights W_j:

$$d_j = \| X - W_j \| = \sqrt{\sum_{i=1}^{M} (x_i - w_{ij})^2} \tag{4}$$

Being the winning neuron, the one with the smallest distance between them. The weights of this winning neuron will therefore be adjusted, in the direction of the input vector, according to the expression:

$$W_J(t+1) = W_j(t) + \eta(t) \cdot h_j(t) \cdot (X - W_j(t)) \tag{5}$$

where η is the learning rate, and therefore the update of the weight vector W_J $(t + 1)$ for time $t + 1$, and h_j (t) is the neighborhood function for time t.

With this adjustment, each node of the output layer develops the ability to recognize future input vectors that are presented to the network and are similar to it, grouping them in this environment according to a self-organizing process, which gives them propensity for grouping (clustering) [44]. More detailed information for the Kohonen networks and theirs mathematical foundations can be found in Kohonen (2001) [45].

2.3. Pre-Processing of Data

On the initial data set consisting of 17,942 trials, an exploratory and descriptive analysis phase of the data was carried out, conditioned especially by the general parameter *RF* and *CF*, and which allowed abundant discarding of trials (including all those vectors with missing data). Similarly, some parameters were discarded, based on bibliographic recommendations [24,41], due to their low significance. Among them: all the deep water parameters and some redundant structural parameters. The discards resulted in a dimensional reduction to a total of 23 variables.

2.3.1. Data Escalation

The original data, once those considered anomalous had been discarded, underwent a scaling process based on the fact that they came from two very different sources such as: those from laboratory tests at different scales, and those from prototypes. Maintaining this lack of dimensional coherence would introduce an additional problem when modelling, and would also greatly hinder an elementary descriptive analysis of the data. Therefore, the adoption of a single scale is proposed [46].

The scaling process was carried out by applying a scaling based on the theory of dimensional analysis to represent the general equations of hydraulics and obtain the dimensionless Buckingham-Pi monomials, and specifically those based on Froude similarity [47].

For practical purposes, it is established that $H_{m0\,t}$ acquires a value common to all trials of 1 m [24,25,48].

2.3.2. Debugging the Data

Once the data has been scaled, the sample is debugged in order to use it as an input space in the training of neural networks. To do this it must be ensured that the integrating patterns do not contain outliers, since their presence can be highly detrimental the network training [49]. Another significant aspect that makes the elimination of outliers advisable is the improvement in the performance of the error functions that will determine the goodness of the model, and is especially desirable when we work with the MSE function [50].

For dealing with the problem caused by the presence of outliers in the input data, this work chose from different strategies [51], the one that proposes the early detection of them, later proceeding with their removal from the sample and finally facing the modelling process with a sample clean of outliers.

To check if an input pattern can be considered as a multidimensional outlier, a process based on the Mahalanobis distance is proposed [52].

To apply this test, a detection of those data vectors that present one-dimensional outliers is previously performed, establishing a class that will be compared by means of the Mahalanobis distance test with the other sample that does not contain univariate outliers, established through statistics such as Box Fisher's *F*, or Wilks' lambda statistic and supported by graphics (box-plots).

For practical purposes, the strategy referred to above means that from the univariate and multivariate analysis of the data, it turns out that the data set is reduced to another set made up of 23 variables and a total of 10,097 patterns, which constitute the definitive base, or sample space, available for a subsequent dimensionality reduction by applying diverse techniques. In the following table (Table 1) the main statistical parameters of this debugged data base are summarized.

Table 1. Statistical parameters of the scaled input–output patterns of the database used for training of the ANN, after the debugging process (scaled to $H_{m0\,t}$ = 1 m).

Variable	Min	Max	Mean	Std. Deviation
$H_{m0\,\text{toe}}/L_{m-1}$ (−)	0.004	0.113	0.046	0.014
$H_{m0\,toe}$ (m)	0.280	3.765	1.013	0.142
$T_{p\,toe}$ (s)	2.993	108.138	5.314	3.111
$T_{m-1,0t}$ (s)	2.721	69.942	4.940	2.500
$T_{m\,toe}$ (s)	1.441	20.466	4.299	1.224
$cot\alpha_u$ (−)	−5.000	100.000	2.860	7.946
$cot\alpha_d$ (−)	0.000	7.000	1.872	1.351
$cot\alpha_{incl}$ (−)	−1.347	11.299	2.166	1.692
$cot\alpha_{excl}$ (−)	−1.347	8.144	1.845	1.474
$tan\alpha_B$ (−)	0.000	0.101	0.002	0.010
B_h (m)	0.000	34.188	1.042	2.511
B_t (m)	0.000	19.231	0.743	1.832
B (m)	0.000	34.188	1.055	2.521
B (o)	0.000	80.000	3.126	10.469
h (m)	0.375	30.000	4.251	3.215
h_t (m)	0.375	23.434	3.800	2.952
h_b (m)	−2.652	7.826	0.070	0.675
D (m)	0.000	2.370	0.195	0.249
γ_f (−)	0.380	1.000	0.739	0.277
R_c (m)	0.000	13.675	1.594	1.089
A_c (m)	−5.242	13.675	1.506	1.069
G_c (m)	0.000	15.170	0.988	1.443
q (m^2/s)	0.000	0.320	0.012	0.032

2.3.3. Dimensionality Reduction

The good generalizability of an ANN model is linked to its complexity. The presence of a very high number of features (>30) results in the well-known "curse of dimensionality" [23] to which the ANN are not alien [43]. Avoiding overfitting is one of the most important goals to achieve with dimensionality reduction. Therefore, an attempt is made to reduce the input dimension as much as possible, without loss of information associated with the sample variance.

In the present study, this objective has been achieved by applying, on the one hand, the discriminant analysis technique of Principal Component Analysis (PCA), and on the other hand, the application of Kohonen networks (SOM).

Assuming the matrix X of the sample data with p features and n integrating patterns (vectors) of the sample:

$$X = \begin{pmatrix} x_{11} & x_{12} & \cdots & x_{1p} \\ x_{21} & x_{22} & \cdots & x_{2p} \\ \vdots & \vdots & \ddots & \vdots \\ x_{n1} & x_{n2} & \cdots & x_{np} \end{pmatrix} \tag{6}$$

PCA consists of finding orthogonal transformations of the original features to obtain a new set of uncorrelated ones, called Principal Components. These uncorrelated features are the eigenvectors, and are obtained in decreasing order of importance, this importance being associated with the amount of variance explained by them. As a result, the components are linear combinations of the original features and it is expected that only a few (the first of them) will capture most of the variability of the data, thus obtaining a reduction in size after the transformation that this technique involves [53]. Geometrically, transformation can in fact be explained as a rotation in the p-dimensional space (see Figure 2), looking for the projection that maximizes the information provided by the multivariate pattern in terms of variance.

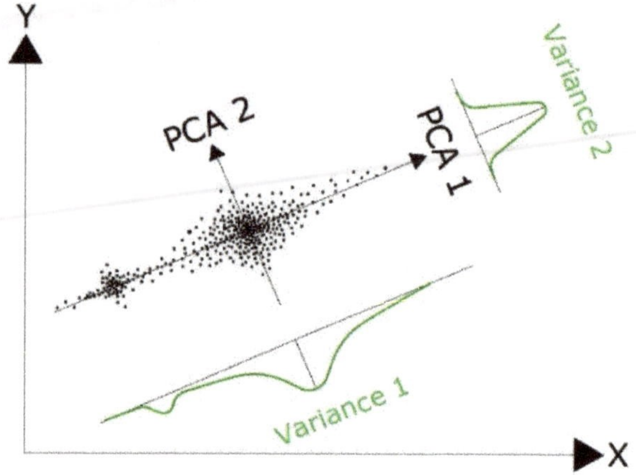

Figure 2. Spatial interpretation of the PCA technique. The projection on the new axes PCA1 and PCA2 maximizes the information provided by the multivariate variable in terms of variance.

The space generated by the first q components is then a q-dimensional vector subspace of the original p-dimensional space. Thus, the principal components of X will be the new variables:

$$Y_j = X \cdot t_j,\ j = 1, \ldots, p \tag{7}$$

for each, the new variable j is constructed from the j-th eigenvector of $S = \text{var}(X)$. Expressing it in a less compact way:

$$\begin{pmatrix} Y_1 \\ Y_2 \\ \vdots \\ Y_p \end{pmatrix} = \begin{pmatrix} t_{11} & t_{12} & \cdots & t_{1p} \\ t_{21} & t_{22} & \cdots & t_{2p} \\ \vdots & \vdots & \ddots & \vdots \\ t_{n1} & t_{n2} & \cdots & t_{np} \end{pmatrix} \cdot \begin{pmatrix} x_1 \\ x_2 \\ \vdots \\ x_p \end{pmatrix} \tag{8}$$

The transformed data matrix is $Y = X \cdot T$, and represents the "observations" of the new variables (principal components) on the n sample patterns.

In another way, the application in this case of a Kohonen network model, also indirectly allows the interpretation of existing relationships at the level of self-similarity. Information that will ultimately allow decisions to be made on reducing the dimensionality of the input space with topology preserving [54], either by interpretation of these detected relationships between the variables, or by confirmation of those already detected using the PCA technique. This is achieved through the interpretation of the distance matrix (U-matrix), and significantly, of the plane components (P-Matrix) [55]. The U-matrix represents in a 2D lattice the Euclidean distance between neighboring nodes. The component planes allow the graphic display, also in 2D, of each of the variables in the data set, by expressing the values of the weight vectors.

Although it should be noted that this procedure has an obvious drawback, which lies in the subjectivity of the criteria for establishing these relationships between variables.

2.4. Proposed Models

Prior to the construction of the first models, a parametric contrast of homogeneity of the joint sample is carried out, demonstrating that there are significant differences between the samples, once they have been scaled, from both tests on reduced models and from prototypes [30], thus breaching one of the premises that motivated the creation of an international and homogeneous base on wave overtopping [56].

It is well known that there are founded differences between overtopping rates determined from conventional scale models of breakwaters (generally based on Froude's law of similarity) [57], especially for rubble mound breakwaters in laboratories, and those measured on similar prototypes [42,58–61]. To confirm this lack of homogeneity, that could significantly influence any approach of ANN models, it was decided to carry out certain hypothesis tests, understanding these as a test of significance to demonstrate whether certain hypotheses that are assumed to be true, are with a certain degree of security. The hypothesis test proposed is a parametric test, checking for its certain population parameters, and also assuming that the distribution of the data is of a certain known type (distributed according to a normality hypothesis).

The parametric tests carried out consist of the Fisher test of Snedecor's *F*-contrast statistic over the sample variance and the bilateral hypothesis test on the equality of the sample means carried out using the Student's *t*-test statistic. Table 2 shows the results of the bilateral contrast carried out over the means and sample variances.

Table 2. *p*-valor results on the bilateral *F*-Snedecor and *t*-Student contrast (confidence level $\alpha = 0.05$). In bold, those values that allow accepting the null hypothesis of equality of means or variances have been highlighted.

Variable	Variances	Means	Variable	Variances	Means
$H_{m0\ toe}/L_{m-1}$	<0.0001	0.005	B	<0.0001	<0.0001
$H_{m0\ toe}$	<0.0001	<0.0001	β	0.027	<0.0001
$T_{p\ toe}$	<0.0001	<0.0001	h	<0.0001	**0.218**
$T_{m-1,0t}$	<0.0001	<0.0001	h_t	<0.0001	**0.286**
$T_{m\ toe}$	**0.974**	<0.0001	h_b	<0.0001	**0.382**
$cot\alpha_u$	<0.0001	**0.286**	D	<0.0001	<0.0001
$cot\alpha_d$	<0.0001	0.003	γ_f	**0.351**	<0.0001
$cot\alpha_{incl}$	0.044	**0.589**	R_c	<0.0001	<0.0001
$cot\alpha_{excl}$	<0.0001	0.005	A_c	<0.0001	<0.0001
$tan\alpha_B$	<0.0001	0.025	G_c	<0.0001	<0.0001
B_h	<0.0001	<0.0001	q	<0.0001	<0.0001
B_t	<0.0001	<0.0001			

The results highlighted in bold in Table 2, finally allow concluding that the two samples are not homogeneous, due to the fact that only five of the variables ($cot\alpha_u$, $cot\alpha_{incl}$, h, h_t, h_b) meet the hypothesis of equality of means, but additionally, none of them meet the hypothesis of equality in variances.

The proven existence of such a lack of homogeneity leads to proposing a differentiating strategy, which implies proposing suitably differentiated models for each sample with its own sampling characteristics. Therefore, two initial models are proposed, that will be trained with a total number of 9997 patterns, preserving an additional number of 100 patterns for extra validation purposes:

- Model I: Corresponds to an ANN model for the definition of which, all the available patterns have been used after the debugging and the dimensionality reduction process.
- Model II: Involves a division of the input pattern space into two distinct groups or clusters. The first of them trained with data from laboratory tests, and the second with tests from prototypes.

Model I will result in obtaining a single ANN, while Model II will involve the construction of 2 different ANNs: Sub model II.1 and sub model II.2. Sub model II.2 implies the application of a Kohonen network as a previous step for the optimized obtaining of the training, verification and test subsets on a very small sample of 171 patterns from prototype tests, according to the methodology described by Bowden and Maier in 2002 [62]. This is not necessary for the rest of models for which this division is randomly conducted. The quoted methodology, which includes a Kohonen network with 10×10 nodes that allows selecting three cases for each of these nodes, where the first of the data will be used for the training subset, the second for the verification subset and finally the third goes to the validation subset. In the case that in one of the nodes of the self-organized map only one pattern existed, it would be mandatorily destined to the training subset, while if there were only two existing patterns, it would be acted so that the first one was destined to the subset of training, while the second would be destined to the verification one. In this way it is possible to reduce the patterns necessary to train the model to a minimum.

Univariate MLP networks are proposed for the construction of the predictive model. For the determination of its structure, it is assumed that more than one hidden layer does not represent an appreciable improvement [63], and on the contrary, would suppose a substantial increase in training time in addition to an increase in the possibility of network overfitting, with a lower capacity for generalization associated [23,43,64]. Therefore, the incremental calculation of the number of neurons is proposed as a valid criterion, by a trial and error procedure, and not the number of layers [65].

Before the training of the networks, the variables of the input patterns will be standardized to the range $[-1, 1]$ by means of the following function:

$$x' = \frac{(U_x - L_x)\cdot(x - x_{min})}{(x_{max} - x_{min})} + L_x \tag{9}$$

where x' is the escalated variable; x is the original variable; x_{max} is the maximum of variable x in the original sample; x_{min} is the minimum of the variable x in the original sample; U_x the transformed value of the maximum of the variable x; L_x is the transformed value of the minimum of the variable x; and where the maximum and minimum escalation ranges are $U_x = 1$, and $L_x = -1$.

This scaling proposal implies that the activation function will be the hyperbolic tangent, which entails the mandatory use of a linear activation function in the output layer [66].

Candidate networks will be trained with MATLAB software (Mathworks ©) according to a cross-verification procedure [43], and using the Levenberg-Marquardt algorithm, as it is better adapted to the characteristics of the available sample space, providing better results, both in terms of error, as well as computational time and stability [67,68].

For practical purposes, from the complete sample of 9997 patterns, 85% will be used in the network training process, which means randomly dividing up that sample into a training subset consisting of 6997 patterns, and 1500 patterns intended for cross-verification process, while the remaining subset consisting of 1500 additional patterns will serve for the model test.

3. Discussion

This study investigates the capabilities of two differentiated ANN models to predict the overtopping rate based on different boundary conditions and in the capacity of the ANN to work optimally in homogeneous sample spaces [69]. The performance of these two models will be carried out based on the statistics of the mean squared error (MSE), and the correlation coefficient (r).

The mean squared error measures the average of the squared errors, that is, the difference between the simulated value ($\hat{q}_i$) and the observed value (q_i) across the range of data (n):

$$MSE = \frac{\sum_{i=1}^{n}(\hat{q}_i - q_i)}{n} \tag{10}$$

The mean squared error is one of the most used functions, with interesting properties that make its use generalized, such as it is easily calculated and penalizes large errors. As a disadvantage, it requires errors to be distributed independently and normally [50].

The correlation coefficient (r) is an indicator of the degree of linear statistical dependence, and is calculated according to:

$$r = \frac{\sum_{i=1}^{n} (\hat{q}_i - \hat{\overline{q}}) \cdot (q_i - \overline{q})}{\sqrt{\sum_{i=1}^{n} (\hat{q}_i - \hat{\overline{q}})^2 \cdot \sum_{i=1}^{n} (q_i - \overline{q})^2}} \quad (11)$$

where $\hat{\overline{q}}$ is the mean value of the simulated variable and $\overline{q}$ is the average value of the observed variable.

3.1. Obtaining the Reduced Dimension of the Input Vector

A first step will be to carry out a correlational analysis, by obtaining the correlation matrix. The information it provides is sufficient to make decisions about discarding some variables, but not enough to others, so it is necessary to resort to more sophisticated techniques such as those that will be used later. The existing correlations of the different explanatory variables with the variable to be predicted are very low (maximum r value of 0.322), which shows the high non-linearity of the process to be analyzed (Table 3).

Table 3. Correlation with explained variable (q).

Variable	Pearson Coef. *r*	Variable	Pearson Coef. *r*
$H_{m0\ toe}/L_{m-1}$	−0.097	$cot\alpha_{excl}$	−0.028
β	−0.100	$cot\alpha_{incl}$	−0.067
h	0.068	γf	0.306
H_{m0toe}	−0.033	D	−0.253
$T_{p\ toe}$	0.007	R_c	−0.322
$T_{m\ toe}$	0.058	B	−0.058
$T_{m1,0t}$	0.031	h_b	0.067
h_t	0.107	$tan\alpha B$	0.043
B_t	−0.098	B_h	−0.059
$cot\alpha_d$	−0.022	A_c	−0.312
$cot\alpha_u$	−0.012	G_c	−0.212

In general, the correlation coefficients between the different variables that make up the set of input variables are low, or very low, reaching only significant correlation values in the following pairs of values:

- B and B_h: referred to the dimensions of the berm (0.999)
- h and h_t: referred to depth at the toe of the structure or the submergence of the toe (0.91)
- $T_{m\ t}$ and $T_{m1\ t}$: period values at the toe of the structure (0.63)
- D and γ_f: variables related to the size and roughness factors (0.777)
- R_c and A_c: variables relative to freeboard (0.860)
- $Cot\alpha_{incl}$, $cot\alpha_{excl}$, $cot\alpha_d$: variables related to the slope geometry (0.828 to 0.921)

Regarding PCA analysis, Table 4 represents the contribution of each variable to the first eight principal components, together with their correspondent eigenvalues and the cumulative variance explained by them. The total variance cumulated by them is higher than 75%, that is one of the criteria accepted in practice for stablishing the contributing limit to an effective model. Another adopted criterion will be that the variance explained by them be major than the mean, so major than one, rule also proportionated by the first eight components.

Table 4. Loadings of the first eight variables.

Variable	F1	F2	F3	F4	F5	F6	F7	F8
$H_{m0\,toe}/L_{m1\,0\,t}$	−0.603	−0.254	0.305	−0.292	−0.319	0.197	−0.117	0.220
β	−0.020	−0.153	−0.058	−0.195	0.302	0.131	−0.107	−0.050
h	0.273	0.631	−0.145	−0.153	0.575	0.259	0.166	−0.041
$H_{m0\,toe}$	0.222	−0.184	−0.388	−0.271	−0.003	0.089	0.017	0.683
$T_{p\,toe}$	0.499	−0.090	−0.358	0.334	−0.115	−0.304	−0.118	−0.073
$T_{m\,toe}$	0.597	−0.096	−0.402	0.319	0.063	−0.325	0.089	0.067
$T_{m1,0t}$	0.563	−0.083	−0.370	0.391	−0.078	−0.352	−0.119	−0.072
h_t	0.338	0.575	0.011	−0.186	0.612	0.201	0.042	−0.060
B_t	−0.127	0.152	−0.446	−0.028	0.001	0.219	0.328	0.158
$cot\alpha_d$	0.617	−0.445	0.414	−0.109	0.197	0.233	−0.273	0.015
$cot\alpha_u$	0.195	−0.273	0.169	0.180	0.003	0.270	0.614	−0.184
$cot\alpha_{excl}$	0.591	−0.589	0.408	−0.025	0.163	0.233	−0.104	0.039
$cot\alpha_{incl}$	0.638	−0.442	0.527	−0.178	−0.017	0.068	−0.013	0.010
γf	0.188	0.457	0.280	0.581	−0.239	0.305	−0.040	0.172
D	−0.012	−0.444	−0.489	−0.532	0.224	−0.195	0.089	0.137
R_c	0.422	0.213	−0.489	−0.185	−0.425	0.431	−0.109	−0.006
B	0.367	0.461	0.379	−0.481	−0.303	−0.329	0.142	−0.029
h_b	0.192	0.528	0.205	−0.346	0.017	−0.205	−0.232	0.208
$tan\alpha_B$	0.247	−0.141	0.277	0.127	−0.233	−0.013	0.589	0.231
B_h	0.360	0.467	0.371	−0.484	−0.297	−0.330	0.128	−0.038
A_c	0.421	0.142	−0.512	−0.175	−0.417	0.433	−0.107	−0.027
G_c	−0.022	−0.410	−0.286	−0.440	0.111	−0.249	0.183	−0.082
q	−0.049	0.183	0.261	0.457	0.303	−0.207	−0.016	0.518
Eigenvalue	3.471	3.116	2.908	2.387	1.786	1.596	1.165	1.021
% Explained	15.093	13.550	12.641	10.380	7.766	6.938	5.063	4.439
Cumulative	15.093	28.643	41.284	51.665	59.431	66.369	71.432	75.871

The first component (F1) explains 15.09% of the variance and is dominated by variables related with the period, slope of the structure and by the wave steepness. The second component (F2) with a similar percentage of the variance explained (13.55%) is dominated by variables related with the submergence and $cot\alpha_{excl}$. And the third component (F3) explains 12.64% of the total variance and is dominated by the variables $cot\alpha_{incl}$ and A_c.

The analysis of the correlation circle, that corresponds to a projection of the initial variables of the first two factors of the PCA onto a two-dimensional plane, provides relevant information that allows observing correlations between the variables and interpreting the axes, or main factors, and thus being able to eliminate correlations that could be redundant and therefore detrimental to the predictability of the model. For the present case, it is shown, in the correlation circle (in the projection of both F1 and F2 axes) (See Figure 3a), that the percentage of variability represented by the first two factors is not particularly high (28.64%). Therefore, to avoid a misinterpretation of the graphics, it also requires a visualization in axes 1 and 3, and interpretation of the influence of presence or absence of certain parameters (See Figure 3b).

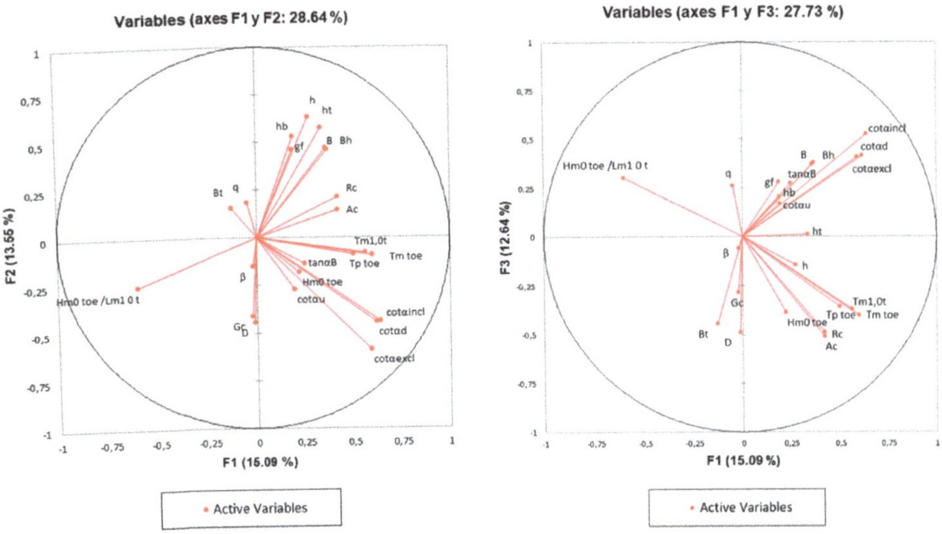

(a) (b)

Figure 3. Dimensionality reduction using the PCA technique: (**a**) Correlation circle on the F1–F2 plane; (**b**) Correlation circle on the F1–F3 plane.

Both plots confirm the results shown in Table 4, without a significant relevance of the variables, accompanied by a lack of clear interpretation of the axes, but on the other hand, with the evidence of certain interesting relationships between the variables.

It can be seen that there is a strong grouping between the variables related to the period ($T_{m1.0t}$; $T_{p\,t}$; $T_{m\,t}$), with a high positive correlation between them, a trivial matter already detected in the correlation matrix, which at least allows reducing their number, in any case keeping only one of them. Another group with a strong positive correlation is composed of those variables related to the geometric characterization of the slope ($cot\alpha_{incl}$; $cot\alpha_{excl}$; $cot\alpha_{d}$) on which it will act in a similar way. The same procedure can be carried out with the variables relative to the width of the berm and its horizontal projection (B; B_h), and from which it is inferred that only variable B will be preserved. The grouping of variables in the correlation circle, in the projection of both axes F1 and F2, seems to determine the lack of correlation between the variables that make up the most obvious groupings, with similar direct cosines, such as those determined by $cot\alpha_{incl}$, $cot\alpha_{excl}$, $cot\alpha$d, and those like: $cot\alpha_u$, $tan\alpha_B$, $H_{m0\,t}$. The foregoing leads to considering that both groupings of variables must be present in the input space, although with the particular restrictions indicated previously for some of them. The spectral wave steepness variable ($H_{m0\,t}/L_{m1\,t}$), negatively correlated with the freeboard variables (R_c, A_c), should be kept as above. Finally, the strong link between the width of the crest and the characteristic size of the protection elements in the breakwater is clearly reflected along with its strong link with the F2 axis.

The projection on the F1 and F3 axes explain a total variability of 27.7%, which is a percentage very similar to that explained by the previous projection (F1 and F2). Additionally, in this projection the correlations established for the first circle of projections are maintained, even the observed groupings are very similar. This robustness in the projection reaffirms the initial idea of finally discarding several of these correlated variables.

A technique that is usually used when the information provided by the PCA analysis carried out on the total of the variables is not very informative, is to consider the contribution of some variables whose contribution is doubted, such as that of supplementary variables studying the effect of their elimination on the projection space. In this case, the input space is censored by considering as supplementary variables all those that have shown an evident correlation in the previous projections. Figure 4 shows the new correlation circle with this elimination step applied.

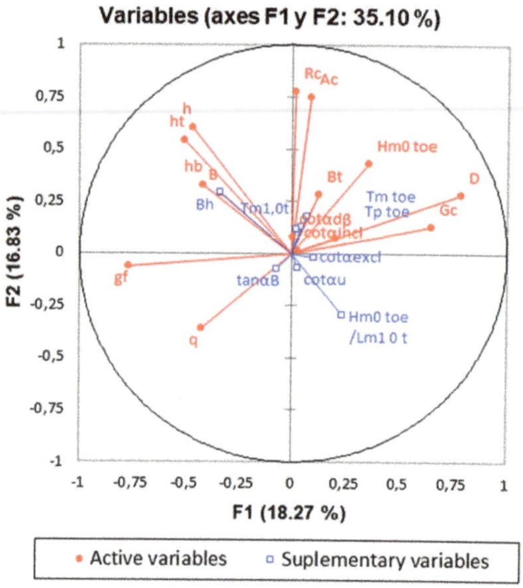

Figure 4. Correlation circle in the first two principal components with variable elimination. In blue colored appear those variables that are candidates to be eliminated.

Figure 4 demonstrates that the variability explained by the first two factors increases slightly with a total of the variance explained equal to 35.10%. Additionally, that there is strong linking between variables relates with freeboard parameters and F1 axis, which could explain the major variance in the sample. The above, and its comparison with the initial projections, indicates that a reduction in dimensionality may be beneficial for the explanation of the problem without a significant loss of information [53], and therefore this reasoning can be valid for the composition of a model with a smaller input dimension.

Alternatively, the application of a Kohonen network model on the same input pattern space, with a dimension of 23 factors (all of which come from the previous pre-processing processes), will indirectly allow the interpretation of the relationships existing at the level of self-similarity between input patterns, in a two-dimensional projection, where every pixel in that 2-dimensional map is characterized by a multidimensional vector. Information that will ultimately allow the reduction of the dimensionality of the entrance space, either by interpretation of these detected relationships, or by confirmation of those already detected using the PCA technique.

For the specific purposes of the present study, the constructed model, built with a Gaussian neighborhood function in every node, has been carried out using the following training scheme, characterized by two different phases [66], where each set is shown 500 times to the SOM. During the training phase, the complete preprocessed data set is shown 500 times to the SOM, built with a Gaussian neighborhood function in every node. The first of these, or rough adjustment phase, is performed with a learning rate with values in the range between 0.9 to 0.1, and with a neighborhood ratio that varies from 2 to 1, and with an extension of training up to 100 times. The second or fine-tuning phase is completed with a unique learning rate of $\eta = 0.01$, with a neighborhood ratio of 0, and with training extension up to 100 epochs. With this, the total length of the training will be:

100 + 100 = 200 times. The dimension of the input tile for this model will be 25 × 25 units (625 total units).

The following figure (Figure 5) shows the different component planes obtained after network training, with a total of 23 units (one for each component variable of the input and output space).

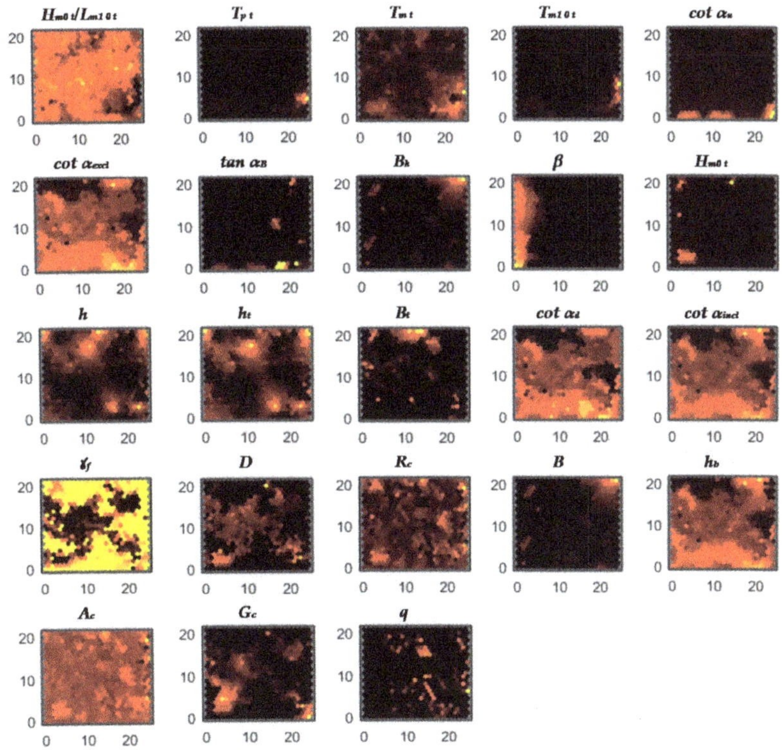

Figure 5. Dimensionality reduction. Self-organizing map (SOM): component planes of 23 variables.

From the analysis of these component planes the existence of several evident relationships between the variables is deduced. The first of these concerns the parameters related to the definition of the slopes of the dykes and which comparatively shows the existence of a direct and significant correlation across the entire range of the data between the variables ($cot\alpha_{excl}$, $cot\alpha_{incl}$, $cot\alpha_{d}$). A reason why the information contributed by them can be redundant, and informs which two of them should be discarded. However, it is noteworthy that another of the variables related to that group of parameters, which refers to the cotangent of the slope of the structure in the part of the slope above the berm ($cot\alpha_{u}$), presents a projection pattern that is notoriously different from the previous ones, but without a distinctive response in the component plane, so it should not be taken into account.

Another very significant relationship detected by the SOM is the one that shows the component planes of the roughness factor variables (γ_f) and the mean diameter (D). The comparison of both planes shows the existence of a negative correlation between them, and thus the greater sizes, the lower the roughness factor. This relationship is evidence in the existing empirical knowledge and taken into account [32,41], but it is comforting to confirm that the ANN is capable of detecting it as well. Both parameters should be preserved a priori.

The next detected relationship is the one between the width of the berm (B) and its projected width (B_h), and that is also preserved across the data range. Therefore, only one of them should be selected, discarding the other.

It could be thought that, based on design criteria, there was a direct relationship between the width of the berm and the width of the toe (B_t), or with the width of the crest, however, this has not been detected at the data base analyzed for the crest width of the structure, so this supposed relationship will be discarded. While the existence of a partial correlation, at least in a region of the projection plane, between the variables of the berm width and the width of the bench (see Figure 5) is detected, which indicates that some of the breakwaters that have been tested have been designed with a theoretical pattern that relates both variables. The foregoing forces not discarding these variables, but to keep them in the input space, since this relationship is partial in the sample space it is necessary to preserve that differentiation.

A last relationship highlighted (Figure 5), and also expected by the existing empirical knowledge, is the one presented by the depth variables at the toe of the structure (h), and the one that define the submergence of it (h_t). In this case, as expected, its correlation is direct or positive. However, the lack of correlation between both variables and the berm (h_b) is also striking, therefore, following the above reasoning, at least two of them should be maintained, discarding the third of them.

In view of the results obtained and interpreted after applying both the PCA technique and the SOM maps, a reduction in the size of the input patterns can be achieved. It results in a final dimension of the input vector of 15 parameters (see Table 5).

Table 5. Selected variables that finally composes the input vector for the ANN modelling.

Parameter [1]	Description	Unit
$H_{m0\ toe}/L_{m-1}$	Dimensionless spectral wave steepness	-
β	Wave incidence angle	o
$T_{m\ t}$	Average period at the toe of the structure	s
$H_{m0\ toe}$	Significant spectral wave height at the foot of the structure	m
h	Water depth at the structure toe	m
h_t	Toe submergence	m
B_t	Toe width	m
$cot\alpha_{incl}$	Average co-tangent of the slope of the structure, considering the contribution of the berm	-
γ_f	Roughness/permeability factor for the structure	-
B	Berm width	m
D	Size of the structure elements along the slope	m
h_b	Berm submergence	m
R_c	Crest freeboard of the structure respect to sea water level (swl)	m
A_c	Armour crest freeboard respect to swl	m
G_c	Crest width	m

[1] Windward side.

3.2. Model Selection

The results obtained after the training process of the different architectures tested for each model, show better performance of the aggregate model (Model I) over the disaggregated model (Model II), both in terms of error and correlation, as shown in Tables 6–8, and Figure 6, in which is possible to distinguish the results for each of the subsets used in the cross-verification process: Training (TR), Verification (V), Test (T) of the better Model I.

Table 6. Model I: Results in the test subset based on the number of neurons in the hidden layer. MSE and *r* statistic.

N° ud.	5	10	15	20	25	30	35
MSE	7.02×10^{-5}	6.75×10^{-5}	5.69×10^{-5}	5.72×10^{-5}	3.82×10^{-5}	5.06×10^{-5}	7.08×10^{-5}
r	96.51	96.28	97.33	97.17	98.17	97.68	97.28

Table 7. Model II.I: Results in the test subset based on the number of neurons in the hidden layer. MSE and *r* statistic.

N° ud.	5	10	15	20	25	30	35
MSE	7.02×10^{-5}	6.71×10^{-5}	8.43×10^{-5}	5.86×10^{-5}	3.82×10^{-5}	6.28×10^{-5}	6.98×10^{-5}
r	96.98	96.24	96.31	97.37	96.13	97.28	96.81

Table 8. Model II.II: Results in the test subset based on the number of neurons in the hidden layer. MSE and *r* statistic.

N° ud.	5	10	15	20	25	30	35
MSE	1.68×10^{-7}	1.82×10^{-7}	3.00×10^{-7}	3.88×10^{-8}	1.45×10^{-8}	1.46×10^{-7}	1.02×10^{-7}
r	84.05	63.10	66.10	84.51	75.33	74.66	74.52

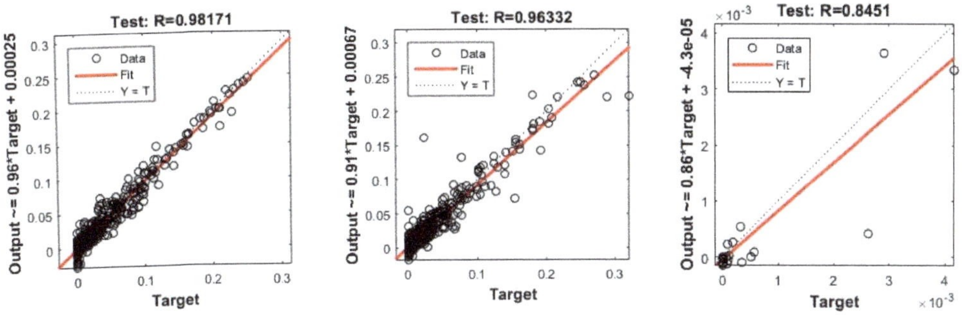

(**a**) (**b**) (**c**)

Figure 6. Results on the proposed models. Correlation graphs are shown for each one and correspond to the test subset: (**a**) Model I; (**b**) Model II.1; (**b**) Model II.2. In each figure the dashed line corresponds to a perfect fit. Additionally, a linear adjustment of the results is shown for each model (only for test subset).

The finally selected architecture for Model I, based on the results obtained, is an MLP network with 15 input variables, 25 neurons in the hidden layer, and a single neuron in the output layer (see Tables 6–8, with the trial results to determine the best architecture in the results of the different models proposed).

The results are shown in the form of correlation plots for test subset in Figure 6. Noting that for the test subset the correlation values are greater than 0.98. Although they are similar to those obtained for sub model II.1 (0.96), they are much higher than those obtained with sub model II.2 (0.84). The results in terms of error (MSE) are similar for both model I (3.85×10^{-5}) and model II (3.82×10^{-5}), with the known exception that the MSE is not an absolute statistic, but a relative one [64].

The analysis of the residuals establishes, as a desirable objective for an ideal model, that its distribution be carried out according to a pattern as close as possible to a normal distribution as a clear indicator of the absence of any hidden trend or bias in the modelling performed. In the present case, a careful analysis of this distribution shows that although it is close to normal, it does not fit significantly to it. This is demonstrated by the chi-square and Kolmogorov-Smirnov fit tests carried out, and that they are presented below (see Table 9), together with their correspondent graphical adjustment (Figure 7a).

Table 9. Results of the different contrast test on the residuals for the selected model.

Statistics	Chi-Squared	Kolmogorov-Smirnov
Empirical Parameter	3.922×10^{15}	0.161
Theoretical Parameter	72.15	
p-value (bilateral)	<0.0001	<0.0001
Confidence interval (α)	0.05	0.05

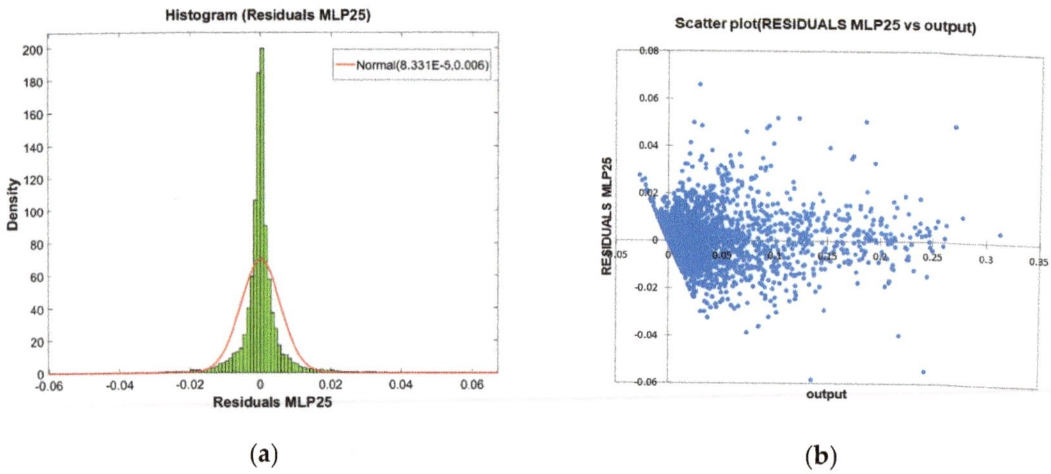

Figure 7. Results for the selected model: (**a**) Graphical fit of the residual histogram to a normal distribution; (**b**) Scatter plot of the model residual versus outputs.

The graphical analysis of the scatter plot of the residuals (Figure 7b) shows adequate behavior across the entire response range, except in the range of low values of overtopping rate, for which it does show a certain tendency towards non-compliance with the hypothesis of constant variance of the residuals. This heteroscedasticity may be linked to scale problems in the tests or introduced by iso-energetic sequences of waves [60,61], since the behavior of the prediction for very low overtopping rates has been associated with high levels of uncertainty [69], or may be due to the need to perform further specific transformations on the input variables beyond those already applied in the present study [25].

An extra validation test performed on an additional sample of 100 patterns, and with the selected ANN, provides good performance, with correlations of 0.98, which could validate its generalizability. However, what is more interesting, after classifying the component patterns into two different classes, the first one corresponding to tests on seawalls, and the second corresponding to sloped breakwaters, they show a similar aspect and is very suitable for the prediction of the overtopping rate on both (see Figure 8). Thus, the results for the typology of seawalls provides a correlation coefficient of 0.996, while for slope breakwaters it provides a similar result of 0.998.

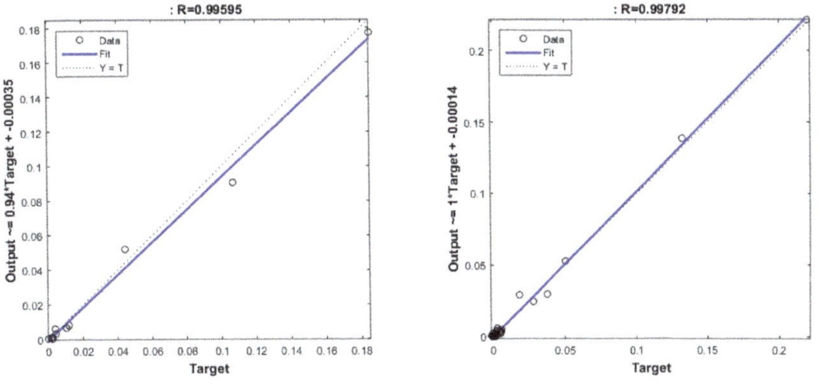

(a) (b)

Figure 8. Extra-validation results on the selected model: (**a**) Correlation graph on the seawall models; (**b**) Correlation graph on the slope breakwaters models.

3.3. Sensitivity Analysis

Finally, a sensitivity analysis is carried out on the selected ANN, specifically on the component parameters of the input vector. This analysis is performed using a pruning technique, and the ratio that was proposed for this purpose:

$$r_s = \frac{er_j}{er_T} \tag{12}$$

where x′ is the sensibility ratio, and er_j the error function value for the trained network. In this case, the MSE is chosen as the error criterion to define the sensitivity ratio.

This procedure is especially useful when the input variables are essentially independent of each other [64], and conversely, the more interdependencies there are between the variables, the less reliable they will be. Hence, among other reasons, the importance of the previously performed dimensionality reduction procedure, which now supports the application of that sensitivity analysis.

It is observed in the Figure 9 that all the variables have a significance ratio greater than 1.05, which according to that criterion implies that all the variables are significant a priori, and that therefore it would be desirable to maintain them for adequate network performance. The above has a clear derivative, which is to suppose that the process of dimensionality reduction has been successful, since any of the existing variables will provide enough relevant information and their elimination may imply worse predictive capacity of the network.

Beyond the previous observations, it is noted that the most influential variable is the freeboard of the wall with respect to swl (R_c), an issue that is confirmed by PCA analysis. It is noteworthy that a variable closely related to it, the other freeboard parameter, the crest height with respect to swl (A_c), is quite far, in terms of significance, from the parameter R_c. This fact is relevant since in some works [59] it has been determined that the scale effect seems to depend a lot on the superior geometry of the breakwater. This results in many more significant associated effects on small overtopping rates, which are incidentally, also the most numerous in the database. Given this, and to try to mitigate these effects as much as possible, some authors propose the dimensionless of these variables [25,30,59].

r_s	
R_c	5.21
$\cot\alpha_{incl}$	3.62
H_{m0}/L_{m-1}	2.32
h_t	1.98
D	1.91
γ_f	1.75
B	1.70
$H_{m0\,toe}$	1.42
B_t	1.32
h_b	1.31
A_c	1.25
β	1.19
G_c	1.17
h	1.15
T_{mt}	1.13

Figure 9. Sensibility analysis. The histogram represents the contribution of each variables of the model input space in terms of significance of the r_s ratio.

Another significantly interesting variable is the average cotangent where the contribution of the berm ($cot\alpha_{incl}$) is considered [42,59]. Similarly of interest, the wave steepness (H_{m0}/L_{m-1}) is highlighted. In addition to these, are both parameters related to roughness (and in essence, to the porosity of the mantle) where their close relationship with overtopping is already known empirically [32], and which in turn have a substantial dependence on the dimensionless freeboard (R_c/H_{m0}).

Overall, the results are consistent in terms of the significance of these parameters with similar studies carried out with different preprocessing techniques [30]. And it should be mentioned that some parameters in this study are may be penalized, due to the fact that they have been poorly represented in the database. For example, this happens with the wave incidence angle parameter (β) that shows a significant lack of data in some ranges of that continuous variable. In the following figure (see Figure 10a) the distribution of the aforementioned parameter with respect to the significant wave height at the toe structure ($H_{mo\,t}$) is presented since, as Van der Meer cites [48], this relationship is strongly related to the overtopping phenomenon, and shows the existence of poor representability in the ranges greater than 50°.

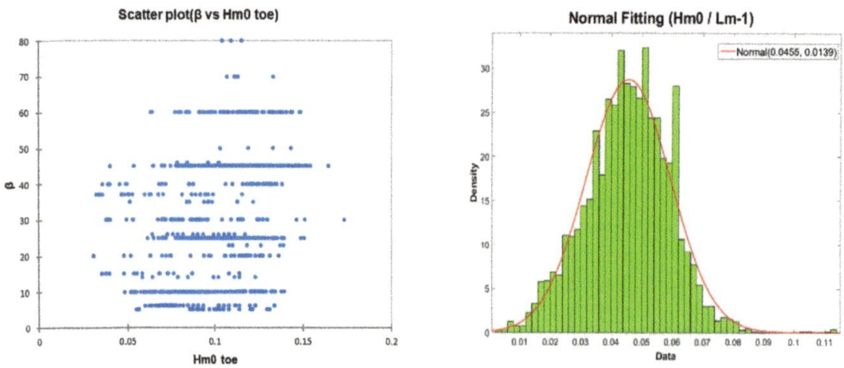

(a) (b)

Figure 10. Analysis of the model sample: (**a**) Scatterplot of the parameter β vs. $H_{mo\,t}$; (**b**) Histogram of the $S_{m-1,0}$ parameter (spectral wave steepness), with adjustment of a normal probability density function.

The importance given to the dimensionless parameter of the wave steepness, particularly for wave overtopping energy conversion [70], which has good representation in the database, both in its distribution and in the quality of that distribution (normalized distribution), should also be highlighted (see Figure 10b). Faced with possible uncertainties associated with scale phenomena [59], although the parameter's existence in the field of validity is remarkable with values over 0.07 that are physically not possible as the wave breaks on steepness [56], the use of wave steepness as a variable is recommended. This also represents the effects induced by local breakage and waves [25], and is therefore strongly related with the overtopping.

Due to the fact that for wave overtopping conversion the maximum overtopping rates correlated with the lower R_c/H_s relationships [71] are highly desirable. These will generally be associated with low crested structures, specifically with R_c/H_s lower than 1. It would be desirable that the training sample be well represented in this range, as does happen, and is shown in the following figure (Figure 11a). Checking the model for those exceedance rates corresponding to the range in the previously mentioned sample of 100 extra cases (corresponding to a total of 57 cases), the result is encouraging, with values of the correlation coefficient greater than 0.98, as shown in the Figure 11b.

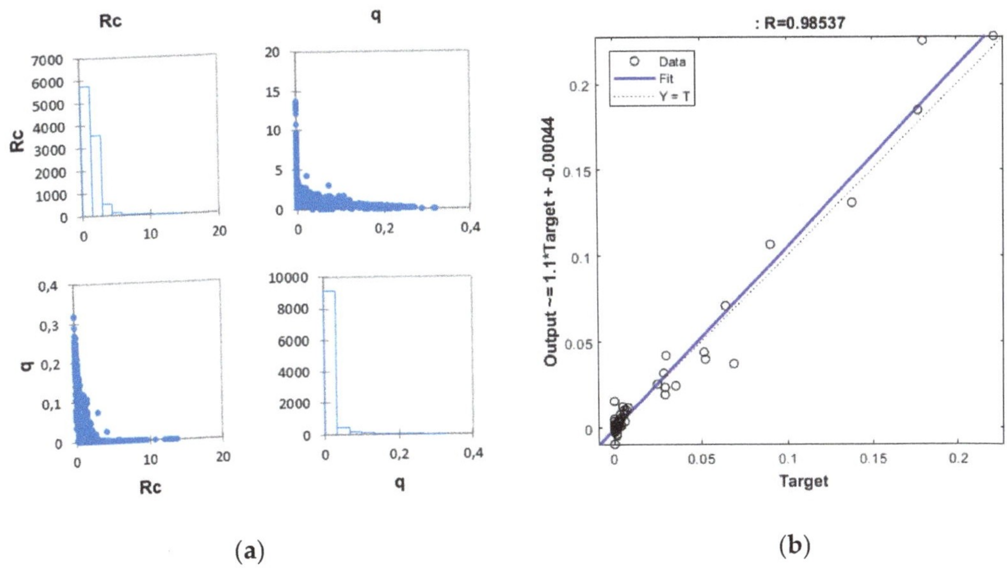

Figure 11. Results on the selected model for the optimum specific range for wave overtopping conversion: (**a**) Vertical freeboard (R_c) vs. overtopping rate relationship in the model sample; (**b**) Correlation graph for the R_c/H_s < 1 specific range.

Thus, and in accordance with the above mentioned, any future improvement in the model should necessarily focus on that data range. This desired approach is in practice the opposite of what is usually done for defense structures.

Another crucial issue related to the generation of the data is the need to make the range of data tested wide enough to include extraordinary events, given that ANNs are usually unable to extrapolate beyond the range of the data used for training [65,72]. This ensures that they always work in the expected range, avoiding poor predictions when the validation data contain values outside of the range of those used for training. In this sense, there is a preponderance of low flow rates that reinforces the idea of a disaggregated approach in future models.

4. Conclusions

As part of a sustainable strategy to take advantage of some existing breakwater infrastructure, and its partial reconversion as a Wave Energy Converter while maintaining its defense purpose, this study is framed in which a model based on artificial neural networks for the overtopping rate forecasting is proposed for a wide range of breakwaters. The adjusted prediction of the overtopping rate constitutes the first step in the study of subsequent modifications to be made to these structures.

To achieve this purpose, existing data from CLASH-EurOtop have been subjected to a preprocessing step, where only the parameters from laboratory tests have been previously scaled according to the Froude model law (to $H_{m0\,toe}$ = 1 m). Subsequently, the entire data base was subjected to an extensive process of exploration, debugging, and dimensionality reduction, until an optimized input pattern in the ANN model was obtained. Using only 15 of the 34 initial features, sufficient relevant information was used to train a model with generalization skills and high predictive efficiency. This preliminary phase derives substantial conclusions such as:

- It is worth noting the lack of homogeneity in the database due to its diverse origin, where the existence of data at different scales forces the adoption of a data scaling procedure, which introduces uncertainty into the model. It is concluded that this lack of homogeneity is masked in the final model by the significant difference in sample size.
- Relevant effects associated to the scale are quoted, especially in what concerns to the superior geometry of the breakwater. WEC devices that are located in existing structures, where power generation capacity is combined with defensive capacity, that have small magnitudes of the overtopping rate, and which incidentally are the most common, are especially sensitive to these effects.
- Linked with the above conclusion, a new and more appropriate transformation of the inputs must be proposed that minimizes the observed heteroscedasticity effects in this range of overtopping rates.
- The present work shows the suitability of multivariate statistical techniques, and specifically the Mahalanobis distance, for the detection of outliers, and also the Principal Component Analysis for the reduction of the dimension of the input vector, a task shared with the Kohonen Self Organizing Maps application.

Several ANN models have been proposed and the architecture finally selected has been an MLP 15-25-1. This was obtained after a cross-verification training process with the Levenberg-Marquardt algorithm, and which corresponds to a model that takes into account both the data from prototypes and small-scale tests.

The results are very encouraging since they allow obtaining predictions with very high correlation coefficients (>0.98) and where the validation process carried out shows that the model is equally suitable for both seawalls and slope breakwaters. This has been justified by the prevalence of those that refer to crest freeboard of the structure with respect to swl over the rest of the parameters, and the average cotangent of the slope of the structure considering the contribution of the berm.

This final conclusion reinforces the belief that subsequent studies, in which an adequate classification criterion of the input parameters will be obtained, will undoubtedly reinforce the good performance of the ANN model. For wave energy conversion, the lower R_c/H_s relationship the higher overtopping rate, therefore this criterion will allow future models to be developed and trained in that specific range of patterns.

Author Contributions: J.M.O. contributed to the work described in this paper by carrying out the analysis and treatment of the data, proposing and analyzing the models and results, and writing—Original Draft preparation. M.D.E. contributed to the work by supervising the methodology and the obtained results and gave the final approval. J.-S.L.-G. contributed with knowledgeable discussion and suggestion. V.N. and M.G.N. contributed with knowledgeable discussion and suggestion. All of the co-authors participated in editing the final paper. All authors have read and agreed to the published version of the manuscript.

Funding: This research received no external funding.

Institutional Review Board Statement: Not applicable.

Informed Consent Statement: Not applicable.

Data Availability Statement: Not applicable.

Acknowledgments: The authors acknowledge the assignment of the EurOtop 2018 Database, specially to Formentin, Z.M., Zanuttigh, B., and Van der Meer, J.W.

Conflicts of Interest: The authors declare no conflict of interest.

References

1. IEA-OES (International Energy Agency-Ocean Energy Systems). *Annual Report 2006*; Technical Report; IEA-OES: Paris, France, 2006.
2. Esteban, M.D.; Espada, J.M.; Ortega, J.M.; López-Gutiérrez, J.S.; Negro, V. What about Marine Renewable Energies in Spain? *J. Mar. Sci. Eng.* **2019**, *7*, 249. [CrossRef]
3. Breeze, P. Marine Power Generation Technologies. In *Power Generation Technologies*, 3rd ed.; Elsevier Ltd.: Oxford, UK, 2019; pp. 323–349.
4. Zhou, Z.; Benbouzid, M.; Charpentier, J.F.; Scuiller, F.; Tang, T. A Review of Energy Storage Technologies for Marine Current Energy Systems. *Renew. Sustain. Energy Rev.* **2013**, *18*, 390–400. [CrossRef]
5. Lewis, M.; Neill, S.; Robins, P.; Hashemi, M. Resource assessment for future generations of tidal-stream energy arrays. *Energy* **2015**, *83*, 403–415. [CrossRef]
6. Lavi, A. Ocean thermal energy conversion: A general introduction. *Energy* **1980**, *5*, 469–480. [CrossRef]
7. Seyfried, C.; Palko, H.; Dubbs, L. Potential local environmental impacts of salinity gradient energy: A review. *Renew. Sustain. Energy Rev.* **2019**, *102*, 111–120. [CrossRef]
8. Falcão, A.F.O. Wave energy utilization: A review of the technologies. *Renew. Energy* **2010**, *14*, 899–918. [CrossRef]
9. Babarit, A.; Bull, D.; Dykes, K.; Malins, R.; Nielsen, K.; Costello, R.; Roberts, J.; Bittencourt Ferreira, C.; Kennedy, B.; Weber, J. Stakeholder requirements for commercially successful wave energy converter farms. *Renew. Energy* **2017**, *113*, 742–755. [CrossRef]
10. IDAE. *Evaluación del Potencial de la Energía de las olas*; Technical Report; IDEA: Madrid, Spain, 2011.
11. Margheritini, L.; Vicinanza, D.; Frigaard, P. SSG wave energy converter: Design, reliability and hydraulic performance of an innovative overtopping device. *Renew. Energy* **2009**, *34*, 1371–1380. [CrossRef]
12. Astariz, S.; Iglesias, G. The economics of wave energy: A review. *Renew. Sustain. Energy Rev.* **2015**, *45*, 397–408. [CrossRef]
13. Reguero, B.G.; Losada, I.J.; Méndez, F.J. A global wave power resource and it seasonal, interannual and long-term variability. *Appl. Energy* **2015**, *148*, 366–380. [CrossRef]
14. Shields, M.A.; Payne, A.I.L. Strategic Sectoral Planning for Offshore Renewable Energy in Scotland. In *Marine Renewable Energy Technology and Environmental Interactions*, 1st ed.; Shields, M.A., Payne, A.I.L., Eds.; Springer: Berlin, Germany, 2014; pp. 141–152.
15. Aderinto, T.; Li, H. Ocean Wave Energy Converters: Status and challenges. *Energies* **2018**, *11*, 1250. [CrossRef]
16. The Wave Dragon Technology. Wave Dragon Web Site. Available online: http://www.wavedragon.net/?option=com_content&task=view&id=4&Itemid=35 (accessed on 1 June 2020).
17. Foteinis, S.; Tsoutsos, T. Strategies to improve sustainability and offset the initial high capital expenditure of wave energy converters (WECs). *Renew. Sustain. Energy Rev.* **2017**, *70*, 775–785. [CrossRef]
18. Contestabile, P.; Iuppa, C.; Di Lauro, E.; Cavallaro, L.; Andersen, T.L.; Vicinanza, D. Wave loadings acting on innovative rubble mound breakwater for overtopping wave energy conversion. *Coast. Eng.* **2017**, *122*, 60–74. [CrossRef]
19. Iuppa, C.; Contestabile, P.; Cavallaro, L.; Foti, E.; Vicinanza, D. Hydraulic Performance of an Innovative Breakwater for Overtopping Wave Energy Conversion. *Sustainnabily* **2016**, *8*, 1226. [CrossRef]
20. Vicinanza, D.; Nørgaard, J.H.; Contestabile, P.; Andersen, T.L. Wave loadings acting on Overtopping Breakwater for Energy Conversion. *J. Coast. Res.* **2013**, *65*, 1669–1674. [CrossRef]
21. Cuadra, L.; Salcedo-Sanz, S.; Nieto-Borge, J.C.; Alexandre, E.; Rodríguez, G. Computational intelligence in wave energy: Comprehensive review and case study. *Renew. Sustain. Energy Rev.* **2016**, *58*, 1223–1246. [CrossRef]
22. Govindaraju, R.S.; Rao, A.R. *Artificial Neural Networks in Hydrology*, 1st ed.; Water Science and Technology Library, Kluwer Academic Publishers: Dordrecht, The Netherlands, 2000; pp. 1–329.
23. Bishop, C.M. Neural Networks. In *Pattern Recognition and Machine Learning*, 3rd ed.; Springer: New York, NY, USA, 2006; pp. 225–290.
24. Van Gent, M.R.A.; Van den Boogaard, H.F.P.; Pozueta, B.; Medina, J.R. Neural network modelling of wave overtopping at coastal structures. *Coast. Eng.* **2007**, *54*, 586–593. [CrossRef]
25. EurOtop Manual. EurOtop-Wave Overtopping of Sea Defenses and Related Structures. An Overtopping Manual Largely Based on European Research, but for Worldwide Application, 2nd ed. 2018. p. 320. Available online: https://www.overtopping-manual.com (accessed on 24 May 2020).
26. Formentin, S.M.; Zanuttigh, B. A methodological approach for the development and verification of artificial neural networks based on application to wave-structure interaction processes. *Coast. Eng. J.* **2018**, *60*, 1–20. [CrossRef]

27. Verhaeghe, H.; De Rouck, J.; Van der Meer, J.W. Combined classifier–quantifier model: A 2-phases neural model for prediction of wave overtopping at coastal structures. *Coast. Eng.* **2008**, *55*, 357–374. [CrossRef]
28. Peixó, J.; Van Oosten, R.P. Wave Transmission at Various Types of Low-Crested Structures Using Neural Networks. Master's Thesis, Delft University of Technology, Delft, The Netherlands, 2005.
29. Panizzo, A.; Briganti, R. Analysis of wave transmission behind low-crested breakwaters using neural networks. *Coast. Eng.* **2007**, *54*, 643–656. [CrossRef]
30. Formentin, S.M.; Zanuttigh, B.; Van der Meer, J.W.A. Neural Network Tool for Predicting Wave Reflection, Overtopping and Transmission. *Coast. Eng.* **2017**, *59*, 1–31. [CrossRef]
31. Zanuttigh, B.; Formentin, S.M.; Briganti, R. A neural network for the prediction of wave reflection from coastal and harbour structures. *Coast. Eng.* **2013**, *80*, 49–67. [CrossRef]
32. Molines, J.; Medina, J.R. Explicit Wave-Overtopping Formula for Mound Breakwaters with Crown Walls Using CLASH Neural Network–Derived Data. *J. Waterw. Port Coast. Ocean Eng.* **2016**, *142*, 1–13. [CrossRef]
33. Deo, M.C. Artificial neural networks in coastal and ocean engineering. *Indian J. Geo-Mar. Sci.* **2010**, *34*, 589–596.
34. Najafzadeh, M.; Barani, G.A.; Hessami-Kermani, M.R. Group method of data handling to predict scour depth around vertical piles under regular waves. *Sci. Iran.* **2013**, *20*, 406–413.
35. Azamathulla, H.M.; Zakaria, N.A. Prediction of scour below submerged pipeline crossing a river using ANN. *Water Sci. Technol.* **2011**, *63*, 2225–2230. [CrossRef]
36. Ayoubloo, M.K.; Etemad-Shahidi, A.; Mahjoobi, J. Evaluation of regular wave scour around a circular pile using data mining approaches. *Appl. Ocean Res.* **2010**, *32*, 34–39. [CrossRef]
37. Bateni, S.M.; Borghei, S.M.; Jeng, D.S. Neural network and neuro-fuzzy assessments for scour depth around bridge piers. *Eng. Appl. Artif. Intell.* **2007**, *20*, 401–414. [CrossRef]
38. López, I.; Aragonés, L.; Villacampa, Y.; Satorre, R. Modelling the cross-shore beach profiles of sandy beaches with Posidonia oceanica using artificial neural networks: Murcia (Spain) as study case. *Appl. Ocean Res.* **2018**, *74*, 205–216. [CrossRef]
39. Negro, V.; Varela, O. Comportamiento funcional, reflexión, transmisión, y amortiguación. Remonte, descenso y rebase. In *Diseño de Diques Rompeolas*, 2nd ed.; Colegio de Ingenieros de Caminos Canales y Puertos: Madrid, Spain, 2010; pp. 267–304.
40. Rodríguez, A.M.; Sánchez, J.F.; Gutiérrez, R.; Negro, V. Overtopping of harbour breakwaters: A comparison of semi-empirical equations, neural networks, and physical model tests. *J. Hydraul. Res.* **2015**, *53*, 1–14. [CrossRef]
41. Verhaeghe, H.; Van der Mer, J.W.; Steendam, G.J.; Besley, P.; Franco, L.; Van Gent, M.R.A. Wave overtopping database as the starting point for a neural network prediction method. In *Coastal Structures*; ASCE: Portland, OR, USA, 2003; pp. 418–429.
42. De Rouck, J.; Geeraerts, J. *CLASH—D46: Final Report*; Full Scientific and Technical Report; Gent University: Gent, Belgium, 2005.
43. Haykin, S. *Neural Networks: A Comprehensive Foundation*, 2nd ed.; Pearson Prentice Hall: Upper Saddle River, NJ, USA, 1999; pp. 1–823.
44. Kohonen, T. Self-Organized Formation of Topologically Correct Feature Maps. *Biol. Cybern.* **1982**, *43*, 59–69. [CrossRef]
45. Kohonen, T. *Self-Organizing Map*, 3rd ed.; Springer: New York, NY, USA, 2001; pp. 1–501.
46. Van der Meer, J.W.; Van Gent, M.R.A.; Pozueta, B.; Steendam, G.J.; Medina, J.R. Applications of a neural network to predict wave overtopping at coastal structures. In Proceedings of the International Conference on Coastlines, Structures and Breakwaters, London, UK, 20–22 April 2005; ICE Thomas Telford: London, UK, 2005; pp. 259–268.
47. Buckingham, E. On Physically Similar Systems: Illustrations of the Use of Dimensional Equations. *Phys. Rev.* **1914**, *4*, 345–376. [CrossRef]
48. Van der Meer, J.W.; Verhaeghe, H.; Steendam, G.J. The new wave overtopping database for coastal structures. *Coast. Eng.* **2009**, *56*, 108–120. [CrossRef]
49. Azme, K.; Zuhaymi, I.; Haron, K. The Effects of Outliers Data on Neural Network Performance. *J. Appl. Sci.* **2005**, *5*, 1394–1398.
50. Maier, H.R.; Dandy, G.C. Neural networks for prediction and forecasting of water resources variables: A review of modelling issues and applications. *Environ. Model. Softw.* **2000**, *15*, 101–124. [CrossRef]
51. Peña, D. *Análisis de Datos Multivariantes*, 1st ed.; McGraw-Hill: Madrid, Spain, 2002; pp. 133–170.
52. Aggarwal, C.C. *Outlier Analysis*, 1st ed.; Springer International Publishing: Cham, Switzerland, 2017; pp. 1–422.
53. Jolliffe, I.T. *Principal Component Analysis*, 2nd ed.; Springer: New York, NY, USA, 2002; pp. 111–147.
54. Vesanto, J.; Alhoniemi, E. Clustering of the Self-Organizing Maps. *IEEE Trans. Neural Netw.* **2000**, *11*, 586–600. [CrossRef]
55. Ultsch, A.; Siemon, H.P. Kohonen's Self Organizing Feature Maps for Exploratory Data Analysis. In Proceedings of the INNC'90, Paris, France, 9–13 July 1990; Kluwer: Dordrecht, The Netherlands, 1990; pp. 305–308.
56. Pozueta, B.; Van Gent, M.; Van der Boogaard, H.; Medina, J. Neural network modelling of wave overtopping at coastal structures. World Scientific. In Proceedings of the 29th International Conference on Coastal Engeniering, Lisbon, Portugal, 19–24 September 2004; pp. 4275–4287.
57. Kortenhaus, A.; Oumeraci, H.; Geeraerts, J.; De Rouck, J.; Medina, J.R.; González-Escrivá, J.A. Laboratory effects and further uncertainties associated with wave overtopping measurements. World Scientific. In Proceedings of the 29th International Conference on Coastal Engeniering, Lisbon, Portugal, 19–24 September 2004; Volume 4, pp. 4456–4468.
58. Franco, L.; Geeraerts, J.; Briganti, R.; Willems, M.; Bellotti, G.; De Rouck, J. Prototype measurements and small-scale model tests of wave overtopping at shallow rubble-mound breakwaters: The Ostia-Rome yacht harbour case. *Coast. Eng.* **2009**, *56*, 154–165. [CrossRef]

59. Andersen, L.; Burcharth, H.T.; Gironella, X. Comparison of new large and small scale overtopping tests for rubble mound breakwaters. *Coast. Eng.* **2011**, *58*, 351–373. [CrossRef]
60. Romano, A.; Bellotti, G.; Briganti, R.; Franco, L. Uncertainties in the physical modelling of the wave overtopping over a rubble mound breakwater: The role of the seeding number and of the test duration. *Coast. Eng.* **2015**, *103*, 15–21. [CrossRef]
61. Williams, H.E.; Briganti, R.; Romano, A.; Dodd, N. Experimental analysis of wave overtopping: A new small scale laboratory dataset for the assessment of uncertainty for smooth sloped and vertical coastal structures. *J. Mar. Sci. Eng.* **2019**, *7*, 217. [CrossRef]
62. Bowden, G.J.; Maier, H.R. Optimal division of data for neural network models in water resources applications. *Water Resour. Res.* **2002**, *3*, 1611–1619. [CrossRef]
63. Hornik, K.; Stinchcombre, M.; White, H. Multilayer feedforward networks are universal approximators. *Neural Netw.* **1989**, *2*, 359–366. [CrossRef]
64. García-Bartual, R.L. *Redes Neuronales Artificiales en Ingeniería Hidráulica y Medio Ambiental: Fundamentos*; Technical University of Valencia: Valencia, Spain, 2005.
65. Minns, A.W.; Hall, M.J. Artificial neural networks as rainfall-runoff models. *Hydrol. Sci. J.* **1996**, *41*, 399–417. [CrossRef]
66. Hagan, M.T.; Demuth, H.B.; Beale, M.H.; De Jesús, O. *Neural Network Design*, 2nd ed.; Martin Hagan: Stillwater, MN, USA, 2014; pp. 36–60.
67. Levenberg, K. A method for the solution of certain problem in least squares. *Q. Appl. Math.* **1944**, *2*, 164–168. [CrossRef]
68. Marquart, D.W. An algorithm for least-squares Estimation of Nonlinear Parameters. *J. Soc. Ind. Appl. Math.* **1963**, *11*, 431–441. [CrossRef]
69. Zanuttigh, B.; Formentin, S.M.; Van der Meer, J.W. Prediction of extreme and tolerable wave overtopping discharges through an advanced neural network. *Ocean Eng.* **2016**, *127*, 7–22. [CrossRef]
70. Kofoed, J.P.; Frigaard, P.; Friis-Madsen, E.; Sørensen, H.C. Prototype testing of the wave energy converter wave dragon. *Renew. Energy* **2006**, *31*, 181–189. [CrossRef]
71. Tedd, J.; Kofoed, J.P.P. Measurements of overtopping flow time series on the Wave Dragon, wave energy converter. *Renew. Energy* **2009**, *34*, 711–717. [CrossRef]
72. Flood, I.; Kartam, N. Neural networks in civil engineering, I: Principles and understanding. *J. Comput. Civil Eng.* **1994**, *8*, 131–148. [CrossRef]

sustainability

MDPI

Article

Probabilistic Assessment of Hybrid Wind-PV Hosting Capacity in Distribution Systems

Dichen Liu [1], Chenxu Wang [1,*], Fei Tang [1,*] and Yixi Zhou [2]

1 School of Electrical Engineering and Automation, Wuhan University, Wuhan 430072, China; dcliu@whu.edu.cn

2 State Grid Hangzhou Electric Power Supply Company, Hangzhou 310000, China; sgcczhouyx@163.com

* Correspondence: wangchenxu@whu.edu.cn (C.W.); tangfei@whu.edu.cn (F.T.)

Received: 14 February 2020; Accepted: 10 March 2020; Published: 11 March 2020

Abstract: In recent years, hybrid wind-photovoltaic (PV) systems are flourishing due to their advantages in the utilization of renewable energy. However, the accurate assessment of the maximum integration of hybrid renewable generation is problematic because of the complex uncertainties of source and demand. To address this issue, we develop a stochastic framework for the quantification of hybrid energy hosting capacity. In the proposed framework, historical data sets are adopted to represent the stochastic nature of production and demand. Moreover, extreme combinations of production and demand are introduced to avoid multiple load flow calculations. The proposed framework is conducted in the IEEE 33-bus system to evaluate both single and hybrid energy hosting capacity. The results demonstrate that the stochastic framework can provide accurate evaluations of hosting capacity while significantly reducing the computational burden. This study provides a comprehensive understanding of hybrid wind-PV hosting capacity and verifies the excellent performance of the hybrid energy system in facilitating integration and energy utilization.

Keywords: Hybrid energy system; wind power; photovoltaic; hosting capacity; distribution system

1. Introduction

Over the past decades, the utilization of renewable energy sources (RES) has grown dramatically due to concerns over the environment and carbon emissions [1]. Wind power and photovoltaic (PV) comprise a significant proportion of RES due to mature technologies, economic advantages, and abundant primary energy. The strategical deployment of RES in distribution systems as distributed generation (DG) can bring serval benefits [2,3], such as voltage profile improvement, power loss reduction, and reliability enhancement. However, the excessive penetration of DG can alter the normal operational behavior of distribution systems and cause various negative impacts, such as overvoltage [4,5], voltage unbalance [6], thermal overloading [7], and harmonic distortions [8,9]. The performances of systems becomes unacceptable when the amount of DG exceeds the hosting capacity [10]. For this reason, it is of paramount importance to system planners to determine the maximum amount of DG that can be connected.

Hosting capacity is defined as the amount of DG that can be connected without endangering the reliability or voltage quality of systems [11]. The idea of hosting capacity was introduced in 2004, and Math Bollen et al. [12] further developed this concept. Throughout the next few years, hosting capacity analysis was regarded as a transparent tool for the planning and design of distribution networks. Furthermore, based on the time-varying nature of generation and demand, Castelo de Oliveira et al. [13] emphasized that the hosting capacity needs to be analyzed throughout different periods, and developed the concept of dynamic hosting capacity. The analysis of dynamic hosting capacity requires time series data, e.g., forecasting the values of generation and demand [14,15]. In this context, the combined dynamic hosting capacity with forecasting data has the potential to predict and facilitate the operation of systems.

In addition to the development of the concept, abundant studies have been conducted using various techniques for the assessment and enhancement of hosting capacity. To provide guidelines for system planners and researchers on various assessment techniques, the authors in [16] presented a comprehensive overview of three different methods for quantifying PV hosting capacity, i.e., deterministic [17], stochastic [18,19], and time series methods [20]. The main differences between the three categories of methods are whether and how uncertainties are considered. Deterministic methods usually adopt the worst-case scenario (e.g., maximum production and minimum load) to evaluate the extreme impact of DG. Therefore, these methods may lead to conservative results. To realistically assess the states of systems, the stochastic and time series methods adopt probability distributions and time-series data, respectively, to represent uncertainties of source and demand.

The accurate analysis of hosting capacity involves, not only the variations of DG production and load demands, but also the uncertainties of DG location and rated power [16]. In this context, Smith et al. [21] developed a Monte Carlo simulation-based (MCS) stochastic method to provide a comprehensive assessment with various uncertainties. This stochastic method takes into account the uncertainties in the size and location of DG by generating thousands of potential DG deployment schemes. Then, load flow calculations are performed to address the stochastic nature associated with DG production and load consumption. For a specific deployment scheme, the hosting capacity result can be obtained if any technological limitations are violated. With the multiple potential DG deployment schemes, the final result is no longer one exact value but a probability distribution.

Based on the stochastic method in [21], some applications and variations have been developed. This method was conducted in [22] to estimate the PV hosting capacity of 16 utility distribution feeders. Compared with the 15% rule adopted by system planners [10] (i.e., the total DG rating should be lower than 15% of the feeder annual peak load), the stochastic analysis provides more realistic estimations of hosting capacity. Moreover, some studies [23,24] adopted the stochastic method to perform the sensitivity analysis of hosting capacity to the characteristics of the feeder (e.g., voltage class, peak load demand, and voltage regulation equipment) and PV generation (e.g., location and power factor). Dubey et al. [25] presented a comprehensive understanding of the PV hosting capacity problem by developing a mathematical formulation and an hourly stochastic analysis framework of hosting capacity assessment. Given that the analysis framework is stochastic, an approach was also established to quantify the accuracy of the obtained results. A simplified MCS-based method was developed in [26] to estimate the PV hosting capacity of 50000 real low-voltage systems from a distribution utility in Brazil. The obtained statistical results provide a risk-based guide to determine strategies to deal with increasing PV penetration. Instead of using load flow calculations, Abad et al. [27] proposed an optimization model of hosting capacity and embedded this model into the stochastic framework to address the variability of production and consumption.

The literature review in the realm of hosting capacity assessment illustrates that abundant methods have been developed for PV hosting capacity problems. These studies have contributed significantly to the understanding of the impact of different variables on PV hosting capacity. However, less research work has been done on the hybrid wind-PV hosting capacity assessment. Wind power and PV, to some extent, are complementary [28,29]. Therefore, the hybrid wind-PV system may have the potential to generate smoother production than single renewable generation. Some studies have reported the advantages of the hybrid wind-PV system in facilitating energy production [30] and enhancing power supply reliability [31]. In this context, if the complementarity can be captured through the hybrid deployment of wind power and PV, distribution systems could have the potential to accommodate more DG capacity. More importantly, hybrid renewable energy can provide more power for demand, which helps achieve the goal of reducing carbon emissions.

Since the output of wind power and PV have different characteristics [20], the methods for hybrid wind-PV hosting capacity assessment can be different compared to those methods for only

PV. The most significant difference is how to consider the uncertainties of DG production. For PV generation, the maximum output usually happens at midday due to solar irradiation. Therefore, only critical periods, in terms of potential limit violations, are evaluated in some studies. For instance, the authors in [23,26] only considered periods 10 am-2 pm and 11 am-1 pm, respectively. The adoption of critical periods can significantly reduce the computational burden while ensuring the accuracy of assessments. However, it is difficult to determine such periods for wind power. Under such circumstances, the evaluation of hybrid wind-PV hosting capacity can be very time-consuming due to the need for a large number of load flow calculations to capture the potential technological violations.

Motivated by the research gap and challenges, this paper aims to present a comprehensive assessment of hybrid wind-PV hosting capacity in distribution systems. To achieve this goal, we develop a stochastic framework that takes into account the uncertainties of load demands, DG production, location, size, and type. Moreover, to avoid multiple load flow calculations, the extreme combinations of DG production and load demands are defined. The proposed framework is implemented to evaluate both single and hybrid DG hosting capacity. Compared with the comprehensive load flow method, both the efficiency and accuracy of the proposed framework are verified. The simulation results provide system planners with an understanding of hosting capacity and energy utilization of the hybrid wind-PV system from the probabilistic perspective.

The paper organization is as follows. Section 2 presents the modeling of time-varying DG production and load demands. Section 3 introduces the stochastic framework for DG hosting capacity assessment. Section 4 performs the proposed framework on the IEEE 33-bus system to evaluate single and hybrid DG hosting capacity. Section 5 presents the discussion on the main features and applications of the proposed framework. Finally, Section 6 summarizes the main conclusions of this study.

2. Time-Varying Renewables and Demand

In hosting capacity assessment, the time-varying characteristics of both generation and demand need to be considered in evaluating the operating state of the systems. In this context, some studies [32,33] adopted the historical data of generation and demand as inputs to perform time series analysis to identify potential constraint violations. In general, time series analysis requires historical data sets with a long time scale and high-resolution. On the one hand, long time scale historical data, such as one year or a few years [16], is needed to capture various combinations of production and demand. On the other hand, a relatively short time interval [34] (e.g., 1-min, 10-min, and 15-min) is critical for capturing the variabilities of these uncertainties. However, the adoption of time series analysis introduces a significant number of power flow calculations into assessments. For instance, a time series analysis on the data set of one year at the interval of 15-min requires 35,040 power flow simulations to cover the complete assessment. Therefore, the significant computational burden makes time series analysis laborious and intractable.

One of the solutions to improve the efficiency of host capacity assessment is to extract the historical data that has the most significant impact on system constraints. In this context, Ochoa et al. [35] proposed a two-step (i.e., discretization and aggregation steps) technique to allocate the historical data of wind power and load demand into a finite number of bins based on their joint probability of occurrence. Subsequently, Sun et al. [30] further developed this technique by addressing the "coincidence" of wind power, PV, and load.

To outline the procedure of the two-step technique, we adopt wind power and load demand as an example. In Figure 1a, the curves present a 10-day snapshot of 15-min wind power and demand data with the values normalized against respective peak values. The discretization process allocates the time-dependent wind power and demand data into a series of bins. For instance, when the width of the bins is set to 0.1 p.u, the data points 0.43 p.u (wind power) and 0.64 p.u (load demand) are allocated into bins (0.40 p.u, 0.50 p.u] and (0.60 p.u, 0.70 p.u], respectively. Then, the aggregation process groups the bins into multiple combinations, e.g., {(0.40 p.u, 0.50 p.u] and (0.60 p.u, 0.70 p.u]}, as shown as the two-dimensional mesh in Figure 1b, using the above processes to sweep all data points and allocate them into corresponding combinations. Since load demand is never below 0.40 p.u, only 52 out of 100 potential combinations have non-zero probabilities of occurrence. These combinations are labeled with "√" in Figure 1c.

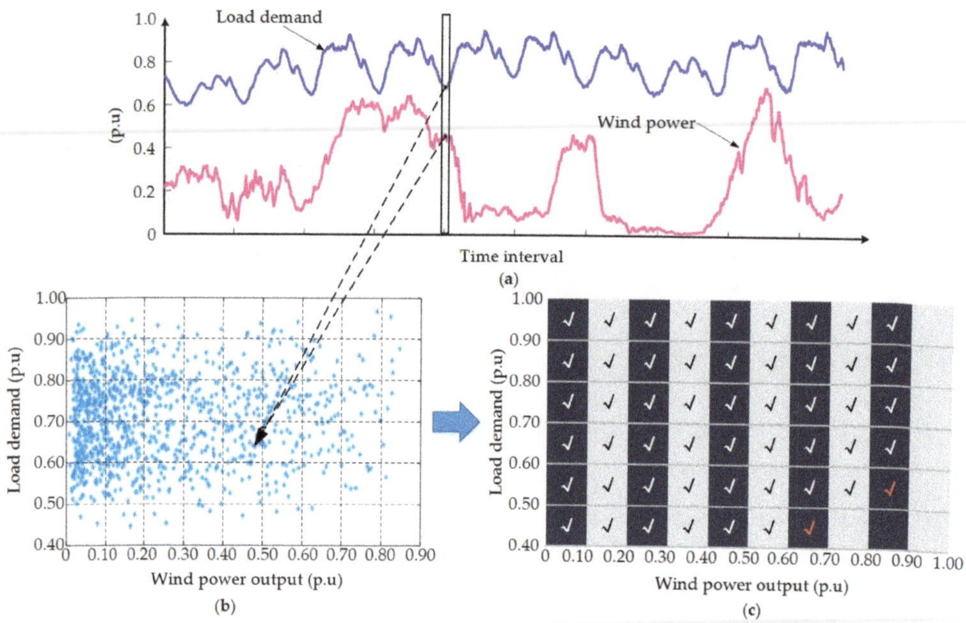

Figure 1. The procedure of discretization and aggregation of historical data: (**a**) historical data of wind power and load; (**b**) scatter plot of wind power and load; (**c**) visualization of wind power-load combinations.

The discretization-aggregation technique allocates the time series data into a finite number of combinations, thus reducing the number of load flow calculations. More importantly, this technique can preserve the interrelationships between production and demand. However, two challenges still need to be addressed when applying this technique in the hosting capacity assessment. The first is to determine which combinations are most critical in driving the system constraints. Second, the selection of bin width is empirical. Therefore, it is imperative to evaluate the impact of different bin widths on hosting capacity results.

For the first challenge, consider the combinations labeled with the red "√" in Figure 1c, which are characterized by relatively high generation and low demand. If technical constraints are not violated in these combinations, they are unlikely to be violated in other combinations. Therefore, these combinations can represent the worst-case scenarios in determining the hosting capacity, and we define them as extreme combinations. For the second challenge, different bin widths need to be considered. Figure 2 presents the combinations of wind power and load demand when the bin width is set to 0.05 p.u. It is evident that the total number of potential combinations increases significantly compared with the results in Figure 1c, but the number of extreme combinations is almost the same. The adoption of a smaller bin width can generate more compact intervals to be evaluated. Therefore, the selection of bin width may affect the accuracy of hosting capacity assessment, and the quantitative analysis of this impact is presented in Section 4.

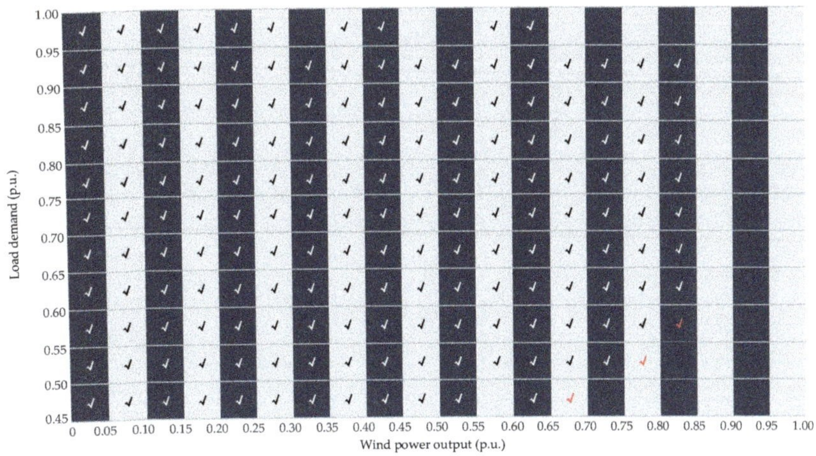

Figure 2. Visualization of wind power-load combinations.

3. Framework for Hosting Capacity Assessment

In this section, we develop a stochastic framework for hosting capacity assessment. The determination of hosting capacity requires clear performance limits as criteria. Note that the most restrictive impact of DG integration is bus overvoltage [23,26]. This paper primarily considers the voltage limits to determine the hosting capacity. The proposed approach can also be extended to consider other criteria, such as conductor thermal capacity and transformer overload [26].

3.1. *Definition of Variables*

Two types of variables need to be considered in hosting capacity assessment. The first category of variables relates to the deployment of DG, including location, type, and the rated power of DG. The second category of variables includes DG production and load consumption. To consider the above variables, we define some terms as follows:

(1) Location penetration (Loc_i): The definition of location penetration is the ratio of the number of selected DG locations N_S to the number of all potential locations N_P. This variable denotes the number of locations for the integration of DG. To fully understand the impact of location penetration on hosting capacity, this variable is increased by a 10% step from 10% to 100%. Let $\boldsymbol{Loc_{pen}}$ be the set of potential location penetration, i.e., $\boldsymbol{Loc_{pen}}$ = {10%, 20%, ... , i×10%, ... ,100%}.
(2) Deployment scheme of DG (S_{ij}): For a specific bus penetration Loc_i, a deployment scheme S_{ij} can be obtained by randomly selecting $Loc_i \times N_P$ buses for DG integration. Let $\boldsymbol{S_i}$ be the set of deployment schemes corresponding to Loc_i. Various potential deployment schemes can be generated using the MCS. Since the deployment of DG is stochastic, the trial number of MCS can affect the accuracy of probabilistic hosting capacity.
(3) DG rated power (DG_{rated}): The rated power of DG at each selected bus is assumed to be proportional to the corresponding peak load consumption.
(4) Type of DG: This variable represents the combination of DG technologies. Two types of DG are considered in this paper, i.e., wind power and PV. As an example, a combination of DG types (50% wind, 50% PV) indicates that the half capacity of DG at the selected buses is wind power, and the other half is PV panels.
(5) Load demands and DG production: A one-year historical data set with a 15-min time resolution is adopted to represent the stochastic nature of generation and consumption. The discretization-aggregation technique in Section 2 is applied to process the historical data and determine the extreme combinations. Since the selection of bin width is empirical, seven different

values are considered to provide comparative results. Let the $\Delta \boldsymbol{d}$ be the set of bin widths, i.e., $\Delta \boldsymbol{d}$ = {0.10 p.u, 0.05 p.u, 0.025 p.u, 0.01 p.u, 0.005 p.u, 0.0025 p.u, 0.001 p.u}.

3.2. Stochastic Analysis Framework

In this section, the computational procedure of the proposed stochastic framework is developed. As shown in Figure 3, the proposed framework has three modules. Module 1 aims to generate various DG deployment schemes. Then, Module 2 conducts the impact analysis to obtain the hosting capacity results of different deployment schemes. Module 3 performs the statistical analysis of the obtained hosting capacity results.

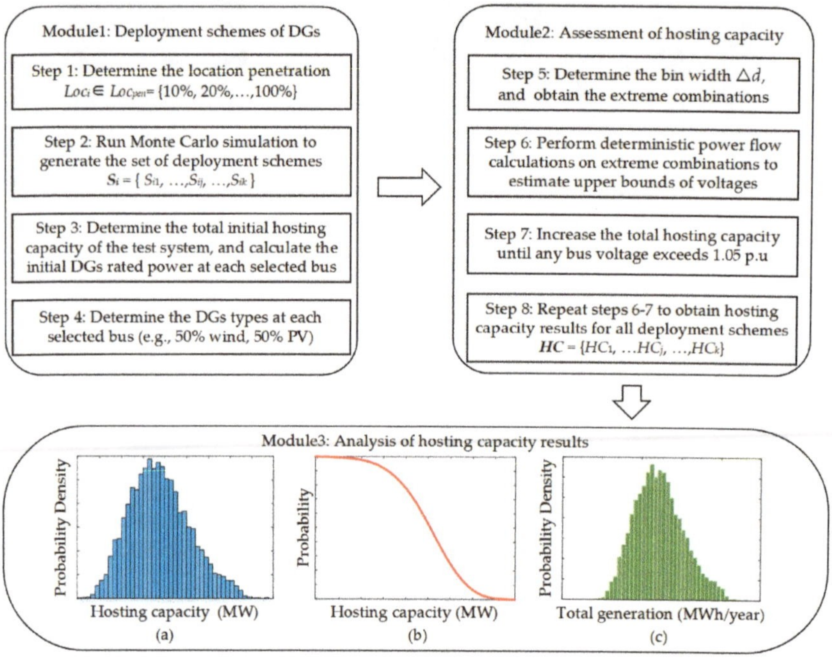

Figure 3. The proposed stochastic framework for hosting capacity analysis.(**a**) Histogram of hosting capacity; (**b**) Cumulative probability curve of hosting capacity; (**c**) Histogram of total generation

(1) Deployment schemes of DG

This module addresses the generation of multiple DG deployment schemes. The variables involved in this module include DG location penetration, rated power, and type. The steps of Module 1 are as follows:

Step 1: Determine the location penetration $Loc_i \in \boldsymbol{Loc_{pen}}$, ($i = 1, 2, \ldots, 10$);

Step 2: Perform MCS to generate k DG deployment schemes with location penetration level Loc_i. The set of deployment schemes is represented as $\boldsymbol{S_i} = \{S_{i1}, \ldots, S_{ij}, \ldots, S_{ik}\}$;

Step 3: Determine the initial total hosting capacity (e.g., 1 MW) of the test system. For each deployment scheme, S_{ij}, the initial DG rated power at each bus is allocated based on the corresponding peak load demand;

Step 4: Determine the DG type at each selected bus (e.g., 50% wind, 50% PV).

(2) Assessment of hosting capacity

This module aims to analyze the impact of DG production on the operating states of systems. The hosting capacity of each deployment scheme is determined by performing deterministic power flow calculations on extreme combinations. Therefore, the variable considered in this module is the bin width $\triangle d$. The steps of Module 2 are as follows:

Step 5: Determine the bin width $\triangle d$, and obtain the extreme combinations to be evaluated;

Step 6: For a specific deployment scheme, S_{ij}, perform deterministic power flow calculations on the extreme combinations to estimate the upper bounds of voltages. For instance, if the extreme combination is {(0.65 p.u, 0.70 p.u], (0.45 p.u, 0.50 p.u]}, the upper bounds of voltages can be obtained with DG generation and load demand set to 0.70 p.u and 0.45 p.u, respectively;

Step 7: If none of the bus voltages exceed the technical constraints, i.e., 1.05 p.u, the total hosting capacity is increased by a fixed step size (e.g., 0.01 MW) and the procedure is repeated from Step 6. Otherwise, the procedure is terminated and the hosting capacity result, HC_j, is obtained for the test system with deployment scheme S_{ij};

Step 8: Repeat steps 6–7 to obtain the hosting capacity results for each deployment scheme in $\boldsymbol{S_i}$. The hosting capacity results are represented as $\boldsymbol{HC} = \{HC_1, \ldots HC_j, \ldots ,HC_k\}$.

(3) Analysis of hosting capacity results

In Module 3, statistical analysis of the obtained hosting capacity results is performed. The results of this module are as follows:

Histogram of hosting capacity: The histogram of hosting capacity can be obtained based on the results $\boldsymbol{HC} = \{HC_1, \ldots HC_j, \ldots ,HC_k\}$. As shown in Figure 3a, the results provide an understanding of the probabilistic hosting capacity of the test system with a specific location penetration.

Cumulative probability curve of hosting capacity: This curve can be obtained based on the histogram of hosting capacity. As shown in Figure 3b, the curve is presented in a descending trend. This curve helps system planners to estimate the probability of having a hosting capacity higher than a specific value.

Histogram of total generation: The total energy generation of DG is a valuable indicator for the utilization of renewable energy. Therefore, the histogram of energy generation in one year, as shown in Figure 3c, is calculated to provide a comparison of the energy utilization of different types of DG.

4. Numerical Results

In this section, numerical studies are carried out on the IEEE 33-bus distribution system to demonstrate the efficacy of the proposed framework. Firstly, brief introductions to the test system and historical data sets of load demand, wind, and PV production are presented. Then, two critical parameters (bin width and trail number) in the stochastic framework are determined through comparisons with the comprehensive assessment. Following that, detailed assessments of both single and hybrid DG hosting capacity are performed.

4.1. Data

The proposed framework is evaluated on the IEEE 33-bus distribution network, whose single line diagram is presented in Figure 4. We set the voltage of the grid supply point to 1.0 p.u, and the lower and upper bounds of voltage for each bus are 0.95 p.u and 1.05 p.u, respectively. The total peak demand for this test system is 3.715 MW. The detailed parameters and configurations of the test system are available in [36].

The historical data of load demand, wind speed, and solar irradiation are derived from a typical distribution system in Hubei province, China. The data set of one year at intervals of 15 min has a total of 35,040 data points. Load demand is normalized against its peak value. Wind speed and solar irradiation are processed and applied to their characteristic curves [37], respectively. Figure 5 presents the variations of load demand, wind power, and PV production in the whole year. It can be seen from the curves that both load demand and PV output have visible seasonal patterns. The load in summer is relatively lower than in winter, while PV output shows the opposite trend. However, the pattern of wind power is less clear than the load and PV output. In the test system, different buses are geographically close. Therefore, it is reasonable to assume that the production of the same type of DG follow the same time series curve.

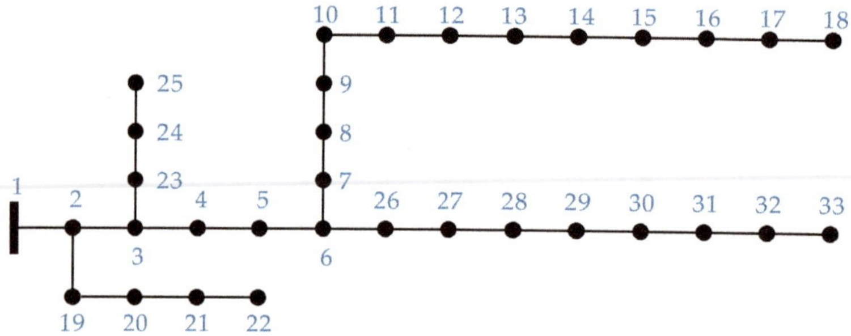

Figure 4. Single line diagram of the IEEE 33-bus distribution system.

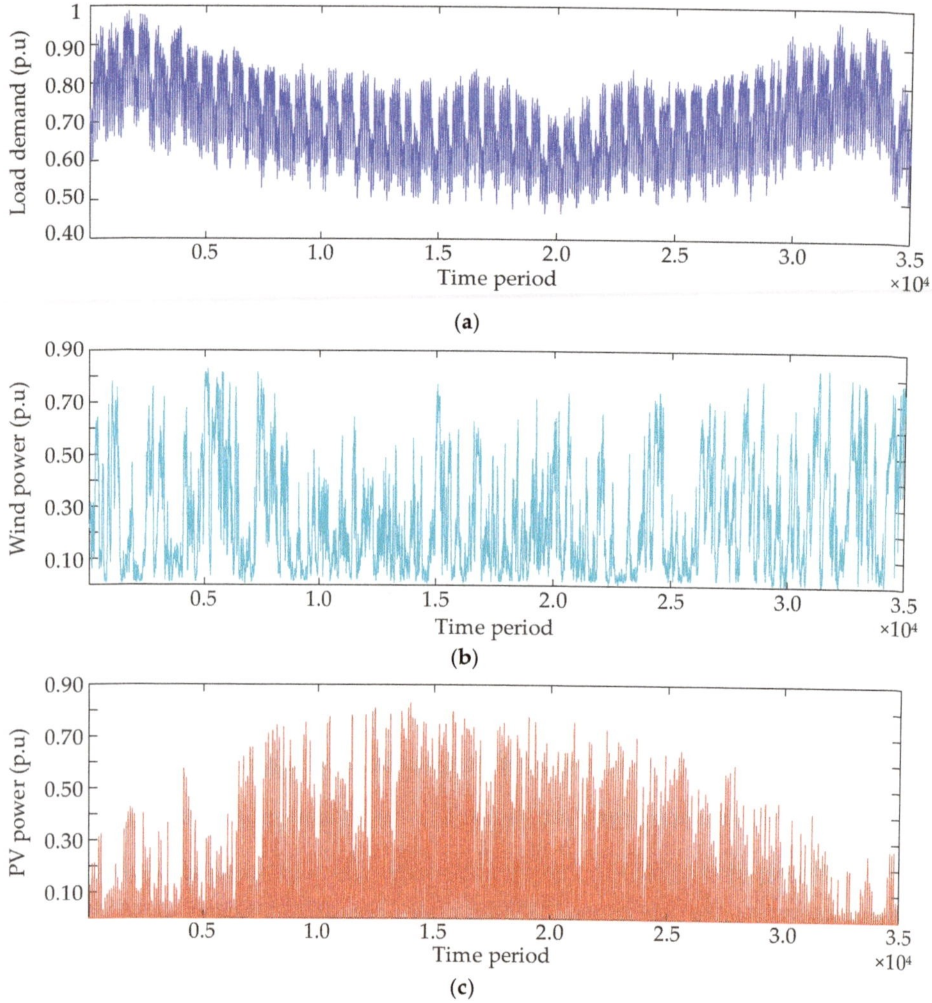

Figure 5. One-year time series data: (**a**) load demand; (**b**) wind power; (**c**) PV power.

The discretization-aggregation technique in Section 2 is adopted to establish the combinations of renewables and load demands. Figure 6 shows the coincident periods for wind power-load, PV-load, and wind power-PV when the bin width is selected as 0.05 p.u. For the 400 potential combinations of wind power-load, only 175 have non-zero probabilities of occurrence. Similarly, only 133 combinations are non-zero in the PV-load case. For the coincidence of renewable energies, there is a negative correlation between wind power and PV, and the correlation coefficient between two resources is −0.162. The PV output mainly depends on solar irradiation. As a result, solar power is negligible during nighttime. In contrast, wind power during daytime is typically less than at nighttime. The complementarity between wind and solar power has the potential to facilitate energy integration.

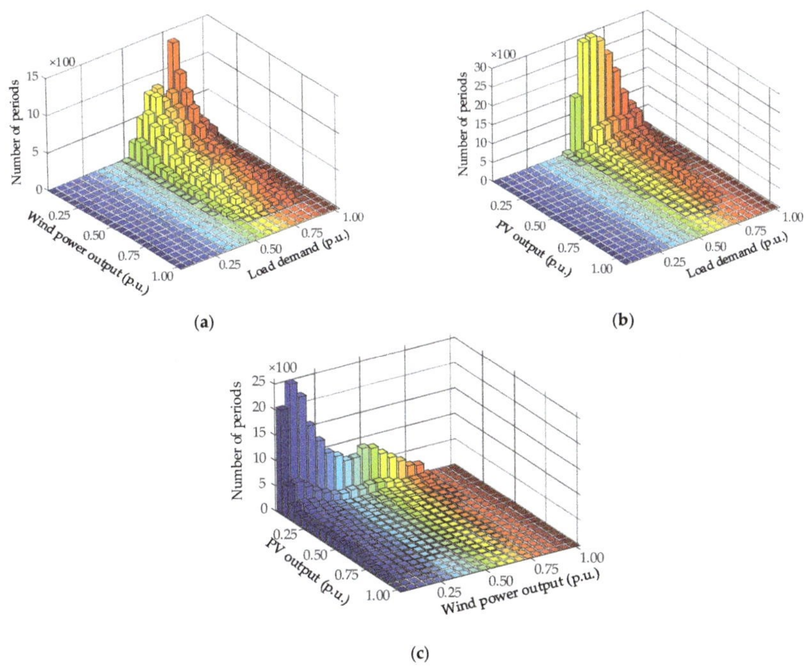

Figure 6. Coincident periods for (**a**) wind power-load; (**b**) PV-load; (**c**) PV-wind power.

In hosting capacity assessment, the occurrences of relatively high production and low demand are critical in driving in system constraints. Table 1 presents the extreme combinations to be evaluated for wind power and PV hosting capacity. For wind power-load and PV-load cases, only three and four extreme combinations need to be considered, respectively. Compared with the historical dataset, the introduction of extreme combinations can significantly reduce the number of load flow calculations.

Table 1. Extreme combinations for wind power-load and PV-load cases.

Renewables and Load	Extreme Combinations {(renewables], (load]} (p.u)
Wind Power-Load	{(0.65 0.70], (0.45 0.50]}, {(0.75 0.80], (0.50 0.55]} {(0.80 0.85], (0.55 0.60]}
PV-Load	{(0.05 0.10], (0.45 0.50]}, {(0.50 0.55], (0.50 0.55]} {(0.55 0.60], (0.55 0.60]}, {(0.80 0.85], (0.60 0.65]}

4.2. Impacts of Simulation Parameters on Hosting Capacity

Before the evaluation of hosting capacity, two critical parameters should be determined. One of them is the bin width, which is used to determine the extreme combinations of generation and demand. This parameter may have an impact on the accuracy of hosting capacity. Another parameter is the trial number of MCS, which represents the number of DG deployment schemes. Since the analysis framework is stochastic, this parameter may affect the accuracy of the probability distribution of the obtained results.

Firstly, the selection of different bin widths is conducted, and only wind power is considered to simplify the analysis. The proposed stochastic framework is implemented with seven different bin widths, Δd, i.e., 0.1 p.u, 0.05 p.u, 0.025 p.u, 0.01 p.u, 0.005 p.u, 0.0025 p.u, and 0.001 p.u. Moreover, the comprehensive load flow method is adopted to render comparative results. The computational procedure of comprehensive assessment is similar to Module 2 of the proposed framework, i.e., increase the total DG capacity until the limitation is violated. Still, the difference is that the former method performs load flow calculations with all historical data of wind power and load to capture the violation of system constraints. Therefore, the results of the comprehensive load flow method can be treated as the exact values.

Taking a specific DG deployment as an example, the locations of wind power integration are Bus 2, 7, 24, and 33. Table 2 shows the number of extreme combinations and hosting capacity results under different bin widths. As a comparison, the result obtained by the comprehensive assessment is 10.87 MW, and the computational time is 894 s. The result obtained by the proposed framework with $\Delta d = 0.1$ p.u is 12.97% less than the exact value. That means the adoption of a relatively large bin width underestimates the hosting capacity and leads to a conservative result. In contrast, when Δd is less than 0.01 p.u, the relative errors are less than 1%. The accurate estimations demonstrate that the proposed extreme combination can provide a suitable alternative to the time series data in hosting capacity assessment when an appropriate bin width is selected.

Table 2. Hosting capacity results with different bin widths.

Width of Bins	Number of Extreme Combinations	Hosting Capacity Results (MW)	Error (%)	Computational Burden (s)
0.1 p.u	2	9.46	12.97	1.19
0.05 p.u	3	10.22	5.98	1.24
0.025 p.u	6	10.63	2.21	1.32
0.01 p.u	10	10.77	0.92	1.47
0.005 p.u	15	10.83	0.37	1.72
0.0025 p.u	21	10.86	0.09	1.94
0.001 p.u	26	10.86	0.09	2.23

The reason for the conservative estimation of hosting capacity can be further explained through the time series results of bus voltage. Figure 7 presents the 1-day snapshot of the time series voltage curves at Bus 18. The blue curve denotes the accurate results obtained by the comprehensive load flow method. Three different Δd, i.e., 0.1 p.u, 0.01 p.u, and 0.001 p.u, are conducted as the comparative cases. For these cases, the corresponding results in Table 2, i.e., 9.46 MW, 10.77 MW, and 10.86 MW, are used to calculate the upper and lower bounds of bus voltage. As described in Section 3.2, the upper bounds of the voltages are calculated using the lower and upper bounds of load and wind power bins, respectively. The lower bounds can be calculated using the opposite rule. It is evident from the results that the accurate voltage curve is always contained by the range of upper and lower bounds in each case. Therefore, when the corresponding hosting capacity in each case is applied to the actual time series data, it can be guaranteed that no voltage violation occurs. The comparison of the upper bounds of the voltages in different cases can explain the conservatism of the obtained hosting capacity. When Δd is relatively large, the upper bound of voltages is overestimated, and the voltage violation can be observed with less DG capacity. In contrast, when Δd is appropriately selected, the interval between the upper and lower bounds of voltage is much narrower than those with larger bin widths. Under such circumstances, the hosting capacity can be accurately estimated.

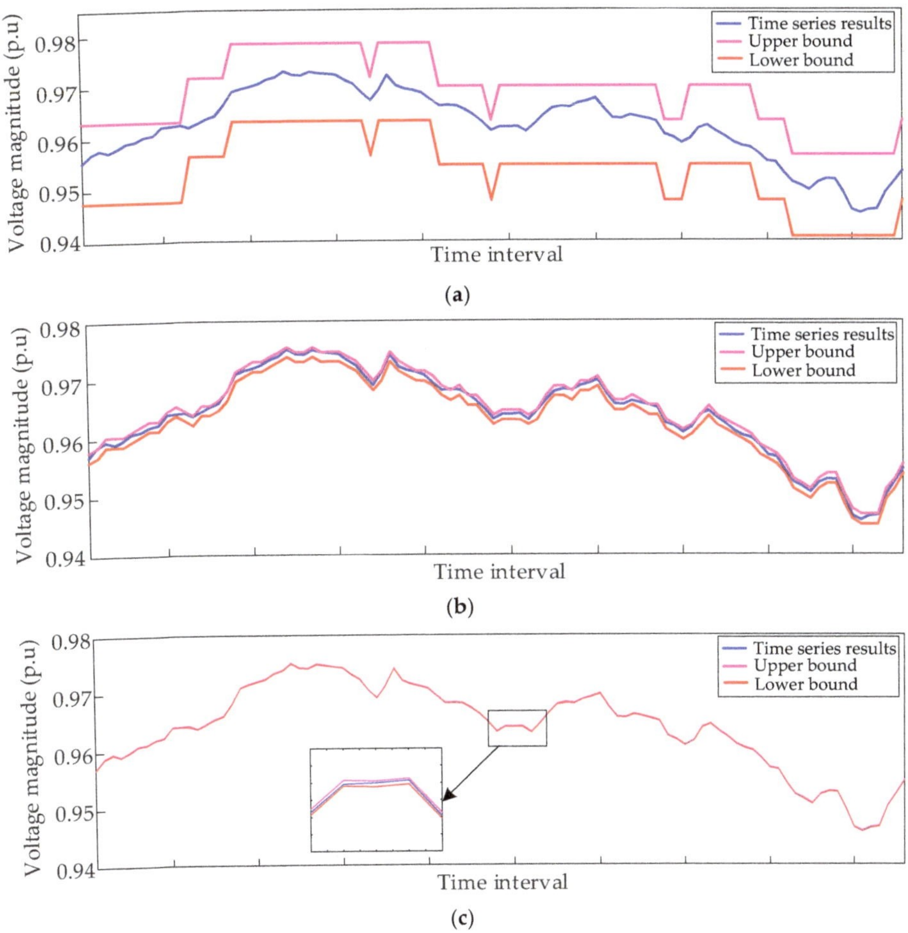

Figure 7. Time-series curves of voltage at Bus 18 with different bin widths: (**a**) $\Delta d = 0.1$ p.u; (**b**) $\Delta d = 0.01$ p.u; (**c**) $\Delta d = 0.001$ p.u.

On the comparison of computational burden, due to the increase of the number of extreme combinations, the computational time required by the proposed framework with smaller Δd increases slightly, as shown in Table 2. Compared with the comprehensive assessment, the proposed framework can provide accurate estimations of hosting capacity with at least 400 times (894/2.23) speed up in efficiency.

In the test system, the number of candidate buses is 32, and each of them has an equal probability of DG integration. For location penetration of 20%, the total possible number of DG deployment schemes is more than 4×10^5. The hosting capacity assessment for such a large number of potential DG deployments results in a significant computational burden. Therefore, MCS is adopted to simulate a relatively small number of scenarios to obtain approximated results. Moreover, the trial number of MCS needs to be determined to achieve a compromise between accuracy and computational burden. In this context, we adopt variance coefficient β [38] of the obtained hosting capacity results as the stopping criteria of MCS. That means the results of MCS are assumed to be accurate if β is less than a specific value.

The variance coefficient, β, should be appropriately selected to ensure the accuracy of the obtained results. Therefore, we consider two different values of β, i.e., 1% and 0.5%, to provide comparative results. Assume that there are four buses in each deployment scheme that are the candidate locations for DG integration. Under such circumstances, the total number of potential deployments can be obtained by the enumeration method (EM). The hosting capacity results of a total of 35,960 deployment scenarios are treated as accurate results. The mean and standard deviation (STD) values of hosting capacity obtained by EM are 6.66 MW and 2.45 MW. As a comparison, the results obtained by MCS are presented in Table 3. The trial number required by MCS with two different β values are 1300 and 5400, respectively. It can be found that the approximated results are more accurate when stopping criteria is relatively small. Figure 8 presents the comparisons between the probability distribution curves. The results indicate that MCS can well approximate the accurate curves obtained by the EM with β set to 0.5%. Therefore, in the following case study, the stopping criterion of MCS is set as $\beta \leq 0.5\%$.

Table 3. Statistical results of hosting capacity with different beta values.

Stopping Criteria β	Mean Values (MW)	Standard Deviation (STD) Values (MW)
1%	6.61	2.34
0.5%	6.67	2.45

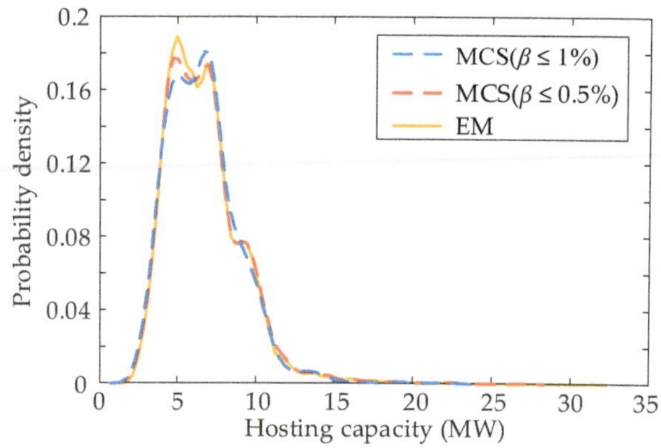

Figure 8. Comparison of approximate results and exact results. MCS: Monte Carlo simulation-based; EM: enumeration method.

4.3. Hosting Capacity Assessment of Single DG Technology

In this section, we perform the stochastic framework to assess the hosting capacity of single DG technology. Based on the analysis in Section 4.2, the parameters bin width, Δd, and variance coefficient, β, are set to 0.001 p.u and 0.5%, respectively, to ensure the accuracy of the results. For each DG technology, location penetration is increased from 10% to 100%, with a step size of 10%.

Table 4 presents the statistical characteristics of wind power and PV hosting capacity results (HC_{WP} and HC_{PV}) under location penetration 20%, 50%, and 80%. It is evident that the mean values of hosting capacity increase with higher location penetration, while the standard deviations present an opposite trend. The reasons for such results are as follows. Firstly, with increased location penetration, the installed DG capacity at each location reduces (the initial total capacity is assumed as 1 MW and increased with a fixed step). In this case, the impact of DG on the system is less severe, and the test system can accommodate more DG capacity. Secondly, higher location penetration alleviates the uncertainties of DG location. Therefore, the obtained hosting capacity results with different deployment schemes are less dispersed.

Table 4. Statistical results of hosting capacity with different location penetration.

Loc_i	Mean Value of HC_{WP} (MW)	STD Value of HC_{WP} (MW)	Mean Value of HC_{PV} (MW)	STD Value of HC_{PV} (MW)
20%	7.05	2.09	7.26	2.12
50%	7.78	1.39	8.05	1.43
80%	7.92	0.72	8.21	0.73

When comparing the hosting capacity of different DG types, it is worth mentioning that the test system has a higher PV hosting capacity than wind power at the same location penetration. For instance, the results for PV is 3.3% higher than that of wind power when location penetration is 50%. In general, the occurrences of high DG production and low demand are of great concern as they promote voltage rise. Therefore, the difference in wind power and PV hosting capacity can be explained using extreme combinations. For the case of wind power, most of the deployment schemes are constrained by the combination {(0.729 p.u, 0.730 p.u], (0.507 p.u, 0.508 p.u]}. While the corresponding combination for the PV case is {(0.749 p.u, 0.750 p.u], (0.590 p.u, 0.591 p.u]}. The former combination can cause a slightly higher voltage rise at the same DG capacity.

In addition to the hosting capacity results, Table 5 presents the comparisons of mean values of total energy production from wind power and PV (E_{WP} and E_{PV}). Although the hosting capacity for PV is slightly higher than wind power, total energy production from PV is less than 50% of wind. The reason for the difference between total energy production is the higher wind power capacity factor in this area. Total generation production can realistically reflect energy utilization. Therefore, in the planning of DG, not only the maximum DG capacity, but also the total generation production should be considered.

Table 5. Mean values of energy production with different location penetration.

Loc_i	Mean Value of E_{WP} (MW·h)	Mean Value of E_{PV} (MW·h)
20%	15,642.71	7367.37
50%	17,254.52	8175.09
80%	17,579.30	8331.41

Figure 9 presents the probability distributions of wind power and PV hosting capacity under 50% location penetration. The results are shown using histograms and cumulative probability curves. It can be found that Gamma distribution can approximate both wind power and PV hosting capacity. Moreover, the obtained results with other location penetrations also follow the same rule. The histograms provide the estimations of extreme values of hosting capacity. For wind power hosting capacity, the obtained minimum and maximum values are 4.10 MW and 12.65 MW, respectively. As a comparison, the corresponding results for PV hosting capacity are 4.26 MW and 13.21 MW, respectively. The difference between the extreme values is partly due to the locations of DG. In general, the relatively high hosting capacity can be obtained with DG located near the first bus. Figure 9b,d presents the cumulative probability curves of hosting capacity. These curves provide system planners with a straightforward tool for estimating the probability of having a hosting capacity higher than a specific value. For example, the probability of having wind power hosting capacity higher than 6 MW is 0.9072, while in the PV case, the result is 0.9353.

Since the obtained hosting capacity results with different location penetration can be approximated by Gamma distributions, it is imperative to establish the relationship between location penetration and the parameters of Gamma distributions. The probability density function of Gamma distribution is given as follows:

$$f(x|\alpha,\beta) = \frac{1}{\beta^{\alpha}\Gamma(\alpha)}x^{\alpha-1}e^{\frac{-x}{\beta}} \quad (1)$$

where $\Gamma(\cdot)$ is the Gamma function; parameters α and β are shape and scale parameters, respectively.

The shape and scale parameters can be calculated as follows:

$$\begin{cases} \alpha = \mu_x^2 / \sigma_x \\ \beta = \sigma_x / \mu_x \end{cases} \tag{2}$$

where μ_x and σ_x are the expected value and standard deviation of variable x, respectively.

Based on the hosting capacity results with different location penetrations, the curves of the parameter of Gamma distribution to location penetration can be obtained and fitted by the fifth-order polynomials, as shown in Figure 10. These curves can provide planners with useful information. For instance, when the location penetration is 35%, system planners can utilize the curve to obtain the shape and scale parameters and estimate the probability distribution of hosting capacity under the target location penetration. Moreover, once such curves are established for a specific distribution system, the probabilistic hosting capacity of the system with any location penetration can be obtained without simulating thousands of DG deployment schemes.

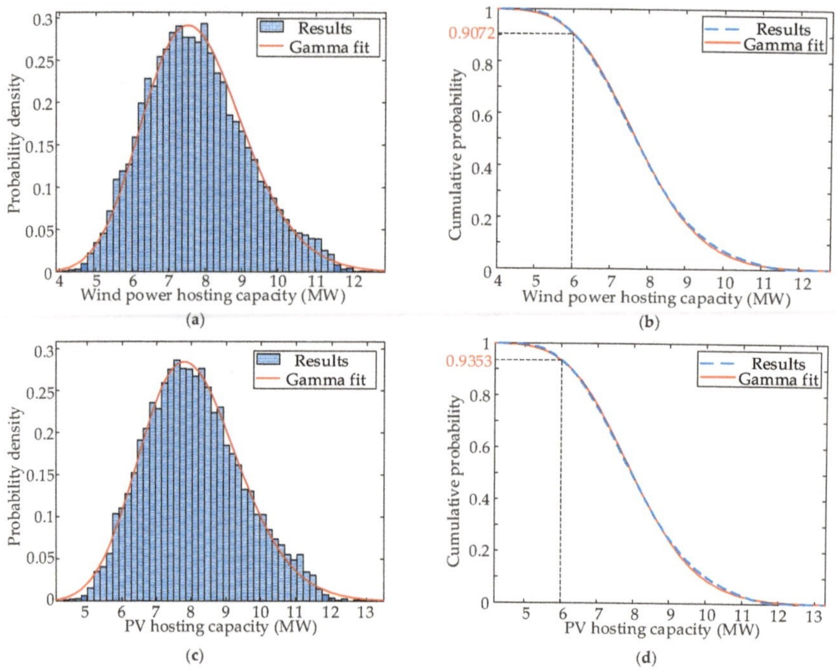

Figure 9. Hosting capacity results of single DG technology: (**a**) and (**b**) probability distribution of wind power hosting capacity; (**c**) and (**d**) probability distribution of PV hosting capacity.

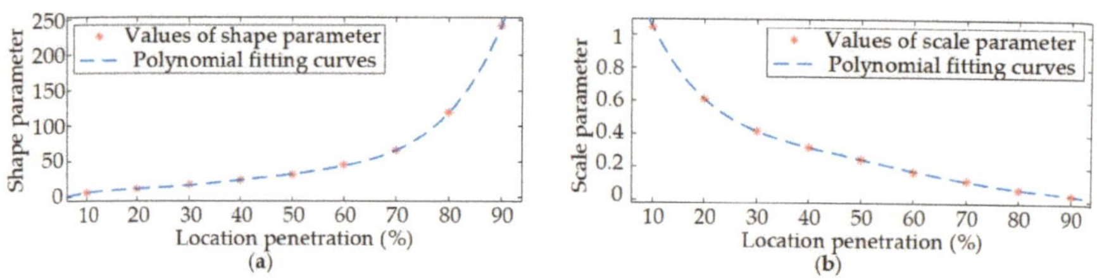

Figure 10. The curves of shape and scale parameters to location penetration: (**a**) shape parameter; (**b**) scale parameter.

4.4. Hosting Capacity Assessment of Hybrid Wind-PV

In this section, the proposed framework is conducted to estimate the hybrid wind power-PV hosting capacity. This section only considers the different ratios of wind power and PV under 50% location penetration to simplify the presentation.

Monte Carlo simulation is performed to generate various DG deployment schemes. For a specific DG deployment scenario, the percentage of wind power generation is increased from 0% to 100% with 5% increment steps. The proposed framework is performed to calculate the corresponding hosting capacity. Figure 11 presents the probability distributions of the obtained hosting capacity results with different wind power percentages. As the ratio of wind power increases, the transition of hybrid DG hosting capacity follows a two-stage process. In the first stage, the probability distributions of hosting capacity shift toward the right until the percentage of wind power reaches 75%. In the second stage, as the proportion of wind power continues to increase, the probability distribution curves of hosting capacity move to the left.

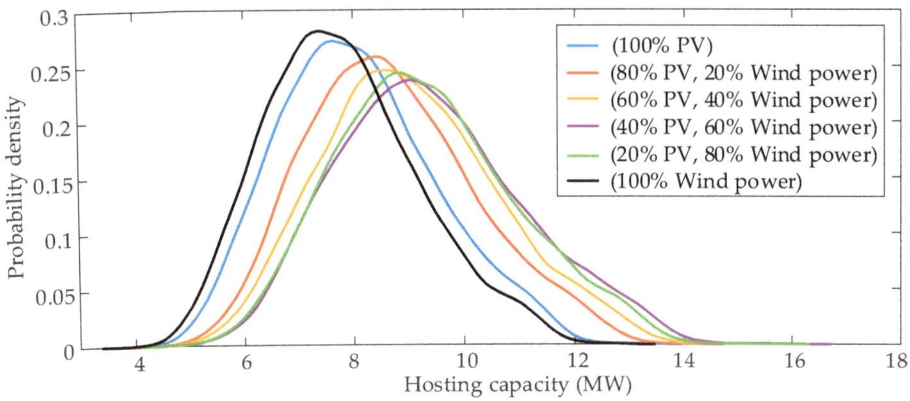

Figure 11. Probability distributions of hosting capacity with different DG technology combinations.

Table 6 presents the mean values of hosting capacity and total energy production under two maximum situations. The hybrid wind power-PV hosting capacity reaches its maximum value when the percentage of wind power is 75%. In this case, the mean value of total hosting capacity is 9.39 MW, of which wind power and PV capacity are 7.04 MW and 2.35 MW, respectively. Compared with the results of single wind power and PV technologies, the total hosting capacity increases by about 1.21 (9.39/7.78) and 1.17 (9.39/8.05) times, respectively. The mean value of total energy production for one year is 17,999.26 MW·h, of which the energy production of wind power accounts for about 87%. The total energy production increases by about 1.04 and 2.20 times when compared with single wind power and PV, respectively. Another important result is the maximum energy production, and this value is reached when the wind power percentage is 85%. In this case, the mean values of hosting capacity and total energy production are 9.12 MW and 18,596.31 MW·h, respectively.

Table 6. Mean values of hosting capacity and energy production.

	Mean Value of HC_{WP} (MW)	Mean Value of HC_{PV} (MW)	Mean Value of E_{WP} (MW·h)	Mean Value of E_{PV} (MW·h)
Maximum hosting capacity	7.04	2.35	15,617.89	2381.37
Maximum energy production	7.75	1.37	17,207.29	1389.02

As can be seen from the above results, wind power has a dominant role in both cases of maximum hosting capacity and energy production. However, PV is also essential as the complementarity generation to increase total hosting capacity and energy production. The results demonstrate that the complementarity between wind power and PV can promote the system to accommodate more DG capacity, thus increasing the utilization of renewable energy. Besides, the difference between the two maximum results demonstrates that hosting capacity and energy production do not reach the optimal values simultaneously. Therefore, considering the investment of different DG types, system planners and DG investors may need to resort to multi-objective optimization to achieve compromise solutions.

Similar to the hosting capacity of single DG technology, the probability distribution of hybrid wind power-PV hosting capacity with different wind power ratios can also be approximated by Gamma distribution. Moreover, the probability distribution curves of hosting capacity in Figure 11 have a similar shape. Using Equation (2) to calculate the parameters of Gamma distribution corresponding to each curve, the obtained mean and STD values of the shape parameters are 32.13 and 0.4, which indicates that the shape parameters of different curves are almost the same. In this context, only the relationship between wind power percentage and scale parameter needs to be established. Moreover, the change of scale parameters can represent the trend of hosting capacity. As can be seen from the curve in Figure 12, the system can accommodate relatively higher DG capacity when the wind power percentage is 65–75%. This result is consistent with the previous analysis in Table 6. This curve has two applications. Firstly, system planners can obtain the probability distributions of hosting capacity with different wind power and photovoltaic combinations based on the curve. Secondly, DG investors can adjust their investment plans for wind power and PV generation based on the curve so that they can maximize the DG capacity and obtain more revenue from energy production.

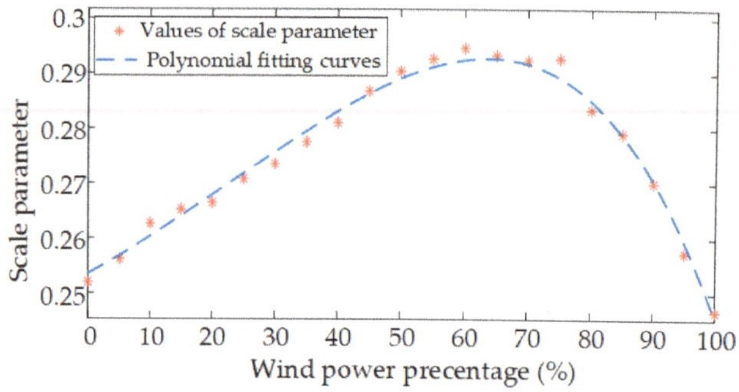

Figure 12. The curves of scale parameters to wind power percentage.

5. Discussion

In this section, we discuss the main features, applications, and future research of this study.

Firstly, the proposed stochastic framework provides system planners with an accurate and efficient tool for hosting capacity assessment. The framework adopts the discretization-aggregation technique to deal with the time series data. Compared with the application of this technique in previous studies [30,35], this study concentrates more on the extreme combinations and the appropriate selection of bin width. Thanks to the introduction of extreme combinations, the number of power flow calculations required by the hosting capacity evaluation is significantly reduced. More importantly, the simulation results demonstrate the necessity of appropriate selection of bin width.

Secondly, this study presents the assessments of wind power, PV, and hybrid wind-PV hosting capacity. The simulation results indicate that the hybrid energy system has the potential to increase hosting capacity and energy production due to the complementarity of wind speed and solar irradiation. The performance of the hybrid wind-PV system in facilitating energy integration and utilization varies with different wind power and PV ratios. Compared with the studies that focused on PV hosting capacity [23–26], this study can contribute a better understanding of hybrid wind-PV hosting capacity from a probabilistic point of view.

Thirdly, the simulation results identify that Gamma distribution can approximate the probability distributions of both single and hybrid DG hosting capacity. These findings can provide system planners and DG investors with serval useful tools, such as the curves in Figures 10 and 12. Such results can be generally established for other networks. These curves can, not only help system planners assess the hosting capacity under different location penetrations, but also assists DG investors in making informed decisions on the investment plan of different DG types.

This study performs the hosting capacity assessment without considering active network managements (ANMs). ANMs, such as on-load tap changers [39] and energy storage [40,41], are verified to be effective ways for the enhancement of hosting capacity. However, due to the need for the knowledge of the succession of operating states (e.g., state of charge of energy storage), the assessment of enhanced hosting capacity can be more challenging when considering ANMs. Therefore, the establishment of a critical time series of DG production and load demand can be one of the future research works.

6. Conclusions

In this paper, a stochastic framework is developed to study the hybrid wind-PV system hosting capacity from a probabilistic view. To reduce the computational burden of assessments, we define the extreme combinations of DG productions and load demand based on historical data sets. The proposed framework is implemented in the IEEE 33-bus system to evaluate wind power, PV, and hybrid energy hosting capacity. The main findings of this study are as follows:

(1) The accuracy of the proposed framework depends on the selection of bin width. Relatively large bin width can lead to a conservative estimation of hosting capacity. Therefore, to obtain accurate results, it is recommended that the bin width needs to be smaller than 0.01 p.u.
(2) Due to the difference in energy resources, the system has slightly higher PV hosting capacity than wind power. Still, the total energy production of PV is much less than that of wind power. Moreover, the probability distributions of wind power and PV hosting capacity with different location penetration can be approximated by Gamma distribution.
(3) Due to the complementarity between wind power and PV, the system can accommodate more DG capacity, thus promoting the utilization of renewable energy. However, the hosting capacity and energy production of the hybrid wind-PV system do not reach optimal values simultaneously. Therefore, the multi-objective optimization method can be used to achieve a compromise between DG capacity and energy production.

The main application of this study is to provide a useful tool for system planners to assess DG hosting capacity. This study can be extended to consider other low carbon technologies, such as electric vehicles and biomass-fueled gas engines. Also, the proposed combination of production and demand can be used in other energy management studies that require extensive power flow calculations, such as analysis of system losses and energy reductions. Regarding future research, more binding constraints, such as voltage unbalance, conductor thermal capacity, and transformer overload, can be incorporated to establish a more comprehensive hosting capacity assessment. Furthermore, a critical time series for DG production and load demand needs to be developed to evaluate the hosting capacity enhanced with ANMs.

Author Contributions: The authors confirm their contributions to the paper as follows: Supervision, D.L.; methodology, software, and writing—original draft preparation, C.W.; project administration, F.T.; visualization and writing—review and editing, Y.Z. All authors have read and agreed to the published version of the manuscript.

Funding: This work was supported by the National Natural Science Foundation of China under grant No. 51977157.

Acknowledgments: This work was supported by the National Natural Science Foundation of China under grant No. 51977157.

Conflicts of Interest: The authors declare no conflict of interest.

References

1. Haque, M.M.; Wolfs, P. A review of high PV penetrations in LV distribution networks: Present status, impacts and mitigation measures. *Renew. Sustain. Energy Rev.* **2016**, *62*, 1195–1208. [CrossRef]
2. Zhang, S.; Cheng, H.; Li, K.; Tai, N.; Wang, D.; Li, F. Multi-objective distributed generation planning in distribution network considering correlations among uncertainties. *Appl. Energy* **2018**, *226*, 743–755. [CrossRef]
3. Rabiee, A.; Mohseni-Bonab, S.M. Maximizing hosting capacity of renewable energy sources in distribution networks: A multi-objective and scenario-based approach. *Energy* **2017**, *120*, 417–430. [CrossRef]
4. Petinrin, J.; Shaaban, M. Impact of renewable generation on voltage control in distribution systems. *Renew. Sustain. Energy Rev.* **2016**, *65*, 770–783. [CrossRef]
5. Olivier, F.; Aristidou, P.; Ernst, D.; Cutsem, T.V. Active management of low-voltage networks for mitigating overvoltages due to photovoltaic units. *IEEE Trans. Smart Grid* **2015**, *7*, 926–936. [CrossRef]
6. Schwanz, D.; Möller, F.; Rönnberg, S.K.; Meyer, J.; Bollen, M.H.J. Stochastic assessment of voltage unbalance due to single-phase-connected solar power. *IEEE Trans. Power Deliv.* **2017**, *32*, 852–861. [CrossRef]
7. Etherden, N.; Bollen, M.H.J. Overload and overvoltage in low-voltage and medium-voltage networks due to renewable energy-Some illustrative case studies. *Electr. Power Syst. Res.* **2014**, *114*, 39–48. [CrossRef]
8. Santos, I.N.; Cuk, V.; Almeida, P.M.; Bollen, M.H.J.; Ribeiro, P.F. Considerations on hosting capacity for harmonic distortions on transmission and distribution systems. *Electr. Power Syst. Res.* **2015**, *119*, 199–206. [CrossRef]
9. Bollen, M.H.J.; Das, R.; Djokic, S.; Ciufo, P.; Meyer, J.; Rönnberg, S.K.; Zavoda, F. Power quality concerns in implementing smart distribution-grid applications. *IEEE Trans. Smart Grid* **2017**, *8*, 391–399. [CrossRef]
10. Ismael, S.M.; Abdel Aleem, S.H.; Abdelaziz, A.Y.; Zobaa, A.F. State-of-the-art of hosting capacity in modern power systems with distributed generation. *Renew. Energy* **2019**, *130*, 1002–1020. [CrossRef]
11. Bollen, M.H.J.; Rönnberg, S.K. Hosting Capacity of the Power Grid for Renewable Electricity Production and New Large Consumption Equipment. *Energies* **2017**, *10*, 1325. [CrossRef]
12. Bollen, M.; Hassan, F. *Integration of Distributed Generation in the Power System*; John Wiley & Sons: Hoboken, NJ, USA, 2011.
13. Castelo de Oliveira, T.E.; Bollen, M.; Ribeiro, P.F.; de Carvalho, P.M.S.; Zambroni, A.C.; Bonatto, B.D. The Concept of Dynamic Hosting Capacity for Distributed Energy Resources: Analytics and Practical Considerations. *Energies* **2019**, *12*, 2576. [CrossRef]
14. Dellino, G.; Laudadio, T.; Mari, R.; Mastronardi, N.; Meloni, C.; Vergura, S. Energy Production Forecasting in a PV plant using Transfer Function Models. In Proceedings of the 2015 IEEE 15th International Conference on Environments and Electrical Engineering (EEIC), Rome, Italy, 10–13 June 2015.
15. Bruno, S.; Dellino, G.; La Scala, M.; Meloni, C. A Microforecasting Module for Energy Management in Residential and Tertiary Buildings †. *Energies* **2019**, *12*, 1006. [CrossRef]
16. Mulenga, E.; Bollen, M.H.J.; Etherden, N. A review of hosting capacity quantification methods for photovoltaics in low-voltage distribution grids. *Int. J. Electr. Power Energy Syst.* **2020**, *115*, 105445. [CrossRef]
17. Ebe, F.; Idlbi, B.; Morris, J.; Heilcher, G.; Meier, F. Evaluation of PV hosting capacity of distribution grids considering a solar roof potential analysis-Comparison of different algorithms. In Proceedings of the 2017 IEEE Manchester PowerTech, Manchester, UK, 18–22 June 2017.
18. Conti, S.; Raiti, S. Probabilistic load flow using Monte Carlo techniques for distribution networks with photovoltaic generators. *Sol. Energy* **2007**, *81*, 1473–1481. [CrossRef]
19. Kabir, M.N.; Mishra, Y.; Bansal, R.C. Probabilistic load flow for distribution systems with uncertain PV generation. *Appl. Energy* **2016**, *163*, 343–351. [CrossRef]
20. Behravesh, V.; Keypour, R.; Foroud, A.A. Stochastic analysis of solar and wind hybrid rooftop generation systems and their impact on voltage behavior in low voltage distribution systems. *Sol. Energy* **2018**, *166*, 317–333. [CrossRef]
21. Smith, J. *Stochastic Analysis to Determine Feeder Hosting Capacity for Distributed Solar PV*; EPRI Technical Update: Knoxville, TN, USA, 2012.

22. Smith, J. *Alternatives to the 15% Rule: Modeling and Hosting Capacity Analysis of 16 Feeders*; Electric Power Research Institute: Palo Alto, CA, USA, 2015.
23. Ding, F.; Mather, B. On Distributed PV Hosting Capacity Estimation, Sensitivity Study, and Improvement. *IEEE Trans. Sustain. Energy* **2017**, *8*, 1010–1020. [CrossRef]
24. Dubey, D.; Santoso, S.; Maitra, A. Understanding photovoltaic hosting capacity of distribution circuits. In Proceedings of the IEEE Power & Energy Society General Meeting, Denver, CO, USA, 26–30 July 2015.
25. Dubey, A.; Santoso, S. On Estimation and Sensitivity Analysis of Distribution Circuit's Photovoltaic Hosting Capacity. *IEEE Trans. Power Syst.* **2017**, *32*, 2779–2789. [CrossRef]
26. Torquato, R.; Salles, D.; Pereira, C.O.; Meira, P.C.M.; Freita, W. A Comprehensive Assessment of PV Hosting Capacity on Low-Voltage Distribution Systems. *IEEE Trans. Power Deliv.* **2018**, *2*, 1002–1012. [CrossRef]
27. Abad, M.S.S.; Ma, J.; Zhang, D.; Ahmadyar, A.S.; Marzooghi, H. Probabilistic assessment of hosting capacity in radial distribution systems. *IEEE Trans. Sustain. Energy* **2018**, *4*, 1935–1947. [CrossRef]
28. Monforti, F.; Huld, T.; Bódis, K.; Vitali, L.; D'Isidoro, M.; Lacal-Arántegui, R. Assessing Complementarity of Wind and Solar Resources for Energy Production in Italy. A Monte Carlo Approach. *Renew. Energy* **2014**, *63*, 576–586. [CrossRef]
29. Ren, G.; Wan, J.; Liu, J.; Yu, D. Spatial and temporal assessments of complementarity for renewable energy resources in China. *Energy* **2019**, *177*, 262–275. [CrossRef]
30. Sun, W.; Harrison, G.P. Wind-solar complementarity and effective use of distribution network capacity. *Appl. Energy* **2019**, *247*, 89–101. [CrossRef]
31. Jurasz, J.; Beluco, A.; Canales, F.A. The impact of complementarity on power supply reliability of small scale hybrid energy systems. *Energy* **2018**, *161*, 737–743. [CrossRef]
32. Ochoa, L.F.; Padilha-Feltrin, A.; Harrison, G.P. Time-series-based maximization of distributed wind power generation integration. *IEEE Trans. Energy Convers.* **2008**, *23*, 968–974. [CrossRef]
33. Khoshkbar-Sadigh, A.; Smedley, K.M. The necessity of time-series simulation for investigation of large-scale solar energy penetration. In Proceedings of the IEEE Power & Energy Society Innovative Smart Grid Technologies Conference (ISGT), Washington, DC, USA, 18–20 February 2015.
34. Abad, M.S.S.; Ma, J.; Zhang, D.; Ahmadyar, A.S.; Marzooghi, H. Sensitivity of Hosting Capacity to Data Resolution and Uncertainty Modeling. In Proceedings of the 2018 Australasian Universities Power Engineering Conference (AUPEC), Auckland, New Zealand, 27–30 November 2018.
35. Ochoa, L.F.; Dent, C.J.; Harrison, G.P. Distribution Network Capacity Assessment: Variable DG and Active Networks. *IEEE Trans. Power Syst.* **2010**, *25*, 87–95. [CrossRef]
36. Baran, M.E.; Wu, F.F. Network reconfiguration in distribution systems for loss reduction and load balancing. *IEEE Trans. Power Deliv.* **1989**, *4*, 1401–1407. [CrossRef]
37. Wang, C.; Liu, C.; Tang, F.; Liu, D.; Zhou, Y. A scenario-based analytical method for probabilistic load flow analysis. *Electr. Power Syst. Res.* **2020**, *181*, 106193. [CrossRef]
38. Prusty, B.R.; Jena, D. A critical review on probabilistic load flow studies in uncertainty constrained power systems with photovoltaic generation and a new approach. *Renew. Sustain. Energy Rev.* **2017**, *69*, 1286–1302. [CrossRef]
39. Wang, S.; Chen, S.; Ge, L.; Wu, L. Distributed generation hosting capacity evaluation for distribution systems considering the robust optimal operation of OLTC and SVC. *IEEE Trans. Sustain. Energy* **2016**, *7*, 1111–1123. [CrossRef]
40. Babacan, O.; Torre, W.; Kleissl, J. Siting and sizing of distributed energy storage to mitigate voltage impact by solar PV in distribution systems. *Sol. Energy* **2017**, *146*, 199–208. [CrossRef]
41. Palacios-Garcia, E.J.; Moreno-Muñoz, A.; Santiago, I.; Moreno-Garcia, I.M.; Milanés-Montero, M.I. PV Hosting Capacity Analysis and Enhancement Using High Resolution Stochastic Modeling. *Energies* **2017**, *10*, 1488. [CrossRef]

sustainability MDPI

Article

Renewable Energy for Sustainable Growth and Development: An Evaluation of Law and Policy of Bangladesh

Mohammad Ershadul Karim [1,*], Ridoan Karim [1,2], Md. Toriqul Islam [1,3], Firdaus Muhammad-Sukki [4,*], Nurul Aini Bani [5] and Mohd Nabil Muhtazaruddin [5]

1 Faculty of Law, University of Malaya, Kuala Lumpur 50603, Malaysia; ridoankarim@um.edu.my (R.K.); toriqul@siswa.um.edu.my (M.T.I.)
2 School of Business Administration, East Delta University, Abdullah Al Noman Road, Noman Society, Mozaffor Nogor, East Nasirabad, Khulshi, Chittagong 4209, Bangladesh
3 Department of Law and Justice, Bangladesh University of Business and Technology, Dhaka 1216, Bangladesh
4 School of Engineering, Robert Gordon University, Garthdee Road, Aberdeen AB10 7GJ Scotland, UK
5 Razak Faculty of Technology and Informatics, Universiti Teknologi Malaysia, Kuala Lumpur 54100, Malaysia; nurulaini.kl@utm.my (N.A.B.); mohdnabil.kl@utm.my (M.N.M.)
* Correspondence: ershadulkarim@um.edu.my (M.E.K.); f.b.muhammad-sukki@rgu.ac.uk (F.M.-S.); Tel.: +60-187-693-629 (M.E.K.)

Received: 23 September 2019; Accepted: 15 October 2019; Published: 17 October 2019

Abstract: Bangladesh's constant growth with an annual 6% plus Gross Domestic Product (GDP) for more than the last two decades and achievements in other socio-economic metrics in recent times is impressive and recognized by various global authoritative bodies. The extent of overwhelming economic ventures in the private sector coupled with the commitments of the government clearly demonstrates the transformation of the country from a primarily agro-based economy to one influenced by the manufacturing and service sectors. Bangladesh is fortunate to have fossil fuel reserves on a limited scale, though these are not enough to run the ongoing massive scale development activities, both in private and public sectors. Thus, the constant and uninterrupted supply of energy at an affordable price remains a serious concern for the successive governments. Therefore, this issue of supply of constant energy has turned to be an important part in the national development agenda. Besides, the country is one of the worst victim nations of the devastating effects of global warming and climate change. As Bangladesh is geographically located in a favorable place in the world map with the availability of plenty of renewable energy sources (RES), the policymakers started to take initiatives leading to exploiting these sources to meet the energy demand of the country. There are both prospects and administrative, legal, technological, socio-cultural and environmental challenges. To address these challenges, it requires comprehensive policy initiatives. A good number of technical and scientific research containing findings and recommendations are available. This paper, which is based on adopting a qualitative research methodology where the contents of secondary sources were analyzed, is an initial attempt to highlight the renewable energy developments in Bangladesh, and subsequently, to evaluate the relevant legal and policy initiatives in the light of international best practices. We advance several recommendations that the stakeholders can consider exploiting RES effectively to attain inclusive, equitable and sustainable development in Bangladesh. These include, *inter alia*: (1) Enhancing government participation to lead the development of renewable energy (RE); (2) ensuring localization of RE technology; (3) reducing the expenses of energy generation through RES and providing assistance in initial investments; (4) introducing comprehensive legal and regulatory policy for the development of RE industry in Bangladesh; and (5) conducting effective public awareness.

Keywords: RE prospects and challenges; RE regulations and policy; RE in Bangladesh

1. Introduction

Energy generated from fossil or traditional fuel sources, such as natural oil, gas and coal, etc., causes negative impacts on our surrounding environment as they produce greenhouse gases (GHGs) which are responsible for climate change and global warming. Climate change has devastating effects and consequences on the existence and survival of human beings, biodiversity and ecosystems on this earth. Therefore, the reduction of emission of GHGs in order to mitigate climate change has turned out to be an important concern for almost everyone. The global community has proposed and adopted various policy initiatives which advocated to shift the energy production from fossil fuels to energy generated from renewable energy sources (RES), such as solar, geothermal, biomass, wind, biogas, hydro-power, etc., since these are reported to be cleaner, more sustainable and relatively less pollutant [1].

Bangladesh's constant growth with an annual 6% plus Gross Domestic Product (GDP) for more than the last two decades and achievements in other socio-economic metrics in recent times is impressive and recognized by various global authoritative bodies. Bangladesh is frequently projected as an agricultural country that encounters various catastrophic natural disasters every year. However, in recent years, it aims to transform from an agro-based nation to an industrial one, and this is reflected through a slow, but massive industrial growth. The extent of overwhelming economic ventures and the commitments of the government clearly demonstrates Bangladesh's economic shift. The country is fortunate to have fossil fuel reserves on a limited scale, and these are not enough to run the ongoing massive scale development activities, both in private and public sectors. Thus, the constant and uninterrupted supply of energy at an affordable price remains a serious concern for the successive governments. Therefore, this issue has been getting importance in the development agenda constantly.

The Government of Bangladesh has promised and has taken initiatives to ensure quality energy production and supply to every door by 2020. In implementing this high ambition, the government set plans to generate 10% of its power's demand through energy generated from RES by 2020 [2]. Fuel diversification seems to be an exclusive approach for ensuring sufficient electricity supply and stimulating economic advancement of the country. Consequently, Bangladesh has adopted this fuel diversification program for extending its renewable energy (RE) industry and ensuring the ever-increasing electricity demand. In 2008, the government formulated the Renewable Energy Policy. Public-private investment is welcomed in promoting the RE ventures to replace the fossil fuel-based energy generation to the RE contributing to the national grid [3]. Government's Renewable Energy Policy presumed to generate 5% of energy from the RES by 2015 and to increase a total of 10% of the total electricity generation by 2020 [4]. Meanwhile, the government has launched a new project "500 MW Solar Power Mission" to satisfy the escalating need for energy consumption [5].

Bangladesh has signed the Paris Agreement to the United Nations Framework Convention on Climate Change (UNFCC) on the 21st September 2016 and ratified the same on the 4th November, 2016. The Paris Agreement is not like the Kyoto Protocol that puts an obligation toward the advanced nations of mitigating emissions of GHG. It is a bottom-up treaty that obliges every country to undertake essential efforts for the protection of the environment and keeping climate stable [6]. Having ratified the Paris Agreement, the country has pledged to reduce its GHG emissions by 5% below 'business-as-usual' level by 2030 using domestic resources regardless of the country's GDP. In such a context, Bangladesh is also responsible to the establishment, development and implementation of the legal framework for addressing the environmental concerns, formulate rules and policies for reducing the GHG emissions, and realize a sustainable economic strategy to generate electricity [6].

Although the government wishes seriously to expand a significant portion of RE contribution in the national grid, such expansion can only be achieved if fundamental challenges encompassing the RE projects are properly addressed. Despite the execution of comprehensive policies and development strategies, which eventually worked as the catalyst for RE penetration in power generation, RE projects and green technologies still face the financial, technical and even political barriers. Even though many scholars, researchers, bodies, and agencies from both public and private sectors have shown interests and great concerns about the scientific and technical challenges encircling the expansion of RE in Bangladesh, there is a dearth of literature on this issue, particularly, from the legal point of view. This paper is primarily legal in nature which purports to share the prospects of RES and problems in exploiting the RES in Bangladesh context. In doing so, it has attempted to identify the key obstacles within the existing legal and regulatory mechanisms and suggest some solutions that are based on international best practices. To this end, this paper is divided into seven parts, including the introduction and conclusion. Part 2 of the paper will discuss the basics of RE and its growing importance, while Part 3 will highlight the prospects of RE in Bangladesh context and Part 4 will share some challenges for RE in Bangladesh context. Part 5 of this paper covers discussion on the legal, regulatory and policy aspects of RE in Bangladesh and Part 6, finally, proposes some suggestions and recommendations.

2. Renewable Energy Basics and its Growing Importance

Energy generated from RES, such as solar, wind, biomass, thermal or hydro, etc., is theoretically considered as renewable as each one could yield an infinite amount of energy [7]. Uses of energy generated via these sources are not new and ancient people successfully utilized these also. For example, the Egyptian Nile civilization used to utilize the wind to move ships, grain-grinding facilities, and boat propellers. Even the Chinese and Japanese started using wind-run water pumping systems in the historical past, leading to cost minimization dramatically [8]. The succeeding civilizations opted to use windmills as alternative sources, since it is accessible and comparatively less expensive [9]. In fact, in recent times, the increasing energy crisis and stringent enforcement of carbon emission laws for reducing the GHG emissions has forced many nations to think about alternative supplies of energy [10]. This segment will highlight some basics about RES, RE, their importance and relevant concerns in exploiting them.

2.1. *Renewable Energy: Definitional Challenges*

Generally, a universally accepted definition of RE and RES are difficult to pinpoint because of divergent understanding of these terms by various stakeholders. As a result, various definitions of RE can be found in both scholarly works and in jurisdiction specific literature. For instance, the definition of RE can be found in Article III of the Statute of the International Renewable Energy Agency (IRENA) [11], the EU Directive 2009/28/EC [12] on the development and progression of the utilization concerning electricity from sustainable resources, the Glossary of Statistical Terms of Organization for Economic Co-operation and Development (OECD) [13], and the International Energy Agency (IEA) in its 2010 Renewable Information Report [14], etc. A cautious look at these definitions will reveal that these have incorporated specific attributes of either RE or the RES, i.e., RE means the electricity produced from solar sources, wind, geothermal, biomass, and hydropower assets, etc.

The notion of the definition is essential not only to get an agreement on a term's meaning and scope, but also to delimit its parameters. Moreover, it is critically significant in matters of jurisprudence and the regulatory frameworks too. The absence of consensus on specific lawful definition of 'RES' creates much confusion and invites more questions. Various definitions of the same term in authoritative literature invites heated debates as the stakeholders may reveal unexpected contradictions, while undertaking RE-related activities [15]. Since the significance of RE has been emerging increasingly as a result of its inherent benefits and prospects, the global community should reach a consensus regarding the relevant definitions for legal and administrative purposes, which is imperative for its ultimate success.

In the domestic level of Bangladesh, according to section 2 of the Sustainable and Renewable Energy Development Authority Act, 2012 (Act No. 48 of 2012) [16], "renewable energy" is defined as the energy and power originated from biomass, biofuel, biogas, hydropower, solar energy, wind power, hydrogen cell, geothermal, tide and wave or energy and power generated from any other sources declared by the government as renewable energy in the official gazette notification. The same section further defines 'sustainable energy' and 'non-renewable energy' too. It is provided that "non-renewable energy" is the energy and power produced from mineral gas, coal, peat coal, natural oil, any other fossil fuel, nuclear power, and any other non-renewable energy resources declared by the government as non-renewable energy in the official notification gazette notification [16].

While defining any term(s) and/or phrase(s), the words either 'means' or 'includes' are used in legal texts. The use of the word 'means' denotes exclusivity, i.e., the definitions are exclusive and cannot be expanded. On the other hand, the word 'includes' is used to mean flexibility to allow to consider similar kinds of things. It is evident that in both the definitions of the words 'renewable energy' and 'non-renewable energy', the word 'means' is used, which indicates that it is not possible to include any things else other than those included in the definition. This legal definition of RE will eventually help to understand the regulatory measures in describing the prospects and challenges of RE of Bangladesh in this paper.

2.2. *Renewable Energy Sources: Importance and Concerns*

RES, due to the utilization of indigenous assets, can possibly give energy with near-zero discharges of both air toxins and GHG emissions [17]. That is why, clean energy, environmentally friendly power energy, sustainable energy, alternative energy, green energy, etc., are used as the synonyms of RE. The importance of this type of energy is documented in many scientific and policy research. Aside from the researchers and specialists, even the legal experts have embraced the significance of RE for the preservation of the global environment and combating climate change [6]. In the case of *Preussen Elektra AG v Schhleswag AG* [2001], *Case C-379/98*, the Court of Justice of the European Union (CJEU) reiterated that the use of sustainable energy resources for generating power is paramount in connection with saving the planet because of its undeniable role in reducing GHGs, which scientists ascribe to be the chief driver of climatic doom [18].

Environmental experts and legal researchers have been campaigning for a shift toward the concentration from finite non-RES to renewables to maintain the development activities, despite the fact that there are some inherent initial concerns [19]. Nevertheless, these concerns are not exceptional in this sector alone; rather, they are generally present whenever any new technological developments are introduced [6]. With the entry of, and advance in every sector, triggered by the scientific advancements, the global community has understood the significance of improvement of the RES in a practical, scalable, and capable ways [20].

In the global context, the first and formal, even though indirect, talk on RE started in the 1970s, in particular, in the United Nations (UN) Stockholm Declaration on the Human Environment 1972 [6]. While sharing on the dangers of depletion of non-renewable materials, the world leaders emphasized on the utilization of RES [21]. Subsequently, the UN and many other global and regional agencies stressed using RE resources, especially, in power generation in an environment-friendly and sustainable manner. Numerous international bodies, forums, non-governmental organizations (NGOs) and instruments, are established and developed to concentrate on promoting RE resources for securing sustainable energy solutions in the context of global warming and climate change.

Above all, the international endeavor proceeded steadily for moving forward through many initiatives. A list of such initiatives include, but not limited to, World Commission on Environment and Development Report, 1987 [22]; Nairobi Program of Action for the Development and Utilization of New and Renewable Sources of Energy, 1981 [23]; Intergovernmental Panel on Climate Change (IPCC), 1988 by United Nations Environment Program (UNEP); World Meteorological Organization (WMO), 1988; UNFCCC (Rio Earth Summit) 1992; the 3rd Conference of the Parties (COP3), 1997; Kyoto Protocol to the UNFCCC; UN Secretary General's declaration, 'Sustainable Energy for All (SE4ALL)' to attain 30% of the global RE target by 2030 [6]; Resolution 65/151, UN General Assembly's declaration on 'International Year of Sustainable Energy for All, (2012)' by 2014–2024 [24]; UN Sustainable Development Goals (SDGs), 2015, especially, the goals No. 7 and 13, and finally, the Paris Agreement to the UNFCCC, 2016.

Since the late 1970s, the worldwide network's dependence on RE has been emerging over 10% every year [25], and starting from 2014, more than 164 nations have embraced the renewables targets [26]. In 2012, the utilization of RES helped to provide up to 13.2% of the worldwide essential energy supply. The same figure rose to 22% of worldwide power usage in 2013. It was estimated that this number is expected to rise to 26% in 2020. To share this in a practical setting, this number is greater than the current overall power demands of Brazil, Russia, India, China and South Africa (BRICS) nations put together [27]. Figure 1 shows the world's total primary consumption from 2007 until 2016, illustrating the rising trend of RE share.

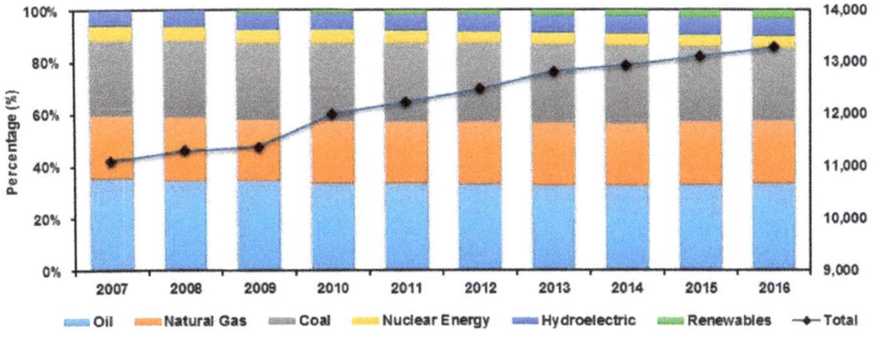

Figure 1. Total primary energy consumption (2007–2016). Adapted from British Petroleum (BP) Statistical Review of World Energy [28–32].

Despite the positive outlook and promises, worldwide demand for RE has been expanding at an underwhelming rate, especially, compared to fossil fuel-based energy demands. About USD 16 billion and USD 17 billion were put for RE resources in 2015 and 2016, respectively [33]. This figure is marginally lower than the normal interest in 2014, which was USD 19 billion. Such a circumstance has urged the private sector entrepreneurs to approach to dedicate resources to creating power utilizing renewables. The dedication of the private sector commitments—including speculation—is additionally obvious as the interest in the RE limit exceeding that of hydrocarbon incumbents demonstrating a steady trend of difference for the fifth year in a row [34].

Experts believe that RE frameworks perform best at little to medium scale and are perfect for rural and geographically disadvantaged regions where it is difficult to cover these territories through traditional fossil-based energy sources [35]. RE frameworks offer an attractive alternative prospect in such regions. Besides, the energy generated utilizing renewables is less vulnerable to the volatility of price and value, which is generally suffered by the oil and gas markets [36]. Thus, buyers can remain certain about the supply of energy and can additionally be profited through—among other things—delivering and exchanging additional energy. On the other hand, a few administrative and policy-oriented difficulties, consumption payments, tax benefits, and other surfeit expenses tend to depress the investors of RE projects compared to investment in fossil fuel-based projects [37].

Asides from the concerns raised above, many years of systemic abuse of the RES have led to a plethora of worries stemming from the lack of adequacy of productivity of the existing power management framework as the energy bills are still quite high. Moreover, private contractors are raking in enormous financial benefits, and the cash spent to help the exercises of various foundations—such as controllers, transport framework administrators, open utilities and universal organizations—further magnify the battle of competing with fossil-based power sources [38]. Disheartening as these worries may be, there are still signs of hope. In recent times, there are signs that the RE projects that have been utilizing the RES judiciously are overcoming these challenges leading to financial improvement [39]. Furthermore, it should additionally be appreciated that the issue of initial outflow relating to the capital expenses is not limited only to the RE, rather any power development project has to incur similar initial high up-front costs.

Renewables are taking a consistently growing pace in the overall energy industry with an evolving number of driving organizations focusing on aggressive inexhaustible power targets [40]. Unfortunately, even such growth cannot decrease the level of GHG emissions that the world desires. Hence, effective, consolidated and synergic strategies and efforts from the stakeholders are required to utilize RES across the board at a quicker pace. Moreover, the significance of an all-inclusive concession to the environment, as advocated by the Paris Agreement, holds guarantees for a feasible eco-accommodating development and progression of the world for the future.

3. Prospects of Renewable Energy in Bangladesh

Fossil-fuel sources, more specifically natural gas, oil and coal, play the lead role in electricity generation in Bangladesh [3]. The electricity demand in the country increased considerably from 0.4 to 1.38 quadrillion between 1997 and 2016 [41] and the country so far could not provide 100% access to electricity, due to *inter alia*, the inadequate volume of energy sources [42]. The total coal assets in Bangladesh were only 1063 million tons, whereas, the natural gas reserve demonstrated to be available with a volume of 9.7 trillion cubic feet (TCF) according to the statistics of 2013 [43]. Besides, to supply fuel in different industries, including energy, Bangladesh imports nearly 1.2 million tons of crude oil and 2.6 million tons of refined petroleum commodities every year [44]. Bangladesh absorbed approximately 2.132 million metric tons of coal, 175.69 kilo barrels of oil per day and 28.37 billion cubic meters of natural gas in 2018 [45].

Energy application in the country has expanded significantly, and the generation altogether relies mainly upon non-RES. Energy production through these ways adds practically 40% of the absolute carbon dioxide (CO_2) discharges by the nation [46]. In 2018, Bangladesh generated energy amounting 7,418 MW, whereas the demand was 11,534 MW. Such a gap between supply and demand results in the impediment of the financial and innovative advancement of the nation. Such a situation has influenced the policymakers to officially undertake activities to utilize the RES to alleviate the energy needs while maintaining the ecological effects.

Bangladesh is blessed to have a generous amount of RES and the effective exploitation of these promises to satisfy the energy need of the country. Among the accessible energy assets, biomass is considered as the significant RES available in the country, which can reduce the utilization of and reliance on non-RES. Solar energy is also very promising as the country gets an immense scale of sunlight, due to its geographical location. It may be pertinent to share that the rural and coastal areas of Bangladesh have been enjoying the benefits of the installed solar photovoltaic (PV) panels. Besides, several government agencies and NGOs undertook some activities for delivering electricity from small scale hydro plants and wind turbines despite the fact that the country is not very suitable to consider hydropower and wind for power generation. Hence, the RE resource, such as biomass and solar can be utilized to ensure energy security.

RES are the assets that are renewed persistently through natural transformations and can be reused because of their inherent properties. The RES incorporate biomass, solar, wind, geothermal and hydropower energy, but exclude conventional nuclear fuels. Figure 2 exhibits a preview of various RES available for electricity or power generation [47].

From the discussion above, it can be revealed that an immense volume of RES is available in Bangladesh. Even though the use of RE technology has become a worldwide trend, the country yet greatly struggling to utilize these sources. From among the list, biogas, biomass, and solar are regarded as the probable, favorable and productive sources for the sustainable energy generation in the country. Table 1 provides a review of RE resources and potentials in Bangladesh [46].

Even with the availability of all these RES, it is a matter of great concern that only 1.0% of the nation's cumulative electricity generation originates from RE resources; however, RE estimates to provide approximately 19% of cumulative global electricity supply [48]. In this given context, the Government of Bangladesh has formulated and adopted some strategies and policies to utilize RES for electricity production up to 10% by 2020 [4]. The following segment will provide a brief survey of the available RES in Bangladesh.

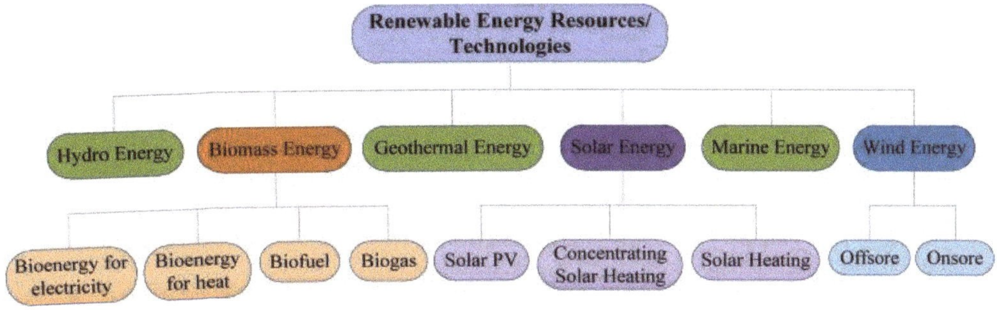

Figure 2. An overview of renewable energy sources (RES) and technologies.

Table 1. Renewable energy (RE) potential in Bangladesh.

Resources	Potential	Entities Involved
Solar	Enormous	Public and private sector
Wind	Resource mapping required	Public sector/PPP
Hydro	Limited potential for micro or mini-hydro (max. 5 MW). Estimated hydro potential approximately 500 MW	Mainly public entities
Domestic biogas system	8.6 million m3 of biogas	Public and private sector
Rice husk-based biomass gasification power plant	300 MW considering 2 kg of husk consumption per kWh	Mainly private sector
Cattle waste-based biogas power plants	350 MW considering 0.752 m^3 of biogas consumption per kWh	Mainly private sector

3.1. Biomass

The natural substance originating from living or dead life forms like the plant, yields, tree and its deposits are considered as biomass. Biomass is contemplated as an outstanding, reliable and sustainable energy source. Biomass absorbs CO_2 for photosynthesis within the sight of sun-based energy to deliver natural mixes required for its development. However, it is well documented that biomass discharges CO_2. Hence, even though biomass is a sustainable energy source, it is not relatively so clean for the environment. Notwithstanding, on the planet, approximately 2.6 billion individuals rely upon biomass for purposes, such as heating or cooking [49].

Being a primarily agricultural country, Bangladesh has an enormous volume of biomass assets that incorporates rice shell, animal waste, crop sediment, timber, municipal waste, jute stock, sugarcane residue and additional complementary springs because of the country's rainforest, the ecological and biological system. In Bangladesh, practically 64% of all-out lands are utilized for agricultural reasons [50]. Along these lines, the nation has a huge extent of agricultural deposits from different harvests, including sugarcane, rice, vegetables, jute, wheat, beets, maize, coconut, cotton groundnut and millet developed all the year round. As a result, practically 70% of individuals, immediately or discursively, depend on biomass energy in Bangladesh [51].

Other than biomass, biogas is created by anaerobic processing that can be utilized for cooking, lighting, and power production and the residue can be utilized for fertilizer, fish feed and compost. Thus, Bangladesh has a tremendous opportunity concerning biogas production from current build-ups and waste assets. It was affirmed that it was possible to utilize 2.91 billion m^3 biogas in Bangladesh in 2012–2013, which was equal to 1.455 billion liters of diesel [52].

Bio-fuel creation is still on the embryonic step in Bangladesh in which ten pyrolysis plants were set up to deliver bio-oil from different biomasses [53]. Be that as it may, the plants are not operating legitimately because of the absence of suitable support and specialized technology [54].

The Renewable Energy Policy of Bangladesh 2008 intends to outfit the possibilities, expansion and utilization of RES. For instance, it is provided for biomass gasification and clean energy advancement, discouraging the use of energy generated from fossil fuel sources. Clean energy from biomass can result in relatively less carbon emission contrasted with the reliance on fossil fuel substances. Therefore, biomass has great potential in Bangladesh for achieving the clean environment goals set by the government.

3.2. Solar

Solar-based electricity is the most popular, infinite and effective energy source which is well-accepted everywhere throughout the world. Concentrating solar power (CSP) and solar PV are very promising advancements, and the solar-based home system can deliver power using solar-oriented radiation. In the case of Bangladesh, there is a tremendous extent of possibilities to use solar-powered radiation because the country is situated in the topographical area [55]. The country gets a normal daily solar irradiation of 4.2–5.5 kWh/m^2 that can create roughly 1,862.5 kWh/m^2 per year (see Figure 3) [56].

Solar oriented PV panels assume a significant job in the worldwide power sector and offer about 0.7% of complete energy production. Solar PV provides approximately 7.8% of annual electricity creation in Italy, 5% in Germany and 6% in Greece [57]. Japan and China are the Asian driving nations creating around 13.6 GW and 20 GW sun-oriented power individually [58]. Similarly, modern technological advancement in solar home-system and solar-based innovations are progressively appealing and compelling in Bangladesh.

Bangladesh has already experienced a few success and fruitful execution of solar-based energy usage [59]. The nation has very nearly 234 MW energy production potential from sunlight-based home frameworks [60]. Bangladesh possesses a capability of 50,174 MW electricity generation from solar PV, as displayed in Table 2 [61]. With about 5 million Solar Home System (SHS), Bangladesh has the world's largest SHS. Be that as it may, the nation is creating and delivering just 3 MW from rooftop top panels to the national grid. Nevertheless, Bangladesh Power Development Board (BPDB) is attempting to additionally introduce joined solar cycle plants and LED road light, in order to reduce the necessary electricity demand. Very recently, the country has successfully inaugurated the largest solar power plant in Teknaf area with a capacity to produce 28 MW, which can feed 20 MW to the local substation. With regards to concentrated solar power (CSP), this is a promising technology for power generation in which the solar radiation is concentrated to generate high temperature for producing steam in a solar thermal power plant [62]. A number of researchers have identified the potential of using CSP in Bangladesh [63–67]. In 2011, the State Minister for Power, Energy and Mineral Resources mentioned that the Asian Development Bank (ADB) was willing to finance a 10 MW to 20 MW capacity CSP plant inside Kaptai Hydro-electric Plant which was expected to be ready by 2016 [68]. In 2014, Reliance Power commissioned a 100 MW grid-connected CSP plant in Rajashtan, the largest CSP that utilized Fresnel technology [69].

Table 2. Solar energy potential in Bangladesh [3].

Technology	Potential Power (MW)
CSP (Concentrated solar power)	100
Grid-connected solar PV	50,174
SHS (Solar home systems)	234

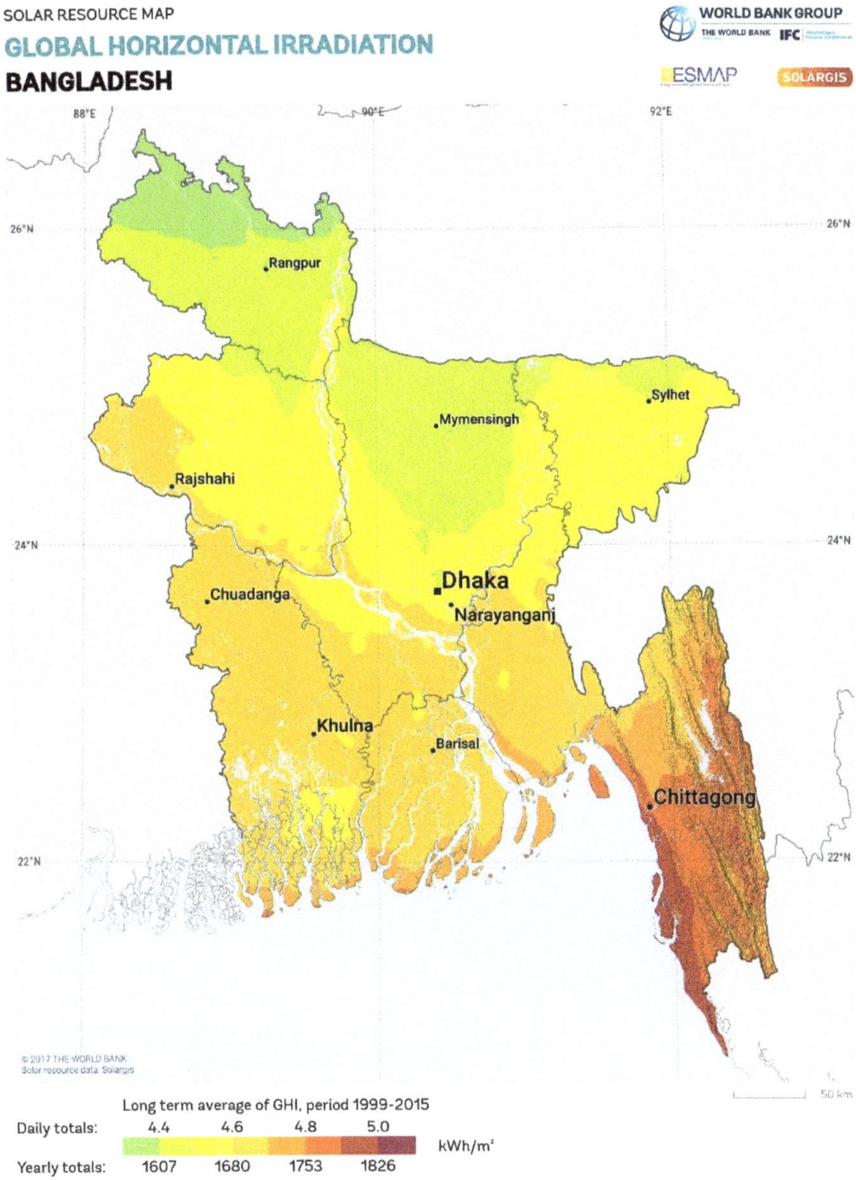

Figure 3. Global horizontal solar irradiation in Bangladesh [56].

3.3. Hydro

Hydro energy implies a type of sustainable power source which utilizes the force of water stream to generate electricity. The massive flow of water transforms its energy into electricity. Bangladesh is blessed with various rivers and waterways; henceforth, the country holds a decent extent of possibilities for hydropower extraction. BPDB and Bangladesh Water Development Board (BWDB) identified some proper micro-hydropower production places where the hydropower installations may provide a satisfying amount of energy in the future. Table 3 provides a summary of such identification of suitable sites.

Table 3. Potential of micro-hydropower sites in Bangladesh identified by Bangladesh Power Development Board (BPDB) and Bangladesh Water Development Board (BWDB) [3].

District	Potential	Entities Involved
Chittagong	Foy's lake	4
	Choto Kumira	15
	Hinguli Chara	12
	Sealock (Chittagong hill tracts)	81
	Lungichara	10
	Budiachara	10
Sylhet	Nikhari Chara	26
	Madhab Chara 1500 ft. from fall	78
	Rangapani Gung	616
Jamalpur	Bhugai-Kongsa at 2 miles U/S. of Nalitabari	69 for 10 months
	Marisi at Dukabad near Jhinaigati	35 for 10 months
Dinajpur	Dahuk at Burabari	24
	Chawai at U/S of Chawai L.L.P	32
	Talam at U/S of Talam L.L.P	24
	Pathraj at Fulbari	32
	Tangon at D/S of Nargun L.L.P	48
	Punarbhaba at Singraban	11
Rangpur	Buri Khora Chikli at Nizbari	32
	Fulkumar at Raiganj Bazar	48

Sustainable Rural Energy additionally ventures and investigates more conceivable destinations for miniaturized scale hydropower plants in Bangladesh. According to their assessment, Chittagong possesses huge potentials with an expected electricity generation of 135 kW from its rivers and lakes [70].

In 1962, BPDP introduced the first hydro plant in Bangladesh which consists of two units of limit 40 MW, and each one was installed in the Karnafuli stream. Three additional units of limit 50 MW were introduced constantly in 1988. The Kaptai power plant is known as the hydroelectric power plant that utilizes a vertical hub Kaplan turbine [71]. Bangladesh owns huge hydropower potential in Karnafuli waterway, Matamuhuri basin, Sangu stream and in Brahmaputra river [72].

3.4. Wind

Wind energy is known for its eco-accommodating nature and considered as the best sustainable power source that outfits for future energy solutions. The dynamic electricity is created from the moving air as the kinetic energy of wind provides the turbine shaft. The geographical situation of Bangladesh makes the country suitable for little scale wind turbines [73]. BPDB introduced four units of the primary wind power plant at the Muhuri dam zone of Sonagazi in Feni district with a generation capacity of 0.90 MW. Likewise, BPDB introduced another 1 MW wind power plant at Kutubdia Island in the year 2008 which comprises 50 wind turbines of 20 kW limit each [74].

In addition, various government agencies and non-government organizations have introduced some initiatives to establish more wind plants in the country [75]. The country has also consented to an arrangement for a joint project with US-DK Green Energy (BD) Ltd. Which proposes the assistance of the USA and Denmark to introduce the nation's biggest ever wind control plant of limit 60 MW at Cox's Bazar [3].

3.5. Other Renewable Resources

Bangladesh can consider utilizing the prospects of other RES, e.g., tidal electricity, oceanic wave and geothermal energy for power generation in the future in addition to biomass, wind, solar and hydro energy [3]. Ocean waves and tides can be considered as a few tornados hit in the Bay of Bengal consistently. Nevertheless, the prospects of these resources are still under scrutiny, and pragmatic endeavors need to be taken in this regard. All these options are yet to be utilized, due to the absence of appropriate information, innovation, and instruments [76].

4. Renewable Energy in Bangladesh: Issues and Challenges

Bangladesh has a high potential to move towards a better and more sustainable country with greener energy without compromising the country's economic efficiency, human dignity, standards of living, and financial development. Nevertheless, considerable development activities, including a sustainable power source and green arrangements in Bangladesh have had limited accomplishments and are confronting various challenges which are mostly financial, technical, and regulatory in nature [77]. It may be pertinent to share here that these challenges are not unique in Bangladesh's context, rather these present in most jurisdictions.

Since RE projects are more complex, unpredictable and full of unforeseen risks and dangers, the investors in this field may face serious financial challenges having impacts on future development and commercialization of the projects and technologies [78]. For the investors, it is difficult to convince the financial institutions and other financers to get the necessary support. Since they use new technologies that are uncertain, this may cause the payback period to exceed their expectations [79]. Hence, it acts as an impediment to innovative projects. This vulnerability results in high financing expenses for research, improvement and preparation in this sector. Thus, this deceptively raises the cost of clean energy generation, postponing their full assimilation into the energy market.

Sometimes, different RE related projects require huge initial investments. This is again very challenging for the entrepreneurs, since there are already available standard, but cheaper alternatives in the market [80,81]. Besides, there are geographical factors that have effects on the performance of RE projects. Hence, it can be concluded that an already successful RE project in one country probably will not be realistic to another nation for the presence of sunlight-based variables, biomass, wind and sea between nations in the tropical and other areas.

Usually, RE project entrepreneurs are small companies with limited resources [82]. The eagerness for their undertakings depends on the capacity to support the improvement at an insignificant expense contrasted with the gigantic Independent Power Producers (IPP) [83]. In any case, the output of RE projects run by the small companies is not necessarily the same when compared to financially feasible and promising projects, run by the IPPs. Thus, the financial strength of the company is important. All these above-mentioned challenges are crucial for Bangladesh to develop and promote RE in the future. Without addressing these financial barriers, Bangladesh cannot achieve its national goals relating to RE. Some other relevant challenges are discussed as follows.

4.1. Employment of Advanced RE Technologies

Technological constraints plague the RE sector worldwide, and this phenomenon is not limited to Bangladesh alone [84]. Hence, these constraints are hindering the growth of innovation within the sustainable energy sector.

To begin with, the unreliable power supply with respect to Bangladesh's geographical variables makes it difficult and vulnerability for RE's sustainable advancement [85]. Likewise, uncertain technological innovation ensures the development of the RE industry with negative intensity, whereas, the orthodox methods of energy generation still offers a financially feasible choice to the energy stakeholders [86]. Furthermore, there is restricted expertise when it comes to productive practices and hardware aspects of RE [87]. The deficiency of qualified labour and skilled workers with adequate engineering or technical training additionally blocks the pace of the innovative improvement in the sustainable energy industry in Bangladesh. In particular, the inefficiency of human resources with engineering and technical skills in handling the equipment hinders security issues on the RE supply. To prepare the industry for these challenges, it is necessary to train and equip the workforce with necessary facilities though such facilities add additional costs to the RE projects. Moreover, due to the technical issues, there has been ample evidence of time-delays in generating sufficient energy through renewable means, and in such cases, the financers lose the interest to invest in RE projects.

The insecurity of available renewable source supplies for long-term and the price instability is making the situations more vulnerable for the biomass energy projects in Bangladesh [88]. Besides, although the government policy promised to introduce tax incentives in the development of RE projects, however, other than the feed-in-tariff (FiT) system provided by the Bangladesh Energy Regulatory Commission Act, 2003 (Act No. 13 of 2003), no other financial relieve system or tax incentive measures have been formulated so far [83]. Such a position also indirectly discourages entrepreneurs in investing in this sector.

4.2. Policy Related Institutional Barriers

Although the government's intention to promote RE through institutional, legal and regulatory frameworks is appreciable, the policies relating to RE are not firm and comprehensive [89]. The financing bodies in RE projects are more concerned with the heavy return of profits, while the Government of Bangladesh is worried about the allocation of subsidies to bear to achieve the ultimate objective of fuel diversification policy [90]. Hence, the government has to deal with different and conflicting interests with the potential industry players [3]. Such disparities in the RE industry create negative impacts on the investment of RE projects.

Additionally, financial investors find themselves in a continuous dilemma because of the unreliable fuel supply in the industry. Moreover, the stakeholders face several investment barriers, due to lack of practice and governance-related discrepancies, because the regulation of the energy sector in Bangladesh, like in most jurisdictions, is both fragmented and inconsistent [91]. Regulatory expertise is lacking too at times, due to the dearth of technical know-how. In practice, the arrangements relating to RE development and strategy execution in Bangladesh are viewed as isolated exercises as there are more than one authority that deal with this issue and no single authority is made responsible for monitoring the implementation of the Policy.

In general, the sustainable and RE policies are formulated at Ministerial and Parliamentary political process, and, then, such policy is conveyed and executed by the local governments and other agencies which are equipped with specialized technical tools, administrative arrangements, and regulatory structures for proper implementation. However, this top-down methodology seems non-functional in Bangladesh.

The participants and stakeholders in the energy business in Bangladesh additionally seem, by all accounts, to be less organized. There is no proper forum that listens to their perceptions. Even though they are heard occasionally by way of a public hearing before taking any policy initiative by the regulators, their inputs are less counted in the policy-making process, and the whole process is turned as a routine exercise without any significant impact. There is no other better alternative to develop without addressing the issues of active engagements of the potential stakeholders of the RE industry. Additionally, inputs from experts, producers, consumers, and engineers should also be counted to upgrade the ability of RES.

4.3. Legal and Regulatory Concerns

With respect to the legal and regulatory barriers, genuine and significant initiatives are missing toward the advancement and extension of RE in Bangladesh. Without addressing the regulatory concerns, the policy related to fuel diversification can never be achieved. The Bangladesh Energy Regulatory Commission Act, 2003 (Act No. 13 of 2003) and subsequent amendments directly or indirectly support to implement a FiT scheme to develop the RE industry in the country. However, such laws have limited enforcement and impact in the practical field as a FiT scheme functions in a project-to-project basis. Hence, there is no positive output on such regulations. Additionally, the government needs to consider several other incentives in the RE industry, such as higher selling tariffs and tax reductions or tax relief, etc. Associated administrative and regulatory bodies should reallocate the subsidies from conventional energy generation to RES to sponsor the existing efforts of the RE industry. Such initiatives may progressively reduce the burden of different stakeholders and play a significant role in the growth of the sustainable energy industry.

In the case of regulatory and administrative difficulties to cope with the complex issues of RE, two remedies may be considered. It is important to note that although the Government of Bangladesh initiated several policies to support RE, such policies are not found successful and effective, due to the fact, *inter alia*, that Bangladeshi enterprises still do not consider the RE industry as a suitable place for investment [92]. Hence, the first task of the regulatory bodies is to review and evaluate the policies comprehensively and make necessary practical amendments to refocus the country's target of achieving energy fuel mix to offer significance to the sustainable energy source.

The regulators need to consider distributing subsidies for RE utilization. The subsidies for traditional fossil fuel source ought to be periodically eradicated, as well as converted and reallocated to RE assets to develop the sustainable energy industry establishments [93].

Furthermore, the second task of the regulatory bodies is to address the issues relating to the institutional framework relating to the industry. The absence of an efficient working institutional framework on RE must be overwhelmed by empowering joint exertion between government organizations and private establishments with the ultimate objective to explore the financial and technical viability of RE generation. Enhancing the institutional network between government leaders, enterprises and utilities promise to implement carefully designed RE policies. In addition, a portion of the policies and strategic activities need reviews or explicit clarifications on how the existing or newly incorporated legal and regulatory system and standards would approach and direct the execution of such adopted policies.

4.4. Lack of Awareness

There are likewise various social difficulties relating to RES and green innovations. There is a lamentable, but noticeable absence of public awareness and participation with respect to feasible advancements in the energy industry in Bangladesh. The public participation programs, particularly in the rural areas relating to RE development within the country, is not satisfactory [50]. Public participation and awareness about the energy sector promise to deliver the required advancement and development of the commercialization process of the energy industry. Hence, the viability of public awareness programs needs to be increased and boosted with the goal to raise public support for the advancement of RE, which promises to prompt effective usage of sustainable power policies and green approaches ultimately [94]. Relevant government organizations can also come forward to help and guide the potential beneficiaries of RE to motivate them to engage and disseminate positive information relating to the industry.

5. Renewable Energy in Bangladesh: Legal, Regulatory and Policy Aspects

It has already been shared that energy production, management, and supply to the people have become a serious concern for the Government of Bangladesh. Therefore, the government has formulated various policies and adopted regulatory measures [42]. In this context, the contribution of RES in the energy mix can be an exceptional accomplishment that would influence the economic development of the nation significantly.

In 2008, the National Renewable Energy Policy was introduced to promote RE by encouraging distinctive private and public platforms concerning the investment in the RE industry in place of the fossil fuel-based power sources. Since then, the uses, development, preparation, and research on the domestic RE industry have been accelerating; though for exploiting the commercial benefits of these, it will still require a comprehensive strategy.

In recent times, the domestic RE generation and the government's participation in expanding electricity generation through RE increased approximately 560 MW [95]. Bangladesh attempted to increase the electricity supply up to 16,000 MW by 2016 and subsequently, set the target to obtain up to 39,000 MW by 2030 according to the adopted the Power System Master Plan 2010 (PSMP) [96]. The electricity generation target as encapsulated in the PSMP is presented in the following Table 4.

Table 4. Power System Master Plan 2010 [97].

Year	MW
2016	16,000
2021	24,000
2030	40,000

In order to meet these targets, the Government of Bangladesh initiated multiple projects, including the utilization of coal as a commanding energy source for electricity generation and supply in the future [98]. Accordingly, two large coal energy plants with a target to produce 1320 MW are under construction at Khulna and Chittagong. Moreover, the government has undertaken major policy decisions to set-up nuclear energy plants to ensure reliable and environment-friendly electricity for the future [99]. The atomic power plant is anticipated to add another 2000 MW by 2020, and the government wishes to increase to 5000 MW by 2030. In implementing those wishes, the government initiated the process of installation of nuclear power plants in the country. Accordingly, the long-cherished nuclear power plant, with the technological support of Russia, is under construction at Rooppur in the Pabna district [8].

Despite the adoption of so many energy policies, it seems that the government may not be successful in achieving the desired objectives unless it exploits the RES to produce energy. The Government of Bangladesh, hence, concentrated on electricity production up to 500 MW by RE within 2015; but failed to succeed.

Fortunately, the government recognized the reality, revised the existing policies and strategies and attempted to formulate pragmatic strategies for the sector. Consequently, the policymakers of Bangladesh aim to achieve its objectives by producing 10% of its total electricity demand by 2020 from RES [46]. Now Bangladesh is producing around 560 MW of power from renewables which contributes 2.95% of the cumulative energy generation of the country. Specialists figure that achieving 10% of the aggregate supply with around 2000 MW of electricity by the next two years will be a challenge for the country. In accomplishing the goals, the government has initiated numerous endeavors particularly, to achieve the energy-specific objectives.

The government attempts to fostering the cooperative approaches within private and public sectors in reaching that milestone. Accordingly, different government organizations, such as BPDB, Bangladesh Rural Electrification Board, Bangladesh Council of Scientific and Industrial Research and Local Government Engineering Department, started executing numerous RE related activities. Infrastructure Development Company Limited, on the other hand, is a private NGO, is trying to develop the market competitiveness by commercializing RE in rural areas of Bangladesh.

Securing access to electricity for all is one of the prime visions of Bangladesh since its independence, which has been evidenced by incorporating the State's promise in the Fundamental Principles of the State Policy (FPSP). Rural development and agricultural revolution are important FPSP as enshrined in article 16 of the Constitution of the People's Republic of Bangladesh, 1972. In this article, Bangladesh has given words of undertaking adequate measures to bring radical change in rural areas by introducing a revolution in agriculture, ensuring rural electrification, developing cottage and other industries, and improving the education system. In addition to these, it is further provided that the State shall endeavor to develop the communication and public health systems with a view to removing the disparity of living standards between rural and urban settings. The provision of affordable and continuous energy is instrumental to achieve this.

Nonetheless, since independence to date, Bangladesh is suffering in ensuring 100% access to electricity for all. In addition, the findings of various scientific studies have revealed that domestic fossil fuel sources are either depleting or will naturally be depleting in the course of time. Therefore, the promotion of RE has been taken as an alternative source in the Renewable Energy Policy, 2008 for the long-term energy sustainability, pollution control and electricity generation. Because fuel diversification contributes to sustainable electricity production [100], and consumption within the socio-cultural and economic parameters, Bangladesh is in great need to promote energy production through renewable means.

By assessing the international environmental obligations, the country made in the international forum, Bangladesh should design the national policies and regulatory frameworks for RE. From various initiatives taken by the Government of Bangladesh, it is evident that the government truly values sustainable development [101]. For example, the Government of Bangladesh has signed and ratified most of the international environmental law instruments and have been submitting the required state reports regularly. Additionally, the government has enacted national laws and taken policy initiatives to give effect to the provisions of these international instruments.

In order to understand the energy policies of the country, it will be relevant to share here energy-related legal and policy instruments. In the undivided India when present Bangladesh was part of it, the Electricity Act, 1910 (Act No. IX of 1910) was enacted containing provisions, *inter alia*, on grant of licenses for the supply, transmission and use of electricity. That law provides for punishments for a number of offences such as dishonest abstraction of energy, installation of artificial means, maliciously wasting energy or injuring works, theft of line materials, tower members, equipment, etc., from any electric supply system, dishonestly receiving stolen property, unauthorized supply of energy by non-licensees, illegal or defective supply or for non-compliance with order, illegal transmission or use of energy, etc. After more than a century, the government repealed the law and enacted the Electricity Act, 2018 (Act No. VII of 2018) containing similar but updated provisions on these matters to meet the continuously increasing demands of electricity. It will be pertinent to mention here that the law does not contain provisions on the sources of energy, i.e., energy generated from fossil-fuel sources or RES, rather the main focus of the law is to ensure the supply, transmission and use of electricity.

After the independence of the country, the government enacted the Rural Electrification Board Ordinance, 1977 (Ordinance No. LI of 1977) to establish the Rural Electrification Board that will primarily be responsible to take measures for effective use of electrical power for development of the rural economy of the country. This Ordinance of 1977 was also repealed, and the government enacted the Rural Electrification Board Act, 2013 (Act No. 57 of 2013). Same as with the Electricity Act, 2018 (Act No. VII of 2018), the provisions of the law are mainly concerned about the supply, use and transmission of electricity in the rural area and the sources used to produce electricity is not the primary concern. The law provides to establish a body known as Bangladesh Rural Electrification Board (BREB) to this end. Most importantly, this Board has been playing a pivotal role in the promotion and utilization of RES in the rural area of the country as it has taken various initiatives [42], including the introduction of SHS for the first time in the country in 1993 through the project 'Diffusion of Renewable Energy technologies' with the financial assistance of France. In the agro-based economy, BREB is now working to install 2000 solar irrigation pumps.

Moreover, a list of such instruments include: Policy Guidelines for Small Power Plants in the Private Sector, 1996, Private Sector Power Generation Policy, 1996, the Bangladesh Energy Regulatory Commission Act, 2003, Import Duty Exemptions for Solar and Wind of Bangladesh (Statutory

Regulatory Order), 2004, Renewable Energy Policy, 2008, the Sustainable and Renewable Energy Development Authority Act, 2012, Scaling Up Renewable Energy Program for Bangladesh, 2015(SREP Bangladesh), Bangladesh Energy Regulatory Commission (Tariff for Roof Top Solar PV Electricity) Regulations, 2016 (Draft), etc. Some of these are evaluated and discussed below-

5.1. *Policy Guidelines for Small Power Plants in the Private Sector (1996)*

In fulfilling the promises encompassing energy supply as enshrined in the Constitution, the Government of Bangladesh has taken numerous policies on power production, supply and consumption. Even though Bangladesh has to depend on large-scale nationalized efforts to produce sufficient electricity for the national grid, it has also been encouraging privatized endeavors. In line with the aims, the Government has formulated a Policy Guidelines for Small Power Plants in the Private Sector, 1996 allowing private investors to install Small Power Plants (SPP) at the earliest possible time for generating electricity on a commercial basis. Accordingly, private parties could generate power for their own and sell the residue, if any, to anybody else. Initially, the plant size was fixed for producing up to 10 MW and allowing, even more, based on necessity, demand and loading capacity [102].

To encourage private entrepreneurs in power production, the Policy suggests that Petrobangla, a government-owned national oil company of Bangladesh, may supply natural gas to those power plants on a commercial basis with a usual rate if the SPP is not located too far from the gas supply reticulation (section 3). To ease the business, the Policy also provides that the sponsors can use the extant transmission and distribution systems if there is no problem in terms of capacity though the owner of SPP or the sponsors will have to pay a mutually set wheeling charge for using transmission/distribution facilities (section 4). Moreover, it is also provided that the government would not interfere in pricing; rather, it would be fixed on negotiation between the sponsor and the consumers (section 5).

The provisions of 'fiscal and other incentives' for national and foreign investors are included under the heading of "Captive Independent Power Producer' and 'Captive Power Generation' policies in sections 5 and 6, respectively. Moreover, the Policy provisions clarified that the government has no obligation of purchasing the power produced by SPP. Nevertheless, the government may, if it considers essential, purchase the power from the SPP (section 9). Globally, community-based energy production and supply systems are gaining momentum. Many scholars suggest that Bangladesh can consider even the household level system of energy production and distribution in resolving its energy crisis though which it may add sufficient power into the national grid [103].

5.2. *Private Sector Power Generation Policy (1996)*

The Government of Bangladesh formulated the Private Sector Power Generation Policy (PSPGP) in 1996 with a view to ensuring the participation of the private bodies in the power generation leading to promoting the economic growth of the country. The PSPGP set several specific objectives, such as ensuring the access to electricity for all; increasing the annual per capita generation of power; reaching and sustaining minimum 6-7% annual GDP growth; achieving desired socio-economic progress by alleviating poverty; securing adequate electricity supply at an affordable cost by expanding the production of electricity as a whole.

Starting from 11666 GWh, the Policy set a target of generating 16500 GWh power within 2000 and 24160 GWh by 2005, which figured out that an average of 300 MW of power has to be generated more annually. It was estimated that a total of US$ 6.6 billion would be required for expanding, reinforcing, transmitting and distributing electricity (section 1.3).

Other provisions of the said Policy include: Formation of a power cell for facilitating the promotion, improvement, execution, commissioning and operations of private energy production schemes (section 2.0); modality for implementation of private power projects (section 3.0); financing arrangements (section 3.3); security package (section 3.4); tariff for bulk purchase of power at busbar (section 4.0); fiscal incentives (section 5.0); other facilities and incentives for foreigners (section 6.0), and right of interpretation (section 8.0) [104]. Taking the advantages of the Policy, the first private power plant was installed in October 1998, having a capacity of producing 110 MW power to add in the national grid [105].

In Bangladesh, there was no engagement of the private bodies in the energy generation and distribution sectors in the early 1990s. Since 1996, the Government of Bangladesh undertook several reforms plans under the National Energy Policy to offer competition, bring foreign investment with a view to increasing the energy generation and distribution. One of the key Policies was the Policy Guidelines for SPP in Private Sector, 1996 discussed above [106]. During the period, some private energy generation schemes were established as the IPPs selling power to the BPDB. However, due to the friendly investment policy of the government, now private enterprises dominate the energy sector contributing a total of 54.35% in the national power generation [107].

5.3. The Bangladesh Energy Regulatory Commission Act, 2003

With a view to achieving some broad goals, especially, making provisions for the establishment of a self-governing and fair energy regulatory commission, the Parliament of Bangladesh enacted the Bangladesh Energy Regulatory Commission Act, 2003 (Act No.13 of 2003). One of the prime objectives of the Act was to establish the national regulator Bangladesh Energy Regulatory Commission (BERC) for regulating the electricity, gas, and petroleum commodities in Bangladesh. The vision of the BERC was to foster an amicable environment to establish an efficient, properly organized and sustainable energy sector in Bangladesh for ensuring energy at a fair and sensible price, and to secure the interest and satisfaction of the customers by fair practice [108].

Major provisions of the Bangladesh Energy Regulatory Commission Act, 2003 are as follows: The establishment of the BERC as a legal person (section 4); functions of the BERC (section 22); emergency power of the government to control energy use (section 25); settlement of disputes (section 26); issuance of license by the BERC (section 28); renewal, revision and cancellation of license (section 30); tariff (section 34); emergency provision (in favor of the BERC) (section 36); restrictions on publishing information (section 39;) arbitration-settlement by BERC (section 40); appeal against the decision of inspector (section 41); penalty (section 42); penalty for stealing energy (section 44); penalty for obstruction of the construction during the installation or repair of electric line or gas pipeline, etc. (section 45); jurisdiction of trial court (section 50); collection of fee, fine and charges (section 56); power to make rules (section 58) and regulations (section 59), and provisions for issuance of license during transitional period (section 66).

The duties and responsibilities of the Commission include, but not limited to: Creating a conducive atmosphere for the private investors in power generation activities; ensuring transport, transmission, and promoting the petroleum products; maintaining transparency in every sphere of the sector, and finally, to protect the rights of the consumers by creating the competitive market [109]. However, the BERC is conducting widespread activities in line with its missions and visions as a major regulatory body since its inception in 2004.

5.4. Bangladesh (Statutory Regulatory Order), 2004

The government framed the Import Duty Exemptions for Solar and Wind of Bangladesh (Statutory Regulatory Order), 2004 [110] targeting certain resources, such as solar, SPV, solar thermal, solar heat, wind in the electricity, heating and cooling sector. It is applicable for both small- and large-scale ventures of RE project. Through the provisions of this Order, the National Board of Revenue and the regulatory code 155, an exemption in import duties for certain RE products, in particular, the solar cells, modules and lanterns are offered. Additionally, according to this instrument, the rate of duties imposed for other RE resources is comparatively inexpensive too. Moreover, these duty exemptions are not limited to the national companies only, rather extended to foreign companies also. As a result, the duly registered foreign companies also enjoy the same facilities as are enjoyed by the locally owned companies [110].

5.5. The Renewable Energy Policy of Bangladesh, 2008

The Power Division, Ministry of Power, Energy and Mineral Resources, Bangladesh formulated the Renewable Energy Policy on 18 December 2008 [4]. The Policy identified the following three reasons of global crisis encompassing the energy, e.g., the gradual decline of fossil fuel, and the consequent price fluctuation because of the gap between the demand and supply; the necessity of reducing global emissions (up to 80% by 2050) in responding to the climate change, and the continuous demand for energy security.

The objectives of the Policy are to put to use the potentials of RE resources and technologies everywhere; encourage and facilitate both public and private sector investors in RE sectors; scaling up the power generation; conducting training facilitating the use of RE in each unit of energy usage; providing encouragement, legal and environment assistance in the use of RE, etc. In particular, the ultimate goal was to produce 5% of the total electricity demand utilizing RES by 2015 and to increase the portion up to 10% by 2020 [4], which means that 2000 MW has to be generated from RES. Some of the main provisions of the Policy include: Institutional arrangements (section 3); provisions regarding the resource, technology and program development (section 4); investment and fiscal incentives (section 5); regulatory policy (section 6), and the right to interpretation (section 7), etc.

It may appear that the production costs of RE are comparatively higher than the fossil fuels; however, they can be economically viable if all other ancillaries (eco-friendliness, beneficial to health, low operating costs) are taken into consideration. In recent years, Bangladesh is witnessing commendable progress in the RE sector because of the implementation of several policies, particularly, the Renewable Energy Policy of Bangladesh, 2008. Recent statistics show that the country is generating a total of 404 MW of electricity from the RES (see Table 5). Meanwhile, the solar home system has appeared as a success story and getting popularity gradually, principally in the off-grid areas [111].

Table 5. Contribution of the RES in Bangladesh [5].

Method	MW
Installation of Solar Home System (3.5 million)	150
Installation of Rooftop PV at Government/Semi Government offices	3
Installation of PVs on commercial buildings and shopping centers	1
Installation of PVs by the consumer during new electricity connections	11
Installation of Wind-based power plants	2
Installation of Biomass-based power plants	1
Installation of Biogas-based power plants	5
Solar Irrigation	1
Hydro Electric power generation	230
Total	**404**

5.6. The Sustainable and Renewable Energy Development Authority Act, 2012

To secure the energy security, the government established the Sustainable and Renewable Energy Development Authority (SREDA) though the enactment of the law—the Sustainable and Renewable Energy Development Authority Act, 2012 (Act No. 48 of 2012). While enacting the law, the policymakers of Bangladesh realized the need to control global warming, prevent the misuse of energy, reduce the hazard of natural disasters and gradually reduce the dependence of fossil fuel in power generation by promoting the use of RE.

According to this Act, energy refers to power generated using both the renewable and non-RE sources [section 2(5)]. The law bestows legal personality to the authority, i.e., SREDA [section 2(2)]. Some other relevant provisions of the law include: Responsibilities and functions of the authority (section 6); the power of the authority to impose fees (section 7); the constitution of the board of directors (section 9); fund, budget, and accounts and audit (sections 19–21); power to make rules and regulations (sections 26–27), and abolition of Energy Audit Cell, etc. (section 28).

In spite of having diverse plans, so far, Bangladesh achieved a limited success encircling the RE sectors because of numerous reasons, and principally, the lack of project execution experience; lack of strong institutional capability; financial challenges; land constraint; project development challenges, etc. Keeping all these in context, the Government of Bangladesh established the SREDA in 2014 to promote the RE sector and achieve energy efficiency [112]. Since its inception, the SREDA is actively operating multiple action plans with a view to ensuring electricity for all within 2020 by increasing the RE share 1.5% to an increase of 50% in grid supply [113].

5.7. *Bangladesh Energy Regulatory Commission (Tariff for Roof Top Solar PV Electricity) Regulations, 2016 (Draft)*

The government has recently drafted the Bangladesh Energy Regulatory Commission (Tariff for Roof Top Solar PV Electricity) Regulations, 2016 [114]. Once approved, the provisions of the Regulations shall apply to all new RE power plants operated for generating and promoting the sale of electricity produced in such RE power plants. It is also provided that in the Draft that in the current RE power plants, all terms and conditions, tariffs, etc. shall be administered by the prevailing notifications, of course, with the prior approval of the BERC. All terms and conditions required for getting a license are governed as per the provisions of the Bangladesh Energy Regulatory Commission Act, 2003, Renewable Energy Policy, 2008, and the subsequent amendments thereof. In addition to these, no license is required for a plant installed for producing up to 5 MW, and in such a case, a waiver certificate shall have to be taken subject to the fulfillment of the conditions determined by the BERC [114].

BERC remains as authority to govern specific tariffs concerning RE production. The BERC has been working to conceive similar laws and regulations pertaining to the specific FiT system, especially for solar and wind energy ventures. The government in their several policies and laws has mentioned that a broad range of RE projects may serve the country with prosperity that can additionally address four key intentions, i.e., Efficient Energy Access, Energy Security, Industrial Advancement and Environmental Protection.

The draft Regulations imply to undertake FiTs as the primary common essential measure to promote RE in Bangladesh. Such a system is assumed to provide an outstanding result relating to the innovative investments in the RE industry. To improve the RE industry, and enhance the grid-connected electricity production, it is essential to have an efficient administration and the adoption of recent innovative technologies. Such a contribution of RE can deliver advantages concerning every stakeholder in the energy industry of Bangladesh.

The Draft Regulations of 2016 is a substantial foundation to initiate supplementary innovations for reliable and affordable energy solutions. The draft Regulations proposed for a universal FiT concerning the special model of RE technology managed by BERC on a specific basis that is presumed to initiate assistance for new investors. Such a universal approach is much effective than the existing project-by-project strategy. This approach concerning universal FiT for diverse RE exercises has produced concrete outcomes in several nations. Such a method is also capable of reducing excessive profits of private energy enterprises. Nevertheless, the regulations have been in the draft stage since 2016, and have not been enforced yet.

In achieving the desired goals, the Government of Bangladesh has planned to generate electricity by the following means and approaches:

- Development of domestic primary fuels;
- Energy efficiency improvement;
- Private and joint venture participation;
- Coal as a main source of energy;

- Use of alternative energy;
- Use of nuclear energy;
- Cross-border power trade;
- Fuel diversification;
- Construction of effective and efficient infrastructure;
- Low carbon emission;
- Construction of effective and efficient infrastructure.

Net energy metering (NEM), also known as net metering, which allows the prosumers to use the produced electricity anytime of personal need, is gaining popularity all over the world. Such an initiative encourages the prosumers to produce more energy using RES. It is a matter of great hope that the Government of Bangladesh has actively been considering to introduce the NEM system in the country and has released the draft of the Guidelines with the aim to support the government to develop a net metering policy for individual energy generation. It is anticipated that such an initiative, when successfully be implemented, will encourage the consumers to use their self-produced electricity reducing the dependency on the grid power. Thus, such an initiative offers various benefits - the consumers will need to spend less on their electricity bill, emission of GHGs will be reduced, government needs to spend less on the import of fossil-fuel sources for energy production and the so save national money can be used for other socio-economic development activities.

Recently, the BERC has formulated the Electricity Grid Code, 2018 to govern the boundary between the licensee and users, and to establish the procedures for operations of facilities that will use the transmission system. This Grid Code specifies criteria, guidelines, basic rules, procedures, responsibilities, standards and obligations for the operation, maintenance and development of the electricity transmission system. It is anticipated that this Code will help to ensure a transparent, non-discriminatory and economic access and use of the grid, whilst maintaining a safe, reliable and efficient operation which will facilitate to provide a quality and secure electricity supply as reasonably as practicable. Most importantly, this Code contains provisions on RE.

Finally, it can be submitted that the policies, such as policies on fuel diversification, energy efficiency improvement and low carbon emission, can never be achieved without promoting RE in the country. Hence, the government is firmly committed to bringing every possible solution to increase the contribution of the RE industry in the national energy mix and thus, has taken various initiatives. While the initiatives taken by the government can be applauded, there are still avenues to improve the situation and therefore, we have advanced some suggestions and recommendations in the following segment.

6. Suggestions and Recommendations

6.1. Government Leadership

Achieving green and RE goals require the presence and coordination of synergic move by the stakeholders under the auspices of the appropriate regulatory watch. The Government authority is fundamental in planning proper RE approaches. Bangladesh has organized different activities to advance a sustainable power source [115]. However, there is correspondingly a need to give arrangements and support to outstanding business people, enterprises and the business networks in supporting the practicable power drive. This requires a wide understanding and contribution at different dimensions.

Bangladesh still battles to keep its international promises to enhance the usage of renewable resources to promote a green environment. The heavily anticipated Renewable Energy Policy leads Bangladesh's change into a green country and latches onto the prospects of practical improvement. Nevertheless, although the policies attempted to develop the RE industry in Bangladesh and achieve its sustainable goals, there remain many failures on the part of the government agencies in exploiting RES to generate energy.

The awareness regarding environmental sustainability and the need for creative alternative sources for the energy sector are a global phenomenon. Whatever the case, it is imperative for the policymakers of Bangladesh to categorically execute the propelled tasks and plans successfully. It has been seen from historical encounters that there remains an immense mismatch in accomplishing the goals of power designs. Along these lines, the Government of Bangladesh should take the lead position with the end goal to advance RE and execute the undertakings that create RE inside its regional area for the benefit of its citizens.

As the economy of Bangladesh is shifting toward industrialization and the shifting into industry-based economy produces increased earnings to the citizens and stimulated urbanization. Fast accelerating commercial ventures, including increasing income levels, produced an immense need for transportation settings, particularly in immediately developing metropolitan cities. Thus, the country is also increasing its contribution to polluting the environment within the region.

While the economy is fast developing, a difficult assignment remains for the Government of Bangladesh, i.e., to fulfill its expanding need for electricity. It is a fact that the costs of fossil fuel sources far and wide are unpredictable. The lack of adequate natural gas reserves in the country intensified the situation further. In total, these two forces increase Bangladesh's electricity demands. At present, Bangladesh supplies her electricity prerequisites overwhelmingly through gaseous petrol, additionally supported by a bunch of coal and oil plants. Consequently, because of the shortage of gas, coal and oil assets, sustainable power source surfaces as a precise vital alternative for Bangladesh's future advancement plan. Nonetheless, a productive, sustainable power source program involves a broad framework. Therefore, the Government of Bangladesh should take a leadership position in order to promote RE and implement the projects that help develop RE within its territory.

Other than the traditional approaches, the Government should also focus on the modern technical and regulatory approaches to overcome the challenges that they have been facing. Building codes, energy rating schemes, supportive grid connection arrangements, etc. can also be very useful tools in promoting the RE industry. Building codes are the significant mechanism which provides indirect incentive to promote RES. For instance, Australia has been using energy rating schemes that provide criteria for calculating the credit system for PV and solar water heaters to empower new buildings or substantial renovations with the capacity to generate electricity [116]. Such set up in commercial buildings generally implement solar PV to achieve higher energy ratings, which then helps them to achieve higher rentals. It cannot be ignored that for such construction with RE tools, the owners had to make the huge initial investment; however, such appearance helps the construction firms with tax incentives from the government and add to high resale value, especially in the countries where the requirement of energy rating on building sales is mandatory. Hence, building codes and rating schemes can help to promote a green environment and reduce electricity needs with limited public investment. Additionally, regulating little categorized generators like residential PV linking with grid connection can enhance the energy efficiency of the country. Such specific regulation can be a significant step to reduce the skepticism and danger encountered by private sector investors and also reduces management expenses of the energy industry.

6.2. *Ensuring Renewable Energy Development through Technological Means*

To achieve public procurement on RE and to enhance the output of technology, it is important to encourage research, especially when the government intended for the localization of foreign imported technologies. Advancement of confined clarifications and explanations on innovative technologies or methods can frame civic perspectives, including public participation. Additionally, RE establishments in Bangladesh are designed by regional corporations, including external specialists. Information and technology directions are prerequisites to encourage the potential raising in the RE industry and to support the progressive growth for long-term goals even though such technology is imported by foreign ventures and experts.

The idea of RE and energy efficiency could be brought into the curricular exercises in conventional educational and vocational institutions, such as schools and colleges. One of the major challenges relating to RE is the scarcity of sufficient human resource with relevant skill, expertise and technical knowledge in regards to, *inter alia*, handling of equipment [115]. This is further exacerbated by inefficient energy practices, the reluctance of administrative and regulatory bodies, and lack of culture embracing innovations.

Hence, instruction developed and programs preparation on sustainable energy at the tertiary training level promise to facilitate the advancement of RE in the country. Such advancements and developments seem pre-requisites for achieving the sustainable goals of Bangladesh.

6.3. Minimization of Renewable Power Generation Cost and Promoting Market Competitiveness

The costs of generating and transmitting power in the modern age despite all the technological advancements are staggering [117]. The matter is manifold complicated for the RE sector as the industry yet does not benefit from economies of scale. When RE is compared with existing fuel-based energy sources, the economic benefit is prospectively and considerably diminutive [118]. Hence, the major challenge relating to the RE is the difficulties in anchoring the investment in the industry.

In any case, the bankers and financial investors seem to have a lack of confidence to make required speculation about the RE industry. Additionally, the long process of granting bank loans for RE demotivates the participants to engage in business [119]. In this manner, the Government of Bangladesh needs to build up a worthwhile, practical, and controllable subsidized legal, regulatory or political instrument for RE projects.

Because of the absence of such instruments, both the financial specialists and industry players consider it monetarily not lucrative to invest in the RES activities and advancements. Consequently, lack of coordination and non-responsive attitude on the part of the administration decreases the chances of prospective foreign or local investments on RE projects.

Other project motivators, for example, tax incentives, research funding or any other technological support from the part of government can bring down the expense of RE innovation [120]. European nations have embraced such sponsorships and motivating forces as a major aspect of their advanced techniques to build such a complex industry like RE [121]. Financial help, dedicated lessons and embracing the RE experts to create a comprehensive RE study syllabus for universities, colleges and even in schools might catalyze the discourse about advancement, exhibition and systematic awareness-raising campaigns related to the sustainable energy source and green innovations.

In Bangladesh, until now, RE significant ventures have been actualized with the project-to-project base assessment. Specific separate investors invested, and the government fixed the tariff-based on power purchase agreements (PPAs). The countries which have long experiences of RE production, such as Germany, India, China or Spain, have never followed project-by-project contracts. Instead, a market approach was established to drive the considerations and regulation of energy generation stipulations. With the establishments of FiT regime, private sectors of the said countries plan for their investments, based on the offerings of RE premiums (the extra amount in addition to the market price for electricity) or RE PPAs. Furthermore, the outcome of certain applications has been enormous, covering the preceding ten years concerning the penetration of the RE industry as reducing the expenses associated with it.

6.4. Creating a Comprehensive Policy, Legal and Regulatory System

The regulatory framework plays a significant role in the advancement and development of RE technologies within a country's existing energy formation. The regulatory direction is necessary to enhance the competitiveness of the energy market. Such a regulatory framework is also vital to ensure that the stakeholders comply with the established rules and regulations for energy trading. Any deficiency in the regulatory structure may cause a threat to the market liberalization process.

Likewise, the regulatory framework ought to increase public awareness about RE technologies and incentive mechanisms to promote investment in the RE industry.

World Bank's 'Turn Down the Heat', and The Intergovernmental Panel on Climate Change's 'Fifth Assessment Report' were two reports which uncover large haul suggestions for Bangladesh and its kin from likely cataclysmic effects of environmental change. The two reports draw a terrible situation to face in the future if the environmental changes are not addressed.

The Government of Bangladesh has been allocating funds in the research targeting the control of environmental degradation. Regardless of these activities, as all examinations call attention to, Bangladesh will remain helpless, and its kin will confront extreme monetary difficulties from environmental consequences in the future. Accordingly, Bangladesh should approach all countries to target a low-carbon economy dependent on sustainable power source supply. Such a request can only be considered positive, given that Bangladesh itself change approaching towards environment-friendly ventures.

The Government of Bangladesh has schemed different innovative approaches, for example, the Renewable Energy Policy 2008. Be that as it may, there must have a subsequent execution of the components set up by the Policy. Other than that, there must have severe courses of events for each strategy for achieving administrative success.

Financial experts urged that a fruitful sustainable power source framework needs long haul security to recuperate the costs of investors and to support their revenues. The draft Regulation released by BERC in 2016, includes remarks and commitments from various partners, which appears coordinating and immensely conducive in preparing an appropriate structure. Nevertheless, the most critical part of any policy for achieving success is its unwavering quality and dependability. Ventures need a considerate long-haul connection between proprietors of these offices in one hand, and the circulation or transmission matrix between administrators is necessary to deal with the price of electricity produced, on the other. This dependability must be demonstrated in the policies also and ought not to be influenced by political insecurity or regime change. Policies, in general, should incorporate every statutory and legal term relating to the grid technology, including energy reliability, security and essential expansions.

Grid access, including a flexible licensing system, for RE is vital concerning advancement in the treatment of solar PV and wind power. Additionally, complementary measures for the RE generation, such as land acquisition, preparation of human resources, tax exemptions or tariff system, etc. should be comfortable to procure. The draft Regulation of 2016 containing provisions on FiT is promising as the provision contains no fees for licensing in the installations of RE plants with the capacity up to 1 MW. Notwithstanding, a special provision ought to additionally render the guidance on the usage of lands, tenancy contracts, operational standards, availability of equipment and the technological expectancy.

Furthermore, to improve the monetary appeal of RE, there should be no retroactive reduction of tariffs or modification of practices for existing plants under the agreement. To identify every single installation correctly, a registry should be created, preferably by SREDA containing the publicly available information on the type of equipment, year of installation/grid connection, the expiry date of FiT contract, etc. Such registry should be declared "protective" against retroactive changes by law or regulation. Concerning the vulnerable situation in Bangladesh for the energy industry, advanced policy and comprehensive regulatory framework are essential to obtain all possible solutions.

6.5. *Increasing Judicious Public Relation Work to Spur Public Awareness*

The government's decision to tackle climate change crises and developing energy in a sustainable fashion is receiving more traction. From nearly two decades onwards, the Government of Bangladesh has adopted various policies and action plans to promote implementing greener and RE as an alternative source. In spite of continual strides made within this field, developing true renewable sources is not yet able to fulfill its utmost potential. Such a sluggish response is holding back the achievement of broader energy goals [122].

The tepid acceptance of renewable alternatives by the general public is another important factor for diffusion and solidification of RE development [8]. This is why a certain necessity exists for raising the general public's awareness regarding the pros and cons of using fossil fuel generated energies and the benefits of using sustainable and RE. For this to happen, appropriate campaigns and programs should be designed with the help of modern information and technological communication tools. Furthermore, education and material propagation tied to RE should be made as part of the curriculum and education course design at all levels of schooling. Without a proper introduction to the costs and benefits of RE resources and technologies, the public cannot be expected to be conscious about achieving the nation's broader goals.

This process can be helped through government-subsidized programs and initiatives. Besides, once the public comes to know the pivotal role played by RE, research and development programs will most likely be receiving greater funding leading to the development of more efficient and cheaper technology. In this regard, the Government of Bangladesh may want to prioritize budget allocations for the research and development of technologies in engineering and relevant environmental science disciplines [123].

7. Conclusions

The planet-wide impacts of environmental change and the nature of global warming urge to adopt 'hard' International law and relevant arrangements so as to advance electricity generation through a sustainable manner. Notwithstanding, there is no urgent or explicit authoritative decision that binds the countries legally to advance the local utilization of RES to generate electricity [6], though the provisions of the recently adopted Paris Agreement to UNFCCC encourages the use of RE. The scarcity of universal legal arrangements on RE does not, in any case, undermine the significance of its utilization to limit environmental degradation. Henceforth, a binding international instrument, alongside the positive activities and mutual cooperation of the stakeholders worldwide, regional and municipal players in terms of financial and technological information exchange, may assume an instrumental job in the advancement of RE [6].

Since independence, Bangladesh remained extremely reliant on fossil fuel sources to generate electricity, and these sources are depleting naturally. Continuous reduction of these fossil fuels sources, increasing expenses of such sources, energy security and international attention on GHG emissions drive Bangladesh, like many other countries, to consider sustainable policies for the electricity industry. The Government of Bangladesh is constitutionally obliged to eliminate specific inequality in living standards within the metropolitan and rural areas. Moreover, the government is constitutionally bound to achieve a specific end of development by ensuring rural electrification [124]. Therefore, the government has taken various initiatives, including the formulation of the Renewable Energy Policy 2008, aiming at promoting RE. Besides, the government has adopted policies to establish the SREDA with the prime purpose of obtaining sustainable advancement and improvement of the RE industry. Besides, Bangladesh adopted other policies, *inter alia*, to advance the industry by providing subsidies and tax incentives to draw investors into the industry. The investment relating to RE are commonly expensive in comparison with fossil fuel-based power plants, though RE plants can be profitable if one considers every external opportunity that RE offers, such as environmental protection, sustainability and inclusive social development, etc.

Achieving the goals of sustainable and greener development of energy infrastructure in Bangladesh is constrained by several issues in terms of the development and frameworks. A practical breakthrough is necessary to surmount the high upfront costs borne by developing a fresh market to let the RE market growth. This will have the effect of scaling up the renewable sector and drives down the cost of technology. In order for Bangladesh to make headway in building a low carbon society, legal instruments on RE are in dire need of institutional patronage to ensure level playing fields for all key players, the stakeholders, and the general public. Since it is undeniable that energy is the key antecedent for wide-scale development in the global arena, ensuring this development is sustainable and counters inequality and remains inclusive for all segments of the population is equally relevant [125]. The government agencies are not the only ones who should make efforts. The private sector should also be more corporate, socially responsible and make compromises by accepting longer payback periods in RE projects.

Globally, energy is imperative to pursue any type of development, and for attaining sustainable and all-inclusive development [6]. In addition, creating a friendly green climate for the future remains a challenge for every country, not particularly, for Bangladesh [126]. To mitigate climate degradation and to achieve the SDG goals, Bangladesh should lead its energy sector into renewable sources. Providing an equitable opportunity to its citizens and to promote inclusive development for the society, choosing RE is one of the most strategic options that Bangladesh has in its hand now.

Author Contributions: Conceptualization, M.E.K. and R.K.; Data curation, M.E.K. and R.K.; Formal analysis, M.E.K., R.K., M.T.I. and F.M.-S.; Funding acquisition, F.M.-S., N.A.B. and M.N.M.; Investigation, M.E.K., R.K., M.T.I. and F.M.-S.; Methodology, M.E.K. and R.K.; Project administration, M.E.K. and F.M.-S.; Resources, M.E.K.; Software, F.M.-S.; Supervision, M.E.K.; Validation, M.E.K.; Visualization, F.M.-S.; Writing—original draft, M.E.K., R.K., M.T.I. and F.M.-S.; Writing—review and editing, M.E.K., R.K., M.T.I., F.M.-S., N.A.B. and M.N.M.

Funding: This research was partly funded by Universiti Teknologi Malaysia (UTM) and Ministry of Education, grant number Q.K 130000.3556.06G43 and R.K 130000.7756.4J380.

Conflicts of Interest: The authors declare no conflict of interest.

References

1. Barua, A.; Narain, V.; Vij, S. *Climate Change Governance and Adaptation: Case Studies from South Asia*; CRC Press: Boca Raton, FL, USA, 2019; ISBN 9781138054509.
2. Hasan, M.; Tanvir, A.A.; Siddiquee, S.M.S.; Zubair, A. Efficient Hybrid Renewable Energy System for Industrial Sector with On-Grid Time Management. In Proceedings of the IEEE 3rd International Conference on Green Energy and Technology (ICGET), Dhaka, Bangladesh, 11–12 September 2015; pp. 1–6.
3. Halder, P.K.; Paul, N.; Joardder, M.U.H.; Sarker, M. Energy scarcity and potential of renewable energy in Bangladesh. *Renew. Sustain. Energy Rev.* **2015**, *51*, 1636–1649. [CrossRef]
4. Ministry of Power, Energy and Mineral Resources (MPEMR). *Renewable Energy Policy of Bangladesh*; MPEMR: Dhaka, India, 2008.
5. Islam Sharif, S.; Anisur Rahman Anik, M.; Al-Amin, M.; Abu Bakr Siddique, M. The prospect of renewable energy resources in Bangladesh: A study to achieve the national power demand. *Energy Power* **2018**, *8*, 1–6. [CrossRef]
6. Karim, M.E.; Munir, A.B.; Karim, M.A.; Muhammad-Sukki, F.; Abu-Bakar, S.H.; Sellami, N.; Bani, N.A.; Hassan, M.Z. Energy revolution for our common future: An evaluation of the emerging international renewable energy law. *Energies* **2018**, *11*, 1769. [CrossRef]
7. Demirbas, A. Potential applications of renewable energy sources, biomass combustion problems in boiler power systems and combustion related environmental issues. *Prog. Energy Combust. Sci.* **2005**, *31*, 171–192. [CrossRef]
8. Karim, R.; Karim, M.E.; Muhammad-Sukki, F.; Abu-Bakar, S.H.; Bani, N.A.; Munir, A.B.; Kabir, A.I.; Ardila-Rey, J.A.; Mas'ud, A.A. Nuclear energy development in Bangladesh: A study of opportunities and challenges. *Energies* **2018**, *11*, 1672. [CrossRef]
9. Sørensen, B. A history of renewable energy technology. *Energy Policy* **1991**, *19*, 8–12. [CrossRef]
10. Fräss-Ehrfeld, C. *Renewable Energy Sources: A Chance to Combat Climate Change*; Wolters Kluwer Law & Business: The Hague, Netherlands, 2009; ISBN 9041128700.
11. International Renewable Energy Agency (IRENA). *Statute of the International Renewable Energy Agency (IRENA)*; IRENA: Masdar City, UAE, 2011; pp. 1–18.
12. The European Parliament and The Council of The European Union. *European Directive 2009/28/EC*; European Parliament: Brussels, Belgium, 2009; pp. 1–47.
13. Organisation for Economic Co-operation and Development (OECD). The OECD Glossary of Statistical Terms. Available online: https://stats.oecd.org/glossary/ (accessed on 30 August 2019).
14. International Energy Agency (IEA). *Renewables Information 2010*; IEA: Paris, France, 2010.

15. Chen, Y. Promotion of Renewable Energy Globally: Based on Johannesburg Follow-up. *TemaNord NV—2004531* **2004**, 1–72. [CrossRef]
16. Government of the People's Republic of Bangladesh. *The Sustainable and Renewable Energy Development Authority Act 2012*; Ministry of Law, Justice and Parliamentary Affairs: Dhaka, Bangladesh, 2012.
17. Mahlia, T.M.I.; Saidur, R.; Memon, L.A.; Zulkifli, N.W.M.; Masjuki, H.H. A review on fuel economy standard for motor vehicles with the implementation possibilities in Malaysia. *Renew. Sustain. Energy Rev.* **2010**, *14*, 3092–3099. [CrossRef]
18. Court of Justice of the European Union. *PreussenElektra AG v Schhleswag AG [2001]. Case C-379/98*; CJEU: Kirchberg, Luxembourg, 2001; Case C-379.
19. Roseland, M. Sustainable community development: Integrating environmental, economic, and social objectives. *Prog. Plann.* **2000**, *54*, 73–132. [CrossRef]
20. Omer, A.M. Energy, environment and sustainable development. *Renew. Sustain. Energy Rev.* **2008**, *12*, 2265–2300. [CrossRef]
21. United Nations (UN). United Nations Conference on the Human Environment—A/RES/2994(XXVII). In *Proceedings of the United Nations General Assembly—27th Session*; United Nations: Stockholm, Sweden, 1972.
22. Borowy, I. *Defining Sustainable Development for Our Common Future: A History of the World Commission on Environment and Development (Brundtland Commission)*; Routledge: Abingdon, UK, 2013; ISBN 9780415825504.
23. United Nations (UN). United Nations Conference on New and Renewable Sources of Energy—A/RES/36/193. In *Proceedings of the United Nations General Assembly—36th Session*; United Nations: Stockholm, Sweden, 1981.
24. United Nations (UN). United Nations General Assembly Declares 2014–2024 Decade of Sustainable Energy for All—GA/11333: EN/274. In *Proceedings of the Press Releases: United Nations General Assembly—67th Session*; United Nations: Stockholm, Sweden, 2012.
25. Brundtland Commission. *El desarrollo sostenible, una guía sobre nuestro futuro común: El informe de la Comisión Mundial sobre el Medio Ambiente y el Desarrollo*; Oxford Paperback Reference; Oxford University Press: Oxford, UK, 1987; ISBN 9780192820808.
26. International Renewable Energy Agency (IRENA). *Renewable Capacity Statistics 2018*; IRENA: Abu Dhabi, UAE, 2018.
27. International Energy Agency (IEA). *Renewable Energy: Medium-Term Market Report 2015*; IEA: Paris, France, 2015.
28. British Petroleum (BP).
29. British Petroleum (BP). *BP Statistical Review of World Energy*; BP: London, UK, 2011.
30. British Petroleum (BP). *BP Statistical Review of World Energy*; BP: London, UK, 2013.
31. British Petroleum (BP). *BP Statistical Review of World Energy*; BP: London, UK, 2015.
32. British Petroleum (BP). *BP Statistical Review of World Energy*; BP: London, UK, 2017.
33. International Renewable Energy Agency (IRENA). *Renewable Energy Highlights*; IRENA: Abu Dhabi, AE, 2017.
34. Frankfurt School-UNEP Centre. *Global Trends in Renewable Energy Investment 2017*; Frankfurt School-UNEP Centre: Frankfurt, Germany, 2017.
35. Kaundinya, D.P.; Balachandra, P.; Ravindranath, N.H. Grid-connected versus stand-alone energy systems for decentralized power—A review of literature. *Renew. Sustain. Energy Rev.* **2009**, *13*, 2041–2050. [CrossRef]
36. Boyle, G. *Renewable Energy*, 3rd ed.; Oxford University Press: Oxford, UK, 2004.
37. Klass, A.B.; Wilson, E.J. Interstate transmission challenges for renewable energy: A federalism mismatch. *Vanderbilt Law Rev.* **2012**, *65*, 1801–1873.
38. Michalena, E.; Hills, J.M. Introduction: Renewable Energy Governance: Is it Blocking the Technically Feasible? In *Renewable Energy Governance: Complexities and Challenges*; Michalena, E., Hills, J.M., Eds.; Springer: London, UK, 2013; pp. 3–8.
39. Wüstenhagen, R.; Wolsink, M.; Bürer, M.J. Social acceptance of renewable energy innovation: An introduction to the concept. *Energy Policy* **2007**, *35*, 2683–2691. [CrossRef]
40. Fried, L.; Shukla, S.; Sawyer, S. Growth Trends and the Future of Wind Energy. In *Wind Energy Engineering*; Tretcher, T.M.L., Ed.; Academic Press: London, UK, 2017; pp. 559–586. ISBN 9780128094518.
41. KNOEMA. Bangladesh Primary Energy Consumption, 1973-2018Hannah2. Available online: https://knoema.com/atlas/Bangladesh/Primary-energy-consumption (accessed on 31 August 2019).
42. Mollik, S.; Rashid, M.M.; Hasanuzzaman, M.; Karim, M.E.; Hosenuzzaman, M. Prospects, progress, policies, and effects of rural electrification in Bangladesh. *Renew. Sustain. Energy Rev.* **2016**, *65*, 553–567. [CrossRef]

43. British Petroleum (BP). *BP Statistical Review of World Energy*; BP: London, UK, 2014.
44. IEEJ. *Energy Scenario of of Bangladesh*; IEEJ: Tokyo, Japan, 2012.
45. British Petroleum (BP). *BP Statistical Review of World Energy Report*; BP: London, UK, 2019.
46. Das, A.; Halder, A.; Mazumder, R.; Saini, V.K.; Parikh, J.; Parikh, K.S. Bangladesh power supply scenarios on renewables and electricity import. *Energy* **2018**, *155*, 651–667. [CrossRef]
47. Brown, A.; Müller, S.; Dobrotková, Z. *Renewable Energy: Markets and Prospects by Technology*; IEA: Paris, France, 2011.
48. Renewable Energy Policy Network for the 21st Century (REN21). *Renewables 2014 Global Status Report*; REN21: Paris, France, 2014.
49. Berhanu, M.; Jabasingh, S.A.; Kifile, Z. Expanding sustenance in Ethiopia based on renewable energy resources—A comprehensive review. *Renew. Sustain. Energy Rev.* **2017**, *75*, 1035–1045. [CrossRef]
50. Rahman, M.A.; Møller, H.B.; Alam, M.M. Assessing the energy potential of agricultural residues and an approach to meet the rural energy demand: The Bangladesh perspective. *Biomass Convers. Biorefinery* **2018**, *8*, 925–934. [CrossRef]
51. Baul, T.K.; Datta, D.; Alam, A. A comparative study on household level energy consumption and related emissions from renewable (biomass) and non-renewable energy sources in Bangladesh. *Energy Policy* **2018**, *114*, 598–608. [CrossRef]
52. Halder, P.K.; Paul, N.; Beg, M.R.A. Assessment of biomass energy resources and related technologies practice in Bangladesh. *Renew. Sustain. Energy Rev.* **2014**, *39*, 444–460. [CrossRef]
53. Halder, P.K.; Paul, N.; Joardder, M.U.H.; Khan, M.Z.H.; Sarker, M. Feasibility analysis of implementing anaerobic digestion as a potential energy source in Bangladesh. *Renew. Sustain. Energy Rev.* **2016**, *65*, 124–134. [CrossRef]
54. Nikolakakis, T.; Chattopadhyay, D.; Bazilian, M. A review of renewable investment and power system operational issues in Bangladesh. *Renew. Sustain. Energy Rev.* **2017**, *68*, 650–658. [CrossRef]
55. Rahman, M.M.; Islam, A.K.M.S.; Salehin, S.; Al-Matin, M.A. Development of a model for techno-economic assessment of a stand-alone off-grid solar photovoltaic system in Bangladesh. *Int. J. Renew. Energy Res.* **2016**, *6*, 140–149.
56. Solargis Solar Resource Maps of Bangladesh. Available online: https://solargis.com/maps-and-gis-data/download/bangladesh (accessed on 9 October 2019).
57. Kannan, N.; Vakeesan, D. Solar energy for future world: A review. *Renew. Sustain. Energy Rev.* **2016**, *62*, 1092–1105. [CrossRef]
58. Yu, H.J.J.; Popiolek, N.; Geoffron, P. Solar photovoltaic energy policy and globalization: A multiperspective approach with case studies of Germany, Japan, and China. *Prog. Photovolt. Res. Appl.* **2016**, *24*, 458–476. [CrossRef]
59. Chowdhury, P.; Jenkins, A.; Islam, Z.S.; Jenkins, A.; Islam, Z.S. Feasibility of solar-biomass hybrid cold storage for unelectrified rural areas of Bangladesh. In *The Environmental Sustainable Development Goals in Bangladesh*; Selim, S.A., Saha, S.K., Sultana, R., Roberts, C., Eds.; Routledge: London, UK, 2018; pp. 45–55.
60. Khan, I. Power generation expansion plan and sustainability in a developing country: A multi-criteria decision analysis. *J. Clean. Prod.* **2019**, *220*, 707–720. [CrossRef]
61. Faraz, T. Benefits of Concentrating Solar Power over Solar Photovoltaic for Power Generation in Bangladesh. In Proceedings of the IEEE 2nd International Conference on the Developments in Renewable Energy Technology (ICDRET 2012), Dhaka, Bangladesh, 5–7 January 2012; pp. 1–5.
62. Muhammad-Sukki, F.; Ramirez-Iniguez, R.; Abu-Bakar, S.H.; McMeekin, S.G.; Stewart, B.G. An evaluation of the installation of solar photovoltaic in residential houses in Malaysia: Past, present, and future. *Energy Policy* **2011**, *39*, 7975–7987. [CrossRef]
63. Roni, M.M.; Hoque, I.U.; Ahmed, T. Comparative Study of Levelized Cost of Electricity (LCOE) for Concentrating Solar Power (CSP) and Photovoltaic (PV) Plant in the Southeastern Region of Bangladesh. In Proceedings of the IEEE International Conference on Electrical, Computer and Communication Engineering (ECCE), Cox's Bazar, Bangladesh, 7–9 February 2019; pp. 1–6.
64. Noor, N.; Muneer, S. Concentrating Solar Power (CSP) and its prospect in Bangladesh. In Proceedings of the IEEE 1st International Conference on the Developments in Renewable Energy Technology (ICDRET), Dhaka, Bangladesh, 17–19 December 2009; pp. 1–5.

65. Offer, G.; Meah, N.; Coke, A. *Enabling a Transition to Low Carbon Economies in Developing Countries. Case Study: Bangladesh*; Institute of Electrical and Electronics Engineers (IEEE): London, UK, 2011.
66. Shiraishi, K.; Shirley, R.; Kammen, D.M.; Huq, S.; Rahman, F. *Identifying High Priority Clean Energy Investment Opportunities for Bangladesh*; ICCCAD: Dhaka, Bangladesh, 2018.
67. Lipu, M.S.H.; Jamal, T. Techno-economic Analysis of Solar Concentrating Power (CSP) in Bangladesh. *Int. J. Adv. Renew. Energy Res.* **2013**, *2*, 750–762.
68. EnergyBangla Bangladesh—Concentrated Solar Energy Plant on Cards by 2016—HELIOSCSP. Available online: http://helioscsp.com/bangladesh-concentrated-solar-energy-plant-on-cards-by-2016/ (accessed on 9 October 2019).
69. Buckley, T.; Nicholas, S.; Ahmed, S.J. *Bangladesh Electricity Transition: Diverse, Secure and Deflationary Way Forward*; Institute for Energy Economics and Financial Analysis: Cleveland, OH, USA, 2016.
70. Uddin, M.N.; Rahman, M.A.; Mofijur, M.; Taweekun, J.; Techato, K.; Rasul, M.G. Renewable energy in Bangladesh: Status and prospects. *Energy Procedia* **2019**, *160*, 655–661. [CrossRef]
71. Islam, A.K.M.S.; Islam, M.; Rahman, T. Effective renewable energy activities in Bangladesh. *Renew. Energy* **2006**, *31*, 677–688. [CrossRef]
72. Mondal, M.A.H.; Denich, M. Assessment of renewable energy resources potential for electricity generation in Bangladesh. *Renew. Sustain. Energy Rev.* **2010**, *14*, 2401–2413. [CrossRef]
73. Nandi, S.K.; Ghosh, H.R. Prospect of wind–PV-battery hybrid power system as an alternative to grid extension in Bangladesh. *Energy* **2010**, *35*, 3040–3047. [CrossRef]
74. Ullah, H.; Hoque, T.; Hasib, M. Current status of renewable energy sector in Bangladesh and a proposed grid connected hybrid renewable energy system. *Int. J. Adv. Renew. Energy Res.* **2012**, *1*, 618–627.
75. Baten, M.Z.; Amin, E.M.; Sharin, A.; Islam, R.; Chowdhury, S.A. Renewable energy scenario of Bangladesh: Physical perspective. In Proceedings of the IEEE 1st International Conference on the Developments in Renewable Energy Technology (ICDRET), Dhaka, Bangladesh, 17–19 December 2009; pp. 1–5.
76. Guha, D.K.; Henkel, H.; Imam, B. Geothermal Potential in Bangladesh-Results from Investigations of Abandoned Deep Wells. In Proceedings of the World Geothermal Congress, Bali, Indonesia, 25–30 April 2010; pp. 25–29.
77. Tarik-ul-Islam, M.; Ferdousi, S. Renewable Energy Development—Challenges for Bangladesh. *Energy Environ.* **2007**, *18*, 421–430. [CrossRef]
78. Quek, A.; Ee, A.; Ng, A.; Wah, T.Y. Challenges in Environmental Sustainability of renewable energy options in Singapore. *Energy Policy* **2018**, *122*, 388–394. [CrossRef]
79. Shukla, A.K.; Sudhakar, K.; Baredar, P. Renewable energy resources in South Asian countries: Challenges, policy and recommendations. *Resour. Technol.* **2017**, *3*, 342–346. [CrossRef]
80. Williams, N.J.; Jaramillo, P.; Taneja, J.; Ustun, T.S. Enabling private sector investment in microgrid-based rural electrification in developing countries: A review. *Renew. Sustain. Energy Rev.* **2015**, *52*, 1268–1281. [CrossRef]
81. Blazquez, J.; Fuentes-Bracamontes, R.; Bollino, C.A.; Nezamuddin, N. The renewable energy policy Paradox. *Renew. Sustain. Energy Rev.* **2018**, *82*, 1–5. [CrossRef]
82. Sovacool, B.K.; Drupady, I.M. *Energy Access, Poverty, and Development: The Governance of Small-Scale Renewable Energy in Developing Asia*, 1st ed.; Ashgate Publishing: Abingdon, UK, 2012; ISBN 9781409441137.
83. Naznin, N.S.; Trisha, F.R.; Haque, M. Exploring the perspective of entrepreneurship in the energy sector: Case study on renewable energy sector of Bangladesh. *Glob. J. Manag. Bus. Res. C Financ.* **2018**, *18*, 1–9.
84. Shafiullah, G.M.; Arif, M.T.; Oo, A.M.T. Mitigation strategies to minimize potential technical challenges of renewable energy integration. *Sustain. Energy Technol. Assess.* **2018**, *25*, 24–42. [CrossRef]
85. Amin, S.; Islam, S.; Kamal, T.; Mithila, N. Prospects and Constraints of Renewable Energy Sector in Bangladesh: An Analytical Exercise. *World J. Soc. Sci.* **2016**, *6*, 1–12.
86. Islam, N.; Bloemink, J. Bangladesh's Energy Crisis: A Summary of Challenges and Smart Grid-Based Solutions. In Proceedings of the IEEE 2nd International Conference on Smart Grid and Smart Cities (ICSGSC), Kuala Lumpur, Malaysia, 12–14 August 2018; pp. 111–116.
87. Amin, S.B.; Rahman, S. The Impact of Skill Development in Bangladesh Energy Sector. In *Energy Resources in Bangladesh*; Amin, S.B., Rahman, S., Eds.; Springer International Publishing: Cham, Switzerland, 2019; pp. 43–47.
88. Eleftheriadis, I.M.; Anagnostopoulou, E.G. Identifying barriers in the diffusion of renewable energy sources. *Energy Policy* **2015**, *80*, 153–164. [CrossRef]

89. Yaqoot, M.; Diwan, P.; Kandpal, T.C. Review of barriers to the dissemination of decentralized renewable energy systems. *Renew. Sustain. Energy Rev.* **2016**, *58*, 477–490. [CrossRef]
90. Mitra, S.; Sarkar, M.H.; Majumder, A.K. A review of potential renewable energy preference in rural area of Bangladesh. *J. Energy Nat. Resour.* **2017**, *6*, 64–68. [CrossRef]
91. Khandker, L.; Amin, S.B.; Khandker, L.L.; Khan, F. Renewable energy consumption and foreign direct investment: Reports from Bangladesh. *J. Account. Financ. Econ.* **2018**, *8*, 72–87.
92. Alam, M.J.; Ahmed, M.; Begum, I.A. Nexus between non-renewable energy demand and economic growth in Bangladesh: Application of Maximum Entropy Bootstrap approach. *Renew. Sustain. Energy Rev.* **2017**, *72*, 399–406. [CrossRef]
93. Islam, S.; Khan, M.Z.R. A Review of Energy Sector of Bangladesh. *Energy Procedia* **2017**, *110*, 611–618. [CrossRef]
94. Alam Hossain Mondal, M.; Kamp, L.M.; Pachova, N.I. Drivers, barriers, and strategies for implementation of renewable energy technologies in rural areas in Bangladesh—An innovation system analysis. *Energy Policy* **2010**, *38*, 4626–4634. [CrossRef]
95. Nabi, M.S. Can Bangladesh meet its 10% renewable energy target by 2020? *Dhaka Tribune*, 12 January 2019.
96. Wares, S.M.; Hasan, M.; Islam, M.; Saleh, S.T. *Effect of Climate Change on Electricity Demand and Power Generation of Bangladesh*; BRAC University: Dhaka, Bangladesh, 2018.
97. Japan International Cooperation Agency (JICA) & Tokyo Electric Power Company (TEPCO). *Power System Master Plan (PSMP) 2010*; JICA: Dhaka, Bangladesh, 2011.
98. Zaman, R.; Brudermann, T.; Kumar, S.; Islam, N. A multi-criteria analysis of coal-based power generation in Bangladesh. *Energy Policy* **2018**, *116*, 182–192. [CrossRef]
99. Karim, R.; Karim, M.E.; Munir, A.B.; Newaz, M.S. Social, Economic and Political Implications of Nuclear Power Plant in Bangladesh. In Proceedings of the Social Sciences Postgraduate International Seminar, Penang, Malayasia, 29 November 2017; pp. 11–17.
100. Dincer, I. Renewable energy and sustainable development: A crucial review. *Renew. Sustain. Energy Rev.* **2000**, *4*, 157–175. [CrossRef]
101. Rahman, M.M.; Paatero, J.V.; Lahdelma, R.A.; Wahid, M. Multicriteria-based decision aiding technique for assessing energy policy elements-demonstration to a case in Bangladesh. *Appl. Energy* **2016**, *164*, 237–244. [CrossRef]
102. Ministry of Power, Energy and Mineral Resources (MPEMR). *Policy Guidelines for Small Power Plants in Private Sector*; MPEMR: Dhaka, Bangladesh, 1996.
103. Nexant. *Subsidizing Rural Electrification in South Asia: An Introductory Guide*; Nexant: New Delhi, India, 2004.
104. Ministry of Power, Energy and Mineral Resources (MPEMR). *Private Sector Power Generation Policy of Bangladesh*; MPEMR: Dhaka, Bangladesh, 1996.
105. Ministry of Finance (MOF). *Power and Energy Sector Road Map: An Update*; MOF: Dhaka, Bangladesh, 2011.
106. United News of Bangladesh (UNB). Private sector dominates power generation with 54.35% contribution. *Dhaka Tribune*, 1 December 2018.
107. Asian Development Bank (ADB). *Power System Expansion and Efficiency Improvement Investment Program (RRP BAN 42378)*; Asian Development Bank (ADB): Frankfurt, Germany, 2012.
108. Bangladesh Energy Regulatory Commission. Available online: http://www.berc.org.bd/site/page/ef59f071-9770-4095-95b3-b4f1a1f9ba4f/ (accessed on 1 September 2019).
109. Bangladesh Gazette. *Authentic English Text of the Act No 13 of 2003*; Bangladesh Gazette: Dhaka, Bangladesh, 2003; pp. 1–27.
110. National Board of Revenue (NBR). *Import Duty Exemptions for Solar and Wind of Bangladesh (Statutory Regulatory Order)*; NBR: Dhaka, Bangladesh, 2004.
111. Mohammad, H. Achieving sustainable energy targets in Bangladesh. *UN Chron.* **2013**, *52*, 36–39. [CrossRef]
112. The World Bank. *Combined Project Information Documents/Integrated Safeguards Datasheet (PID/ISDS)*; The World Bank: Washington, DC, USA, 2018.
113. Sustainable and Renewable Energy Development Authority (SREDA). *Scaling up Renewable Energy in Low Income Countries (SREP) Investment Plan for Bangladesh Renewable Energy Development Authority*; SREDA: Dhaka, Bangladesh, 2015.

114. Bangladesh Energy Regulatory Commission. *Bangladesh Energy Regulatory Commission (Tariff for Roof Top Solar PV Electricity) Regulations, 2016 (Draft)*; Bangladesh Energy Regulatory Commission: Dhaka, Bangladesh, 2016.
115. Ahmed, S.; Islam, M.T.; Karim, M.A.; Karim, N.M. Exploitation of renewable energy for sustainable development and overcoming power crisis in Bangladesh. *Renew. Energy* **2014**, *72*, 223–235. [CrossRef]
116. Preston, B.; Orgill, T. Adapting to a sustainable energy future: Part 1—The localisation of sustainable energy generation under the New South Wales planning law regime. *Env. Plan. Law J.* **2017**, *34*, 418–435.
117. Kumar, A.; Sah, B.; Singh, A.R.; Deng, Y.; He, X.; Kumar, P.; Bansal, R.C. A review of multi criteria decision making (MCDM) towards sustainable renewable energy development. *Renew. Sustain. Energy Rev.* **2017**, *69*, 596–609. [CrossRef]
118. Rob, H.-L.; Islam, S. *Energy Efficiency and Energy Auditing in Bangladesh*; Economic Dialogue on Green Growth (EDGG): Geneva, Switzerland, 2018.
119. Siddiqui, F.; Newman, P. Grameen Shakti: Financing Renewable Energy in Bangladesh. In *Sustainable Banking: The Greening of Finance*; Bouma, J.J., Jeucken, M., Klinkers, L., Eds.; Greenleaf Publishing Limited: Sheffield, UK, 2001; pp. 88–95. ISBN 1874719381.
120. Hossain, M.F.; Hossain, S.; Uddin, M.J. Renewable energy: Prospects and trends in Bangladesh. *Renew. Sustain. Energy Rev.* **2017**, *70*, 44–49. [CrossRef]
121. Haas, R.; Panzer, C.; Resch, G.; Ragwitz, M.; Reece, G.; Held, A. A historical review of promotion strategies for electricity from renewable energy sources in EU countries. *Renew. Sustain. Energy Rev.* **2011**, *15*, 1003–1034. [CrossRef]
122. Sulaiman, F.; Abdullah, N.; Gerhauser, H.; Shariff, A. An outlook of Malaysian energy, oil palm industry and its utilization of wastes as useful resources. *Biomass Bioenergy* **2011**, *35*, 3775–3786. [CrossRef]
123. Hellsmark, H.; Frishammar, J.; Söderholm, P.; Ylinenpää, H. The role of pilot and demonstration plants in technology development and innovation policy. *Res. Policy* **2016**, *45*, 1743–1761. [CrossRef]
124. Karim, R.; Muhammad-Sukki, F.; Karim, M.E.; Munir, A.B.; Sifat, I.M.; Abu-Bakar, S.H.; Bani, N.A.; Muhtazaruddin, M.N. Legal and regulatory development of nuclear energy in Bangladesh. *Energies* **2018**, *11*, 2847. [CrossRef]
125. Li, W.; Long, R.; Chen, H.; Geng, J. A review of factors influencing consumer intentions to adopt battery electric vehicles. *Renew. Sustain. Energy Rev.* **2017**, *78*, 318–328. [CrossRef]
126. Amin, S.B.; Rahman, S. (Eds.) Renewable Energy Practices in Bangladesh. In *Energy Resources in Bangladesh: Trends and Contemporary Issues*; Springer International Publishing: Cham, Switzerland, 2019; pp. 93–96.

sustainability MDPI

Article

Residential Energy-Related CO_2 Emissions in China's Less Developed Regions: A Case Study of Jiangxi

Yong Yang [1,2], Junsong Jia [1,2,*] and Chundi Chen [3]

1 Key Laboratory of Poyang Lake Wetland and Watershed Research, Ministry of Education, Jiangxi Normal University, Nanchang 330022, China; 18607045927@163.com
2 School of Geography and Environment, Jiangxi Normal University, Nanchang 330022, China
3 College of Architecture and Urban Planning, Tongji University, Shanghai 200092, China; chundichen@tongji.edu.cn
* Correspondence: jiaaniu@jxnu.edu.cn

Received: 26 January 2020; Accepted: 28 February 2020; Published: 5 March 2020

Abstract: The residential sector is the second-largest consumer of energy in China. However, little attention has been paid to reducing the residential CO_2 emissions of China's less developed or undeveloped regions. Taking Jiangxi as a case study, this paper thus aims at fully analyzing the difference of the residential energy-related CO_2 emissions between urban and rural regions based on the Log-Mean Divisia Index (LMDI) and Tapio decoupling model. The main results are showed as follows: (1) Since 2008, residential energy-related CO_2 emissions have increased rapidly in both urban and rural Jiangxi. From 2000 to 2017, the residential energy-related CO_2 emissions per capita in rural regions rapidly increased and exceeded that in urban regions after 2015. Furthermore, the residential energy structures had become multiple in both urban and rural regions, but rural regions still had room to optimize its energy structure. (2) Over the study period, consumption expenditure per capita played the dominant role in increasing the residential energy-related CO_2 emissions in both urban and rural regions, followed by energy demand and energy structure. Energy price had the most important effect on decreasing the urban and rural residential energy-related CO_2 emissions, followed by the carbon emission coefficient. However, urbanization increased the urban residential energy-related CO_2 emissions but decreased the CO_2 emissions in rural regions. Population made marginal and the most stable contribution to increase the residential energy-related CO_2 emissions both in urban and rural regions. (3) Overall, the decoupling status showed the weak decoupling (0.1) and expansive negative decoupling (1.21) in urban and rural regions, respectively.

Keywords: residential energy-related CO_2 emissions; less developed regions; urban and rural regions; LMDI; Tapio decoupling; Jiangxi province

1. Introduction

The Intergovernmental Panel on Climate Change (IPCC) suggests that climate change is a great threat for human survival and development owing to the anthropogenic greenhouse gas (GHG) emissions [1,2]. Since 2006, China has become the largest CO_2 emitter globally [3]. Currently, its emissions account for approximate one third of the global total emissions [4,5]. In this context, China has made commitments and formulated a series of policies to reduce emissions. Meanwhile, there is a consensus that reducing the fast increase of residential CO_2 emissions is a significant pathway to achieve energy-saving and emission-reduction targets in China [6,7]. For China, the residential sector is the second-largest energy consumer category as well as primary CO_2 emissions source [8,9]. Specifically, the residential sector is responsible for 11.7% of energy consumption and 12.6% of CO_2 emissions in 2015 [10]. On one hand, after 40 years of "reform and opening up", China steps into the new development phase. In addition, with its further development of urbanization and industrialization, the residential CO_2 emissions are expected to increase continuously. Thus, it is necessary and significant to study the key driving forces to help raise reasonable and effective mitigation policies to combat climate change [9]. On the other hand, it is widely appreciated that investment, consumption and export are the troika for economic growth, and it is not wise to simply restrain consumption to mitigate the CO_2 emissions [11]. In addition, as reported by a recent study that China's economic prosperity had been coupled with environmental degradation, and suggested that decoupling economic growth from ecological impact had been central to achieve the national sustainable development [12]. Therefore, it is also necessary and imperative for China to discover a green, low-carbon and sustainable path to coordinate the link between developmental and environmental protection [13].

In reality, plenty of studies have been conducted to respond to the above concerns for decades in China. They often focused on exploring influencing factors of the CO_2 emissions or energy consumption in the residential sector, based on national or regional perspectives, or analyzing the relationship between economic growth and CO_2 emissions from residential sector. When exploring the drivers affecting residential CO_2 emissions or energy consumption, many factors such as income [14–16], population [9,11,17], urbanization [18,19], energy intensity (or energy price) [10,20,21], energy structure [6,22,23], floor space (per capita) [24,25], education level [26], etc., were discussed. For example, based on structural decomposition analysis (SDA), Zhu, et al. [11] studied the indirect residential CO_2 emissions during the period 1992–2005. They argued that population size was not the main reason for the emissions growth anymore, although it promoted the indirect emissions to a certain extent. Using Log-Mean Divisia Index (LMDI), Zhao, et al. [27] decomposed the residential energy consumption in urban China from 1998–2007 at a disaggregated product level. The results demonstrated that energy price had the positive effect on the emission reduction, while population and income played the key role in the growth of the CO_2 emissions. Simultaneously, more scholars were also awareness of the urban–rural gap or dual society in China, thereby they attempted to research Chinese residential CO_2 emissions or energy consumption based on the perspective. Based on Sato–Vartia index, Liu, et al. [17] analyzed the impact of China's increased urban and rural residential consumption on CO_2 emissions from 1992–2007. Furthermore, Fan, et al. [6] studied the residential CO_2 emissions evolutions in urban–rural divided China and explored the underlying driving forces from the perspective of end-use and behavior by applying Adaptive Weighting Divisia (AWD) decomposition. With the further study on the issue, some researchers extended their scopes and took into account Chinese regional disparity in their related studies. Taking Guangdong, the most developed province in China, as a case study, Wang, et al. [21] decomposed the influencing factors of direct residential energy-related CO_2 emissions into eight factors by using LMDI. The results showed that residential living standard had the largest contribution to the increase of CO_2 emissions, while energy price was the first inhibiting factor. Similarly, taking Liaoning, a coastal province in northeastern China, as an example, Tian, et al. [9] explored the driving forces of the residential CO_2 emissions from 2002–2007 based on LMDI. The results suggested that population and per-capita consumption were the main factors to the increase of residential CO_2 emissions, whereas carbon intensity had the negative effect on the residential CO_2 emissions growth. Moreover, based on spatial–regional level, Yuan, et al. [22] applied LMDI to analyze the drivers of urban and rural residential CO_2 emissions in China's 30 provinces between 2007 and 2012. They found that population and income-per-capita effects were the main drivers of high urban residential CO_2 emissions in most of the coastal provinces. There were so many studies conducted in many other regions of China. The explanations for these studies can be similarly illustrated, here; we condense some explanations and list some presentative references in Table 1 to save space.

In the analysis of the pathway to coordinate CO_2 emissions and economic growth, numerous studies were performed using the decoupling analysis. Generally, there are two main decoupling methods, i.e., OECD indicator and Tapio decoupling indicator [28,29]. Compared with OECD indicator, the latter was more accurate and flexible, which greatly improved OECD indicator. Therefore, Tapio decoupling indicator was more popular and widely used to analyze the relationship between CO_2

emissions and economic growth. In addition, in China, most of these previous studies focused on the main emission sectors: industry [13,30]; transportation [31,32]; construction [33,34]; agriculture [35,36]. To the best of our knowledge, a few scholars applied the Tapio decoupling indicator to analyze the nexus between CO_2 emissions and economic growth from the residential sector. Based on the Tapio decoupling indicator, Ye, et al. [20] conducted the study to analyze the relationship between residential CO_2 emissions and economic growth in China from 1994–2012, which ignored the regional disparity and urban–rural gap. Using the OECD indicator, Yuan, et al. [22] studied the decoupling status in China's 30 provinces from the residential sector with little content.

Therefore, from Table 1, we can conclude that to date, researchers have mainly focused on developed regions especially coastal provinces and cities such as Guangdong, Liaoning, Jiangsu, Shandong and urban agglomerations, or on national and its 30 provinces when they conducted the studies on exploring the driving forces of CO_2 emissions in the residential sector. Moreover, few studies were conducted on decoupling analysis of the relationship between CO_2 emissions and economic growth from the residential sector compared with other main sectors. The existing related decoupling analysis studies on the residential sector were far from straightforward. As we all know, there is a consensus that China is a multiple-regional and heterogeneous country with the regional disparity and urban-gap [10,23], such as socio-economic, consumption pattern, urbanization, natural resource endowment, energy use, etc. Indeed, China's mitigation targets need all regions' efforts. However, this country's less developed and underdeveloped regions have not yet received academic attention, and few findings can be found in the existing studies worldwide. It is thus crucial to study these less developed and undeveloped regions. This can complete CO_2 emissions profile and help those regions' decision-makers make reasonable mitigation policies on residential sector by fully considering the local situations.

Table 1. Summary of representative studies on investigating the driving forces of Chinese residential sector.

Literature	Region	Scale	Urban–Rural Disparity	Methods
Zha, et al. [14]	China	Nationwide	√	LMDI
Liu, et al. [17]	China	Nationwide	√	Sato–Vartia index
Zhu, et al. [11]	China	Nationwide	×	SDA
Fan, et al. [6]	China	Nationwide	√	AWD
Wang, et al. [21]	Guangdong	Provincial	√	LMDI
Tian, et al. [9]	Liaoning	Provincial	√	LMDI
Bai, et al. [18]	64 cities of Chinese urban agglomerations	City	×	IPAT
Shi, et al. [10]	China and its 30 provinces	Nationwide and Provincial	×	Temporal and spatial LMDI
Yuan, et al. [22]	China's 30 provinces	Nationwide and Provincial	√	Spatial LMDI

Under such a circumstance, this paper aims to make up for the above-mentioned deficiencies by fully considering urban–rural gap and regional disparity; we take Jiangxi, a typical less developed province in central China, as a case study. Figure 1 shows its location. Jiangxi is in central China with an administrative area of 166,900 km^2. Statistically, its total gross domestic product (GDP) reached 2001 billion RenMinBi (RMB) in 2017 [37], which ranked 16 among all 31 regions of China. To the best of our knowledge, its capital city, Nanchang, was assigned the task of pursuing a low-carbon economic transformation in China's first "Low-Carbon Pilot Cities" national project in 2010 [9,38]. More recently, Jiangxi, Fujian, and Guizhou are selected as the first national ecological civilization test beds in 2016, aiming to set up the relatively sound ecological mechanism for China. According to these facts, we can know that Jiangxi had the huge potential of increasing its CO_2 emissions on residential sector with its further development. However, little academic attention was paid to Jiangxi, especially its residential sector. Thus, based on data availability, we first attempt to overview the situation of residential energy-related CO_2 emissions in Jiangxi urban and rural regions. Then, the driving forces affecting the residential energy-related CO_2 emissions in urban and rural regions are explored by using LMDI. In addition, the Tapio decoupling model was applied to analyze the relationship between the residential energy-related CO_2 emissions and economic growth (consumption expenditure). We believe this study has some innovative significance to help this province make mitigation policies and provide important insight for China's other less developed or undeveloped regions to reduce emissions.

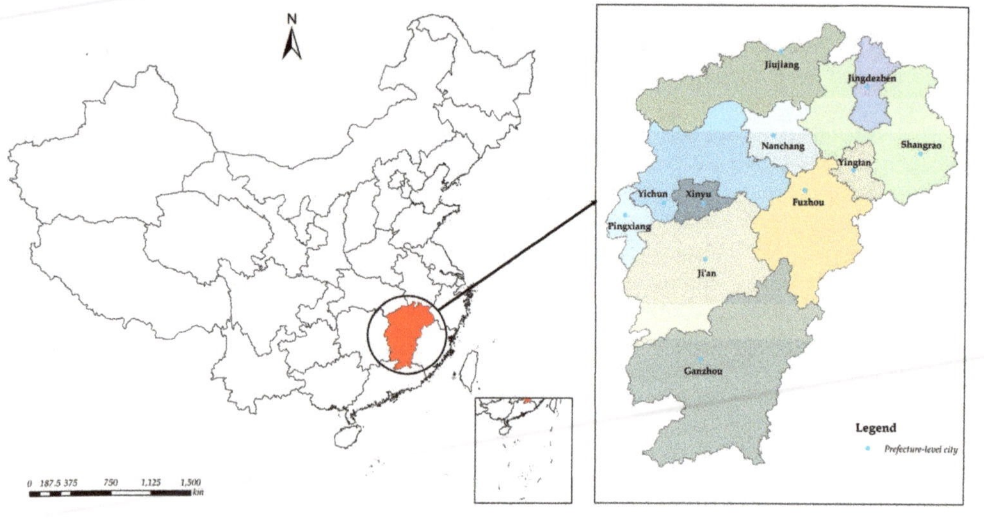

Figure 1. Location of Jiangxi.

The remainder of this paper is organized as follows: Section 2 depicts the methodology and data description. Results and discussions are given in Section 3. The main conclusions and corresponding policy implications are provided in Section 4.

2. Methods and Data Description

2.1. *Estimation of the Residential Energy-Related CO_2 Emissions*

The residential energy-related CO_2 emissions was caused by the residential energy consumption, e.g., lighting, cooking, heating, using household appliances and private transport, in addition, including electricity and heat [9]. Along with these activities, quantities of fossil fuels (e.g., coal, oil, natural gas) and secondary energy sources (e.g., electricity, heat) were consumed, which resulted in CO_2 emissions directly. According to IPCC [39], the residential energy-related CO_2 emissions can be calculated by the following equation:

$$C = \sum_i \sum_j C_{ij} = \sum_i \sum_j E_{ij} \cdot f_j \tag{1}$$

where i represents the resident type, i.e., urban residents and rural residents. j represents the fuel type. It should be noted that all kinds of fuel types consumed by residents are subdivided into five categories in this study based on China Energy Statistical Yearbook [40], i.e., coal, oil, natural gas, electricity, and heat. Specifically, coal includes raw coal, briquettes, coke oven gas, and other gases. Oil includes gasoline, kerosene, diesel oil, lubricants, liquefied petroleum gas (LPG) and other petroleum products. Natural gas includes natural gas and liquefied natural gas (LNG). E_{ij} denotes energy consumption of the i type resident, f_j denotes the carbon emission coefficient of j fuel type. With reference to Kennedy, et al. [41], carbon emission coefficients of different energy types are listed in Table A2. In addition, it is difficult to acquire the carbon emission coefficients of electricity and heat in Jiangxi. In the light of the data of heat can be only found from 2000–2002, which has small influence on the whole results. For this reason, we use China's corresponding carbon emission coefficients to substitute for Jiangxi's carbon emission coefficients of heat. With references to Jia, et al. [38], the carbon emission coefficient of heat can be measured as 0.11 t-CO_2/GJ by the equivalent calorific value. As for the carbon emission coefficient of electricity, here, we calculate and list the coefficient in Table A1.

2.2. Decomposition Method

Decomposition analysis has been widely applied to explore the impact factors affecting the changes of CO_2 emissions in the economic and environmental field. To date, there are two mainstream factor decomposition methods [25,36], viz., SDA and index decomposition analysis (IDA). Generally, SDA depends on input–output (IO) table and produces more accurate decomposition results. However, the IO tables are not available every year so that SDA has the limitation to the annual analysis. Inversely, IDA is easier to acquire data and more alterative to use aggregated data to analyze any years' changes [10]. Hence, IDA is widely applied to decompose the driving factors of energy-related CO_2 emissions. In addition, IDA provides many different indexes to choose [38]. In particular, among these indexes, LMDI has the incomparable advantages for its ease of use and no unexplainable residuals [42]. Moreover, it also provides eight effective strategies to handle zero values problem [43]. In view of these advantages, LMDI has been regarded as the most perfect decomposition method and widely used by many researchers [44,45]. Thus, in this paper, we apply the LMDI to explore the influencing factors affecting the residential energy-related CO_2 emissions in Jiangxi urban and rural regions.

The Kaya identity was widely used to reveal the influencing factors of CO_2 emissions. With the further development of the studies, more influencing factors are discussed. According to Kaya [46], the extended determinants of the residential energy-related CO_2 emissions can be expressed as:

$$C = \sum_i \sum_j C_{ij} = \sum_i \sum_j \frac{C_{ij}}{E_{ij}} \cdot \frac{E_{ij}}{E_i} \cdot \frac{E_i}{Y_{ie}} \cdot \frac{Y_{ie}}{Y_i} \cdot \frac{Y_i}{P_i} \cdot \frac{P_i}{P} \cdot P \tag{2}$$

Let $K_{ij} = \frac{C_{ij}}{E_{ij}}$, $ES_{ij} = \frac{E_{ij}}{E_i}$, $EP_i = \frac{E_i}{Y_{ie}}$, $ED_i = \frac{Y_{ie}}{Y_i}$, $CP_i = \frac{Y_i}{P_i}$, $U_i = \frac{P_i}{P}$, $P = P$ in Equation (2) can be written as:

$$C = \sum_i \sum_j K_{ij} \cdot ES_{ij} \cdot EP_i \cdot ED_i \cdot CP_i \cdot U_i \cdot P \tag{3}$$

where the variables are defined in Table 2.

Table 2. Definition of variables.

Variables	Definition
C	Total residential energy-related CO_2 emissions
C_{ij}	Residential energy-related CO_2 emissions of energy j by resident i
E_{ij}	Residential energy consumption of energy j by resident i
E_i	Residential energy consumption of resident i
Y_{ie}	Residence expenditure of resident i
Y_i	Consumption expenditure of resident i
P_i	Population of resident i
P	Total population of Jiangxi province
K_{ij}	Carbon emission coefficient of energy j by resident i
ES_{ij}	Share of energy j in residential energy consumption by resident i
EP_i	Residential energy consumption per unit of residence expenditure for resident i
ED_i	Share of residence expenditure to consumption expenditure of resident i
CP_i	Consumption expenditure per capita of resident i
Ui	Share of population of resident i to total population

Thereafter, to explore the contribution of the influencing factors to the total CO_2 emissions, next, we use LMDI, and it has two formations, i.e., additive LMDI (LMDI-I) and multiplicative LMDI (LMDI-II). Based on the additive LMDI, the changes of the total CO_2 emissions (C_{TOT}) from the baseline period (C_0) to the target period (C_T) can be decomposed as the following seven effects:

$$\Delta C_{TOT} = C_T - C_0 = \Delta C_{K_{ij}} + \Delta C_{ES_{ij}} + \Delta C_{EP_i} + \Delta C_{ED_i} + \Delta C_{CP_i} + \Delta C_{U_i} + \Delta C_P \quad (4)$$

Each effect can be calculated as follows:

$$\begin{aligned}
\Delta C_{K_{ij}} &= \sum_i \sum_j (C_{ij,T} - C_{ij,0}) / (\ln C_{ij,T} - \ln C_{ij,0}) \cdot (\ln K_{ij,T} - \ln K_{ij,0}) \\
\Delta C_{ES_{ij}} &= \sum_i \sum_j (C_{ij,T} - C_{ij,0}) / (\ln C_{ij,T} - \ln C_{ij,0}) \cdot (\ln ES_{ij,T} - \ln ES_{ij,0}) \\
\Delta C_{EP_i} &= \sum_i \sum_j (C_{ij,T} - C_{ij,0}) / (\ln C_{ij,T} - \ln C_{ij,0}) \cdot (\ln EP_{i,T} - \ln EP_{i,0}) \\
\Delta C_{ED_i} &= \sum_i \sum_j (C_{ij,T} - C_{ij,0}) / (\ln C_{ij,T} - \ln C_{ij,0}) \cdot (\ln ED_{i,T} - \ln ED_{i,0}) \\
\Delta C_{CP_i} &= \sum_i \sum_j (C_{ij,T} - C_{ij,0}) / (\ln C_{ij,T} - \ln C_{ij,0}) \cdot (\ln CP_{i,T} - \ln CP_{i,0}) \\
\Delta C_{U_i} &= \sum_i \sum_j (C_{ij,T} - C_{ij,0}) / (\ln C_{ij,T} - \ln C_{ij,0}) \cdot (\ln U_{i,T} - \ln U_{i,0}) \\
\Delta C_P &= \sum_i \sum_j (C_{ij,T} - C_{ij,0}) / (\ln C_{ij,T} - \ln C_{ij,0}) \cdot (\ln P_T - \ln P_0)
\end{aligned} \quad (5)$$

$$L(a,b) = \begin{cases} (a-b)/(\ln a - \ln b), a \neq b(a > 0, b > 0) \\ 0, a = b(a > 0, b > 0) \end{cases} \quad (6)$$

In addition, to present the decomposition results clearer, here, we also use the multiplicative LMDI. According to Shao, et al. [47], the corresponding multiplicative LMDI can be written as:

$$\Psi C_{TOT} = \frac{C_T}{C_0} = \Psi C_{K_{ij}} \cdot \Psi C_{ES_{ij}} \cdot \Psi C_{EP_i} \cdot \Psi C_{ED_i} \cdot \Psi C_{CP_i} \cdot \Psi C_{U_i} \cdot \Psi C_P \quad (7)$$

where $\Psi C_w = \exp\left(\sum_i \frac{\ln \omega_T}{\ln \omega_0} \cdot \frac{(C_{ij,T} - C_{ij,0})/(\ln C_{ij,T} - \ln C_{ij,0})}{(C_T - C_0)/(\ln C_T - \ln C_0)} \right)$, and ω means K_{ij}, ES_{ij}, EP_i, ED_i, CP_i, U_i and P. Ψ_{TOT} means the changes of the total CO_2 emissions between the baseline period and the target period with the corresponding multiplicative formation of the LMDI.

To the best of our knowledge, the CO_2 emission coefficient is usually assumed as a constant value, which has no contributions to the changes of CO_2 emissions. However, in this paper, the CO_2 emission coefficient of electricity changes every year, and it will lead to the changes of CO_2 emissions. Therefore, we finally analyze the following seven effects: the carbon emission coefficient effect ($\Delta C_{K_{ij}}$ and $\Psi C_{K_{ij}}$), energy structure effect ($\Delta C_{ES_{ij}}$ and $\Psi C_{ES_{ij}}$), energy price effect (ΔC_{EP_i} and ΨC_{EP_i}), energy demand effect (ΔC_{ED_i} and ΨC_{ED_i}), consumption expenditure per-capita effect (ΔC_{CP_i} and ΨC_{CP_i}), urbanization effect (ΔC_{U_i} and ΨC_{U_i}) and population effect (ΔC_P and ΨC_P).

2.3. Decoupling Model

The decoupling theory initially originated from the field of physics, which denotes the de-linkage relationship between two or more variables. Organization for Co-operation and Development (OECD) first categorized the decoupling indicators into relative decoupling and absolute decoupling [28]. However, there are some limitations and shortcomings of the OECD decoupling theory. Specifically, it is liable to be affected by the decoupling elasticity and lacked obvious criteria for choosing appropriate factors [36]. In view of this, Tapio [29] subdivided decoupling indicators into eight sub-categories to analyze the relationship between the road traffic and CO_2 emissions in the EU 15 countries from 1970–2001, which greatly improved the framework of OECD decoupling indicators. Since then, the Tapio decoupling model (see Table 3) was widely used to explore the dynamics nexus between economic growth and environmental protection. However, based on the context above, the decoupling analysis of the CO_2 emissions in the residential sector was far from straightforward in China, especially in less developed regions. Therefore, in this paper, the Tapio decoupling model was selected to analyze the decoupling status and the underlying drivers of the residential consumption expenditure from the residential energy-related CO_2 emissions in Jiangxi urban and rural regions.

According to Tapio [29], the decoupling indicator of residential energy-related CO_2 emissions of the *i* type resident from residential consumption expenditure can be expressed as:

$$D_i = \frac{C_i\%}{Y_i\%} = \frac{\Delta C_i / C_i{}^0}{\Delta Y_i / Y_i{}^0} \quad (8)$$

Table 3. Framework of the Tapio decoupling indicators.

Decoupling State	Abbreviation	$\Delta C_i/C_i^0$	$\Delta Y_i/Y_i^0$	D_i
Strong decoupling	SD	<0	>0	$(-\infty, 0)$
Weak decoupling	WD	>0	>0	(0, 0.8)
Recessive decoupling	RD	<0	<0	$(1.2, +\infty)$
Strong negative decoupling	SND	>0	<0	$(-\infty, 0)$
Weak negative decoupling	WND	<0	<0	(0, 0.8)
Expansive negative decoupling	END	>0	>0	$(1.2, +\infty)$
Expansive coupling	EC	>0	>0	(0.8, 1.2)
Recessive coupling	RC	<0	<0	(0.8, 1.2)

2.4. Data Description

Based on data availability, in this paper, the study period ranges from 2000 to 2017. To the best of our knowledge, since 1953, with a gap from 1963 to 1965, Chinese government has formulated plans for national economic and social development every five years, namely "Five-Year Plan (FYP)". To make the related results and analysis clearer, here, the study period is consistent with the FYP and subdivided into four stages, i.e., 2000–2005 (10th FYP), 2005–2010 (11th FYP), 2010–2015 (12th FYP) and 2015–2017 (13th FYP). In addition, it is noteworthy that Chinese government proposed a new stage, called "the new normal", of China's economic development during the 12th FYP, which was coincided with the last two stages.

All residential energy data are collected from Energy Balance Sheet of Jiangxi Province in the China Energy Statistical Yearbook (2001–2018) [40]. The related consumption expenditure per capita and population data directly derive from the Jiangxi Statistical Yearbook (JSY) (2001–2018) [37]. It should be noticed that consumption expenditure is subdivided into eight categories (e.g., food, cigarettes and wine, clothing, residence, household appliances, and services) in the JSY [37]. Additionally, among these categories, the data of residence expenditure (including the expenditure of housing, electricity, water, fuel and others) is the main source of the residential energy-related CO_2 emissions [21]. Thus, we split the residence expenditure from the consumption expenditure to make the analyses clearer. The share of residence expenditure to consumption expenditure is defined the residential energy demand in this study.

3. Results and Discussion

3.1. Overview the Situation of Residential Energy-Related CO_2 Emissions in Urban and Rural Jiangxi

3.1.1. The Trends of Residential Energy-Related CO_2 Emissions

As shown in Figure 2, the urban residential energy-related CO_2 emissions grew from 540.38 $\times 10^4$ t in 2000 to 1112.84 $\times 10^4$ t in 2017, with an annual average increase amount of 33.67 $\times 10^4$ t and a growth rate of 4.34%. The trend of urban residential energy-related CO_2 emissions could be subdivided into three phases: a fluctuating phase (2000–2005), a slow growth phase (2005–2008) and a rapid increase phase (2008–2017). From 2000–2005, it could be easily seen that the urban residential energy-related CO_2 emissions was from 540.38 × 10^4 t to 548.28 × 10^4 t, with an annual growth rate of only 0.29%. This phenomenon might be mainly attributed to the large-scale adjustment of the energy structure and cut down the share of coal use. From 2005–2008, because the global economic crisis took place around 2008, people restricted and reduced all kinds of expenditure in daily life, but the consumption of electricity could not fall for people's habits and customs [38]. Specifically, the urban residential energy-related CO_2 emissions were not more than 600 × 10^4 t. After 2008, the urban residential energy-related CO_2 emissions grew to 615.17 × 10^4 t in 2009, then it steadily and quickly increased to 1112.84 × 10^4 t, with an annual growth rate of 7.47%. The rapid growth of the residential energy-related CO_2 emissions might be explained by the recovery from the crisis and the improvement of economy and living standard for urban residents. This situation was mainly caused by the growing demand of high-carbon appliances and private transport (Table 4), consuming large quantities of oil and electricity [23]. For example, the number of family cars, air conditioners and computers in 2017 reached 40-fold, 1.9-fold and 2.3-fold than that in 2005, respectively.

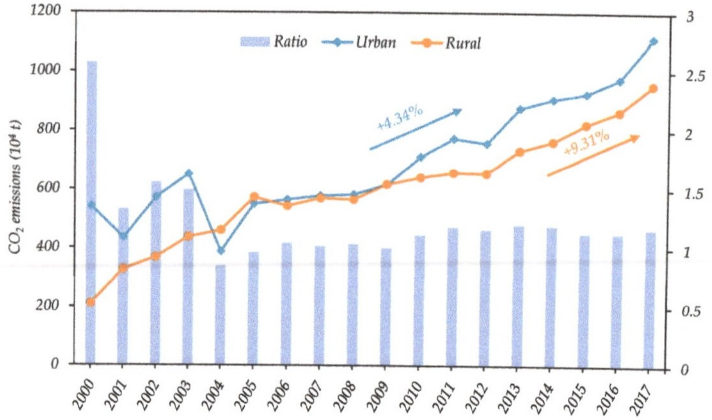

Figure 2. Trends of residential energy-related CO_2 emissions in Jiangxi urban and rural regions.

By contrast, the rural residential energy-related CO_2 emissions grew from 210.13 × 10^4 t to 954.82 × 10^4 t between 2000 and 2017 (Figure 2), with an annual growth amount of 43.81 × 10^4 t and a growth rate of 9.31%. However, the growth rate of rural residential energy-related CO_2 emissions was quicker than that of urban Jiangxi. Here, we also subdivided the rural residential energy-related CO_2 emissions into three phases: a rapid growth phase (2000–2005), a slight decrease phase (2005–2008) and a steadily growth phase (2008–2017). From 2000–2005, the rural residential energy-related CO_2 emissions grew from 210.13 × 10^4 t to 570.94 × 10^4 t, with an annual growth rate of 22.13%. This rapid growth may be explained by the increase of coal and electricity use, it was worth noting that coal still played the dominant role in rural residents' energy use (Figure 3b), leading to 154.64 × 10^4 t in 2000 and 331.08 × 10^4 t CO_2 in 2005, respectively. Analogous to the urban residents, the rural residents were also affected by the economic crisis from 2005–2008, making the corresponding CO_2 emissions declined by −1.27%. From 2008–2017, the rural residential energy-related CO_2 emissions increased from 563.69 × 10^4 t to 954.82 × 10^4 t, with an annual growth rate of 6.03%.

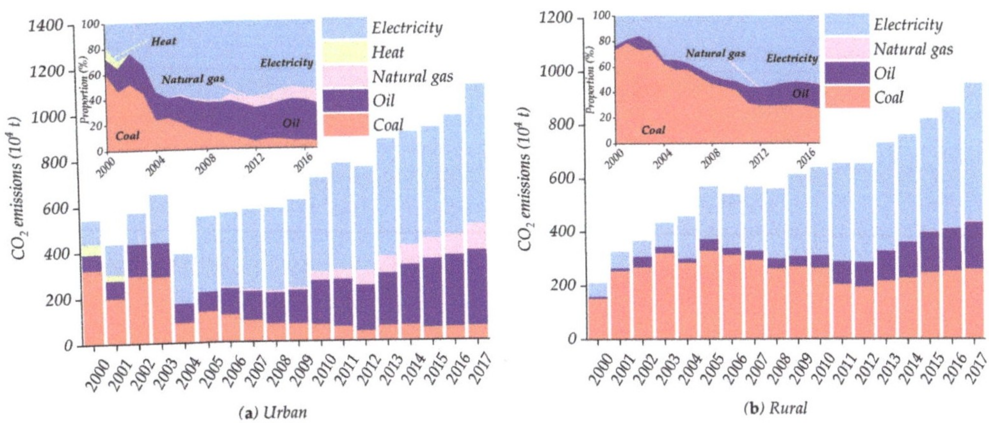

Figure 3. Changes of residential energy-related CO_2 emissions structure in Jiangxi urban and rural regions from 2000–2017. (**a**) Urban. (**b**) Rural.

Table 4. Urban: Ownership of major durable consumer goods per 100 urban households at year-end.

Item (Unit)	2000	2005	2010	2015	2017
Motorcycle	12.96	24.38	20.77	30.67	29.04
Family car	0.39	0.73	5.31	20.02	29.21
Washing machine	80.15	95.29	93.84	90.68	94.21
Refrigerator	75.82	90.66	96.57	96.64	98.6
Color TV set	106.01	139.31	148	139.13	136.29
Computer	4.56	32.03	59.91	73.98	74.33
Air conditioner	17.07	72.41	107.67	124.31	137.65
Mobile telephone	14.37	136.26	181.18	226.16	235.73
Shower heater	58.02	81.77	92.28	91.99	95.9

Figure 2 also showed the ratio of the residential energy-related CO_2 emissions for urban and rural regions widened and narrowed, felling from 2.57 in 2000 to 1.17 in 2017. Over the study period, the urban–rural ratio of residential energy-related CO_2 emission most remained the interval 1.0–1.2 except 2000–2005. Specifically, the ratio was more than 1.3 between 2000–2003. To the best of our knowledge, urban residents owned more energy equipment and consumed more commercial energy than that of rural residents, leading to more residential energy-related CO_2 emissions [17,48]. The ratio was 0.84 and 0.96 between 2003 and 2004, respectively. The reason might be that the large-scale reduction of coal use in urban regions as well as the growing popularity of appliances in rural regions (Table 5). From 2006–2017, the ratio of urban–rural always over 1.0, which mainly resulted from the development of urbanization.

Table 5. Rural: Ownership of major durable consumer goods per 100 rural households at year-end.

Item (Unit)	2000	2005	2010	2015	2017
Motorcycle	17.47	43.39	60.49	77.41	75.6
Family car	–	–	–	10.59	16.54
Washing machine	3.55	7.02	14.08	43.05	55.25
Refrigerator	3.63	10.53	45.84	84.68	91.42
Color TV set	30.16	82.33	106.86	127.13	132.12
Computer	–	2	5.22	24.19	24.59
Air conditioner	0.04	2	10.24	42.41	56.48
Mobile telephone	1.43	64.82	140.98	233.19	249.52
Shower heater	1.14	4.12	16.33	56.25	68.01

3.1.2. The Changes of Residential Energy-Related CO_2 Emissions Per Capita

The changes of residential energy-related CO_2 emissions per capita in Jiangxi urban and rural regions were presented in Figure 4. The urban residential energy-related CO_2 emissions per capita decreased from 470.41 kg-CO_2 in 2000 to 440.96 kg-CO_2 in 2017, with an annual average decrease amount of 1.73 kg-CO_2 and a decrease rate of 0.38%. The trends of urban residential energy-related CO_2 emissions per capita could be also subdivided into three stages: a fluctuating decrease stage (2000–2005), a slow growth stage (2005–2010) and a slightly more rapid growth stage (2010–2017). The urban residential energy-related CO_2 emissions per capita experienced a waved decrease from 470.41 kg-CO_2 in 2000 to 342.79 kg-CO_2 in 2005, then it slowly increased to 361.35 kg-CO_2 in 2010. The decrease of the urban residential energy-related CO_2 emissions per capita could be explained by the following reasons. The proportion of high-quality energy types, such as oil, natural gas, and electricity greatly improved due to the shift of energy structure and the improvement of appliances' energy efficiency [24]. Additionally, urban residents reduced some unnecessary energy consumption when the financial crisis occurred. The urban residential energy-related CO_2 emissions per capita grew from 361.35 kg-CO_2 to 440.96 kg-CO_2 from 2010–2017, this might be explained by the increase of appliances and private transport (Table 4).

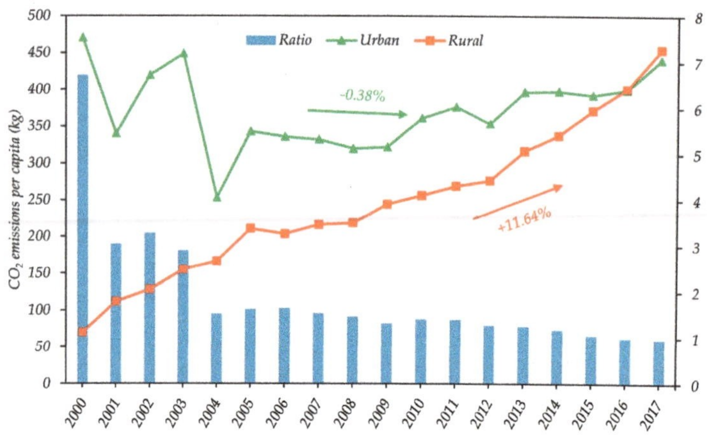

Figure 4. Changes of residential energy-related CO_2 emissions per capita in Jiangxi urban and rural regions from 2000–2017.

By contrast, the rural residential energy-related CO_2 emissions per capita rapidly increased from 70.05 kg-CO_2 to 455.02 kg-CO_2 from 2000–2017 (Figure 4), with an annual average increase amount of 22.65 kg-CO_2 and a growth rate of 11.64%. In particular, the rural energy-related CO_2 emissions per capita were 401.41 kg-CO_2 and 455.02 kg-CO_2 from 2016–2017 which exceeded that of urban regions. Compare to urban regions, the rural residential energy-related CO_2 emissions presented a tedious upward trend. This situation could be explained by the traditional biomass was gradually outpaced by the commercial energy. Moreover, with the improvement of living standards, rural residents had the capacity to pursuit high-quality life, such as buying more energy-intensive appliances and private transport tools (Table 5). However, many of them lacked the consciousness of environmental protection, resulting in more residential CO_2 emissions [49].

As presented in Figure 4, the gap in residential energy-related CO_2 emissions per capita between Jiangxi urban and rural regions decreased from 6.72 to 0.97 from 2000–2017. In addition, it could be divided into two phases: a rapid decrease phase in 2000–2005 and a slow decrease phase between 2005–2017. The approximate reason might be owing to the development of urbanization and the rural economy [23].

3.1.3. Analysis on the Residential Energy-Related CO_2 Emissions Structure

As shown in Figure 3a, the urban residential energy-related CO_2 emissions caused by coal waned from 324.57 × 10^4 t in 2000 to 65.64 × 10^4 t in 2007 and its corresponding share decreased from 60.06% to 5.90%, thereby indicating the reduction of coal use in urban regions. In addition, the residential energy-related CO_2 emissions caused by heat also presented a decrease tendency (46.86 × 10^4 t in 2000 and 2.01 × 10^4 t in 2017). The reason was that Jiangxi reduced the supply of heat, eventually, cut it off after 2002. See other energy types, electricity, oil, and natural gas overall presented an upward trend. The residential energy-related CO_2 emissions caused by electricity increased from 101.81 × 10^4 t to 605.55 × 10^4 t from 2000–2017, the corresponding share increased from 18.84% to 54.42%. It should be noteworthy that the residential energy-related CO_2 emissions induced by electricity exceeded coal and ranked first, which became the most important energy for urban residents after 2004. Since 2007, the residential energy-related CO_2 emissions induced by oil also exceeded the coal and became the second-largest energy type, and its corresponding share grew from 12.42% in 2001 to 29.24% in 2017. It is likely that the private transport increased rapidly (Table 4). However, starting from 2005, urban residents began to use natural gas. The share of residential energy-related CO_2 emissions induced by natural gas increased quickly from 1.09 × 10^4 t in 2005 to 116.29 × 10^4 t in 2007, with an annual growth rate of 47.57%, the corresponding share was from 0.2% to 10.45%. Since 2012, the share of residential energy-related CO_2 emissions induced by natural gas exceeded coal. At present, the residential energy-related CO_2 emissions structure was dominated by electricity, followed by oil, natural gas, and coal in urban regions.

The residential energy-related CO_2 emissions caused by heat was not found in rural regions according to JSY [37]. Here, we only considered coal, oil, natural gas, and electricity. As shown in Figure 3b, the residential energy-related CO_2 emissions for rural residents mainly induced by coal, oil, and electricity, which accounted for over 99% over the study period. In 2000, the residential energy-related CO_2 emissions induced by coal ranked the first, followed by electricity and oil. The share of residential energy-related CO_2 emissions induced by coal decreased from 73.59% to 27.52% from 2000–2007. The residential energy-related CO_2 emissions induced by electricity rapidly increased from 49.02 × 10^4 t to 513.47 × 10^4 t over the study period, with an annual growth rate of 14.82%. In 2009, the residential energy-related CO_2 emissions induced by electricity exceeded the coal, which might be explained by the growing increase of appliances. From 2009–2017, electricity became the most important energy type of the residential energy-related CO_2 emission for rural residents. For oil, the residential energy-related CO_2 emissions induced by oil occupied from 3.08% in 2000 to 18.25% in 2017. In addition, its share was still lower than the coal, although the share of coal decreased. Therefore, it could be deduced that the energy structure had further space to optimize for rural regions. The residential energy-related CO_2 emissions induced by natural gas was minimal among other energy types, which increased from 0.66 × 10^4 t to 4.37 × 10^4 t from 2008–2017. It should be noted that Jiangxi belonged to the "second pipeline of West–East Natural Gas transmission Project (To provide clean energy and shift energy structure in central and south China, the second pipeline of "West–East Natural Gas transmission Project" was launched in February 2008. Specifically, the pipeline went through Xinjiang, Gansu, Ningxia, Shaanxi, Henan, Hubei, Hunan, Jiangxi, Anhui, Jiangsu, Shanghai, Zhejiang, Guangdong, Guangxi, and H.K.)", thereby leading to the use of natural gas after 2008. With the implementation of the "coal to gas" project, the demand of natural gas of rural regions in Jiangxi might increase rapidly and expand the of share of residential energy-related CO_2 emissions induced by natural gas in the foreseeable future. Correspondingly, the share of the residential energy-related CO_2 emissions induced by coal would fall further. To the best of our knowledge, coal was a type of low-quality energy; the direct combustion of coal indoors was harmful to human health for rural residents, especially for women and children [17], which in turn led to air pollution even aggregated climate change. From the perspective of socio-development, if the energy structure were outdated, it would be bad for the development of rural economy and education due to the vulnerability of Chinese rural regions. To date, the share of residential energy-related CO_2 emissions structure was electricity, coal, oil, and natural gas in sequence for rural regions.

3.2. Decomposition Analysis of Residential Energy-Related CO_2 Emissions at Four Stages

As shown that the urban residential energy-related CO_2 emissions increased by only 7.91×10^4 t with a growth rate of 1.46% during the first stage (Figures 5a and 6a), and then rapidly grew by 162.15×10^4 t with a growth rate of 29.58% in the second stage (2005-2010) (Figures 5a and 6b). During the third stage, the CO_2 emissions increased by 215.42×10^4 t at a rate of 30.32% (Figures 5a and 6c) and grew by 186.98×10^4 t at a rate of 20.19% during the fourth stage (Figures 5a and 6d). Therefore, the conclusion could be drawn that the urban residential energy-related CO_2 emissions presented a sequentially upward trend with a total growth rate of 105.94% over the study period (Figure 5a).

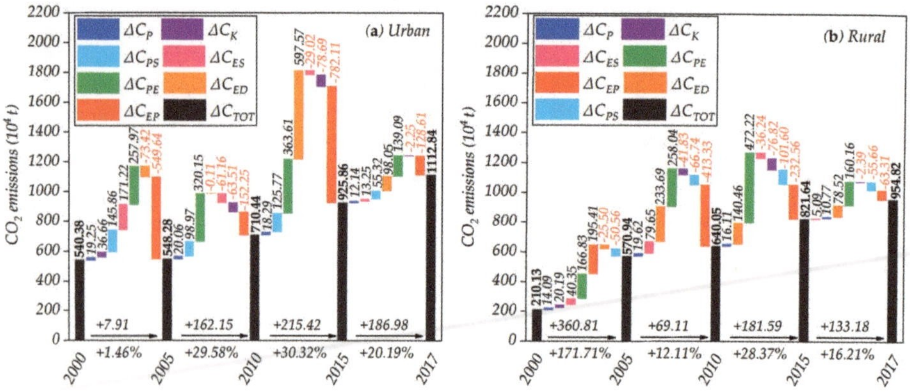

Figure 5. Factors affecting the residential energy-related CO_2 emissions in Jiangxi urban and rural regions from 2000–2017. (**a**) Urban. (**b**) Rural.

With respect to rural regions, it could be easily observed that the rural residential energy-related CO_2 emissions grew by 360.81×10^4 t at a rate of 171.71% in the first stage (Figures 5b and 6a), and then increased by 69.11×10^4 t at a rate of 12.11% in the second stage (Figures 5b and 6b). During the third and fourth stage, the rural residential CO_2 emissions increased by 181.59×10^4 t with a growth rate of 28.37% (Figures 5b and 6c) and grew by 133.18×10^4 t with a growth rate of 16.21% (Figures 5b and 6d), respectively. Analogous to the urban regions, it could be easily noticed that the rural residential energy-related CO_2 emissions also exhibited a sequentially increasing trend at a rate of 354.39% (Figure 5b).

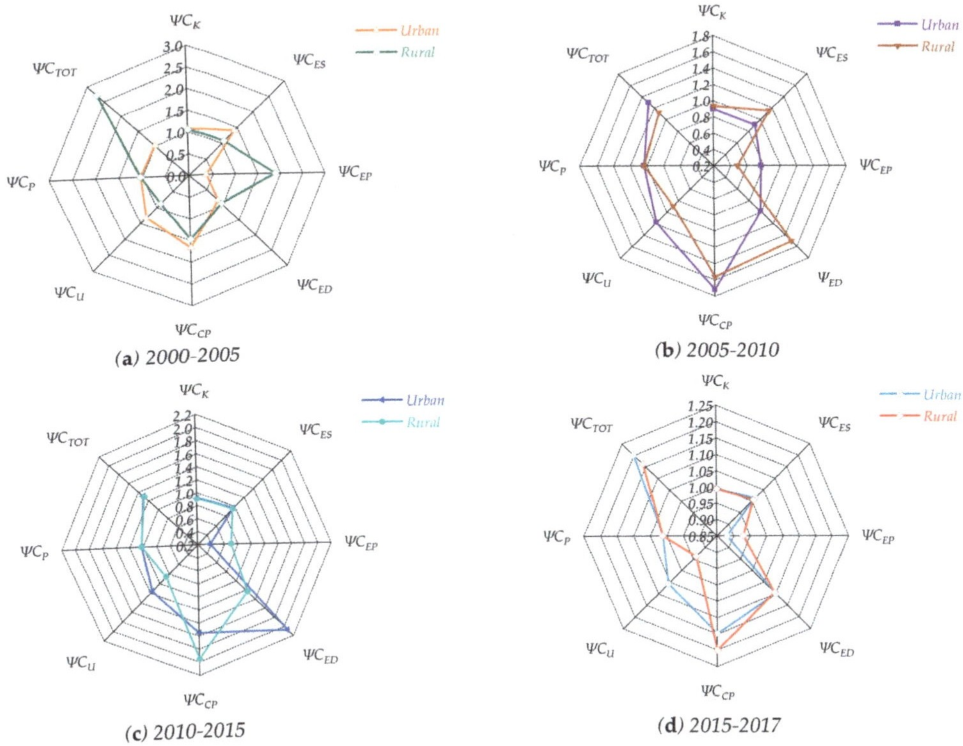

Figure 6. Multiplicative decomposition results of the residential energy-related CO_2 emissions in Jiangxi urban and rural regions in the four stages (**a**) 2000–2005. (**b**) 2005–2010. (**c**) 2010–2015. (**d**) 2015–2017.

3.2.1. Consumption Expenditure Per Capita (CE)

To the best of our knowledge, consumption expenditure per capita could reflect folks' living standards of a region to some extent. As shown in Table A5, it could be intuitively seen that consumption expenditure per capita always exerted a positive influence on the urban residential energy-related CO_2 emissions, which was in consonance with the results in Figure 5a and Table 6. Specifically, the additive decomposition effect of consumption expenditure per capita was 257.97×10^4 t in 2000–2005 (Figure 5a and Table A3). Then, from 2005–2010, 2010–2015, and 2015–2017, the decomposition effects were 162.15×10^4 t, 215.42×10^4 t and 186.98×10^4 t (Figure 5a and Table A3). In addition, the consumption expenditure per-capita effect contributed to the largest annual growth rate of the CO_2 emissions (11.77%), and the corresponding average annual contribution rates at four stages were 9.55%, 11.68%, 10.24% and 7.51% (Table 6), respectively. Thus, the conclusion could be drawn that the consumption expenditure per-capita effect played the most important role in promoting the urban residential energy-related CO_2 emissions.

For rural regions, the consumption expenditure per capita also always had a positive effect on the rural residential energy-related CO_2 emissions (Figure 5b and Table A6). At different four stages, the effect contributed to 166×10^4 t, 258.04×10^4 t, 472.22×10^4 t and 160.16×10^4 t, and the corresponding average annual contribution rates were 15.88%, 9.04%, 14.76% and 9.75% (Figure 5b and Table 7), respectively. Overall, the average annual led to the largest annual contribution rate (29.60%) over the studied decade, suggesting the effect of consumption expenditure per capita was also the most critical promoting factor to the rural residential energy-related CO_2 emissions.

Therefore, simply stated, these results implicated that the effect of consumption expenditure was the first promoting factor to residential energy-related CO_2 emissions for both urban and rural residents, which was in line with the previous studies [9,17]. In fact, with the development of Jiangxi's economy, both urban and rural residents' living standards had greatly improved. Specifically, the consumption expenditure per capita of urban and rural residents increased from 3623.52 RMB and 1642.66 RMB in 2000 to 19244.46 RMB and 9230.21 RMB in 2017 (Figure 7), with a growth rate of 10.32% and 10.69%, respectively. As a result, people had the capacity to shift their consumption patterns to buy more appliances and transport tools, resulting in more energy consumption and CO_2 emissions. According to Pachauri [50], future increases in expenditure levels will lead to further increases in household energy requirement, in turn, a further increase in the urban and rural residential energy-related CO_2 emissions caused by the steady increase of the consumption expenditure per capita in Jiangxi in the future.

3.2.2. Energy Demand (ED)

It could be intuitively observed that the additive decomposition effect of energy demand was -73.42×10^4 t and -0.11×10^4 t in urban regions (Figure 5a and Table A3), exerting a negative effect on decreasing the urban residential energy-related CO_2 emissions during the first and second stages. However, the negative effect weakened year by year. This situation was mainly because the energy consumption growth lagged behind the growth of the consumption expenditure. However, during the third and fourth stages, the decomposition effect increased to 597.57×10^4 t and 98.05×10^4 t (Figure 5a and Table A3), respectively. This could be attributable to the rapid growth of the energy demand, which increased from 10.45% in 2010 to 21.64% in 2015, and then to 23.85% in 2017 (Figure 7). In total, the effect of energy demand made the second-largest contribution to the total annual average rate of urban residential CO_2 emissions (6.77%), and for each of the four stages, the corresponding average annual contribution rates were –2.72%, –0.01%, 16.82% and 5.29%, respectively (Table 6). Overall, the energy demand effect played a key role in increasing the urban residential energy-related CO_2 emissions although it slightly mitigated the CO_2 emissions in Stage 1 and 2 (Figure 6 and Table A5).

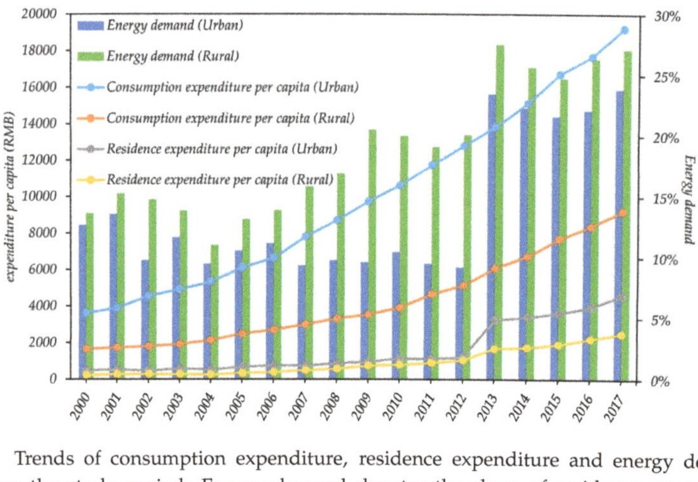

Figure 7. Trends of consumption expenditure, residence expenditure and energy demand in Jiangxi over the study period; Energy demand denotes the share of residence expenditure to consumption expenditure.

Similarly, energy demand also had a positive effect on promoting the rural residential energy-related CO_2 emissions over the study period (Figure 6 and Table A6). In total, it contributed to 427.17×10^4 t CO_2 emissions, with an annual average contribution rate of 11.96% (Figure 5b and Table 7). Compared to urban regions, the energy demand exerted a positive effect on increasing the rural residential energy-related CO_2 emissions except Stage 1 (-25.50×10^4 t), which contributed to 233.69×10^4 t, 140.46×10^4 t and 78.52×10^4 t in Stage 2, 3 and 4 (Figure 5b and Table A4), respectively.

This was because the energy demand decreased from 13.63% in 2000 to 13.13% in 2005, then it increased to 20.01% in 2010 and 27.12% in 2017 (Figure 7).

Table 6. Urban: Types and trends of various effects at different stages and average annual contribution rate in Jiangxi urban regions.

Effects	Average Annual Contribution Rate (%)					Trend
	Stage1 [a]	Stage2 [a]	Stage 3 [a]	Stage 4 [a]	Whole	
Total	0.29 [b]	5.92 [b]	6.06 [b]	10.10 [b]	6.23 [b]	+++++
Carbon emissions coefficient	1.36	−2.32	−2.22	−0.12	−1.17	+−−−−
Energy structure	6.34	−2.23	−0.82	0.72	1.03	+−−++
Energy price	−20.34	−5.55	−22.02	−6.95	−17.55	−−−−−
Energy demand	−2.72	0.00	16.82	5.29	6.77	−−+++
Consumption expenditure per capita	9.55	11.68	10.24	7.51	11.77	+++++
Urbanization	5.40	3.61	3.54	2.99	4.64	+++++
Population	0.71	0.73	0.51	0.66	0.76	+++++

Note: [a] Stage 1, 2, 3 and 4 refer to 2000–2005, 2005–2010, 2010–2015, and 2015–2017, respectively; [b] is the average annual contribution rate of the total urban residential energy-related CO_2 emissions; + and − stand for positive and negative effect on the CO_2 emissions, respectively.

Table 7. Rural: Types and trends of various effects at different stages and average annual contribution rate in Jiangxi rural regions.

Effects	Average Annual Contribution Rate (%)					Trend
	Stage1 [a]	Stage2 [a]	Stage 3 [a]	Stage 4 [a]	Whole	
Total	34.34 [c]	2.42 [c]	5.67 [c]	8.10 [c]	20.85 [c]	+++++
Carbon emissions coefficient	1.92	−1.47	−2.40	−0.15	−2.82	+−−−−
Energy structure	3.84	2.79	−1.13	0.31	2.49	++−++
Energy price	18.60	−14.48	−7.27	−3.85	−14.38	+−−−−
Energy demand	−2.43	8.19	4.39	4.78	11.96	−++++
Consumption expenditure per capita	15.88	9.04	14.76	9.75	29.60	+++++
Urbanization	−4.81	−2.34	−3.17	−3.39	−7.69	−−−−−
Population	1.34	0.69	0.50	0.66	1.70	+++++

Note: [a] Stage 1, 2, 3 and 4 refer to 2000–2005, 2005–2010, 2010–2015, and 2015–2017, respectively; [c] is the average annual contribution rate of the total rural residential energy-related CO_2 emissions; + and − stand for positive and negative effect on the CO_2 emissions, respectively.

3.2.3. Population (P)

It was apparent the effect of population was the most stable and negligible factor to promote the urban and rural residential energy-related CO_2 emissions (Tables A5 and A6), which was consistent with the previous studies [11,20]. Specifically, the average annual promoting impact of population effect (0.76% for urban, 1.70% for rural) was much lower than those of consumption expenditure per-capita effect and energy demand effect (Tables 6 and 7). See from urban and rural, at each of four stages, the additive decomposition effects were 19.25×10^4 t, 20.06×10^4 t, 18.29×10^4 t and 12.14×10^4 t for urban residents (Figure 5a and Table A3), similarly, 14.09×10^4 t, 19.62×10^4 t, 16.11×10^4 t and 10.77×10^4 t for rural regions (Figure 5b and Table A4). Over the study period, it could be found the population effect presented an inverted U-shape trend in urban and rural regions, implying the population effect would weaken in the future. The total population increased from 41.49 million to 46.22 million, with an annual growth rate of only 0.64% in Jiangxi over this 17-year period (Figure 8). In fact, to the best of our knowledge, population was a stable element for a region during a certain period, i.e., population would increase with a stable growth rate.

3.2.4. Urbanization (U)

The additive decomposition results of urbanization in urban regions were 145.86×10^4 t, 98.97×10^4 t, 125.77×10^4 t and 55.32×10^4 t at each of four stages (Figure 5a and Table A3), indicating the effect of urbanization always had a positive influence on increasing the urban residential energy-related CO_2 emissions. The results were consonance with the results in Table A5. In total, the effect of urbanization made a specific contribution to the total urban residential energy-related CO_2 emissions growth (4.64%) (Table 6). Compare to urban regions, the additive decomposition results of urbanization effect in rural regions were –50.56 × 10^4 t, –66.74 × 10^4 t, –101.60 × 10^4 t and –55.66 × 10^4 t during the four stages (Figure 5b and Table A4), respectively. Eventually, the urbanization effect led to the total rural residential energy-related CO_2 emissions decrease at an annual average rate of 7.69% (Table 7). Overall, the results showed the urbanization effect was the inhibiting factor to the rural residential energy-related CO_2 emissions (Table A6). The difference between urban and rural regions could be explained by the followings.

As depicted in Figure 8, the population of urban residents increased from 11.49 million to 25.24 million with an annual growth rate of 4.74%, and the population of rural residents decreased from 30.00 million to 20.98 million with an annual change of –2.08%. Correspondingly, the urbanization rate steadily increased from 27.69% to 54.6% from 2000–2017. In light of the context mentioned above, the urbanization effect increased the urban residential energy-related CO_2 emissions, but decreased the rural residential energy-related CO_2 emissions. On average, when the urbanization increased by 1%, the urban residential energy-related CO_2 emissions might have a growth of about 15.83 × 10^4 t, while the rural residential energy-related CO_2 emissions might have a reduction of about 10.20 × 10^4 t. See from each of four stages, the annual average contribution rates were 5.40%, 3.61%, 3.54% and 2.99% in urban regions (Table 6), and –4.81%, –2.34%, –3.17% and –3.39% in rural regions (Table 7). The urbanization effect presented a decline trend on the whole. In the future, this effect might be offset with the further development of Jiangxi's urbanization. It was demonstrated by the contribution rate of the urbanization effect in Guangdong, the most developed province of China, experienced an increase trend to 53.6% in 2003, then declined to 17.3% in 2012 with a diminishing trend over time [21]. Furthermore, Fan, et al. [19] also pointed out the urbanization contributed to the increase to Chinese residential energy consumption but with a diminishing trend over time.

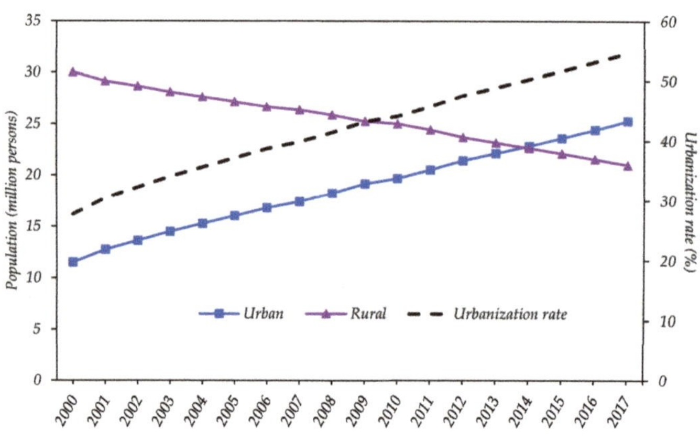

Figure 8. Trends of urban and rural population in Jiangxi over the study period.

3.2.5. Energy Structure (ES)

Energy structure adjustment played a minor role in increasing the residential energy-related CO_2 emissions in urban and rural regions (Tables A5 and A6), which was consistent with most relevant studies [21,51]. Over the study period, the aggregate changes of the energy structure effect were 94.29 × 10^4 t and 88.84 × 10^4 t (Figure 5), and the corresponding average annual contribution rates were 1.03% and 2.49%, respectively. Specifically, in the four stages, the corresponding annual average contribution rates were 6.34%, –2.23%, –0.82%, 0.72% for urban regions (Table 6), and 3.84%, 2.79%,

–1.13% and 0.31% for rural regions (Table 7). The usage of coal steadily decreased in both urban and rural regions, resulting in the share of the residential energy-related CO_2 emissions decreased to 5.90%, 27.52% in 2017 from 60.06%, 73.59% in 2000 (Figure 4). However, the share of residential energy-related CO_2 emissions caused by oil and electricity rapidly increased, especially electricity. Furthermore, the share of natural gas was still low in Jiangxi, especially in rural regions. It was worth mentioning that the source endowment of China (including Jiangxi) was dominated by coal in the long term, and the electricity generation was mainly from coal [9,52]. The renewable and sustainable energy, such as wind, solar, nuclear, and biomass energy, had not been widely used [36]. Thus, it still required a longer time to thoroughly shift the energy structure to the low-carbon pattern to reduce the residential energy-related CO_2 emissions.

3.2.6. Energy Price (EP)

The energy price effect had the most important impact lowering the residential CO_2 emissions in urban and rural regions over the study period (Tables A5 and A6), which was consistent with the previous studies [10,21]. Specifically, the additive decomposition effects were -549.64×10^4 t, -152.25×10^4 t, -782.11×10^4 t and -128.61×10^4 t for urban regions (Figure 5a and Table A3), the corresponding average annual contribution rates were –20.34%, –5.55%, –22.02% and –6.95%, respectively (Table 6). Similarly, in rural regions, the energy price effect led to 195.41×10^4 t CO_2 emissions from 2000–2005. However, the decomposition effect decreased greatly to -413.33×10^4 t, -232.56×10^4 t and -63.31×10^4 t during Stage 2, 3 and 4 (Figure 5b and Table A4), respectively, and for each of the four stages, the average annual contribution rates were 18.60%, –14.48%, –7.27% and –3.85%, respectively (Table 7).

Overall, the energy price effect made the greatest contribution to the total annual rates of urban and rural residential energy-related CO_2 emissions mitigation (–17.55% for urban and –14.38% for rural) (Tables 6 and 7). This was mainly because the energy price increased from 2410.53 RMB/ ton-standard coal equivalent (SCE), 9855.43 RMB/ ton SCE in 2000 to 28078.53 RMB/ ton SCE, 16900.13 RMB/ ton SCE in 2017 for urban and rural regions, respectively (Figure 9), restraining the willing of people's consumption. In particular, it should be noted that the energy price rapidly grew from 9730.42 RMB/ ton SCE in 2012 to 24690.65 RMB/ ton SCE in 2015 because of the implementation of the "ladder electricity price plan (In November 2011, the National Development and Reform Commission (NDRC) issued the "residential ladder electricity price policy" to promote energy conservation and emission reduction as well as social equity. Specifically, residential electricity consumption was divided into three grades and a free grade for low-income households.)" in Jiangxi urban regions, making the energy price reached the peak value (-782.11×10^4 RMB/ton SCE). Moreover, during Stage 1, the energy price decreased from 9855.43 RMB/ ton SCE to 5226.89 RMB/ ton SCE, which stimulated the consumption of rural residents.

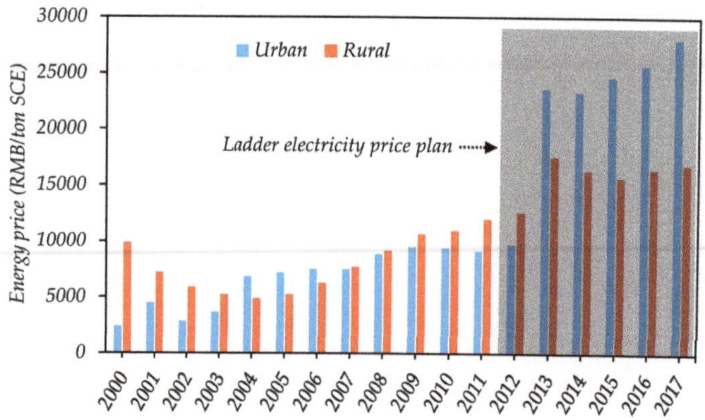

Figure 9. Energy price in urban and rural Jiangxi.

3.2.7. Carbon Emission Coefficient (K)

As mentioned above, in this article, the carbon emission coefficient of electricity was not a constant value. The carbon emission coefficient effect mainly derived from the change of the carbon emission coefficient of power generation (Table A1). In addition, according to Ren, et al. [53], if the contribution of the energy emission factor was negative, the residential energy-related CO_2 emissions decreased. Otherwise, the contribution of the energy emission factor was positive and the residential energy-related increased.

The carbon emission coefficient effect had the least influence on inhibiting the residential energy-related CO_2 emissions in both urban and rural regions (Tables A5 and A6). In total, the average annual contribution rates were –1.17% and –2.82% for urban and rural regions (Tables 6 and 7), respectively. See from each of four stages, the corresponding contribution rates were 1.36% and 1.92% in urban and rural regions from 2000–2005, indicating the carbon emission coefficient effect had a minor influence on increasing the CO_2 emissions. This situation was likely that the carbon emission coefficient of electricity increased from 6.49 t-CO_2/10^4 kwh to 7.99 t-CO_2/10^4 kwh in Jiangxi (Table A1). However, in 2005–2010, 2010–2015, and 2015–2017, the average annual contribution rates decreased to –2.32%, –2.22%, –0.12% for urban regions, and –1.47%, –2.40%, –0.15% for rural regions, respectively. The results might be due to the improvement of power generation technology, updated generation equipment and other factors [53]. Specifically, the carbon emission coefficient of electricity decreased from 7.99 t-CO_2/10^4 kwh to 4.84 t-CO_2/10^4 kwh in 2017. On the other hand, the carbon emission coefficient effect represented an overall downward trend, resulting from the long-term coal-dependent power generation in Jiangxi (Average=80.91%) (Figure 10). Hence, the carbon emission coefficient effect eventually had a relatively marginal negative contribution to the residential energy-related CO_2 emissions reduction. To effectively enhance the effect of energy emission factor to mitigate CO_2 emissions, it is necessary to boost the share of cleaner energy types such as hydropower, wind power, and solar power in Jiangxi's electricity generation structure.

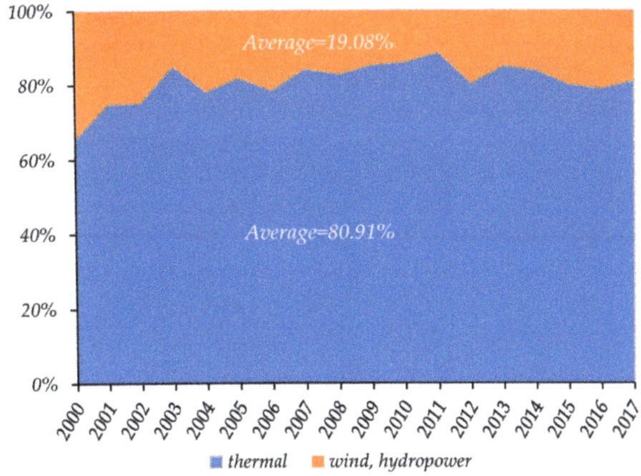

Figure 10. Energy types of power generation in Jiangxi.

3.3. Decoupling Analysis of Residential Energy-Related CO_2 Emissions at Four Stages

As shown in Table 8, four decoupling states occurred in Jiangxi urban regions, namely SD, END, WD, and EC. From 2000–2005, the decoupling states switched every year, thereby implying that an unstable relationship between urban residential energy-related CO_2 emissions and consumption expenditure. Eventually, the decoupling state showed WD, that is to say, the consumption expenditure increased a little faster than the CO_2 emissions. After 2005, the decoupling states converged on the WD in most years of the period 2005–2016, except 2009–2010 and 2012–2013. Furthermore, the decoupling state was EC in the end, suggesting that there is no significant relationship between the CO_2 emissions and consumption expenditure. Overall, the decoupling states were the WD, WD, and EC from 2005–2010, 2010–2015, and 2015–2017, respectively. Over the study period, the decoupling state was WD in Jiangxi urban regions.

Table 8 also presented the decoupling state of Jiangxi rural regions. Similarly, there are also four types of decoupling states: END, SD, WD, and EC. However, there are some differences. Specifically, the decoupling states were stable and converged on END from 2000–2005. Meanwhile most years of the period 2005–2015, the decoupling states mainly converged on WD and SD, except 2008–2009 (END). This caused the decoupling state shift from END in 2000–2005 to WD in 2005–2010 and 2010–2015. In the later period (2015–2017), the decoupling state was EC, resulting from the unsatisfactory decoupling states in 2015–2016 (EC) and 2016–2017 (END). Simply stated, the decoupling state in Jiangxi rural regions showed the END over the study period. Compared to the urban regions, the decoupling process needed further to accelerate in Jiangxi rural regions.

During Stage 1 (2000–2005), both urban and rural residents' income increased with the development of economy. Correspondingly, the consumption expenditure grew by 135% and 37% (Table 8), respectively. To ensure the quick development of economy, the demand for urban residents was restrained, which caused the residential energy-related CO_2 emissions increased by only 1%. Thus, the consumption expenditure increased much greater than the CO_2 emissions, leading to an overall WD (0.01) in Stage 1 although EC and END occurred in 2001–2003 and 2004–2005 (Table 8). See from rural, decline of energy price stimulated the residents' consumption desire which made the energy demand increase (Figure 7). Meanwhile, it caused the CO_2 emissions rose by 172% in 2000–2005 which was greatly higher than that of consumption expenditure. Consequently, the decoupling state was END in this stage.

During Stage 2 (2005–2010), the consumption expenditure rose by 114% (Table 8), which made urban residents pursuit more high-quality life and buy more high-carbon appliance and family cars (Table 4). Thus, the CO_2 emissions grew by 30%, presenting the WD (0.26) (Table 8) in 2005–2010. As to rural residents, on one hand, rural residential consumption expenditure further rose by 45%. On the other hand, traditional outdated energy gradually placed by the commercial energy in Jiangxi rural regions and the CO_2 emissions increased by 12%. As a result, the decoupling state shifted from END to WD in Stage 2.

During Stage 3 (2010–2015), both urban and rural regions presented WD. The growth rate of the consumption expenditure continued to slow down and grew by 89% in urban regions. However, it continued to rise and grew by 74% in rural regions (Table 8). This could be explained by the long-term inequality and lag between urban and rural regions in China's economic development (including Jiangxi), namely duality of urban–rural. Moreover, in order to save energy and reduce emissions, Chinese government (Jiangxi included) formulated and implemented the "ladder electricity price policy" after 2012, which effectively restrained residential energy consumption, especially in urban regions. Thus, the change rate of the CO_2 emissions in urban regions nearly had no obvious increase (30% from 2010–2015), and it grew by 28% in rural regions. Consequently, WD occurred.

Table 8. The decoupling state between residential energy-related CO_2 emissions and consumption expenditure in Jiangxi urban and rural regions over the study period.

Time Period	Urban				Rural			
	C%	Y%	D	State	C%	Y%	D	State
2000–2001	−0.20	0.19	−1.04	SD	0.55	0.02	32.89	END
2001–2002	0.32	0.25	1.29	END	0.13	0.02	6.36	END
2002–2003	0.14	0.15	0.92	EC	0.19	0.05	3.90	END
2003–2004	−0.41	0.14	−2.82	SD	0.05	0.10	0.55	WD
2004–2005	0.42	0.20	2.09	END	0.25	0.15	1.66	END
2005–2006	0.03	0.14	0.19	WD	−0.05	0.06	−0.83	SD
2006–2007	0.02	0.22	0.11	WD	0.05	0.10	0.49	WD
2007–2008	0.01	0.17	0.05	WD	−0.01	0.08	−0.09	SD
2008–2009	0.06	0.17	0.33	WD	0.09	0.04	2.19	END
2009–2010	0.15	0.12	1.29	END	0.04	0.10	0.41	WD
2010–2011	0.09	0.15	0.56	WD	0.02	0.16	0.15	WD
2011–2012	−0.02	0.13	−0.15	SD	0.00	0.07	−0.05	SD
2012–2013	0.16	0.12	1.34	END	0.12	0.16	0.75	WD
2013–2014	0.03	0.13	0.27	WD	0.04	0.08	0.52	WD
2014–2015	0.02	0.14	0.15	WD	0.08	0.12	0.66	WD
2015–2016	0.05	0.09	0.57	WD	0.05	0.06	0.82	EC
2016–2017	0.14	0.13	1.12	EC	0.10	0.07	1.45	END
2000–2005	0.01	1.35	0.01	WD	1.72	0.37	4.68	END
2005–2010	0.30	1.14	0.26	WD	0.12	0.45	0.27	WD
2010–2015	0.30	0.89	0.34	WD	0.28	0.74	0.38	WD
2015–2017	0.20	0.23	0.87	EC	0.16	0.14	1.15	EC
2000–2017	1.06	10.67	0.10	WD	3.54	2.93	1.21	END

Note: C% and Y% denote change rates of residential energy-related CO_2 emissions and consumption expenditure, respectively; D denotes the results of decoupling indicators.

During Stage 4 (2015–2017), the relationship between consumption expenditure and residential energy-related CO_2 emissions transformed from WD to EC in both urban and rural regions. Chinese economic development embarked on a new stage called "the new normal". Specifically, in this stage, China (including Jiangxi) shifted its development pattern from rapid growth to more inclusive and sustainable growth, including higher living standards, cleaner production [54]. Consequently, the growth rate of consumption expenditure and the CO_2 emissions declined, to some extent. Consumption expenditure rose by 23% and 14% in Jiangxi urban and rural regions (Table 8), respectively, which was slower than that of other periods. Additionally, the CO_2 emissions also showed a decline trend and grew by 20% and 16% in Jiangxi urban and rural regions, resulting from the improvement of energy efficiency. Therefore, they both presented EC in this stage, indicating there was insignificant decoupling between consumption expenditure and the CO_2 emissions.

4. Conclusions and Policy Implications

4.1. Main Conclusions

Presently, residential energy consumption has become the second-largest sector to the industrial energy consumption in China. However, little attention was paid to research the difference of the residential energy-related CO_2 emissions between urban and rural areas in the less developed and undeveloped regions. Thus, based on the urban–rural duality, this paper aims at exploring the difference of the residential energy-related CO_2 emissions in Jiangxi urban and rural regions from 2000–2017. We first overviewed the changes of the residential energy-related CO_2 emissions in Jiangxi between urban and rural regions. Then, the LMDI method was introduced to distinguish the major factors affecting the residential energy-related CO_2 emissions. In addition, the Tapio decoupling model was used to analyze the relationship between residential energy-related CO_2 emissions and consumption expenditure. The main results were acquired as follows.

The residential energy-related CO_2 emissions rapidly increased after 2008 in both Jiangxi urban and rural regions. However, the annual growth rate of urban regions (7.47%) was faster than rural regions (6.03%). In addition, the gap between urban and rural regions became narrowed and stable. The energy structures of both urban and rural residential energy-related CO_2 emissions had shifted from coal-dominant to a multiple structure which consisted of oil, natural gas, and electricity. Specifically, the energy-related CO_2 emissions structure was electricity, oil, natural gas, and coal in sequence for urban; it was electricity, coal, oil, and natural gas in sequence for rural. As for the residential energy-related CO_2 emissions per capita, it showed a decline trend in urban regions, while an increasing tendency in rural regions over the study period. The gap in residential energy-related CO_2 emissions for urban and rural residents narrowed from 2000–2007.

The energy price effect played the most important role in reducing the residential energy-related CO_2 emissions in both urban and rural regions. In addition, the carbon emission coefficient had a minor effect on decreasing urban and rural residential energy-related CO_2 emissions. The results showed that urbanization exerted a positive effect on increasing the urban residential energy-related CO_2 emissions, but decreasing the rural residential energy-related CO_2 emissions. Consumption expenditure-per-capita effect was the most important factor increasing the residential energy-related CO_2 emissions in both urban and rural regions, followed by energy demand. From 2000–2017, population effect and energy structure played a minor role in increasing the urban and rural residential energy-related CO_2 emissions.

Over the study period, four decoupling status (END, SD, WD, and EC) occurred in both urban and rural regions when analyzed the relationship between consumption expenditure and residential energy-related CO_2 emissions. Overall, the decoupling state for urban and rural regions were the WD and END. This showed that residential energy-related CO_2 emissions in urban Jiangxi was less depend on consumption expenditure, but rural residential energy-related CO_2 emissions still depended on consumption expenditure.

4.2. Policy Implications

First, since the residential energy-related CO_2 emissions caused by coal was still high in rural regions, which accounted for 27.52% and was second only to electricity (53.78%) even in 2017. However, the corresponding proportion of natural gas was always low and took up less than 1% over the study period. Thus, the energy structure of rural regions had more space to optimize, and more efforts should be paid to further adjust the rural energy structure. For instance, promoting the proportion of natural gas and encouraging the use of renewable and clean energy in rural regions.

Second, since the rural residential energy-related CO_2 emissions per capita had exceeded that of the urban region after 2015, thereby suggesting that the lifestyles in rural regions were more extensive and lower energy efficiency compared with urban regions. On one hand, it is urgent and imperative for the government at all levels to strengthen related public education and guide residents to shift their high-carbon lifestyles to a green and low-carbon pattern, especially for rural residents. In addition, the government should accelerate the elimination of backward production capacity to improve the efficiency in rural regions. On the other hand, residents themselves should also initiatively adapt to the transformation and cultivate the awareness of environmental protection.

Third, consumption expenditure-per-capita effect became the most factor increasing the residential energy-related CO_2 emissions in both urban and rural regions. With the improvement of economic growth and living standards, people had the desire and capacity to pursuit the quality of life and buy more energy-intensive electricity appliances and private transport tools. However, it is not wise to reduce the CO_2 emissions via restraining the consumption, this was because the consumption was a critical impetus of socio-economic development. Thus, the government could apply the fiscal and tax policy to cut down the residential energy-related CO_2 emissions and accelerate the decoupling. Specifically, the government should continue to impose the "Home Appliances Subsidy Program" and monitor its proper extent of implementation. For the high-carbon and extensive goods and services, the government could add the tax to restrain the CO_2-intensive commodities' consumption. Meanwhile for the energy-saving appliances, reducing the tax could stimulate the low-carbon consumption. Moreover, more investment and efforts were needed to improve public transport infrastructures and the energy conservation technology.

Fourth, considering energy price played the most important role in decreasing the residential energy-related CO_2 emissions in both urban and rural regions, indicating tiered energy price was the effective and scientific pathway to achieve energy savings and emissions reduction. In light of the fact, the government needs further to uphold and ameliorate the energy price policy in the future. In particular, when applying the energy price mechanism, the duality as well as inequality of income level and economic development should be comprehensively considered in urban and rural regions.

Last, to the best of our knowledge, residential building energy consumption accounted for a large proportion of the total residential energy consumption, leading to enormous CO_2 emissions. Therefore, it is necessary to improve residential building energy efficiency, especially for rural regions. For urban regions, the intelligent energy management technology in household could be popularized and applied to achieve energy conservation.

5. Limitations and Further Perspectives

There are limitations to this study. Based on the perspective of end-use of energy, only the residential energy-related CO_2 emissions were estimated, while the indirect or embodied residential CO_2 emissions from production perspective were excluded. This could likely underestimate the residential CO_2 emissions in Jiangxi urban and rural regions. Therefore, the embodied residential CO_2 emissions need further study. Moreover, this paper only explored the factors affecting the residential energy-related CO_2 emissions and the decoupling relationship between consumption expenditure and residential energy-related CO_2 emissions in Jiangxi urban and rural regions, specifically carbon emission coefficient, energy structure, consumption expenditure per capita, energy demand, energy price, urbanization, and population. However, family size, floor space of residential buildings and different income levels (an aspect of social inequality) also had the nonnegligible effect on the residential CO_2 emissions. Thus, these influencing factors to the residential CO_2 emissions should be paid more attention in our future work.

Author Contributions: Y.Y., J.J. and C.C. conceived and designed this study; Y.Y. collected and analyzed the data; Y.Y. and J.J. wrote and revised the paper; C.C. contributed to progress of research idea. All authors read and approved the final manuscript.

Funding: This research was funded by Chinese National Science Foundation (Grant No. 71473113), and the Research Project of Humanities and Social Sciences in Jiangxi's Universities (Grant No. GL19225).

Acknowledgments: We thank the anonymous reviewers and editor for their constructive comments and suggestions to improve the quality of this article.

Conflicts of Interest: The authors declare no conflict of interest.

Abbreviation

IPCC	Intergovernmental Panel on Climate Change
OECD	Organization for Economic Co-operation and Development
SDA	Structural decomposition analysis
IDA	Index decomposition analysis
LMDI	Log-mean Divisia index
AWD	Adaptive Weighting Divisia
IPAT	Impact of Population, Affluence, and Technology
GDP	Gross domestic product
FYP	Five-Year Plan
RMB	RenMinBi
SCE	Standard coal equivalent
E	Residential energy consumption
C	Residential energy-related CO_2 emissions
K	Carbon emission coefficient
ES	Energy structure
EP	Energy price
ED	Energy demand
CP	Consumption expenditure per capita
U	Urbanization
P	Population

Appendix A

Calculation of the CO_2 emission coefficient of electricity: Based on Wang, et al. [21], the CO_2 emission coefficient of electricity can be calculated as follows:

$$f_e = \frac{\sum_k \sum_i E_{ik} \cdot f_{ik}}{T_e} = \frac{\sum_i E_i \cdot f_i}{T_e} \tag{A1}$$

where the subscript i denotes the fuel type consumed in electric power generation, i.e., coal, oil and natural gas; k denotes the type of electricity; T_e denotes the total amount of electric power generation; E_i denotes the energy consumption consumed in electric power generation of i fuel type; f_e denotes the CO_2 emission coefficient of electricity; f_i denotes the CO_2 emission coefficient of i type fuel type.

According to Electricity Balance Sheet of JSY (2001–2018) [37], to date, the types of electric power generation include thermal power, hydropower, and wind power. Among these types of electric power generation, thermal power is the main source of CO_2 emissions, and other contribute negligible CO_2 emissions. Thus, here, we only consider the thermal power when calculating the CO_2 emission coefficient of electricity in Jiangxi. In addition, CO_2 emission coefficients of coal, oil, and natural are 2.7412, 2.1358 and 1.626 kg-CO_2/ton-SCE, respectively. The results are listed in Table A1.

Table A1. CO_2 emission coefficient of electricity (t-CO_2/10^4 kwh).

Year	2000	2001	2002	2003	2004	2005	2006	2007	2008
Coefficient	6.493	7.385	6.886	7.117	7.880	7.992	6.958	7.037	6.909
Year	**2009**	**2010**	**2011**	**2012**	**2013**	**2014**	**2015**	**2016**	**2017**
Coefficient	6.640	6.621	6.623	5.561	5.525	5.265	4.898	4.590	4.844

Table A2. CO_2 emission coefficients of different fuel types investigated in this study.

Fuel Type	Carbon Content Unit: (kg/GJ) [a]	Carbon Oxidation Unit: (%) [a]	Net Calorific Value Unit: (TJ/Gg) [a]	Emission Coefficient Unit: (t-CO_2/t)
Raw coal	25.8	100	20.9	1.977
Briquettes	26.6	100	17.6	1.717
Coke oven gas	12.1	100	16726 [b]	7.421 [c]
Other gases	12.1	100	16726 [b]	7.421 [c]
Gasoline	20.2	100	43	3.185
Kerosene	19.5	100	44.1	3.153
Diesel oil	20.2	100	43	3.185
Lubricants	20.0	100	40.2	2.948
Other petroleum products	20.0	100	40.2	2.948
Liquefied petroleum gas	17.2	100	47.3	2.983
Natural gas	15.3	100	38931 [b]	21.84 [c]
Liquefied natural gas	17.5	100	44.2	2.836

Note: [a] The value is the IPCC recommended value; [b] The unit is KJ/m^3; [c] The unit is t-CO_2/10^4 m^3.

Appendix B

Table A3. Urban, LMDI additive: Detailed additive decomposition results of the residential energy-related CO_2 emissions of urban residents in Jiangxi (10^4 t).

Periods	ΔC_{TOT}	ΔC_K	ΔC_{ES}	ΔC_{EP}	ΔC_{ED}	ΔC_{CP}	ΔC_U	ΔC_P
2000–2001	−107.69	14.76	54.10	−289.23	31.01	33.70	43.79	4.17
2001–2002	137.97	−9.21	−21.85	221.16	−160.11	75.83	27.90	4.25
2002–2003	79.07	5.58	32.87	−151.71	108.15	46.53	33.12	4.52
2003–2004	−263.66	21.64	59.71	−308.42	−102.68	40.65	22.06	3.38
2004–2005	162.23	3.89	46.39	−21.44	50.22	61.27	18.98	2.92
2005–2006	14.70	−44.88	−18.83	−25.26	30.59	46.48	23.04	3.56
2006–2007	13.93	3.77	2.03	−0.89	−101.78	90.94	16.07	3.79
2007–2008	4.81	−6.62	−10.58	−90.45	25.43	61.47	21.52	4.05
2008–2009	33.44	−14.68	5.27	−41.69	−9.63	64.78	25.15	4.24
2009–2010	95.26	−1.10	−39.05	6.03	55.29	56.48	13.19	4.42
2010–2011	61.74	0.14	3.04	24.16	−71.44	74.54	26.97	4.32
2011–2012	−15.90	−79.59	41.95	−49.91	−24.32	63.81	29.53	2.62
2012–2013	120.53	59.95	−71.47	−721.79	762.32	65.28	22.96	3.28
2013–2014	30.29	−24.03	−20.46	12.03	−44.37	79.17	24.06	3.90
2014–2015	18.77	−35.16	17.92	−46.60	−24.62	80.81	22.25	4.17
2015–2016	49.32	−32.32	14.89	−39.66	21.32	52.90	26.70	5.49
2016–2017	137.65	30.07	−1.64	−88.95	76.72	86.19	28.61	6.64
2000–2005	7.91	36.66	171.22	−549.64	−73.42	257.97	145.86	19.25
2005–2010	162.15	−63.51	−61.16	−152.25	−0.11	320.15	98.97	20.06
2010–2015	215.42	−78.69	−29.02	−782.11	597.57	363.61	125.77	18.29
2015–2017	186.98	−2.25	13.25	−128.61	98.05	139.09	55.32	12.14
2000–2017	572.46	−107.79	94.29	−1612.61	622.09	1080.82	425.92	69.74

Table A4. Rural, LMDI additive: Detailed additive decomposition results of the residential energy-related CO_2 emissions of rural residents in Jiangxi (10^4 t).

Periods	ΔC_{TOT}	ΔC_K	ΔC_{ES}	ΔC_{EP}	ΔC_{ED}	ΔC_{CP}	ΔC_U	ΔC_P
2000–2001	115.71	7.00	−7.15	82.19	29.35	11.99	−9.99	2.33
2001–2002	40.89	−4.12	−18.94	68.94	−11.64	12.55	−8.87	2.97
2002–2003	68.72	2.41	23.17	49.52	−25.06	26.52	−10.83	2.99
2003–2004	22.88	12.32	41.35	26.57	−96.27	46.15	−10.15	2.92
2004–2005	112.61	2.58	1.92	−31.81	78.12	69.63	−10.72	2.89
2005–2006	−29.35	−27.58	32.88	−99.08	30.99	43.95	−14.09	3.57
2006–2007	26.62	2.46	13.72	−106.71	67.78	55.39	−9.48	3.46
2007–2008	−4.51	−4.55	15.02	−97.22	36.53	56.45	−14.83	4.08
2008–2009	51.68	−11.23	15.51	−91.10	114.32	38.48	−18.57	4.27
2009–2010	24.69	−0.93	2.51	−19.22	−15.93	63.78	−9.77	4.23
2010–2011	15.68	0.11	5.22	−55.89	−30.75	112.32	−19.09	3.75
2011–2012	−2.01	−63.95	16.55	−31.14	33.69	62.72	−22.14	2.25
2012–2013	77.76	35.91	−46.11	−231.22	217.66	116.89	−18.14	2.79
2013–2014	31.19	−19.24	−11.31	54.47	−52.03	75.98	−19.97	3.30
2014–2015	58.97	−29.65	−0.58	31.21	−28.11	104.32	−22.26	4.02
2015–2016	42.90	−28.26	7.12	−40.17	52.19	73.26	−26.13	4.89
2016–2017	90.28	25.87	−2.03	−23.13	26.33	86.90	−29.53	5.88
2000–2005	360.81	20.19	40.35	195.41	−25.50	166.83	−50.56	14.09
2005–2010	69.11	−41.83	79.65	−413.33	233.69	258.04	−66.74	19.62
2010–2015	181.59	−76.82	−36.24	−232.56	140.46	472.22	−101.60	16.11
2015–2017	133.18	−2.39	5.09	−63.31	78.52	160.16	−55.66	10.77
2000–2017	744.69	−100.84	88.84	−513.78	427.17	1057.26	−274.55	60.59

Table A5. Urban, LMDI multiplicative: Detailed multiplicative decomposition results of the residential energy-related CO_2 emissions of urban residents in Jiangxi.

Periods	ΨC_{TOT}	ΨC_K	ΨC_{ES}	ΨC_{EP}	ΨC_{ED}	ΨC_{CP}	ΨC_U	ΨC_P
2000–2001	0.80	1.03	1.12	0.55	1.07	1.07	1.09	1.01
2001–2002	1.32	0.98	0.96	1.56	0.73	1.16	1.06	1.01
2002–2003	1.14	1.01	1.06	0.78	1.19	1.08	1.06	1.01
2003–2004	0.59	1.04	1.13	0.54	0.82	1.08	1.04	1.01
2004–2005	1.42	1.01	1.11	0.95	1.11	1.14	1.04	1.01
2005–2006	1.03	0.92	0.97	0.96	1.06	1.09	1.04	1.01
2006–2007	1.02	1.01	1.00	1.00	0.84	1.17	1.03	1.01
2007–2008	1.01	0.99	0.98	0.86	1.04	1.11	1.04	1.01
2008–2009	1.06	0.98	1.01	0.93	0.98	1.11	1.04	1.01
2009–2010	1.15	1.00	0.94	1.01	1.09	1.09	1.02	1.01
2010–2011	1.09	1.00	1.00	1.03	0.91	1.11	1.04	1.01
2011–2012	0.98	0.90	1.06	0.94	0.97	1.09	1.04	1.00
2012–2013	1.16	1.08	0.92	0.41	2.55	1.08	1.03	1.00
2013–2014	1.03	0.97	0.98	1.01	0.95	1.09	1.03	1.00
2014–2015	1.02	0.96	1.02	0.95	0.97	1.09	1.02	1.00
2015–2016	1.05	0.97	1.02	0.96	1.02	1.06	1.03	1.01
2016–2017	1.14	1.03	1.00	0.92	1.08	1.09	1.03	1.01
2000–2005	1.01	1.07	1.40	0.35	0.84	1.67	1.33	1.04
2005–2010	1.30	0.89	0.91	0.77	0.99	1.72	1.18	1.03
2010–2015	1.30	0.91	0.97	0.38	2.08	1.55	1.17	1.02
2015–2017	1.20	0.99	1.01	0.88	1.10	1.15	1.06	1.01
2000–2017	2.06	0.87	1.25	0.09	1.90	5.12	1.94	1.11

Table A6. Rural, LMDI multiplicative: Detailed multiplicative decomposition results of the residential energy-related CO_2 emissions of rural residents in Jiangxi.

Periods	ΨC_{TOT}	ΨC_K	ΨC_{ES}	ΨC_{EP}	ΨC_{ED}	ΨC_{CP}	ΨC_U	ΨC_P
2000–2001	1.55	1.03	0.97	1.37	1.12	1.05	0.96	1.01
2001–2002	1.13	0.99	0.95	1.22	0.97	1.04	0.97	1.01
2002–2003	1.19	1.01	1.06	1.13	0.94	1.07	0.97	1.01
2003–2004	1.05	1.03	1.10	1.06	0.81	1.11	0.98	1.01
2004–2005	1.25	1.01	1.00	0.94	1.16	1.15	0.98	1.01
2005–2006	0.95	0.95	1.06	0.84	1.06	1.08	0.97	1.01
2006–2007	1.05	1.00	1.03	0.83	1.13	1.10	0.98	1.01
2007–2008	0.99	0.99	1.03	0.84	1.07	1.10	0.97	1.01
2008–2009	1.09	0.98	1.03	0.86	1.21	1.07	0.97	1.01
2009–2010	1.04	1.00	1.00	0.97	0.97	1.11	0.98	1.01
2010–2011	1.02	1.00	1.01	0.92	0.95	1.19	0.97	1.01
2011–2012	1.00	0.91	1.03	0.95	1.05	1.10	0.97	1.00
2012–2013	1.12	1.05	0.94	0.72	1.37	1.18	0.97	1.00
2013–2014	1.04	0.97	0.98	1.08	0.93	1.11	0.97	1.00
2014–2015	1.08	0.96	1.00	1.04	0.97	1.14	0.97	1.01
2015–2016	1.05	0.97	1.01	0.95	1.06	1.09	0.97	1.01
2016–2017	1.10	1.03	1.00	0.97	1.03	1.10	0.97	1.01
2000–2005	2.72	1.05	1.08	1.88	0.95	1.47	0.87	1.04
2005–2010	1.12	0.93	1.15	0.48	1.51	1.56	0.89	1.03
2010–2015	1.28	0.90	0.95	0.70	1.24	1.96	0.87	1.02
2015–2017	1.16	0.99	1.01	0.93	1.10	1.20	0.94	1.01
2000–2017	4.54	0.87	1.19	0.59	1.95	5.40	0.63	1.11

References

1. Fan, J.-L.; Yu, H.; Wei, Y.-M. Residential energy-related carbon emissions in urban and rural China during 1996–2012: From the perspective of five end-use activities. *Energy Build.* **2015**, *96*, 201–209. [CrossRef]
2. IPCC. *Climate Change 2007: IPCC Fourth Assessment Report*; Cambridge University Press: Cambridge, UK, 2007.
3. Mi, Z.; Meng, J.; Guan, D.; Shan, Y.; Liu, Z.; Wang, Y.; Feng, K.; Wei, Y.-M. Pattern changes in determinants of Chinese emissions. *Environ. Res. Lett.* **2017**, *12*, 074003. [CrossRef]
4. Guan, D.; Meng, J.; Reiner, D.M.; Zhang, N.; Shan, Y.; Mi, Z.; Shao, S.; Liu, Z.; Zhang, Q.; Davis, S.J. Structural decline in China's CO_2 emissions through transitions in industry and energy systems. *Nat. Geosci.* **2018**, *11*, 551–555. [CrossRef]
5. Lin, B.; Liu, K. Using LMDI to analyze the decoupling of carbon dioxide emissions from China's heavy industry. *Sustainability* **2017**, *9*, 1198.
6. Fan, J.-L.; Liao, H.; Liang, Q.-M.; Tatano, H.; Liu, C.-F.; Wei, Y.-M. Residential carbon emission evolutions in urban–rural divided China: An end-use and behavior analysis. *Appl. Energy* **2013**, *101*, 323–332. [CrossRef]
7. Zhang, X.; Luo, L.; Skitmore, M. Household carbon emission research: An analytical review of measurement, influencing factors and mitigation prospects. *J. Clean. Prod.* **2015**, *103*, 873–883. [CrossRef]
8. Huang, R.; Zhang, S.; Liu, C. Comparing urban and rural household CO_2 emissions—Case from China's four megacities: Beijing, Tianjin, Shanghai, and Chongqing. *Energies* **2018**, *11*, 1257. [CrossRef]
9. Tian, X.; Geng, Y.; Dong, H.; Dong, L.; Fujita, T.; Wang, Y.; Zhao, H.; Wu, R.; Liu, Z.; Sun, L. Regional household carbon footprint in China: A case of Liaoning province. *J. Clean. Prod.* **2016**, *114*, 401–411. [CrossRef]
10. Shi, Y.; Han, B.; Han, L.; Wei, Z. Uncovering the national and regional household carbon emissions in China using temporal and spatial decomposition analysis models. *J. Clean. Prod.* **2019**, *232*, 966–979. [CrossRef]
11. Zhu, Q.; Peng, X.; Wu, K. Calculation and decomposition of indirect carbon emissions from residential consumption in China based on the input–output model. *Energy Policy* **2012**, *48*, 618–626. [CrossRef]
12. Lu, Y.; Zhang, Y.; Cao, X.; Wang, C.; Wang, Y.; Zhang, M.; Ferrier, R.C.; Jenkins, A.; Yuan, J.; Bailey, M.J.; et al. Forty years of reform and opening up: China's progress toward a sustainable path. *Sci. Adv.* **2019**, *5*, 9413–9422. [CrossRef] [PubMed]

13. Yang, L.; Yang, Y.; Zhang, X.; Tang, K. Whether China's industrial sectors make efforts to reduce CO_2 emissions from production?—A decomposed decoupling analysis. *Energy* **2018**, *160*, 796–809. [CrossRef]
14. Zha, D.; Zhou, D.; Zhou, P. Driving forces of residential CO_2 emissions in urban and rural China: An index decomposition analysis. *Energy Policy* **2010**, *38*, 3377–3383.
15. Fan, J.; Ran, A.; Li, X. A study on the factors affecting China's direct household carbon emission and comparison of regional differences. *Sustainability* **2019**, *11*, 4919. [CrossRef]
16. Feng, Z.-H.; Zou, L.-L.; Wei, Y.-M. The impact of household consumption on energy use and CO_2 emissions in China. *Energy* **2011**, *36*, 656–670. [CrossRef]
17. Liu, L.-C.; Wu, G.; Wang, J.-N.; Wei, Y.-M. China's carbon emissions from urban and rural households during 1992–2007. *J. Clean. Prod.* **2011**, *19*, 1754–1762. [CrossRef]
18. Bai, Y.; Deng, X.; Gibson, J.; Zhao, Z.; Xu, H. How does urbanization affect residential CO_2 emissions? An analysis on urban agglomerations of China. *J. Clean. Prod.* **2019**, *209*, 876–885. [CrossRef]
19. Fan, J.-L.; Zhang, Y.-J.; Wang, B. The impact of urbanization on residential energy consumption in China: An aggregated and disaggregated analysis. *Renew. Sust. Energ. Rev.* **2017**, *75*, 220–233. [CrossRef]
20. Ye, Y.; Shi, X.-Q.; Zou, L.-L. Decoupling economic growth from CO_2 emissions: A decomposition analysis of China's household energy consumption. *Adv. Clim. Chang. Res.* **2016**, *7*, 192–200.
21. Wang, W.; Zhao, D.; Kuang, Y. Decomposition analysis on influence factors of direct household energy-related carbon emission in Guangdong province-Based on extended Kaya identity. *Environ. Prog. Sustain. Energy* **2016**, *35*, 298–307. [CrossRef]
22. Yuan, R.; Rodrigues, J.F.D.; Behrens, P. Driving forces of household carbon emissions in China: A spatial decomposition analysis. *J. Clean. Prod.* **2019**, *233*, 932–945. [CrossRef]
23. Zhang, M.; Bai, C. Exploring the influencing factors and decoupling state of residential energy consumption in Shandong. *J. Clean. Prod.* **2018**, *194*, 253–262. [CrossRef]
24. Nie, H.; Kemp, R. Index decomposition analysis of residential energy consumption in China: 2002–2010. *Appl. Energy* **2014**, *121*, 10–19. [CrossRef]
25. Zhang, M.; Song, Y.; Li, P.; Li, H. Study on affecting factors of residential energy consumption in urban and rural Jiangsu. *Renew. Sust. Energ. Rev.* **2016**, *53*, 330–337. [CrossRef]
26. Ye, H.; Ren, Q.; Hu, X.; Lin, T.; Xu, L.; Li, X.; Zhang, G.; Shi, L.; Pan, B. Low-carbon behavior approaches for reducing direct carbon emissions: Household energy use in a coastal city. *J. Clean. Prod.* **2017**, *141*, 128–136. [CrossRef]
27. Zhao, X.; Li, N.; Ma, C. Residential energy consumption in urban China: A decomposition analysis. *Energy Policy* **2012**, *41*, 644–653. [CrossRef]
28. OECD. *Indicators to Measure Decoupling of Environmental Pressure from Economic Growth*; Organization for Economic Cooperation and Development (OECD): Paris, France, 2002.
29. Tapio, P. Towards a theory of decoupling: Degrees of decoupling in the EU and the case of road traffic in Finland between 1970 and 2001. *Transp. Policy* **2005**, *12*, 137–151. [CrossRef]
30. Wang, Q.; Li, R.; Jiang, R. Decoupling and decomposition analysis of carbon emissions from industry: A case study from China. *Sustainability* **2016**, *8*, 1059. [CrossRef]
31. Zhang, L.; Kou, C.; Zheng, J.; Li, Y. Decoupling analysis of CO_2 emissions in transportation sector from economic growth during 1995–2015 for six cities in Hebei, China. *Sustainability* **2018**, *10*, 4149. [CrossRef]
32. Wang, Y.; Xie, T.; Yang, S. Carbon emission and its decoupling research of transportation in Jiangsu Province. *J. Clean. Prod.* **2017**, *142*, 907–914. [CrossRef]
33. Jiang, R.; Li, R. Decomposition and decoupling analysis of life-cycle carbon emission in China's building Sector. *Sustainability* **2017**, *9*, 886. [CrossRef]
34. Wang, Q.; Zhao, M.; Li, R. Decoupling sectoral economic output from carbon emissions on city level: A comparative study of Beijing and Shanghai, China. *J. Clean. Prod.* **2019**, *209*, 126–133. [CrossRef]
35. Luo, Y.; Long, X.; Wu, C.; Zhang, J. Decoupling CO_2 emissions from economic growth in agricultural sector across 30 Chinese provinces from 1997 to 2014. *J. Clean. Prod.* **2017**, *159*, 220–228. [CrossRef]
36. Zhao, X.; Zhang, X.; Li, N.; Shao, S.; Geng, Y. Decoupling economic growth from carbon dioxide emissions in China: A sectoral factor decomposition analysis. *J. Clean. Prod.* **2017**, *142*, 3500–3516. [CrossRef]
37. Statistic Bureau of Jiangxi. *Jiangxi Statistical Yearbook*; China Statistics Press: Beijing, China, 2001–2018. (In Chinese)

38. Jia, J.; Gong, Z.; Xie, D.; Chen, J.; Chen, C. Analysis of drivers and policy implications of carbon dioxide emissions of industrial energy consumption in an underdeveloped city: The case of Nanchang, China. *J. Clean. Prod.* **2018**, *183*, 843–857. [CrossRef]
39. IPCC. *Gudielines for National Greenhouse Gas Inventories*; Cambridge University Press: Cambridge, UK, 2006.
40. National Bureau of Statistics of China. *China Energy Statistical Yearbook*; China Statistics Press: Beijing, China, 2001–2018. (In Chinese)
41. Kennedy, C.; Steinberger, J.; Gasson, B.; Hansen, Y.; Hillman, T.; Havránek, M.; Pataki, D.; Phdungsilp, A.; Ramaswami, A.; Mendez, G.V. Greenhouse gas emissions from global cities. *Environ. Sci. Technol.* **2009**, *43*, 7297–7302. [CrossRef]
42. Ang, B.W. The LMDI approach to decomposition analysis: A practical guide. *Energy Policy* **2005**, *33*, 867–871. [CrossRef]
43. Ang, B.W.; Liu, N. Handling zero values in the logarithmic mean Divisia index decomposition approach. *Energy Policy* **2007**, *35*, 238–246. [CrossRef]
44. Ang, B.W. Decomposition analysis for policymaking in energy. *Energy Policy* **2004**, *32*, 1131–1139. [CrossRef]
45. Ang, B.W.; Liu, F.L.; Chew, E.P. Perfect decomposition techniques in energy and environmental analysis. *Energy Policy* **2003**, *31*, 1561–1566. [CrossRef]
46. Kaya, Y. *Impact of Carbon Dioxide Emission Control on GNP Growth: Interpretation of Proposed Scenarios*; IPCC Energy and Industry Subgroup, Response Strategies Working Group: Paris, France, 1990.
47. Shao, S.; Yang, L.; Gan, C.; Cao, J.; Geng, Y.; Guan, D. Using an extended LMDI model to explore techno-economic drivers of energy-related industrial CO_2 emission changes: A case study for Shanghai (China). *Renew. Sust. Energ. Rev.* **2016**, *55*, 516–536. [CrossRef]
48. Omer, A.M. Energy, environment and sustainable development. *Renew. Sust. Energ. Rev.* **2008**, *12*, 2265–2300. [CrossRef]
49. Tian, X.; Chang, M.; Lin, C.; Tanikawa, H. China's carbon footprint: A regional perspective on the effect of transitions in consumption and production patterns. *Appl. Energy* **2014**, *123*, 19–28. [CrossRef]
50. Pachauri, S. Household electricity access a trivial contributor to CO_2 emissions growth in India. *Nat. Clim. Chang.* **2014**, *4*, 1073–1076. [CrossRef]
51. Shao, S.; Yang, L.; Yu, M.; Yu, M. Estimation, characteristics, and determinants of energy-related industrial CO_2 emissions in Shanghai (China), 1994–2009. *Energy Policy* **2011**, *39*, 6476–6494. [CrossRef]
52. Wei, Y.-M.; Liu, L.-C.; Fan, Y.; Wu, G. *China Energy Report (2008): CO_2 Emission Research*; Science Press: Beijing, China, 2008. (In Chinese)
53. Ren, S.; Fu, X.; Chen, X. Regional variation of energy-related industrial CO_2 emissions mitigation in China. *China Econ. Rev.* **2012**, *23*, 1134–1145. [CrossRef]
54. Grubb, M.; Sha, F.; Spencer, T.; Hughes, N.; Zhang, Z.; Agnolucci, P. A review of Chinese CO_2 emission projections to 2030: The role of economic structure and policy. *Clim. Policy* **2015**, *15*, S7–S39. [CrossRef]

sustainability

MDPI

Article

Revisiting the Relation between Renewable Electricity and Economic Growth: A Renewable–Growth Hypothesis

Minyoung Yang and Jinsoo Kim *

Department of Earth Resources and Environmental Engineering, Hanyang University, 222 Wangsimni-ro, Seongdong-gu, Seoul 04763, Korea; yminy94@hanyang.ac.kr
* Correspondence: jinsookim@hanyang.ac.kr; Tel.: +82-2-2220-2241

Received: 13 March 2020; Accepted: 10 April 2020; Published: 13 April 2020

Abstract: Global concern about the climate crisis has incited movements for switching to renewable electricity. Renewable electricity can contribute to economic growth as an input factor (electricity generation) and also as an industry (renewable manufacturing). We introduce a new hypothesis, the renewable–growth hypothesis, to investigate the role of the renewable manufacturing industry in the energy–growth nexus study. To test the hypothesis, we select a target country group using the market share of the renewable manufacturing industry and conduct the Granger causality test for solar photovoltaic and wind power. The autoregressive distributed lag bounds testing approach is applied for the causality test. The results show that renewable electricity Granger causes economic growth in target countries, which supports the renewable–growth hypothesis. However, the hypothesis did not hold in countries that export renewable power facilities more than they install them for domestic demand. We believe that the renewable–growth hypothesis would be secured soon if renewable electricity expands broadly over the world.

Keywords: renewable–growth hypothesis; renewable electricity; economic growth; renewable manufacturing; energy–growth nexus

---

1. Introduction

International attention to global warming comprises the global effort for carbon reduction. However, if we keep our pledges at the current level, the world's temperature will increase to almost twice the limit referred to in the Paris Agreement by the end of the century. Climate change is now referred to as the "climate crisis" [1]. This phenomenon also draws attention to the topic of energy, particularly in the electricity industry. As part of that, many countries have displayed their transition to renewable energy from fossil fuels, which are considered the main source of carbon emissions. However, many governments, especially those in Asia, are still using coal-fired power stations. Even worse, the demand growth for gas as an alternative to coal is emerging [2]. Transitioning to renewable energy is a means for solving problems caused by the climate crisis.

Indeed, renewable energy accounted for an estimated 18.1% of total final energy consumption (TFEC) in 2017. Modern renewables composed 10.6% of TFEC, with an estimated 4.4% growth in demand compared to 2016. Particularly in the power sector, renewable energy has grown to account for more than 33% of the world's total installed power-generating capacity in 2018. Solar photovoltaics (PV) comprises 55% of renewable capacity, after the additional installation of around 100 gigawatts in 2018, and is followed by wind power (28%) [3].

Economic development and sustainability are important not only for carbon reduction but also for the global trend promoting renewable energies [4–6]. Developed economies promote renewable sources to strengthen the energy security of supply and control their greenhouse gas emissions [7]. Similarly, understanding the causal relationship between renewable energy and economic growth is significant for a country's economic development and as the basis for policymakers. Therefore, extensive research has been conducted on the relationship between them. In the 1980s, many studies investigated the relationship between energy consumption and economic growth [8]. "Energy–growth nexus" is a term referring to the link between energy consumption and economic growth. The energy–growth nexus has reached a more disaggregated level over time [9].

An analysis of renewable energy growth is needed considering the described circumstances. The facility cost of renewable electricity accounts for a higher percentage in power generation costs compared to non-renewable energy [10]. This means that, if we increase solar facilities, the demand for all sectors related to them also increases. This might lead to positive effects on economic growth. Furthermore, renewable electricity is capital intensive and has a value-added effect. According to Ernst & Young [11], the solar industry in the European Union 28 represented more than 81,000 full-time equivalents and more than EUR 4600 million gross value-added (GVA). They mentioned the installed capacity has a significant impact on job and GVA creation because there is a direct impact on manufacturing and services needed. Thus, the increasing demand for renewable electricity has a positive effect on such industries and is related to economic growth.

However, if a country generally imports equipment and produces its own electricity, renewable electricity demand has less or no positive impact on economic growth. Given the cost structure of the renewable electricity industry, growth is driven by demand growth in a related facility. It has a different growth mechanism when a country simply purchases equipment. If so, it is likely to have a positive impact on the economy of the country producing that facility. This means that if demand for renewable electricity increases, there is no induced effect like great value-added or additional demand in other industries or there is only an impact on the country's imports concerning the overall facility. Therefore, the relationship between renewable electricity demand and economic growth may not be evident in these countries.

There are four hypotheses on the energy–growth nexus considering the number of cases that could be the result of the analysis [12]. First, the "neutrality hypothesis" implies the absence of a causal relationship between energy consumption and GDP. The second one, the "conservation hypothesis," suggests that energy conservation policies may be implemented with little or no adverse effect on economic growth. In the third one, the "growth hypothesis" means that energy consumption is important for economic growth. The last one that implies bidirectional causality between the two factors is the "feedback hypothesis." It is possible to derive energy policy through an analysis of the above hypothesis. The energy conservation policy means reducing energy consumption for economic growth. If a specific country's energy–growth nexus follows the growth hypothesis, energy conservation policies may have an adverse impact on economic growth. However, if under the neutrality hypothesis, energy conservation policies may not have much impact. The main purpose of the energy–growth nexus study is to examine which hypothesis is investigated in a specific country or group of countries. The results of the energy–growth nexus study have yielded mixed results within hypotheses. Existing hypotheses are focused on energy itself and do not consider specific industries. Therefore, the existing hypothesis is not enough to cover the impact of renewable energy manufacturing.

Due to the feature originating from the renewable manufacturing industry, renewable electricity and economic growth will show a positive relationship. Thus, this study tries to fill the gap with a new hypothesis, which is the "renewable–growth hypothesis." We analyze the time-series data of countries to confirm the renewable–growth relationship. The main contributions of this study are as follows. First, this study investigates if the development of the renewable energy generation sector has a positive effect on economic growth considering the features of the renewable power sector. Second, a new perspective of renewable–growth hypothesis is proposed, and the national group supporting the analysis is established. This makes it possible to draw implications for the policy direction of fostering the renewable electricity industry. Furthermore, this study presents the issues to promote further studies.

The paper is organized as follows. Section 2 provides a review of the literature. Section 3 describes the countries to be analyzed and the data to be used for analysis and explains the model and methodology. Section 4 provides empirical results. Section 5 concludes the paper and provides remarks and policy implications.

2. Literature Review

2.1. Overview of Energy–Growth Nexus

The energy–growth nexus has been studied in order to confirm the direction of the causality and its hypothesis has been well developed via various studies (see Table 1). They can be divided into country-specific studies [13–16] and multi-country studies [17–19]. Some of the studies attempted to examine the causal relationship in the energy–growth nexus in both developed and developing countries or by using different period data [8–10,18,20–22]. A study can be conducted for a specific research purpose. Pao and Fu [23] tried to investigate the relationship between economic growth and energy consumption in Brazil. Contrary to the previous study, this study covered various types of energy. They found mixed results: A conservation hypothesis between non-renewable energy consumption and economic growth, a growth hypothesis between non-hydroelectric renewable energy consumption and economic growth, and a feedback hypothesis between economic growth and total renewable energy consumption. Others focused more on the causal relationship between economic growth and energy consumption [24,25]. Apergis and Danuletiu [24] examined the relationship between economic growth and renewable energy consumption for 80 countries using the long-run causality test. They concluded that there is a bidirectional causality between renewable energy consumption and economic growth in the long run. Kazar and Kazar [25] investigated the relationship between development and renewable electricity net generation values for 154 countries with a panel analysis. They found the presence of bidirectional causality in the short run and that the causal relationship differs both in the short run and long run depending on the human development level.

2.2. Electricity and Economic Growth

The energy–growth nexus has been studied in various countries and on a more disaggregated level [9]. In addition to the interest in climate change, many countries and policymakers have tried to implement an electricity conservation policy. The confirmed results of research have been used as a basis for implementing such a policy and to establish the right policy direction for the country. The electricity–growth nexus for a single country has been studied and developed for that reason. Ramcharran [26] studied the electricity–growth nexus in Jamaica and found that the country is energy dependent. Ghosh [27] and Narayan and Smyth [28] investigated energy consumption and economic growth in India and Australia, respectively. They found Granger causality from economic growth to electricity consumption in both countries. In addition, research is being conducted in various countries such as Korea [29], Bangladesh [30], Cyprus [31], China [32], Turkey [33], Malaysia [34], Lebanon [35], and Nigeria [36]. However, due to the omitted variable bias, the studies that use only energy consumption and economic growth as variables should be interpreted with caution. In that context, some studies have attempted to make econometric transformations by adding employment variables to the bivariate model [37] or setting up the model reflecting the structural beaks [38]. Lorde et al. [39] constructed a multivariate model using the neoclassical production model to examine the economic theory. In various studies, bivariate and multivariate models were analyzed, in order or simultaneously. The impact on economic growth has been studied for the implementation of national power-related policies until recently [40–43].

An electricity–growth nexus may be established by setting up a group of countries depending on the research purpose. Yoo [44] analyzed the relationship between electricity consumption and economic growth in ASEAN countries for similar characteristics in the electricity sector. This study tried to confirm the relationship between electricity consumption and economic growth within a similar-featured group but with some differences in terms of investment in the power sector. Each of the four countries showed similar results with the two countries, respectively. The Organization of the Petroleum Exporting Countries (OPEC) is expected to be greatly influenced by energy conservation due to the characteristics of oil-producing countries. However, despite the obvious similarities, the results vary depending on the model and the time period [45,46]. This "no consensus" feature of the electricity–growth nexus was observed among the countries in the group even when the econometric method was modified. Acaravci and Ozturk [47] emphasized that they have results that conflict with the existing literature through the identification of the Granger causality of panel data. At the same time, they stated and emphasized that this issue deserves more attention and needs further research.

Some studies approach the characteristics of a country from an economic perspective. Apergis and Payne [48] analyzed a total of 88 countries using the panel vector error correction models based on the level of economic income. The study showed different results depending on whether a country was assessed in terms of the short run or the long run and also depending on its development level. In the long run, bidirectional causality exists in both high-income and upper-middle-income panels and lower-middle-income country panels. The growth hypothesis is satisfied in lower-middle-income country panels and low-income country panels in the short run. Even in the same group, mixed results were observed according to the short-run and long-run views [49], but some studies show a consensus between studies analyzing the same country [50].

Over time, research has been further disaggregated in various aspects, with national concerns shifting to energy transitions rather than just energy conservation. As a result, studies distinguishing renewable and non-renewable electricity from total electricity have begun [51–57]. Ibrahiem [51] mentions the limitations of the structure of the Egyptian electricity market, such as the crude-oil shortage, the need for renewable growth, or the transition on power mix. Apergis and Payne [52] conducted the panel causality test for the emerging market, including Egypt. Apergis and Payne [53] extend their work by examining the causal dynamics between renewable and non-renewable electricity consumption and economic growth in Central America. They show the negative bidirectional causality between renewable and non-renewable electricity consumption and conclude that the cause of this is that imported petroleum products raise concerns about the security of the region's energy supply. One of their contributions is their investigation of the negative bidirectional causality between renewable and non-renewable electricity consumption. They believe that this is due to imported petroleum products raising concerns about the security of the region's energy supply.

The analysis of renewable electricity is relevant to climate change and policies for global warming mitigation. Based on the results of the study, each country is recommended to increase its investment in renewable energy projects and vice versa (Table 1). According to the confirmed results of Al-mulali et al. [54], they recommended that Latin American countries should encourage not only the investment for renewable energy projects but also reduction in the role of non-renewable sources in electricity consumption. In recent years, these recommendations have been implemented in a way analysis of non-standard Granger causality [55–57]. In this way, it is possible to confirm the effect of specific energy sources [55] or conduct the analysis not only on aggregate electricity models but also those disaggregated into renewable and non-renewable models [57]. Despite the development of the research scope by the economic and econometric approach, the field of energy–growth nexus must still be investigated not only in terms of methodology but also in economics or policy.

Table 1. The previous study-related electricity growth nexus.

Author	Country	Period	Methodology	Finding
Ramcharran (1990) [26]	Jamaica	1970–1986	Granger causality	$EC \rightarrow Y$
Ghosh (2002) [27]	India	1950–1997	Granger causality	$Y \rightarrow EC$
Narayan and Smyth (2005) [28]	Australia	1966–1999	Multivariate Granger causality	$Y \rightarrow EC$
Yoo (2005) [29]	Korea	1970–2002	Cointegration VECM (vector error correction model) Brown parameter stability test	$Y \Leftrightarrow EC$
Yoo (2006) [44]	4 countries	1971–2002	Hsiao's version of Granger causality Standard Granger causality test	$Y \rightarrow EC$ (Thailand, Indonesia) $EC \rightarrow Y$ (Singapore, Malaysia)
Mozumder and Marathe (2007) [30]	Bangladesh	1971–1999	Cointegration VECM	$Y \rightarrow EC$
Zachariadis and Pashourtidou (2007) [31]	Cyprus	1960–2004	Cointegration Granger causality VECM	$Y \Leftrightarrow EC$
Yuan et al., (2007) [32]	China	1978–2004	Cointegration Hodrick–Prescott (HP) filter Granger causality	$EC \rightarrow Y$
Squalli (2007) [45]	All OPEC members	1980–2003	Cointegration ARDL (autoregressive distributed lag) Bounds Test Toda and Yamamoto causality test	$Y \rightarrow EC$ (Indonesia, Libya, Iraq, Algeria,) $EC \rightarrow Y$ (Kuwait, Venezuela) $Y \Leftrightarrow EC$ (Iran, Venezuela, Qatar) Mixed outcomes with different models (Nigeria, Saudi Arabia, Indonesia, Kuwait, and UAE)
Halicioglu (2007) [33]	Turkey	1968–2005	ARDL bounds test Granger causality	$Y \rightarrow EC$
Tang (2008) [34]	Malaysia	1972–2003	ARDL bounds test Toda and Yamamoto causality test Brown parameter stability test	$EC \rightarrow Y$
Abosedra et al., (2009) [35]	Lebanon	1995–2005	Granger causality	$EC \rightarrow Y$
Akinlo (2009) [36]	Nigeria	1980–2006	Cointegration Granger causality VECM	$EC \rightarrow Y$
Ghosh (2009) [37]	India	1970–2006	ARDL bounds test Cointegration VECM	$Y \rightarrow$Electricity supply (short-run)

Table 1. *Cont.*

Author	Country	Period	Methodology	Finding
Acaravci (2010) [38]	Turkey	1968–2005	Cointegration Granger causality VECM	EC → Y
Lorde et al., (2010) [39]	Barbados	1980–2006	Granger causality VECM VAR (vector auto regressive)	EC → Y (short-run) EC ←→ Y (long-run)
Yoo and Kwak (2010) [46]	Argentina Brazil Chile Columbia Ecuador Peru Venezuela	1975–2006	Johansen cointegration Hsiao's Granger causality	EC → Y (Argentina, Brazil, Chile, Columbia, Ecuador) Y ⇔ EC (Venezuela) Y == EC (Peru)
Chandran et al. (2010) [40]	Malaysia	1971–2003	ARDL bounds test Granger causality	EC → Y (short-run) EC → Y (long-run)
Acaravci and Ozturk (2010) [47]	15 transition countries	1990–2006	Pedroni cointegration Granger causality	Y == EC
Apergis and Payne (2011a) [48]	88 countries	1990–2006	Panel cointegration test	Y ⇔ EC (high income and upper-middle-income country panels) EC → Y (short-run, lower-middle-income country panel) Y ⇔ EC (long-run, lower-middle-income country panel) EC → Y (short-run, low income-country panel)
Ozturk and Acravci (2011) [49]	11 Middle East and North Africa (MENA) countries	1971–2006	ARDL bounds test Granger causality	Y → EC (short-run, Israel, Oman) EC → Y (long-run, Egypt and Saudi Arabia) Y == EC (Iran, Jordan, Morocco, Syria)
Bekhet and Othman (2011) [41]	Malaysia	1970–2009	Cointegration Granger causality	EC → Y (long-run)
Shahbaz et al. (2011) [42]	Portugal	1971–2009	ARDL bounds test Granger causality VECM	EC → Y (short-run) Y ⇔ EC (long-run)
Apergis and Payne (2011b) [52]	16 emerging market economies	1990–2007	Panel cointegration Panel Granger causality	Y → REC (short-run) Y ⇔ REC (long-run) Y ⇔ NREC
Apergis and Payne (2012) [53]	6 Central American countries	1990–2007	Panel cointegration test	REC → Y (short-run) Y ⇔ REC (long-run) Y ⇔ NREC
Al-mulali et al. (2014) [54]	18 Latin American countries	1980–2010	Panel cointegration test	Y ⇔ REC (long-run) NREC → Y (short-run) Y ⇔ NREC (long-run)

Table 1. *Cont.*

Author	Country	Period	Methodology	Finding
Halkos and Tzeremes (2014) [55]	36 countries	1990–2011	Nonparametric analysis	Based on Growth hypothesis
Ibrahiem (2015) [51]	Egypt	1980–2011	ARDL bounds testing approach	Y ⇔ REC (long-run)
Kumari and Sharma (2016) [43]	India	1974–2014	Cointegration Granger Causality	Y → EC
Atems and Hotaling (2018) [56]	174 countries	1980–2012	The system generalized method of moments (GMM) approach	Positive relationship between Y, renewable electricity generation and non-renewable electricity generation
Al-Mulali et al. (2018) [50]	Gulf Cooperation Council (GCC) member	1980–2014	Panel cointegration test Panel Granger causality test	Y → EC (short-run) Y ⇔ EC (long-run)
Aydin (2019) [57]	26 OECD countries	1980–2015	Cross-sectional dependence test Panel unit root test Dumitrescu-Hurlin (DH) panel causality test Panel frequency domain causality	Y == EC (DH) Y ⇔ REC, NREC (panel frequency)

Note: Y, EC, REC, NREC mean economic growth, electricity consumption, renewable electricity consumption, and non-renewable electricity consumption, respectively. EC → Y, growth hypothesis; Y → EC, conservation hypothesis; Y ⇔ EC, feedback hypothesis; Y == EC, neutrality hypothesis.

2.3. Revisited Nexus Study

This study tries to investigate the causal relationship between renewable electricity and economic growth. Given what previous studies have referred to as further studies [9,47,56], we "revisited" the nexus study. As mentioned in the previous "Literature Review" section, many studies refer to the existence of a consensus in the nexus field. Nowadays, to try to clarify the related points, several revisited studies are being conducted. Andrew and Bothwell [58] mentioned the limitation of using a bivariate or trivariate model within South Africa. They try to revisit the electricity–growth nexus by using a multivariate model considering the economic aspects of a country like export, employment, and consumer price index. Zortuk and Karacan [59] revisited the energy–growth nexus by selecting countries included in panel data. The ex-Soviet countries located in Central and Eastern Europe and the Caucasus are now in transition into free market economics. Therefore, analysis panel data containing the related countries can clarify the direction of the causal relationship. We also try to revisit the electricity–growth nexus in the same country as the previous study, but we will be using a different methodology and have a different purpose to them.

In this study, we have three distinctions from the previous study. First, the study's purpose is to confirm the renewable electricity–growth nexus by considering the characteristics of renewable energy. This feature is discussed in detail in the next section. Second, to reflect the characteristics that vary depending on the source, we built a model at the renewable source level. The differences in the characteristics of energy sources can affect the investigated findings [56]. There are cases in which previous studies have considered disaggregated level like sources [26,55]; however, only specific sources (steam, hydro, diesel, and gas turbine) were used for data availability and analytical purposes [26]. Halkos and Tzeremes [55] conducted their analysis using wind power, biomass, solar power, and geothermal data. However, the availability of the data used differed from country to country, and a non-parametric technique was applied. The last difference is that generation data is used for the application of the first and second implications mentioned above. The various differences between generation and consumption could affect interpretation in this study, which aimed at reflecting renewable characteristics rather than electricity [40,56]. This study is meaningful in that it analyzes through a revisiting flow in which the implication of the study is made by both the setting of the country and the variables.

3. Data and Methodology

3.1. Renewable–Growth Hypothesis

In this study, we analyze the renewable electricity and economic growth nexus considering renewable manufacturing industries. As mentioned above, the cost structure of renewable electricity is quite different from the conventional thermal power. In the renewable generation sector, much of the cost goes to equipment or facilities, which means a capital expenditure is much higher than an operational expense. On the other hand, the thermal power plants that use fossil fuels have substantial operational costs, although their capital cost is also considerable. Promoting renewable electricity could thus have a positive effect on economic growth through the production of renewable manufacturing industries such as wind turbine and solar panel manufacturing. In other words, renewable electricity can contribute to economic growth through industrial outputs as well as an energy input. We want to call this mechanism the 'renewable–growth' hypothesis in the context of the energy–growth nexus.

To test the renewable–growth hypothesis, we choose a country in which renewable electricity might lead to the development of related industries. We confirm that the renewable–growth nexus in a country includes companies that have developed in the renewable electricity sector. This study attempts to examine the positive relationship between economic growth and renewable electricity demand in such countries based on the market share of renewable manufacturing companies. At the moment, we are analyzing solar PV and wind power, which are expected to grow quickly [3]. We might expect to see a causal relationship by identifying the more disaggregated sectors.

Based on sales volume or revenue, we selected a global renewable energy company in solar [60] and wind [61] power. According to the collected data, we identified the country the company belongs to for analytical purposes. The analysis is conducted on solar and wind power, separately. Tables A1 and A2 show the companies and corresponding countries with the highest sales volume. For the solar PV, we have included the component in the plant and the market share of module companies. Four countries are analyzed for the solar PV: China (including Hong Kong), Canada, the USA, and Korea. A key component of the wind manufacturing industry is the turbine, which accounts for 60% of the total capital expenditure [62]. Thus, the standard for selecting the country included in the wind model is a global wind-turbine company [Table A2]. Wind models include six countries: China, the USA, Denmark, Germany, India, and Spain.

3.2. Data and Estimation Procedure

Consistent with the previous literature, we use real GDP as the dependent variable. Independent variables are electricity generation, gross fixed capital, and labor force. Due to the features of renewables, to transmission, to distribution losses, or even to theft, bias can exist in the investigated result [56]. This feature was also affected by country development level [40,63]. Therefore, we consider electricity generation data. Considering the connectivity to the power grid and intermittent characteristics of renewable energy, the detailed consumption data of each source is not available. To include this, the previous study used aggregate level consumption data. However, this study's purpose is to identify the effects of the nexus depending on the presence of manufacturing companies reflecting the manufacturing feature of each source. Reflecting on the characteristics of the energy source means it must be analyzed at a disaggregated level. Therefore, generation data is used for analysis, and solar PV and wind are configured separately.

We adopt the annual time series data of each country from 1980 to 2017. The data are obtained from the World Bank Development Indicators, World Energy Balances, and US Energy Information Administration and defined as follows: Real GDP (Y) in billion 2010 USD using purchasing power parity, fossil fuel electricity generation (NRE) defined in kilowatt hours, electricity generation from solar PV (RES) defined in kilowatt hours, electricity generation from wind (REW) defined in kilowatt hours, real gross fixed capital formation in current US dollars (K), and total labor force (L) in millions. We convert the unit of real gross fixed capital formation into constant 2010 US dollars using the GDP deflator from WDI. Additionally, all variables are converted to per capita and the natural logarithms are transformed.

3.3. Estimation Strategy

We assume that, due to the characteristics of renewables, renewable electricity generation and economic growth have a relationship. To confirm the causal relationship between them, we must check the stationarity of time series data. If data are not stationary, the estimated model might be a spurious regression and results may lack robustness. Common methods applied to unit root tests are the augmented Dickey–Fuller (ADF) [64], Phillips–Perron (PP) [65], and Kwiatkowski–Phillips–Schmidt–Shin (KPSS) [66]. If the null hypothesis of the ADF and PP tests are rejected, we conclude that time series data has a unit root. The KPSS test is applied to investigate the stationarity, and, if the null hypothesis is rejected, the data may be nonstationary time series data. The KPSS test is more suitable for testing small samples due to the lower lag truncation parameter, and it might complement the limitation of the ADF and PP tests [67]. Therefore, as a tool for cross-checking and for the robustness of results, we conduct ADF, PP, and KPSS tests together.

Before the Granger causality analysis, it is necessary to test for cointegration. There are several methodologies related to cointegration, and the Johansen test [68] is the most common method as it is more generally applicable than the Engle–Granger test [69]. Since this test is common and well known, we provide just a brief overview of this method. Johansen [68] modeled time series as a reduced rank regression, and we can trace test and maximum eigenvalue. The model is given as the following:

$$\Delta Z_t = \omega + \sum_{i=1}^{q-1} \Delta Z_{t-i} + \Pi \Delta Z_{t-1} + \varepsilon_t \tag{1}$$

The Z_t is a vector of p variables consisting of an (n × 1) column, and ω is an (n × 1) vector of constant terms. Γ means coefficient matrices, and Δ is a difference operator. The ε_t follows distribution as $N(0, \Sigma)$. The Π means the coefficient is known as the impact matrix and contains information about the long-run relationships. By employing this method, it is possible to determine the number of the cointegrating vectors of the model.

We can also employ the cointegration test with the autoregressive distributed lag (ARDL) bounds test approach [70]. By using ARDL bounds testing, we can expect to take advantage of the following. First, it is possible to apply such a method irrespective of whether the regressors are I (0) or I (1). Furthermore, compared to the Johansen cointegration techniques, estimating with a smaller sample size is possible [71]. According to Narayan [72], the ARDL bounds test gives a reasonable critical value if the number of samples is between 30 and 80. Furthermore, the ARDL model can be derived in the form of an error correction model. Thi means it is possible to confirm that the long-run and short-run causality is the same as in VECM. In this study, we can establish the ARDL model as the following:

$$\begin{aligned} \Delta GDP_t = \alpha_0 + \sum_{i=1}^{q} \alpha_1 \Delta GDP_{t-i} + \sum_{i=1}^{q} \alpha_2 \Delta NRE_{t-i} + \sum_{i=1}^{q} \alpha_3 \Delta RE_{t-i} + \sum_{i=1}^{q} \alpha_4 \Delta K_{t-i} + \\ \lambda_1 GDP_{t-1} + \lambda_2 NRE_{t-1} + \lambda_3 RE_{t-1} + \lambda_4 K_{t-1} + \mu_t \end{aligned} \tag{2}$$

$$\begin{aligned} \Delta RE_t = \beta_0 + \sum_{i=1}^{q} \beta_1 \Delta RE_{t-i} + \sum_{i=1}^{q} \beta_2 \Delta NRE_{t-i} + \sum_{i=1}^{q} \beta_3 \Delta GDP_{t-i} + \sum_{i=1}^{q} \beta_4 \Delta K_{t-i} + \\ \zeta_1 RE_{t-1} + \zeta_2 NRE_{t-1} + \zeta_3 GDP_{t-1} + \zeta_4 K_{t-1} + \mu_{2t} \end{aligned} \tag{3}$$

where GDP, NRE, RE, and K denote the logarithm form of real GDP, fossil fuel electricity generation, solar PV or wind electricity generation, and gross domestic capital formation, respectively. The parameters α, β are the short-run dynamic coefficient and λ, ζ are the corresponding long-run multipliers of each ARDL model. The μ represents the white noise error term.

To determine the existence of cointegration, we test the lagged value jointly by using the F test. However, as the F statistics from [70] are for several samples, we usually use the critical value from Narayan in the small sample size analysis [72]. The null hypotheses of each model are $H_o: \lambda_1 = \lambda_2 = \lambda_3 = 0$ and $H_o: \zeta_1 = \zeta_2 = \zeta_3 = 0$. If the computed F statistics exceed the upper bound value, the null hypothesis is rejected. However, if the F statistics fall below the lower bound, we can conclude that we cannot reject the no cointegration hypothesis. However, if the value exists between the upper bound and lower bound, the results are inconclusive. Through the cointegration test, if we find the evidence for a long-run relationship between variables, we can conduct the test to check the existence of Granger causality.

According to the results from the cointegration test, there are two cases we must consider. If our variables are not cointegrated, we perform the test as a vector autoregressive (VAR) model in first differenced variable form. If we can confirm the existence of a long-run relationship, we can conduct the model with the error correction term. Thus, the model has a cointegration, and the Granger causality relationships are written as the vector error correction models (VECM) given below:

$$\begin{aligned} \Delta GDP_t = \alpha_0 + \sum_{i=1}^{q} \alpha_{1i} \Delta GDP_{t-i} + \sum_{i=1}^{q} \alpha_{2i} \Delta NRE_{t-i} + \sum_{i=1}^{q} \alpha_{3i} \Delta RE_{t-i} + \sum_{i=1}^{q} \alpha_{4i} \Delta K_{t-i} + \\ \lambda ECT_{t-1} + \mu_{1t} \end{aligned} \tag{4}$$

$$\Delta RE_t = \beta_0 + \sum_{i=1}^{q} \beta_{1i} \Delta RE_{t-i} + \sum_{i=1}^{q} \beta_{2i} \Delta NRE_{t-i} + \sum_{i=1}^{q} \beta_{3i} \Delta GDP_{t-i} + \sum_{i=1}^{q} \beta_{4i} \Delta K_{t-i} + \zeta ECT_{t-1} + \mu_{2t} \tag{5}$$

The null hypothesis of the Granger causality from renewable electricity generation to GDP is $H_o: \alpha_1 = \alpha_2 = \ldots \alpha_q = 0$. To test the short-run causality of renewable electricity to GDP, we impose restrictions on all the lagged renewable electricity generation data using the F test. This is equivalent to the test of lagged first differences of the Granger causality from renewable electricity generation to GDP.

Whether or not they are integrated, the economic variables can be integrated into different orders. In that case, the Wald test will not have an asymptotic chi-square distribution, and VECM cannot be applied for the Granger causality test. Toda and Yamamoto [73] suggest the procedure for solving this problem. According to the standard stationary test, we can determine the order of the variables. Let m be the maximum order of integration, and l be the appropriate maximum lag length in VAR. Then take the preferred VAR model and add the m additional lags of each of the variables. We can test the Granger causality using that model. However, we must test the hypothesis with only the coefficients of the first l lagged values. Rejection of the null hypothesis implies the existence of Granger causality.

4. Empirical Results

4.1. Unit Root Test Results

Tables A3 and A4 present the result of a stationarity test for the solar PV case and the wind power case from ADF, PP, and KPSS test. We use Stata 14.0 for the analysis. The definition of GDP is real GDP, FOG is fossil-fuel electricity generation, RE is renewable electricity generation, and FXC is fixed capital formation. According to the model, RE is divided into solar PV (RES) and wind power (REW). In the process of examining stationarity, the maximum lag length was set to 4 considering the characteristics of the economic variables.

As mentioned above, ADF, PP, and KPSS tests were conducted simultaneously for robustness. If the differenced form does not reject the null hypothesis in more than two tests, the second differenced variable is tested. The first differenced variable is then analyzed based on the statistics considering the trend. Additionally, if we get the result that the first differenced is stationary in at least two tests, we do not write the statistics of the second differenced form in the table. Two or more tests indicate the same order of integration, and we set that value as the integration order of that variable. The results show that most variables in both models are in the form of I (1). Only the GDP variable of Spain in wind power has I (2). Since the order of integration of the variables is mixed, the analysis proceeds with Toda and Yamamoto's procedure.

Therefore, we constructed the VAR model with the variables to find the optimal lag length of the model setting the maximum lag as 4. Lag-order selection is based on Bayesian information criterion (BIC). In each model, the VAR model was constructed according to the optimal time difference obtained in each country, autoregression was checked, and the time difference was adjusted. The order of integration of variables and the optimal lag of the VAR model are summarized in Tables 2 and 3. Table 2 is for the solar PV, and Table 3 is for wind power. To investigate the Granger causality of each country and model, we reconstructed the VAR model with the sum of these two values as the lag length of the new model.

Table 2. Integration order and optimal lag results for solar photovoltaics (PV).

Country	Variables	Order of Integration (m)	Optimal Lag (l)
Canada	GDP FOG RES FXC	I(1) I(1) I(1) I(1)	1
China	GDP FOG RES FXC	I(1) I(1) I(1) I(1)	3
USA	GDP FOG RES FXC	I(1) I(1) I(1) I(1)	4
Korea	GDP FOG RES FXC	I(1) I(1) I(1) I(1)	3

Table 3. Integration order and optimal lag results for wind power.

Country	Variables	Order of Integration (m)	Optimal Lag (p)
China	GDP FOG REW FXC	I(1) I(1) I(1) I(1)	3
USA	GDP FOG REW FXC	I(1) I(1) I(1) I(1)	5
Denmark	GDP FOG REW FXC	I(1) I(1) I(1) I(1)	2
Germany	GDP FOG REW FXC	I(1) I(1) I(1) I(1)	2
India	GDP FOG REW FXC	I(1) I(1) I(1) I(1)	1
Spain	GDP FOG REW FXC	I(2) I(1) I(1) I(1)	2

4.2. Results of the Granger Causality

Tables 4 and 5 summarize the results of the solar PV and wind power, respectively. The country shows that the growth hypothesis presented in the previous nexus study are Canada for the solar PV, and Germany, India, and Spain for wind power. China and the USA were included in both models, satisfying the conservative and feedback hypothesis, respectively. Denmark was the only country with a neutral hypothesis. The renewable–growth relationship is investigated in Canada, USA, and Korea for the solar PV and the USA, Germany, India, and Spain for wind power. China in both models and Denmark in the wind power show different results based on our hypothesis.

Table 4. The results of the Granger causality in solar PV.

Country	Granger Causality	Test Statistics	Hypothesis	
			Conventional	Renewable–Growth
Canada	RES → GDP GDP → RES	3.27 * 0.20	Growth	Yes
China	RES → GDP GDP → RES	3.61 8.70 **	Conservative	No
USA	RES → GDP GDP → RES	20.25 *** 25.74 ***	Feedback	Yes
Korea	RES → GDP GDP → RES	18.14 *** 16.48 ***	Feedback	Yes

Note: *, **, and *** denote the rejection of the null hypothesis of no relationship at the 5% and 1% significance level, respectively.

Table 5. The results of the Granger causality wind power.

Country	Granger Causality	Test Statistics	Hypothesis	
			Conventional	Renewable–Growth
China	REW → GDP GDP → REW	0.79 23.84 ***	Conservative	No
USA	REW → GDP GDP → REW	18.48 *** 17.12 ***	Feedback	Yes
Denmark	REW → GDP GDP → REW	0.09 0.28	Neutral	No
Germany	REW → GDP GDP → REW	14.00 *** 4.25	Growth	Yes
India	REW → GDP GDP → REW	2.75 * 0.04	Growth	Yes
Spain	REW → GDP GDP → REW	6.26 ** 0.71	Growth	Yes

Note: *, **, and *** denote the rejection of the null hypothesis of no relationship at the 5% and 1% significance level, respectively.

In this study, the standard Granger causality test was also conducted to confirm the reliability of the results. The difference of the standard Granger causality is after the stationarity test, which means identifying the cointegration of the model. The Johansen method was used to confirm the cointegration, and the bounds test was also performed considering the small sample size. If cointegration exists, the VECM model is constructed to confirm the causality of the long run and short run. If the results indicate no cointegration in both tests, a VAR model with a differenced variable must be built, and the short run causality is checked using that model. The lag length of each model is the same as that of Toda and Yamamoto. Tables A5 and A6 summarize the standard Granger causality results. Only China, for the solar PV, shows a conflicting result in the cointegration test. The variables in this model are all I (1), so we follow the bounds test result due to the small sample size. Furthermore, because the lag lengths of Canada for solar PV and India for wind power are both 1, the short-run term disappears when constructing the VECM model. Therefore, both cases were analyzed only in the long-run relationship. However, due to the integration order of the variables, we used and analyzed the results using the Toda and Yamamoto procedure.

5. Discussion

The study aims to investigate the renewable–growth hypothesis of the countries with a renewable manufacturing industry. According to the results, the long-run relationship between renewable electricity generation and economic growth is present in all countries. However, in some countries, the results differ from our hypothesis. The results can be attributed to diverse factors like economic or socio-cultural points existing in different countries [45,47,52,74].

In China, the opposite of the renewable–growth hypothesis was found in both solar PV and wind power. China is the largest producer of both solar and wind components. The total installed solar PV was 175 GW in 2018, while the total for wind power was 185 GW in 2018 [75]. However, China's exports of manufactured goods to other countries are much higher than its domestic use [60,61]. Therefore, the causality test with renewable electricity generation in China may not reflect the positive effect of the renewable electricity industry. China's renewable electricity is also a reason for the ambiguous factor of the renewable–growth relationship. The renewable electricity industry induces economic growth in the indirect sectors, such as the development of the manufacturing sector and employment for facilities and the direct impact of the generation itself. Electricity curtailment due to grid issues is a chronic problem in China [76,77]. The resulting loss of generation has a negative effect on economic growth [60].

Denmark also has a similarity with China. In Denmark, renewable energy accounts for more than 50% of total electricity production [78]. However, the increase in installed wind energy capacity is the smallest among countries with wind power [75]. In the absence of increased domestic facilities, it is difficult to identify the growth hypothesis with given variables. Furthermore, Siemens Gamers has manufacturing facilities worldwide [79]. Vestas' manufacturing facilities are also located in other countries [80]. Thus, even if they satisfy the renewable–growth hypothesis, they may not be revealed under the given variables. In the rest of the target countries, we can examine the renewable–growth hypothesis.

The renewable electricity industry of countries showing a growth hypothesis has been more established in recent years. Germany, which shows the renewable–growth relationship between wind power and economic growth, had the largest wind power capacity in 2017. Germany leads the European Union (EU) in terms of power capacity with 56.1 GW, followed by Spain with 23.2 GW, which is also included in our analysis model. The manufacturing facility of Gamesa, which merged with Danish renewable company Siemens in 2016, still exists in Spain, and this could have a positive effect on the country. According to GWEC [81], one of the leading turbine suppliers for the EU is Siemens Gamesa. While installation numbers have slowed, the total capacity of India is 32.8 GW, taking the fourth position in the world's largest wind markets. Because the share of wind-generated electricity in India's total electricity consumption is still 4.35%, there is still growth potential in India's renewable electricity market [81]. The USA has the highest installed capacity after China among our target countries, both for solar PV and wind power. In 2018, the total installed capacity was 51 GW for solar PV and 94 GW for wind power [75]. In Korea, the installed solar PV capacity is 8 GW, but the number of related companies and employment is increasing due to recent aggressive renewable electricity policies. Canada has also installed a solar PV capacity of 3 GW, and while it is not growing much, it is considered to have high potential as the capacity continues to rise.

Under our hypothesis, the driving force of economic growth is the electricity industry itself, and other industries have induced effects from the increase of renewable generation. We used solar PV generation data and wind power generation data as variables to cover these points. However, these variables do not reflect the growth effect of the renewable manufacturing industry effectively in some countries. Development of relevant industries can lead to the growth of domestic products usage. So we set renewable electricity data as a variable in our model. This is a limitation of the study in that it does not consider the characteristics of trade. In the same line, the diverse factors that are inherent in countries need to be considered. The various political, economic, and socio-cultural factors could cause a bias in the nexus study. Therefore, further research to secure the renewable–growth hypothesis is necessary. Developing variables to consider the export-focused characteristic and renewable energy policy of each country could also be the next step of the renewable–growth hypothesis. Furthermore, the existence of renewable growth may not be revealed in the results due to bias arising from the small sample size. It is necessary to overcome this limitation in the future with continuous research on specific energy sources. A country's characteristics may vary and be embodied in the results, so it also becomes the motivation for additional research. Lastly, our study is the first to present the renewable–growth hypothesis. However, it must be noted that while we use an analysis strategy by using Granger causality, it is unknown whether the relationship is positive or negative [82]. A more diversified approach could make it clear whether the hypothesis is correct or not.

6. Conclusions

The increasing concerns about the climate crisis cause trends such as energy transition from fossil fuels to renewable energy. According to the change in the power mix, the renewable electricity industry is affected both directly and indirectly. The difference between conventional electricity and renewable electricity sources lies in the weight of manufacturing-related parts. This is the case with solar PV and wind power, both comprising the world's highest growth in renewable electricity installation. Due to such characteristics, the increase in renewable energy generation has the additional effect of not only leading to advantages in the generation industry but also the development of related industries and increase in employment. As this will have a positive effect on economic growth, we have thus formulated the renewable–growth hypothesis, which is one step further from the growth hypothesis of the existing nexus studies.

In this study, we tried to investigate the relationship between renewable electricity and economic growth. We constructed and analyzed more disaggregated models like solar PV and wind power. Additionally, we established the target group based on global companies in manufacturing. It was expected that the renewable–growth relationship would be represented more accurately in those countries. The variables within the model are real GDP, fossil fuel electricity generation, renewable electricity generation, and fixed capital formation. Renewable electricity generation means the solar PV data for solar PV and wind power data in wind power. Because our purpose is to confirm the renewable–growth hypothesis, we used electricity generation as a variable, considering the difference between generation and consumption data.

The analysis showed that the renewable–growth hypothesis was satisfied except only China for solar PV and China and Denmark for wind power. These results are due to the characteristics of the country's power and power-related manufacturing industries. We can thus conclude that national and industry characteristics can influence the results of the analysis. Furthermore, we can provide the direction for further studies. First, renewable electricity time series data has a relatively small sample size. Therefore, if data is acquired over time, it is necessary to analyze a larger sample size. Doing this would determine the renewable–growth relationship more accurately. Second, considering additional variables to reflect the characteristics of industry or country would be necessary. However, as we can see from the results of other countries, the parameters were appropriate for verifying the hypothesis. Thus, while considering the general case, further research is necessary to reflect the characteristics of each country. The study's main contributions are its presentation of a new perspective in the form of the renewable–growth hypothesis and the establishment and analysis of a target group that reflects the characteristics of the renewable electricity industry.

Author Contributions: Conceptualization, J.K. and M.Y.; Methodology, J.K. and M.Y.; Validation, J.K.; Formal Analysis, M.Y.; Investigation, J.K. and M.Y.; Data Curation, M.Y.; Writing—Original Draft Preparation, M.Y.; Writing—Review & Editing, J.K.; Supervision, J.K.; Funding Acquisition, J.K. All authors have read and agreed to the published version of the manuscript.

Funding: This work was supported by the Human Resources Development program (No. 20194010201860) of the Korea Institute of Energy Technology Evaluation and Planning (KETEP) grant funded by the Korea government Ministry of Trade, Industry and Energy.

Conflicts of Interest: The authors declare no conflict of interest.

Appendix A

Table A1. The highest production company in solar PV (module).

Rank	Company	Country	Production [MW]
1	Jinko Solar	China	9000
2	JA Solar	China	8500
3	Canadian Solar	Canada	8310
4	Hanwha Q CELLS	Korea	8000
5	Trina Solar Ltd.	China	8000
6	Risen	China	6600
7	GCL System	China	5400
8	Talesun	China	4500
9	Suntech/Shunfeng	China	3300
10	Znshine Solar	China	3200
11	Seraphim	China	3000
12	Chint/Astronergy	China	2500
13	First Solar	USA	2200
14	Eging	China	2000
15	BYD	China	1700

Table A2. The market share of global wind power companies.

Rank	Company	Country	Market Share (%, 2014)
1	Vestas	Denmark	16%
2	Siemens Gamesa	Denmark	15%
3	Goldwind	China	12%
4	GE Wind	USA	10%
5	Enercon	Germany	7%
6	Nordex	Germany	6%
7	Envision	China	6%
8	Senvion	Germany (India)	3%
9	Suzlon	India	3%
10	Guodian UP	China	3%
11	Ming Yang	China	2%

Appendix B

Table A3. Unit root test results for solar PV.

Country	Variables		Test Statistics		
			ADF	PP	KPSS
Canada	GDP	Level	−2.144	−6.105	0.225 ***
		First difference	−4.556 ***	−26.242 ***	0.116
	FOG	Level	−1.233	−3.816	0.416 ***
		First difference	−6.374 ***	−40.340 ***	0.0413
	RES	Level	−2.427	−10.865	0.113
		First difference	−3.013 **	−15.118 **	0.0943
	FXC	Level	−2.565	−7.935	0.25 ***
		First difference	−5.095 ***	−31.560 ***	0.0992

Table A3. *Cont.*

Country	Variables		Test Statistics		
			ADF	PP	KPSS
China	GDP	Level	−2.371	−10.118	0.183 **
		First difference	−3.094 **	−15.071 **	0.0984
	FOG	Level	−1.804	−8.698	0.2 **
		First difference	−3.546 **	−17.539 **	0.187 **
	RES	Level	−0.372	−0.427	0.7 ***
		First difference [1]	−4.122 **	−24.522 ***	0.116
	FXC	Level	−1.750	−6.024	0.413 ***
		First difference	−3.875 ***	−17.560 **	0.128 ′
USA	GDP	Level	−0.485	−0.967	0.301 ***
		First difference	−5.216 ***	−30.371 ***	0.244 ***[2]
	FOG	Level	1.072	2.428	0.343 ***
		First difference	−6.544 ***	−41.387 ***	0.158 *[2]
	RES	Level	−2.923	−7.352	0.266 ***
		First difference	−5.637 ***	−35.112 ***	0.0927
	FXC	Level	−2.716	−11.397	0.142 ′
		First difference	−5.800 ***	−34.972 ***	0.0615
Korea	GDP	Level	−3.063	−8.906	0.126 ′
		First difference	−2.654 *	−13.941 **	0.293 ***[2]
	FOG	Level	−1.362	−5.247	0.113
		First difference	−2.902 *	−52.064 ***	0.0957
	RES	Level	−2.557	−11.407	0.111
		First difference	−2.963 *	−46.013 ***	0.112
	FXC	Level	−2.222	−8.004	0.284 ***
		First difference	−5.671 ***	−34.438 ***	0.06

Note: *, **, and *** indicate the level of significance at 10%, 5%, and 1% for ADF, PP. ′, *, **, and *** indicate the level of significance at 10%, 5%, 2.5%, and 1% for KPSS. [1] first difference form using trend when testing ADF and PP. [2] the result of second difference form is stationary.

Table A4. Unit root test results for wind power.

Country	Variables		Test Statistics		
			ADF	PP	KPSS
China	GDP	Level	−2.371	−10.118	0.183 **
		First difference	−3.094 **	−15.071 **	0.0865
	FOG	Level	−1.804	−8.698	0.164 *
		First difference	−3.546 **	−17.539 **	0.158 *[2]
	REW	Level	−2.863	−10.116	0.0846
		First difference	−4.715 ***	−28.629 ***	0.0773
	FXC	Level	−1.750	−6.024	0.413 ***
		First difference	−3.875 ***	−17.560 **	0.128 ′[2]
USA	GDP	Level	−0.485	−0.967	0.301 ***
		First difference	−5.216 ***	−30.371 ***	0.244 ***[2]
	FOG	Level	1.072	2.428	0.343 ***
		First difference	−6.544 ***	−41.387 ***	0.158 *[2]
	REW	Level	−2.515	−10.241	0.305 ***
		First difference [1]	−7.100 ***	−43.660 ***	0.033
	FXC	Level	−2.716	−11.397	0.142 ′
		First difference	−5.800 ***	−34.972 ***	0.0615

Table A4. *Cont.*

Country	Variables		Test Statistics		
			ADF	PP	KPSS
Denmark	GDP	Level	−1.684	−3.320	0.414 ***
		First difference	−4.619 ***	−27.620 ***	0.0744
	FOG	Level	−0.391	−1.660	0.426 ***
		First difference	−8.083 ***	−48.551 ***	0.0307
	REW	Level	−2.446	−1.864	0.448 ***
		First difference	−4.422 ***	−24.953 ***	0.0876
	FXC	Level	−2.696	−13.698	0.175 *
		First difference	−4.836 ***	−28.317 ***	0.102
Germany	GDP	Level	−1.291	−2.180	0.444 ***
		First difference	−4.301 ***	−25.521 ***	0.114
	FOG	Level	−1.094	−6.726	0.225 ***
		First difference	−6.980 ***	−44.514 ***	0.0939
	REW	Level	−0.777	−0.912	0.459 ***
		First difference	−2.665 *	−12.125 *	0.331 ***[2]
	FXC	Level	−3.843 **	−11.353	0.128 ′
		First difference	−2.662 *	−30.106 ***	0.0918
India	GDP	Level	−1.422	−2.319	0.454 ***
		First difference	−3.879 ***	−22.200 ***	0.0682
	FOG	Level	−2.775	−3.588	0.411 ***
		First difference	−4.405 ***	−26.979 ***	0.167 *[2]
	REW	Level	−1.771	−5.858	0.318 ***
		First difference [1]	−5.555 ***	−34.473 ***	0.162 *[2]
	FXC	Level	−1.843	−6.219	0.342 ***
		First difference	−6.157 ***	−38.722 ***	0.0848
Spain	GDP	Level	−2.854	−5.346	0.239 ***
		First difference	−2.704 [1]	−15.080 **	0.225 ***
		Second difference	−8.087 ***	−45.537 ***	0.0283
	FOG	Level	−1.146	−7.526	0.173 *
		First difference	−5.891 ***	−44.595 ***	0.0633
	REW	Level	−0.946	−3.469	0.394 ***
		First difference [1]	−4.827 ***	−29.612 ***	0.328 ***[2]
	FXC	Level	−2.425	−7.233	0.166 *
		First difference	−3.404 **	−22.619 ***	0.117

Note: *, **, and *** indicate the level of significance at 10%, 5%, and 1% for ADF, PP. ′, *, **, and *** indicate the level of significance at 10%, 5%, 2.5%, and 1% for KPSS. [1] first difference form using trend when testing ADF and PP. [2]: the result of second difference form is stationary.

Appendix C

Table A5. The results from the standard Granger causality test in solar PV.

Country	Cointegration		Granger Causality	
	Johansen	Bounds Test	Short-Run	Long-Run
Canada	Rank 1	6.388 ***	- [1]	RES → Y ***
China	Rank 1	1.935	RES → Y **	- [2]
USA	Rank 1	4.867 *	No causality	RES → Y *
Korea	Rank 1	10.649 ***	Y → RES ***	RES → Y *** Y → RES ***

Note: ′, *, **, and *** indicate the level of significance at 10%, 5%, 2.5%, and 1% for bound-testing. *, **, and *** indicate the level of significance at 10%, 5%, and 1% for Granger causality. [1] the lag of this model is 1, so we can investigate the long-run relationship result only. [2] according to the bounds test result, we investigate the short-run relationship using only VAR.

Table A6. The results from the standard Granger causality test in wind power.

Country	Cointegration		Granger Causality	
	Johansen	Bounds Test	Short-Run	Long-Run
China	Rank 1	5.951 ***	No causality	No causality
USA	Rank 1	4.737 ***	Y → REW **	No causality
Denmark	Rank 2	4.041 ′	No causality	REW → Y ***
Germany	Rank 1	8.590 ***	REW → Y * Y → REW *	REW → Y ***
India	Rank 2	5.888 ***	- [1]	REW → Y **
Spain	Rank 3	16.836 ***[3]	REW → Y **	Y → REW *

Note: ′, *, **, and *** indicate the level of significance at 10%, 5%, 2.5%, and 1% for Bound-testing. *, **, and *** indicate the level of significance at 10%, 5%, and 1% for Granger causality. [1] the lag of this model is 1, so we can investigate the long-run relationship result only. [2] this model does not have cointegration, so we can investigate the short-run relationship only. [3] while this model includes the I (2) variable, we applied the bounds test for reference.

References

1. Damian, C. Climate Crisis: 11,000 Scientists Warn of 'Untold Suffering'. *The Guardian*. 5 November 2019. Available online: https://www.theguardian.com/environment/2019/nov/05/climate-crisis-11000-scientists-warn-of-untold-suffering (accessed on 2 January 2020).
2. Pierre, F.; Matthew, W.J.; Michael, O.; Robbie, M.A.; Judith, H.; Glen, P.P.; Wouter, P.; Julia, P.; Stephen, S.; Corinne, L.Q.; et al. Global Carbon Budget 2019. *Earth Syst. Sci. Data* **2019**, *11*, 1783–1838. [CrossRef]
3. REN21 (Renewable Energy Policy Network for the 21st Century). *Renewables 2018: Global Status Report*; REN21 Publications: Paris, France, 2019.
4. Kaygusuz, K. Energy for sustainable development: Key issues and challenges. *Energy Sources Part B Econ. Plan. Policy* **2007**, *11*, 73–83. [CrossRef]
5. Kaygusuz, K.; Yuksek, O.; Sari, A. Renewable energy sources in the European union: Markets and capacity. *Energy Sources Part B Econ. Plan. Policy* **2007**, *2*, 19–29. [CrossRef]
6. Park, S.; Kim, J. The effect of interest in renewable energy on US household electricity consumption: An analysis using Google Trends data. *Renew. Energy* **2018**, *127*, 1004–1010. [CrossRef]
7. Moselle, B. Why Support Renewables? In Proceedings of the EPRG Spring Research Seminar, University of Cambridge, Cambridge, UK, 13 May 2011.
8. Kraft, J.; Kraft, A. Interfuel substitution and energy consumption in the industrial sector. *Appl. Energy* **1980**, *38*, 348–349. [CrossRef]
9. Ozturk, I. A literature survey on energy–growth nexus. *Energy Policy* **2010**, *38*, 340–349. [CrossRef]

10. Harold, A.; Pablo, R.; Michael, T. *Renewable Power Generation Costs in 2018*; International Renewable Energy Agency: Abu Dhabi, UAE, 2019.
11. EY (Ernst & Young). Available online: https://www.solarpowereurope.org/wp-content/uploads/2018/08/Solar-PV-Jobs-Value-Added-in-Europe-November-2017.pdf (accessed on 2 January 2020).
12. Payne, J.E. Survey of the international evidence on the causal relationship on the causal relationship between energy consumption and growth. *J. Econ. Stud.* **2010**, *37*, 53–95. [CrossRef]
13. Yu, E.S.H.; Hwang, B.K. The relationship between energy and GNP: Further results. *Energy Econ.* **1984**, *6*, 186–190. [CrossRef]
14. Payne, J.E. On the dynamics of energy consumption and output in the US. *Appl. Energy* **2009**, *86*, 575–577. [CrossRef]
15. Soytas, U.; Sari, R.; Ozdemir, O. Energy Consumption and GDP Relations in Turkey: A Co-integration and Vector Error Correction Analysis. *Glob. Bus. Technol. Assoc.* **2001**, *1*, 838–844.
16. Soytas, U.; Sari, R. Energy consumption, economic growth, and carbon emissions: Challenges faced by an EU candidate member. *Ecol. Econ.* **2009**, *68*, 1667–1675. [CrossRef]
17. Masih, A.; Masih, R. Energy consumption and real income temporal causality, results for a multi-country study based on cointegration and error-correction techniques. *Energy Econ.* **1996**, *18*, 165–183. [CrossRef]
18. Soytas, U.; Sari, R. Energy consumption and GDP: Causality relationship in G-7 countries and emerging markets. *Energy Econ.* **2003**, *25*, 33–37. [CrossRef]
19. Narayan, P.K.; Smyth, R. Energy consumption and real GDP in G7 countries: New evidence from panel cointegration with structural breaks. *Energy Econ.* **2008**, *30*, 2331–2341. [CrossRef]
20. Apergis, N.; Tang, C.F. Is the energy-led growth hypothesis valid? New evidence from a sample of 85 countries. *Energy Econ.* **2013**, *38*, 24–31. [CrossRef]
21. Cheng, B.S. Energy Consumptions and economic growth in Brazil, Mexico and Venezuela: A time series analysis. *Appl. Energy* **1997**, *4*, 671–674. [CrossRef]
22. Sari, R.; Soytas, U. The growth of income and energy consumption in six developing countries. *Energy Policy* **2007**, *35*, 889–898. [CrossRef]
23. Pao, H.T.; Fu, H.C. Renewable energy, non-renewable energy and economic growth in Brazil. *Renew. Sustain. Energy Rev.* **2013**, *25*, 381–392. [CrossRef]
24. Apergis, N.; Danuletiu, D.C. Renewable energy and economic growth: Evidence from the sign of Panel long-run causality. *Int. J. Energy Econ. Policy* **2014**, *4*, 578–587.
25. Kazar, G.; Kazar, A. The Renewable energy production-economic development nexus. *Int. J. Energy Econ. Policy* **2014**, *4*, 312–319.
26. Ramcharran, H. Electricity consumption and economic growth in Jamaica. *Energy Econ.* **1990**, *12*, 65–70. [CrossRef]
27. Ghosh, S. Electricity consumption and economic growth in India. *Energy Policy* **2002**, *30*, 125–129. [CrossRef]
28. Narayan, P.K.; Smyth, R. Electricity consumption, employment and real income in Australia evidence from multivariate Granger causality tests. *Energy Policy* **2005**, *33*, 1109–1116. [CrossRef]
29. Yoo, S.H. Electricity consumption and economic growth: Evidence from Korea. *Energy Policy* **2005**, *33*, 1627–1632. [CrossRef]
30. Mozumder, P.; Marathe, A. Causality relationship between electricity consumption and GDP in Bangladesh. *Energy Policy* **2007**, *35*, 395–402. [CrossRef]
31. Zachariadis, T.; Pashourtidou, N. An empirical analysis of electricity consumption in Cyprus. *Energy Econ.* **2007**, *29*, 183–198. [CrossRef]
32. Yuan, J.; Zhao, C.; Yu, S.; Hu, Z. Electricity consumption and economic growth in China: Cointegration and co-feature analysis. *Energy Econ.* **2007**, *29*, 1179–1191. [CrossRef]
33. Halicioglu, F. Residential electricity demand dynamics in Turkey. *Energy Econ.* **2007**, *29*, 199–210. [CrossRef]
34. Tang, C.F. A re-examination of the relationship between electricity consumption and economic growth in Malaysia. *Energy Policy* **2008**, *36*, 3077–3085. [CrossRef]
35. Abosedra, S.; Dah, A.; Ghosh, S. Electricity consumption and economic growth, the case of Lebanon. *Appl. Energy* **2009**, *86*, 429–432. [CrossRef]
36. Akinlo, A.E. Electricity consumption and economic growth in Nigeria: Evidence from cointegration and co-feature analysis. *J. Policy Modeling* **2009**, *31*, 681–693. [CrossRef]

37. Ghosh, S. Electricity supply, employment and real GDP in India: Evidence from cointegration and Granger-causality tests. *Energy Policy* **2009**, *37*, 2926–2929. [CrossRef]
38. Acaravci, A. Structural breaks, electricity consumption and economic growth: Evidence from Turkey. *J. Econ. Forecast.* **2010**, *2*, 140–154.
39. Lorde, T.; Waithe, K.; Francis, B. The importance of electrical energy for economic growth in Barbados. *Energy Econ.* **2010**, *32*, 1411–1420. [CrossRef]
40. Chandran, V.G.R.; Sharma, S.; Madhavan, K. Electricity consumption–growth nexus: The case of Malaysia. *Energy Policy* **2010**, *38*, 606–612. [CrossRef]
41. Bekhet, H.A.; Othman, N.S. Causality analysis among electricity consumption, consumer expenditure, gross domestic product (GDP) and foreign direct investment (FDI): Case study of Malaysia. *J. Int. Financ. Econ.* **2011**, *3*, 228–235.
42. Shahbaz, M.; Tang, C.F.; Shabbir, M.S. Electricity consumption and economic growth nexus in Portugal using cointegration and causality approaches. *Energy Policy* **2011**, *39*, 3529–3536. [CrossRef]
43. Kumari, A.; Sharma, A.K. Analyzing the causal relations between electric power consumption and economic growth in India. *Electr. J.* **2016**, *29*, 28–35. [CrossRef]
44. Yoo, S.H.; Kim, Y. Electricity generation and economic growth in Indonesia. *Energy* **2006**, *31*, 2890–2899. [CrossRef]
45. Squalli, J. Electricity consumption and economic growth: Bounds and causality analyses of OPEC members. *Energy Econ.* **2007**, *29*, 1192–1205. [CrossRef]
46. Yoo, S.H.; Kwak, S.Y. Electricity consumption and economic growth in seven South American countries. *Energy Policy* **2010**, *38*, 181–188. [CrossRef]
47. Acaravci, A.; Özturk, I. Electricity consumption–growth nexus: Evidence from panel data for transition countries. *Energy Econ.* **2010**, *32*, 604–608. [CrossRef]
48. Apergis, N.; Payne, J.E. A dynamic panel study of economic development and the electricity consumption–growth nexus. *Energy Econ.* **2011**, *33*, 770–781. [CrossRef]
49. Ozturk, I.; Acaravci, A. Electricity consumption and real GDP causality nexus: Evidence from ARDL bounds testing approach for 11 MENA countries. *Appl. Energy* **2011**, *88*, 2885–2892. [CrossRef]
50. Al-Mulali, U.; Tang, C.F.; Tan, B.W.; Ozturk, I. The nexus of electricity consumption and economic growth in Gulf cooperation council economies: Evidence from non-stationary panel data methods. *Geosyst. Eng.* **2019**, *22*, 40–47. [CrossRef]
51. Ibrahiem, D.M. Renewable electricity consumption, foreign direct investment and economic growth in Egypt: An ARDL approach. *Procedia Econ. Financ.* **2015**, *30*, 313–323. [CrossRef]
52. Apergis, N.; Payne, J.E. Renewable and non-renewable electricity consumption–growth nexus: Evidence from emerging market economies. *Appl. Energy* **2011**, *88*, 5226–5230. [CrossRef]
53. Apergis, N.; Payne, J.E. The electricity consumption–growth nexus: Renewable versus non-renewable electricity in Central America. *Energy Sources Part B Econ. Plan. Policy* **2012**, *7*, 423–431. [CrossRef]
54. Al-mulali, U.; Fereidouni, H.G.; Lee, J.Y. Electricity consumption from renewable and non-renewable sources and economic growth: Evidence from Latin American countries. *Renew. Sustain. Energy Rev.* **2014**, *30*, 290–298. [CrossRef]
55. Halkos, G.E.; Tzeremes, N.G. The effect of electricity consumption from renewable sources on countries' economic growth levels: Evidence from advanced, emerging and developing economies. *Renew. Sustain. Energy Rev.* **2014**, *39*, 166–173. [CrossRef]
56. Atems, B.; Hotaling, C. The effect of renewable and nonrenewable electricity generation on economic growth. *Energy Policy* **2018**, *112*, 111–118. [CrossRef]
57. Aydin, M. Renewable and non-renewable electricity consumption-economic growth nexus: Evidence from OECD countries. *Renew. Energy* **2019**, *136*, 599–606. [CrossRef]
58. Phiri, A.; Nyoni, B. Re-visiting the electricity–growth nexus in South Africa. *Stud. Bus. Econ.* **2016**, *11*, 97–111. [CrossRef]
59. Zortuk, M.; Karacan, S. Energy–growth nexus revisited: An empirical application on transition countries. *Environ. Dev. Sustain.* **2018**, *20*, 605–623. [CrossRef]
60. BNEF (Bloomberg New Energy Finance). *3Q 2018 Global PV Market Outlook*; BNEF Publications: New York, NY, USA, 2018.

61. BNEF (Bloomberg New Energy Finance). *Technical and Financial Service in the Wind Sector*; BNEF Publications: New York, NY, USA, 2018.
62. BNEF (Bloomberg New Energy Finance). *Onshore Wind the Experience Curve Revisited*; BNEF Publications: New York, NY, USA, 2018.
63. Karki, S.K.; Michael, D.M.; Hossein, S. Energy and environment in the ASEAN: Challenges and opportunities. *Energy Policy* **2005**, *33*, 499–509. [CrossRef]
64. Dickey, D.A.; Fuller, W.A. Distribution of the Estimators for Autoregressive Time Series with a Unit Root. *J. Am. Stat. Assoc.* **1979**, *74*, 427–431.
65. Phillips, P.C.B.; Perron, P. Testing for a unit root in time series regression. *Biometrika* **1988**, *75*, 335–346. [CrossRef]
66. Kwiatkowski, D.; Phillips, P.C.B.; Schmidt, P.; Shin, Y. Testing the null hypothesis of stationarity against the alternative of a unit root. *J. Econom.* **1992**, *54*, 159–178. [CrossRef]
67. Lin, B.; Moubarak, M. Renewable energy consumption—Economic growth nexus for China. *Renew. Sustain. Energy Rev.* **2014**, *40*, 111–117. [CrossRef]
68. Johansen, S. Estimation and Hypothesis Testing of Cointegration Vectors in Gaussian Vector Autoregressive Models. *Econometrica* **1991**, *59*, 1551–1580. [CrossRef]
69. Engle, R.F.; Granger, C.W.J. Co-integration and error correction: Representation, estimation, and testing. *Econometrica* **1987**, *55*, 251–276. [CrossRef]
70. Pesaran, M.H.; Shin, Y.; Smith, R. Bounds testing approaches to the analysis of level relationships. *J. Appl. Econom.* **2001**, *16*, 289–326. [CrossRef]
71. Fosu, O.E.; Magnus, F.J. Bound Testing Approach to Cointegration: An Examination of Foreign Direct Investment Trade and Growth Relationships. *Am. J. Appl. Sci.* **2006**, *3*, 2079–2085.
72. Narayan, P.K. The Saving and investment nexus for China: Evidence from cointegration tests. *Appl. Econ.* **2005**, *37*, 1979–1990. [CrossRef]
73. Toda, H.Y.; Yamamoto, T. Statistical inferences in vector auto regressions with possibly integrated processes. *J. Econom.* **1995**, *66*, 225–250. [CrossRef]
74. Jin, T.; Kim, J. Relationship between coal consumption and economic growth for OECD and non-OECD countries. *Geosyst. Eng.* **2016**, *19*, 48–56. [CrossRef]
75. BP (British Petroleum). Available online: https://www.bp.com/content/dam/bp/business-sites/en/global/corporate/pdfs/energy-conomics/statistical-review/bp-stats-review-2019-full-report.pdf (accessed on 17 February 2020).
76. NDRC (National Development and Reform Commission). Available online: https://www.ndrc.gov.cn/xxgk/zcfb/tz/201803/t20180323_962694.html (accessed on 25 February 2020).
77. Lin, B.; Wang, Y. Inconsistency of economic growth and electricity consumption in China: A panel VAR approach. *J. Clean. Prod.* **2019**, *229*, 144–156. [CrossRef]
78. REN21 (Renewable Energy Policy Network for the 21st Century). *Advancing the Global Renewable Energy Transition*; REN21 Publications: Paris, France, 2018.
79. Simens Gamesa. Available online: https://www.siemensgamesa.com/about-us/location-finder (accessed on 27 February 2020).
80. Vestas. Available online: http://us.vestas.com/manufacturing (accessed on 27 February 2020).
81. GWEC (Global Wind Energy Council). Available online: https://gwec.net/wp-content/uploads/vip/GWEC_PRstats2017_EN-003_FINAL.pdf (accessed on 6 February 2020).
82. Taeyoung, J.; Jinsoo, K. Coal Consumption and Economic Growth: Panel Cointegration and Causality Evidence from OECD and Non-OECD Countries. *Sustainability* **2018**, *10*, 660. [CrossRef]

sustainability

MDPI

Article

Strategic Planning of Offshore Wind Farms in Greece

Sofia Spyridonidou [1], Dimitra G. Vagiona [1,*] and Eva Loukogeorgaki [2]

1 Department of Spatial Planning and Development, Aristotle University of Thessaloniki, 54124 Thessaloniki, Greece; sspyrido@plandevel.auth.gr
2 Department of Civil Engineering, Aristotle University of Thessaloniki, 54124 Thessaloniki, Greece; eloukog@civil.auth.gr
* Correspondence: dimvag@plandevel.auth.gr

Received: 9 January 2020; Accepted: 23 January 2020; Published: 26 January 2020

Abstract: In the present article, a new methodological framework for the efficient and sustainable exploitation of offshore wind potential was developed. The proposed integrated strategic plan was implemented for the first time at national spatial planning scale in Greece. The methodological approach is performed through geographical information systems (GIS) and Microsoft Project Server Software and includes five distinct stages: (i) definition of vision/mission, (ii) identification of appropriate areas for offshore wind farms' (OWFs) siting, (iii) determination of the OWFs' layout, (iv) calculation of the OWFs' (projects) total investment cost and, finally, (v) portfolio analysis. The final outcome of the proposed strategic planning is the prioritization of the proposed sixteen offshore wind projects based on their strategic value, as well as the estimation of the overall investment cost of the entire portfolio. High economic, socio-political and environmental benefits could be achieved through the implementation of only 60% of the total investment capital of the proposed strategic plan.

Keywords: strategic planning; site selection process; offshore wind farms; geographic information systems; portfolio analysis; Greece

1. Introduction

In recent years, there has been a growing interest towards the installation of OWFs, due to the existence of multiple benefits related to the siting and operation of wind turbines offshore, such as existence of stronger winds of longer duration, availability of extensive free space for the construction of large-scale projects, reduction, and/or avoidance of noise and visual disturbances caused to the landscape by these structures, etc. Following the installation of the first OWF in Denmark in the early 1990s, a significant increase of the offshore wind industry was noted in the first decade of 2000, with the overall capacity doubling every 2–4 years [1].

On a global scale, according to statistical figures from Global Wind in 2014, over 90% of all offshore wind installations were implemented in European waters. Offshore wind energy in Europe reached the record figure of 3148 megawatt (MW) of total installed capacity in 2017, which corresponds to 560 new offshore wind turbines and 17 OWFs [1,2]. This particular record is two times higher than the figures of 2016 and 4% higher than the previous record of 2015 [2]. In the following year (2018), 409 new offshore wind turbines connected to the electricity grid across 18 offshore wind projects in Europe [3]. The referred amount corresponds to 2649 MW of net additional capacity, which is 15.8% lower than in 2017 [3]. Thus, Europe's cumulative offshore wind capacity reached 18,499 MW at the end of 2018, which corresponds to a total of 4543 grid-connected wind turbines across 11 European countries [3]. At present, the UK has the largest amount of offshore wind capacity in Europe with 44% of all installations in MW, followed by Germany with 34%, Denmark (7%), Belgium (6.4%) and the Netherlands (6%) [3]. Finally, in recent years focus was given on the development of OWFs in deeper waters [4–7], where floating support structures are preferable. For example, useful methodologies have been developed for determining with accuracy all the relevant economic decision variables of floating OWFs [6], for proposing the best technological alternatives [5] and analyzing future wind resources in deeper waters [7].

An important process of the installation of OWFs is the determination of areas suitable for the deployment of the offshore energy systems. The site selection for utilizing OWFs corresponds to a multidimensional decision-making issue. Although several applications in various spatial planning scales can be found in the literature [8–24], the applications of OWF siting at national scale are really handful. Only five out of seventeen [8,10,14,17,19], refer to site suitability analyses on a national spatial planning scale. More specifically, [8] applied five factors and seven constraints using multi-criteria decision making and GIS models to provide a suitability map for offshore wind energy in Egypt. The analysis was conducted at large scale covering the whole of Egypt and its surrounding waters. The large-scale potential of China's offshore wind energy from the perspective of current technical, spatial, and economic constraints and its possible contributions to the nation's energy system was investigated in [10]. With the aid of a GIS-based tool, offshore wind potential was evaluated as a combination of wind resources, technical projections of wind turbines, economic costs and spatial constraints of offshore wind farms. Location-specific levelized production cost and cost supply curves of offshore wind energy were also developed. A two-step decision-making procedure was adopted in [14] to evaluate the locations for offshore wind farms in Greece. Unsuitable locations were initially rejected using GIS and three constraint factors, while the remaining sites were evaluated with the AHP method and five evaluation criteria. A spatial model for the assessment of offshore wind energy potential, production costs, and the identification of suitable areas based on GIS was presented in [17] and applied in a part of the Danish Exclusive Economic Zone (EEZ). Finally, a multi-criteria site selection analysis was performed by considering technical, social, and civil restrictions for finding the most suitable offshore wind farm locations in Turkey among the 55 coastal regions, including their technical power capacities [19].

Considering the tools and techniques that have been applied so far in the OWF siting literature, several researchers have applied multi-criteria techniques to rank OWF siting alternatives (e.g., [8,12–14]), indicating that multi-criteria decision analysis plays a crucial role in OWF siting. GIS presents also an important tool for the identification and the selection of suitable sites for the installation of wind farms either on land or in the marine environment [25]. It has been used globally in many countries for the site selection of offshore wind projects, such as Egypt [8], Ohio-USA [9], China [10], Greece [11–14], South Korea [15], Spain [16], Thailand [26], etc.

What is missing from the current literature and practice is the development of an integrated strategic plan for the efficient and sustainable exploitation of the offshore wind potential and the relevant deployment of OWFs. In the present paper a new Strategic Planning methodology to identify and prioritize suitable areas for offshore wind sites is introduced, which addresses a gap in knowledge in the offshore wind energy field. In order to do this, this work utilizes a countrywide case study (Greece) where the developed methodology is applied. It should be noted that no OWFs have been developed so far in Greece.

The methodology proposed and applied includes the implementation of five distinct stages corresponding to: (i) the definition of vision and mission of the strategic planning, as previously mentioned, (ii) the identification of appropriate areas for OWFs' siting based on specific exclusion criteria, (iii) the determination of the OWFs' layout in the aforementioned areas, (iv) the calculation of the OWFs' (projects) total investment cost by calculating the capital expenditure (CAPEX), the operating expenses (OPEX) and the decommissioning expenses (DECEX) and, finally, (v) a portfolio analysis based on seven assessment criteria.

A critical advantage of the proposed methodology is that it addresses existing gaps on renewable energy sources (RES) siting issues, by: (i) introducing a holistic, step-by-step, OWF siting methodology, which considers all the relevant critical issues that an OWF developer or/and an energy planner should analyze and resolve, (ii) providing a long-term planning approach (25 years, after the plan's implementation), and (iii) recognizing a multi-disciplinary approach, as it considers legal, technical, economic, environmental, societal, and political issues. The final outcome is the assessment and ranking of all OWF project proposals considering their strategic value and cost constraints. The proposed methodology can be easily applied in other regions by following the abovementioned five stages. The novelty of the paper lies both on the integrated methodology itself (strategic spatial planning) and on the tools and criteria used in the analysis.

More specifically, GIS is used in a twofold way in the present paper: (i) for identifying the most suitable areas (SAs) for OWFs in Greece and (ii) for determining the layout and for the first time the precise location coordinates of the wind turbines in each OWF (OWFs' mapping and micro-siting determination) and, therefore, the energy capacity of the projects. The estimation of the precise location coordinates of the wind turbines in an OWF is a critical issue in the planning phase and should be accomplished before the construction phase of such large-scale projects. Moreover, the present paper proposes for the first time a prioritization of OWF projects (and, thus, an identification of optimum sites for OWFs' installation) through portfolio analysis. Portfolio analysis includes a collection of projects/proposals that will compete for selection based on their cost relative to their strategic value. The Microsoft Project Server tool, which integrates multi-criteria evaluation techniques and mathematical optimization, is used for the first time on the subject of OWFs' siting, in order to perform the required portfolio analysis and strategic scenarios in relation to the country' energy needs.

Finally, a total of twenty (exclusion and assessment) criteria are employed in the proposed methodology and the present site suitability support framework almost fully covers the economic, social, political, technical and environmental dimensions of the OWFs' siting problem in a national spatial planning scale. The proposed methodology deploys a number of criteria and restrictions of previous studies (e.g., wind velocity, water depth, distance from protected areas), while it introduces innovative criteria in relation to OWF siting issues. Exclusion criteria such as seismic hazard zones, landscape protection/visual and acoustic disturbance as well as assessment criteria (AC) such as electrical energy demand and distance from military exercise areas (firing fields and exercise locations) are applied for the first time at national planning scale.

This paper is structured in eight sections. Section 2 briefly presents the proposed methodology. Section 3 defines the exclusion criteria and the relevant exclusion zones and presents the sources and the processing method of the required spatial data. In Section 4, the main technical specifications of offshore wind turbines (e.g., most suitable support structure definition) and their layout characteristics considered in this study are presented. Section 5 describes the method applied for estimating CAPEX, OPEX, and DECEX during the life cycle of the projects and, thus, the total investment costs of the portfolio projects. Section 6 includes a description of the criteria used for assessing SA for OWF siting and of the portfolio analysis, while in Section 7 the results of all stages of the methodological framework are presented and discussed. Finally, in Section 8 the conclusions of the present study are cited.

2. Materials and Methods

In order to identify the most appropriate, sustainable, technically and economically viable solutions to site offshore wind projects in Greece the strategic planning methodology shown in Figure 1 is developed and applied in the present paper. The proposed methodological framework consists of five stages, which are analyzed below.

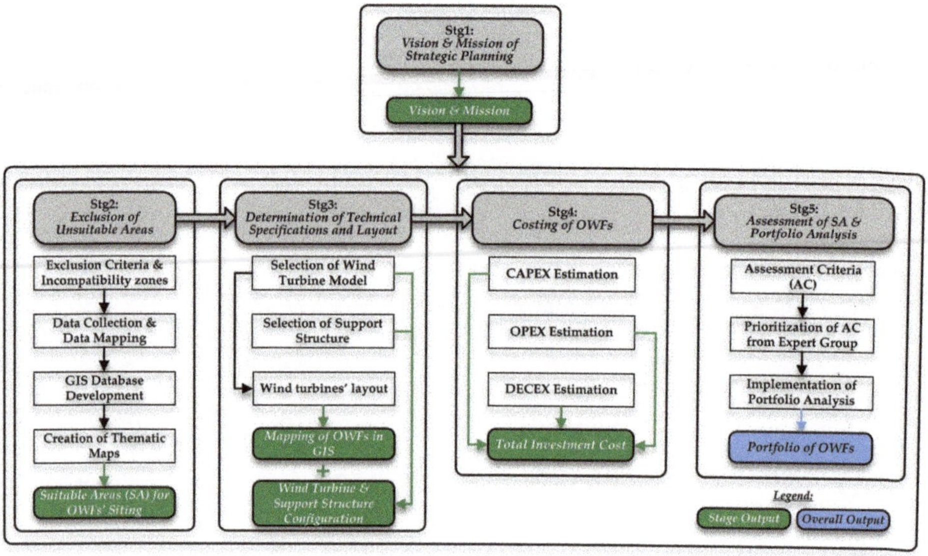

Figure 1. Proposed strategic planning methodology for offshore wind farms (OWFs') siting in Greece.

Stage 1 (Stg1)—Vision and Mission of Strategic Planning

In the first stage, the vision and mission of the strategic planning are defined, on which the next four stages are based. This stage is approached through a combination of proactive and empirical strategy [27]. It takes into account the current situation in the examined country, regarding the issue of energy independence, as well as the future demand for the production of a large number of public commodities, such as electricity, with the ultimate goal to export the latter and, thus, improve the country's current economic status. At this point, it is worth to mention that in Greece by the end of June 2018, the total wind capacity, generated only by onshore wind turbines, was 2690.5 MW, representing an increase of only 1.5% or 39.2 MW compared to the end of 2017 [28]. Thus, the abundant wind potential existing in the Greek marine environment remains still unexploited. In addition, the national effort to reduce GHG emissions is focused on the energy sector. The policy plan for Greece 'National Energy Plan: Roadmap to 2050' was posted by the Ministry for Energy, Environment and Climate Change in 2012 [29]. The roadmap aimed at a reduction of 60% to 70% of CO2 emissions from the energy sector by 2050 compared to 2005, with 85%–100% of electricity coming from RES [29].

Stage 2 (Stg2)—Exclusion of Unsuitable Areas

This particular stage is based on the use of the GIS mapping tool. It includes the exclusion of the areas deemed unsuitable for the siting of OWFs, through the application of various exclusion criteria, resulting to the definition of SA on a national level. The exclusion criteria are defined based on the special characteristics of the examined area, considering also the relevant provisions of the Greek Specific Framework for the Spatial Planning and Sustainable Development for the Renewable Energy Sources (SFSPSD-RES) [30].

Stage 3 (Stg3)—Determination of Technical Specifications and Layout

This stage deals with technical issues related to such projects, such as the selection of the model type of the wind turbine (rotor-nacelle-assembly), the selection of most suitable type of support structure, etc. The required technical specifications are determined by the following elements: (i) the specific characteristics of the suitable sites, that is, wind velocity, wind direction, water depth and the available surface area/shape of the proposed sites, and, also, (ii) studying similar projects that have been completed and are in full or partial operation to this day [31]. In addition, using GIS, the OWFs are sited within the SA identified in Stage 2.

Stage 4 (Stg4)—Costing of OWFs

This stage includes the estimation of CAPEX, OPEX and DECEX of all proposed projects, considering the available data of similar completed projects that are still in operation [31].

Stage 5 (Stg5)—Assessment of SA and Portfolio Analysis

In the final stage of the proposed methodology, the project portfolio is created for the strategic planning of OWFs in Greece and its analysis is carried out. The portfolio decision analysis approach combines multi-criteria evaluation and mathematical optimization and is characterized by the following elements: (i) formation of one portfolio of project proposals (basic goal), taking into account multiple objectives, interactions, and resource constraints; (ii) capturing of the decision makers' preferences regarding the objectives by utilizing a multi-attribute value function; (iii) implementation of integer optimization to obtain the feasible portfolio with the greatest overall strategic value and (iv) potential implementation of interactive "what-if" analyses to examine how the optimal portfolio of actions changes in response to changes in the model parameters or constraints [32]. The objective of this particular stage is to calculate the strategic value of the proposed projects based on specific AC and to prioritize the implementation of the portfolio projects. AC arise mainly from the special characteristics of the SA and their prioritization is achieved with the contribution of an expert group (EG) on issues related to the siting of RES and, more specifically, OWFs, through a questionnaire survey. The portfolio analysis is carried out using Microsoft Project Portfolio Server software. Various energy policy scenarios are formulated for the country based on economic restrictions.

3. Exclusion Criteria and Data Collection/Digitization

3.1. Exclusion Criteria

The study area is defined by the EEZ of Greece and any area outside of it, is legally excluded. The EEZ of a country or otherwise the National Territorial Waters, particularly in recent years, is formally taken into consideration, as a siting criterion, [16,20,33–35]. The exclusion criteria considered in this paper are discussed below.

Wind Velocity

Wind velocity is a significant criterion for the site selection of an OWF, as it is directly linked to the economic feasibility of the project. Therefore, an accurate and detailed analysis of wind data is crucial for a potential wind energy assessment of the proposed suitable sites. In this study, wind velocity data are provided based on measurements made at the height of 80 m on an hourly basis and includes measurements for 10 years (2009–2018). In the present site suitability analysis, marine areas, where annual average wind velocity is smaller than 6 m/s at a height of 80 m above the mean water level, are considered unsuitable for the siting of OWFs [16].

Water Depth

Water depth is one of the key criteria for OWFs' siting, as it significantly contributes to the determination of the investment cost of such projects [1]. Specifically, the water depth affects the selection of the wind turbine's support structure, as well as the CAPEX and OPEX of an OWF project, which increase significantly in deeper waters. For example, according to [36], it can be assumed that with water depth the costs increase due to mooring, anchoring, and cabling costs in deeper waters. In the present investigation, the maximum limit of water depth is set to 500 m [16,36,37].

Military Zones

These marine areas are officially used by the National Army either for training purposes or as firing fields and therefore cannot be considered for any other use. The present criterion is taken into consideration by [8,15,33,34,37].

Seismic Hazard Zones

The seismic hazard factor should be considered generally in the site selection process to reduce construction cost. Greece corresponds to one of the most seismically active countries worldwide. Therefore, all infrastructures should be adequately designed against earthquake. In the case of OWFs, this fact may lead to special designs of the wind turbines' support structure and, therefore, to larger construction costs. Thus, the areas belonging to the Seismic Hazard Zone III (0.36 g) in Greece are excluded. The present criterion has not been considered so far in any other study of OWFs' siting internationally, whereas it has been proposed as a criterion for selecting sites suitable for OWF developments in the South Korea by [38], but it was not considered as a site selection criterion in their study.

Underwater Cables

This exclusion criterion refers to the cables that already exist on the seafloor and serve either for electricity transmission or for telecommunication purposes (e.g., [8,12,33,34]). It is important to consider the underwater routes of those cables, in order to avoid any damage to them during the installation process of OWF developments.

Distance from Ports

The distance of an OWF project from a port presents an important factor affecting the total investment cost, since it has a direct impact on the installation costs, the operation and maintenance costs, as well as the decommission costs of the OWF [39]. Specifically, the total investment cost decreases as the location of an OWF is closer to an existing port, while, moreover, the proximity of the installation area to a port simplifies the overall project management (e.g., no need to install a substation within the marine environment). This criterion has been considered in the site selection of hybrid offshore wind and wave energy systems in Greece [37]. The selected limit of the distance from a domestic port is set in the present paper at 100 km and marine areas that are further away from 100 km are excluded.

Distance from High Voltage Electricity Grid

The distance of an OWF from the national electricity grid and particularly from a high voltage grid is extremely important for technical and economic reasons. A connection to the high voltage grid is selected, because in the opposite case (connection to a medium or low voltage grid) there might be a serious risk of cable destruction due to overloading of the electricity grid [20,26,36]. There are studies that set the distance of the candidate siting areas from the electricity grid at a limit of 200 km [8,39], while there are others that reduce this limit to 60 km [20] or even to 40 km [26]. In the present paper, the limit of 100 km from the existing and the potential officially approved high voltage electricity grid was selected. Moreover, one of the most important factors for the development of OWFs is the evaluation of capacity of the grid. In Greece, the Independent Power Transmission Operator (IPTO) S.A. undertakes the role of transmission system operator for the Hellenic Electricity Transmission System (medium and high voltage grid). In 2018, IPTO published an approved future plan for the spatial development of medium and high voltage grids in Greece as a target for 2027 [40]. In this plan, the majority of the islands that are both close and far away of the mainland lack medium and high voltage grids. Therefore, it is impossible and economically not viable to find locations for the development of OWFs in a long distance from the mainland in the near future. A more detailed analysis of the capacity of grid could be useful for the OWFs' development, but this analysis is out of the scope of the present study.

Landscape Protection/Visual and Acoustic Disturbance

The present criterion is related to the distance of an OWF from the coast and it has been used to ensure landscape protection, avoid visual and acoustic disturbances, and ensure the social acceptance of an OWF [15,17]. In the present study, marine areas with a distance from the coast smaller than 20 km are considered unsuitable for OWFs' siting and are excluded from further analysis. This limit is defined based on [36]. Moreover, the 20 km ensure a distance of at least 130 times of the total height of the selected offshore wind turbine, in order to avoid the visual and acoustic impacts of the project.

Distance from Shipping Routes

The existence of safe navigation routes that connect the plethora of Greek islands with the mainland is an extremely important issue. In order to ensure the protection of shipping movement either for trade or tourism, a safety distance of approximately 5 km (3 miles) from the referred routes is selected [9,13].

Distance from Marine Protected Areas

In this paper, marine protected areas correspond to Sites of Community Importance (SCI) of Natura 2000, national marine environmental parks, coastal bathing waters monitored and assessed in the framework of the Monitoring Programme of Bathing Water Quality according to the provisions of the Directive 2006/7/EC and swimming beaches awarded with the Blue Flag. In this paper, the minimum distance from marine protected areas is selected equal to 2 km, as according to previous studies [11,13,17] the relevant distance limit is set at 1–2 km.

Distance from Wildlife Refugees and Migration Corridors

This criterion includes migration corridors and wetlands of international importance, as defined according to the Ramsar Convention. The specific criterion is considered in order to reduce the potential risk of birds' collision on the wind turbines, mainly during the migratory period. The installation of OWFs should be avoided within the boundaries of the referred areas which are hosting a variety of birds. An exclusion zone of 3 km is taken into account [10,13].

Distance from Residential Network

According to the national legislative framework (SFSDSP-RES) [30] minimum distances from residential settlements and from traditional settlements equal to 1 km and 1.5 km respectively are taken into consideration.

Based on all the above, Table 1 summarizes the exclusion criteria considered in the present paper and their incompatibility zones.

Table 1. Exclusion criteria and incompatibility zones.

No.	Exclusion Criterion	Factor	Unsuitable Areas
EC.1	Exclusive Economic Zone	Legal	Outside the boundaries
EC.2	Wind Velocity	Economic	<6 m/s
EC.3	Water Depth	Economic/Technical	>500 m
EC.4	Military Zones	Political/Restrictive	All
EC.5	Seismic Hazard Zones	Protective/Restrictive	Zone III (0.36g)
EC.6	Underwater Cables	Protective/Technical/Restrictive	All
EC.7	Distance from Ports	Economic/Technical	>100 km
EC.8	Distance from High Voltage Electricity Grid	Economic/Technical	>100 km
EC.9	Landscape Protection/Visual and Acoustic Disturbance	Social/Political/Protective	≤20 km
EC.10	Distance from Shipping Routes	Social/Political/Protective	≲5 km (≤3 miles)
EC.11	Distance from Marine Protected Areas	Environmental/Protective	≤2 km
EC.12	Distance from Wildlife Refugees and Migration Corridors	Environmental/Protective	≤3 km
EC.13	Distance from Residential Network	Legal/Social/Protective	≤1 km (non-traditional settlements) ≤1.5 km (traditional settlements)

3.2. Data Collection/Digitization

In order to identify and analyze all the environmental, economic, technical, legal, and political characteristics of the EEZ of Greece, it was essential to collect and appropriately digitize, if necessary, certain geographical information data from national institutes, research centers, services, and official international and national websites that provide officially approved cartographic data.

More specifically, the digital data used in the present study in correspondence with the responsible entity/source are as follows: (i) Water depth data obtained from the Hellenic Navy Hydrographic Service [41]. (ii) Wind velocity data provided by the Hellenic Centre for Marine Research [42]. (iii) Data of the EEZ of Greece, the Mediterranean Sea and Greece gathered from the electronic database of the European Statistical Service [43]. (iv) Data of SCI, national marine environmental parks, coastal bathing waters, swimming beaches and wetlands of international importance obtained from the "GEODATA" official national website, which has been characterized as the national gate of geographical information data of Greece [44]. (v) Data of the underwater telecommunication cables within the EEZ of Greece, which were collected from the electronic database of the official European website "EMODnet" [45].

Except of the above, the following data were identified, collected, and mapped: (i) The verified shipping routes of the whole country were digitized through the basemaps of the cartographic tool ArcGIS, using the same projected coordinate system. (ii) The military zones used for training purposes and as firing fields, which were provided in analog format by the Hellenic Navy Hydrographic Service [41] and they were, then, appropriately digitized. (iii) The migration corridors, which were mapped by obtaining a corresponding map (in image format) from the Hellenic Ornithological Society [46]. (iv) The domestic ports, which were mapped by providing information of their locations and their names [47]. Only the officially designated ports of the country were mapped. (v) The seismic hazard zones of the country, which were digitized through the official seismic hazard map, collected as an image from the Technical Chamber of Greece [48]. (vi) The data related to the underwater cables of the electricity grid, the locations of the 400 kilovolt (kV) high voltage centers and the 150 kV high voltage substations, obtained from the IPTO, through an official map found in [40].

4. Technical Specifications and OWF Siting Layout

4.1. Definition of Wind Turbine Model

In the present study, the generic 5 MW turbine, which was developed by the National Renewable Energy Laboratory (NREL) [49] is selected. This wind turbine model has been also used in several previous studies [39,50–52], while a large number of existing and fully or partially operational OWFs globally, deploy offshore wind turbines with the same nominal power (ten in Europe and six in Asia) [31].

4.2. Selection of Wind Turbines' Support Atructure

The selection of the wind turbines' support structure is related to the water depth of the SA (Stg2, Figure 1). These areas are located at a water depth of over 50 m (see Section 7.2 below) and, therefore, floating platforms as support structures are preferable [16,39,51]. In the present paper, for water depths of 50–200 m, the Tension Leg Buoy (TLB) is selected as the floating support structure of each wind turbine. TLB systems can serve as sustainable solutions for the exploitation of RES [39,53]. In the North Sea, where harsher wind and wave conditions compared to the Greek marine environment exist, it has been demonstrated that TLBs present the most cost-effective systems for water depths from 50 to 200 m [52]. As for water depths of 200 to 500 m, the Hywind concept is selected. The simple design of Hywind offers many important advantages, such as evidence-based technology, simple support structure construction with the possibility of standardization and lower fabrication cost, as well as robustness and suitability in case of harsh environmental conditions [54]. The Hywind floating system, although slightly more expensive than the SWAY system in terms of cost per megawatt hour (MWh) [39], corresponds to a floating platform that has been used with success globally [54,55].

4.3. Definition of the Wind Turbines' Layout Within the Suitable Areas

The layout of the wind turbines depends on various factors (e.g., cost, wake effects, etc.) and its determination corresponds to an optimization problem [56,57]. In this paper, an oriented approach that fulfills the aim of the present investigation is developed which, moreover, takes into consideration the European best practices. Specifically, the distance between two successive turbines at a line parallel and perpendicular to the prevailing wind direction is denoted as dx and dy respectively, while D_{rotor} denotes the rotor diameter. In existing, fully-commissioned European OWFs (e.g., Nysted OWF Denmark, Eneco Luchterduinen OWF Netherlands, Kentish Flats OWF United Kingdom, Belwind OWF Belgium), the values of dx are between $4.6{\sim}12.1D_{rotor}$, while the corresponding range for dy is $3.2{\sim}8D_{rotor}$ [58]. In addition, according to [59] NREL recommends values of dy correspond to $5{\sim}10D_{rotor}$. In the present study, the wind turbines' layout is determined with the use of 'Advanced Editor Tools' in ArcGIS and according to the following elements: (i) main wind direction of each specific site, (ii) D_{rotor} of the selected wind turbine model, (iii) shape of each OWF site and, also, (iv) the referred European standards, in order to minimize the array losses in the proposed OWF projects. Considering all the above, in this study, the defined dx and dy values for the wind turbines layout are $7D_{rotor}$ and $7D_{rotor}$ respectively.

5. Cost of OWFs

5.1. Estimation of CAPEX

The components of CAPEX considered in the present study are: (i) development and consenting, (ii) construction phase insurance, (iii) rotor-nacelle-assembly costs, (iv) production costs (including tower and support structure), (v) mooring costs (including installation for the case of floating wind turbines), (vi) grid costs (including installation), and (vii) installation of the whole wind turbine system [39,51]. According to [16,60–62], CAPEX is estimated taking into account the water depth and the distance from the shore. In addition to these factors, the CAPEX of such projects depends upon: (i) the wind turbine support structure deployed, which can be either fixed to the sea bed or floating [16,39,51,52,60] and (ii) the nominal power of the wind turbine.

5.2. Estimation of OPEX

OPEX of OWFs are a major part of the total investment costs and they are directly linked and largely affected by the distance from ports and by the water depth [39,51]. In addition, according to [16,39,51], OPEX per year for floating support structures correspond to approximately 3% of CAPEX (€/MW). In [39,51] OPEX was considered equal to 3.7% of CAPEX (€/MW) for wind turbines with a TLB floating platform and equal to 3.44% of CAPEX (€/MW) for wind turbines with a Hywind floating platform. After calculating OPEX for all OWFs for their first year of operation, OPEX during their total life cycle, which is defined equal to 25 years [62], can be estimated. For this purpose, the formula of the present value of annually allocated expenses (PVAAE) (Eq 1.) was used [63]:

$$PVAAE = \frac{(1+i)^n - 1}{i(1+i)^n} \tag{1}$$

where n are the years of operation (25 years) and i the interest rate, which is taken equal to 2.5%.

5.3. Estimation of DECEX

DECEX of an OWF project may correspond to 0%–4% of the total investment costs [16,51,62,64]. In the present paper, for each project, DECEX is taken equal to 2% of the corresponding total investment cost, since it is considered that although a significant amount is collected from recycling the salvaged construction materials, this amount does not suffice to cover the DECEX in total.

6. Assessment of Suitable Areas and Portfolio Analysis

6.1. Assessment Criteria

The AC used for the portfolio analysis include: (i) wind velocity (AC.1) [8,12–14,20,26,36], (ii) water depth (AC.2) [8,26,36], (iii) electrical energy demand (AC.3), (iv) distance from ports (with water depth >10 m in terms of draft requirements) (AC.4) [36], (v) distance from high voltage electricity grid (AC.5) [8,9,12], (vi) distance from Marine Protected Areas (AC.6) [12–14] (vii) distance from military exercise areas (firing fields and exercise locations) (AC.7). It is worth to note that "electrical energy demand" and "distance from military exercise areas (firing fields and exercise locations)" have not been used before as AC for the siting of OWFs.

6.2. Prioritisation of AC

The prioritization of AC is based on a suitably designed questionnaire, which was sent via e-mail to an EG on siting of RES, and, more specifically, of OWFs. The group consists of forty experts from universities, institutes, research centers and companies around the world (USA, Europe, and Asia). These experts were carefully selected, considering the different backgrounds of the participants, so that their distinct opinions reflect different strategic orientations of the present RES siting problem and in order to emphasize the complexity of such siting problems. Out of the forty questionnaires sent out, seven were successfully completed, and were answered by professors, experienced scientific researchers and spatial analysts of RES from various European countries (Greece, Spain, Italy, The Netherlands, etc.). The experts prioritized the selected criteria in the questionnaires based on their own high experience and their own different preferences; the majority of the experts have over seven years of experience on such topics. Figure 2 presents the final prioritization of the AC. The consistency ratio calculated by Microsoft Project Server reached 100%.

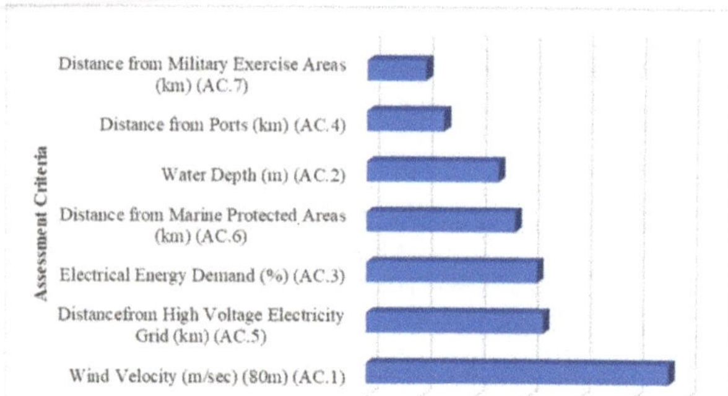

Figure 2. Relevant weights of assessment criteria (AC).

6.3. Portfolio Analysis

This particular step involves the actual selection process, where all project proposals are examined in conjunction with the AC and economic restrictions. More specifically, the portfolio analysis arrives at a prioritization of the project proposals, depending on their performance in relation to the AC. Thus, if a project has a major impact on several parallel AC, it is placed high on the priority list and receives a high strategic value score.

In this paper, the portfolio analysis is carried out using Microsoft Project Server software and include the following main steps: (i) input of AC in the library of the influencing factors, (ii) input of all projects (OWFs) and of their estimated total investment cost in the project center, (iii) assessment of the strategic impact of each OWF, based on the particular features of the projects related to the specific AC using a 5-point scale (Table 2), (iv) input of the relevant weight of each AC (Figure 2) according to the EG and (v) creation of scenarios based on economic restrictions (50%, 60%, and 75% of the total portfolio).

Table 2. Scaling of AC.

AC	Scaling				
	None (0)	Low (1)	Moderate (3)	Strong (6)	Extreme (9)
AC.1	<6	6–7	7–8	8–9	>9
AC.2	>100	>65	50–65	35–50	20–35
AC.3	0	0–5	5–10	10–15	>15
AC.4	0	15–25	25–35	35–45	45–55
AC.5	>500	Up to 500	Up to 400	Up to 300	Up to 200
AC.6	>100	>75	60–75	45–60	30–45
AC.7	0	0–20	20–40	40–60	60–80

7. Results and Discussion

7.1. Strategic Planning Vision and Mission

The Strategic Planning vision and mission involves a clear constant declaration of purpose that describes the values and priorities of the country, as regards the implementation of OWFs and it is addressed to several different stakeholders, people or groups that are directly or indirectly linked to the implementation of the proposed Strategic Plan. In this paper the vision/mission is defined as follows:

"To acquire national energy independence, by considering sustainable development as well as social responsibility and acceptance as a top priority. The main strategic goal is to promote integrated solutions for the siting, technical characteristics and decision-making issues of OWFs".

7.2. Identification of Suitable Areas

For identifying marine areas suitable for OWFs' siting in Greece the 13 exclusion criteria, EC.1~EC.13, (Table 1) are taken into account. Figure 3a–d show indicatively the developed for this purpose thematic maps of the exclusion criteria corresponding to: (a) wind velocity (EC.2), (b) marine protected areas (EC.11), wildlife refuges and migration corridors (EC.12), (c) water depth (EC.3) and (d) seismic hazard zones (EC.5). By creating, editing, and managing three different linear models with the use of geoprocessing tools in the 'ModelBuilder' in ArcGIS software, the SA for OWFs' siting in the Greek marine environment emerge and are shown in Figure 4. Specifically, applying the referred models, all prohibited and unsuitable areas erased of the map, according to the limitations that defined on Stg 2.

The number of SA is sixteen (16). Certain very small sites, less than 2.5 km^2, are not taken into account, since the installation of wind turbines in these small areas does not fulfill any financial purpose. Among the exclusion criteria considered, the water depth and the lack of a high voltage electricity grid represent the two most limiting factors. More specifically, with regard to latter exclusion criterion, there are very few high voltage centers of 400 kV in the south of the country, which are actually situated close to the shore. It should be also noted that for all SA, the water depth is larger than 50 m, advocating the use of floating support structures.

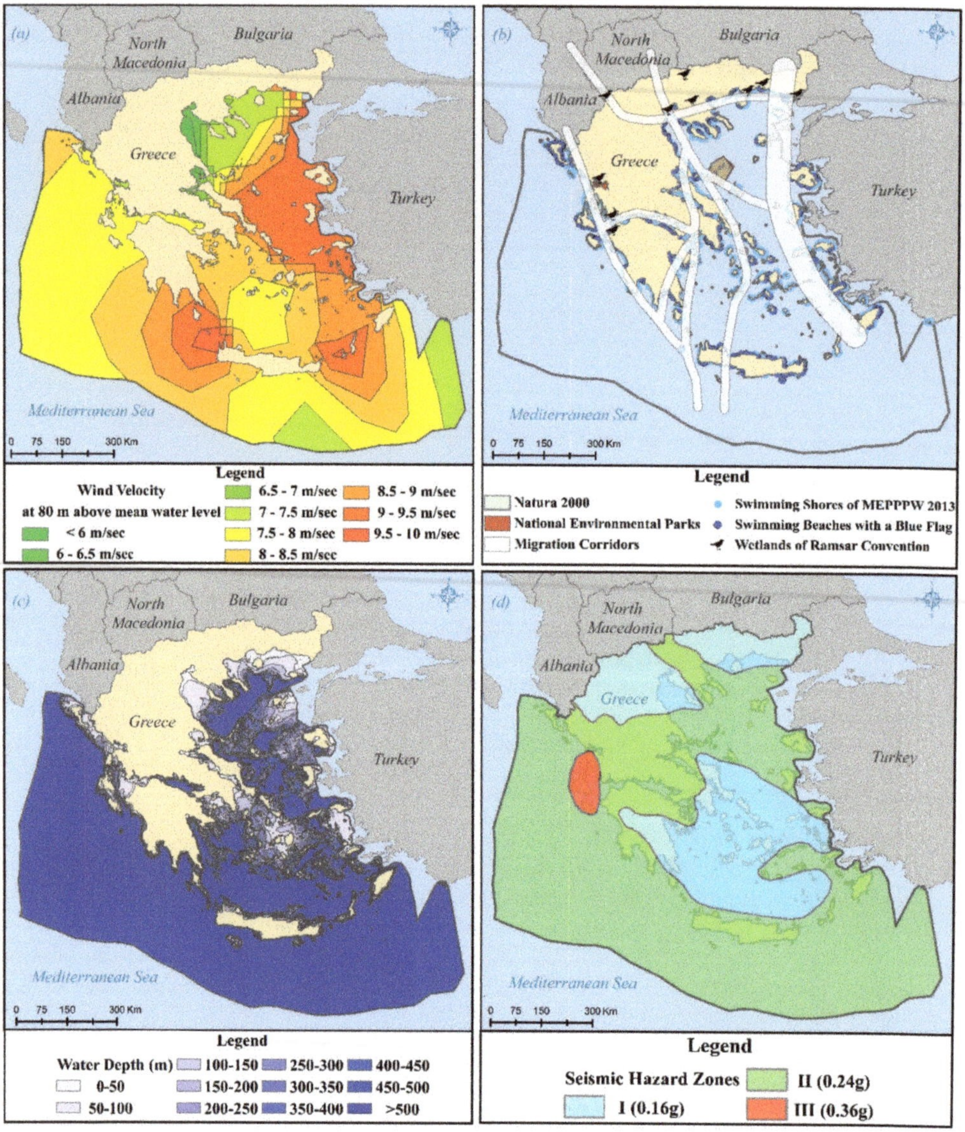

Figure 3. Thematic maps of: (**a**) wind velocity (EC.2), (**b**) marine protected areas (EC.11), wildlife refuges and migration corridors (EC.12), (**c**) water depth (EC.3) and (**d**) seismic hazard zones (EC.5).

7.3. Identification of Suitable Areas

The mapping of the 16 OWFs is carried out, using the 'Advanced Editor Tools' in ArcGIS, and the exact number of 5 MW wind turbines is calculated, along with the capacity of each OWF (Table 3). Indicatively, the micro-siting configuration of OWF15 and OWF4 is presented in Figure 5.

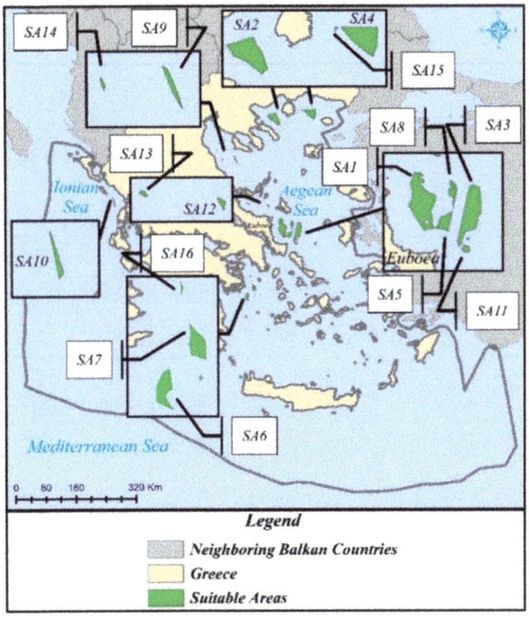

Figure 4. Suitable areas for OWFs siting in Greece.

Table 3. Basic features of the 16 OWFs.

No.	Name of OWF	Location	Area (km^2)	Wind Turbine Support Structure [1]	Project Capacity (MW)
1	"Eolos"	Southeast of Euboea	677.43	TLB and Hywind	4310
2	"Aphrodite"	Southwest of Thasos	552.07	TLB and Hywind	3500
3	"Poseidon"	Southeast of Euboea/North of Andros	542.6	TLB and Hywind	3430
4	"Apollo"	Southeast of Thasos/Northwest of Samothrace	477.46	TLB	3095
5	"Zeus"	Southeast of Euboea	125.63	TLB and Hywind	785
6	"Hera"	South of Hydra	110.29	TLB and Hywind	680
7	"Ares"	East of Hydra	104.46	TLB and Hywind	600
8	"Athena"	Southeast of Euboea	82.03	Hywind	485
9	"Hephaestus"	Northeast of Volos	46.79	Hywind	280
10	"Demeter"	Northwest of Lefkada/West of Preveza	42.7	Hywind	270
11	"Dionysus"	Southeast of Euboea/North of Andros	30.94	Hywind	205
12	"Hestia"	Northwest of Skyros	22.71	Hywind	150
13	"Hermes"	Northeast of Euboea/South of Skopelos	15.6	Hywind	100
14	"Nemesis"	North of Volos	6.9	TLB	50
15	"Artemis"	East of Thasos	4.81	TLB	45
16	"Persephone"	East of Poros	4.45	TLB	40
Total	**Portfolio OWFs**	**EEZ of Greece**	**2846.87**	**TLB and Hywind**	**18,025**

[1] TLB is considered for water depths 50–200 m, while Hywind for water depths 200–500 m (see Section 4.2).

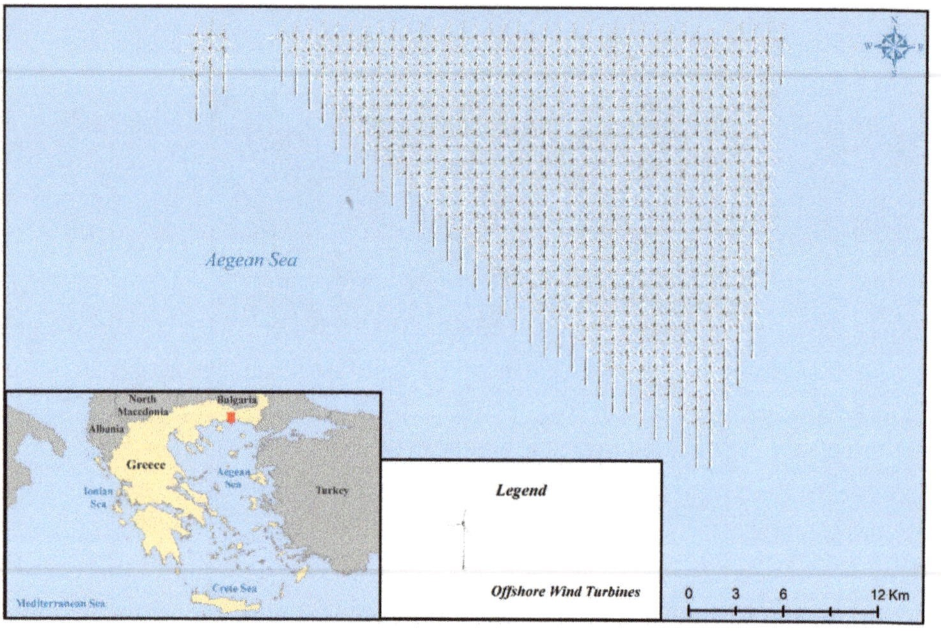

Figure 5. Micro-siting configuration of OWF15 and OWF4.

According to European Network of Transmission System Operators for Electricity [65], the net electricity generating capacity of Greece was in total 16,392.43 MW for 2017 and specifically the amount of 8220.02 MW was produced from RES (25.33% from onshore wind energy, 29.78% from solar energy, 0.73% from biomass energy, 41.35% from renewable hydro energy and 2.81% from other renewable energy sources) and the amount of 8172.41 MW was produced from non-RES. Consequently, with the development and the implementation of the proposed strategic plan, the results from the applied methodology reveal that 45.34% of the electrical energy produced from all the projects in the above table would suffice for the country to achieve energy independence.

In order to keep the total investment costs (e.g., operation and maintenance costs) to minimum and to deploy the suitable marine sites to maximum, the large OWF projects (e.g., OWF1, OWF2) should not be separated into smaller projects. It is suggested that the referred large projects should be implemented in phases, such as 'Gode Wind OWF' in Germany (Project Capacity: 582 MW) [66,67].

7.4. Investment Cost of OWFs

According to [39], CAPEX of offshore 5 MW wind turbines with a TLB platform and Hywind platform, located at a distance of 200 km from the port and at a water depth of 200 m, are estimated equal to 3.537×10^6 €/MW and 3.807×10^6 €/MW respectively. Moreover, it has been shown in [1] that an increase of 10% in water depth or of the distance from the shore will lead to an increase of 1% of investment costs, related mainly to installation and grid connection costs, as well as support structure costs (including installation). By using this rule-of-thumb and considering the aforementioned values of CAPEX per MW as a basis for CAPEX calculations, the CAPEX of OWFs for a given water depth and distance from shore are finally estimated. For example, for a 5 MW wind turbine with a TLB floating platform installed at a water depth of 100 m and at a distance of 50 km from the shore, CAPEX is estimated equal to 3.095×10^6 €/MW (a 12.5% reduction of the 3.537×10^6 €/MW base value has been assumed, due to decrease of both water depth and distance from the shore). Regarding OPEX and DECEX, these quantities are expressed as percentages of CAPEX and of total investment cost respectively, as explained in Sections 5.2 and 5.3 above. Table 4 includes the total investment cost of the

16 OWFs, which is used as input in the portfolio analysis. The proposed projects, shown in Tables 3 and 4, could be financed by European Investment Bank (EIB), as the EIB Board (2015) approved support for strategic infrastructure investment totaling nearly € 7 billion for new OWFs in the UK and Belgian coasts [68]. Moreover, the proposed OWF projects could be financial supported by both private (e.g., Greek private banks) and public funds (e.g., Greek public bank).

Table 4. Investment cost and OWFs' characteristics in relation to the AC.

A/A	Investment Cost (€)	AC.1 (m/s)	AC.2 (m)	AC.3 (%)	AC.4 (km)	AC.5 (km)	AC.6 (km)	AC.7 (km)
OWF1	27.255×10^9	9–9.5	50–500	73.79	47.5	47.0	33.5	22.5
OWF2	20.858×10^9	7–7.5	100–500	59.92	51.5	62.0	30.0	46.5
OWF3	22.173×10^9	9–9.5	150–500	58.72	90.5	47.5	42.0	42.0
OWF4	17.283×10^9	7–8	50–150	52.99	98.0	72.5	28.5	40.5
OWF5	5.092×10^9	9–9.5	150–500	13.44	73.0	29.5	24.5	42.0
OWF6	4.375×10^9	8–8.5	150–500	11.64	52.0	54.5	51.0	4.5
OWF7	3.573×10^9	8–8.5	100–500	10.27	45.0	25.0	34.5	4.0
OWF8	3.338×10^9	9–9.5	450–500	8.30	69.0	57.0	51.5	48.5
OWF9	1.799×10^9	7–7.5	200–400	4.79	85.0	47.0	31.0	32.5
OWF10	1.720×10^9	7.5–8	250–500	4.62	34.0	31.0	29.0	69.5
OWF11	1.382×10^9	9–9.5	150–500	3.51	92.5	29.0	24.0	19.5
OWF12	962.336×10^6	8.5–9.5	300–400	2.57	35.0	81.0	17.5	48.0
OWF13	664.512×10^6	8–8.5	350–500	1.71	39.0	29.5	20.5	77.0
OWF14	272.904×10^6	6–6.5	50–100	0.86	98.5	40.5	19.5	47.5
OWF15	246.342×10^6	7–7.5	50–150	0.77	70.5	46.0	22.5	25.5
OWF16	219.871×10^6	8–8.5	100–200	0.68	30.0	22.0	28.5	0.5

7.5. Results of Portfolio Analysis

The characteristics of all OWFs in relation to the AC required as input in the portfolio analysis are presented in Table 4.

By taking into account the relevant weights of the AC (Figure 2) and the data of Table 4, the portfolio analysis is implemented. Table 5 presents the prioritization of the project portfolio based on the calculated strategic value. The implementation of the whole portfolio achieves a strategic value of 100%, since all proposed projects are implemented, and the investment cost of the whole portfolio amounts to 111.214×10^9 €.

Table 5. Strategic value of portfolio projects.

Ranking	OWF	Strategic Value (%)
1	OWF3	8.46
2	OWF1	8.17
3	OWF5	7.75
4	OWF16	7.44
5	OWF8	7.27
6	OWF7	7.2
7	OWF6	6.69
8	OWF13	6.35
9	OWF11	6.13
10	OWF4	5.84
11	OWF10	5.58
12	OWF2	5.45
13	OWF15	4.84
14	OWF12	4.69
15	OWF14	4.11
16	OWF9	4.03

Next, three possible scenarios for the implementation of portfolio projects are examined. The first scenario corresponds to an investment of 50% of the total portfolio (i.e., 55.607 × 10^9 €), the second one to an investment of 60% of the total portfolio (i.e., 66.729 × 10^9 €), while the third one to an investment of 75% of the total portfolio (i.e., 83.411 × 10^9 €). By implementing the first scenario, eleven OWFs with a 70.08% total strategic value can be realized including OWF3, OWF5, OWF16, OWF8, OWF7, OWF13, OWF11, OWF4, OWF15, OWF12, and OWF14 (in descending order). Most of these OWFs are located Southeast of Euboea (OWF3, OWF5, OWF8, OWF11, Figure 4) and in the Northern Aegean (OWF4, OWF14, OWF15). The total investment cost of these eleven OWFs is equal to 55.207 × 10^9 €. On the other hand, in the case of the second scenario, the following twelve OWFs with a 76.92% strategic value and a total investment cost of 66.181 × 10^9 € are implemented (Figure 4): OWF3, OWF1, OWF16, OWF8, OWF7, OWF6, OWF13, OWF11, OWF10, OWF15, OWF12, and OWF14 (in descending order). Finally, the third scenario leads to the implementation of the following fourteen OWFs with 87.79% strategic value: OWF3, OWF5, OWF16, OWF8, OWF7, OWF6, OWF13, OWF11, OWF4, OWF10, OWF2, OWF15, OWF12, and OWF14 (in descending order). The total investment cost of these projects amounts to 82.160 × 10^9 €, while most of these OWFs are located Southeast of Euboea (OWF3, OWF5, OWF8, OWF11, Figure 4) and in the Northern Aegean (OWF2, OWF4, OWF14, OWF15).

It is worth noting that any of the three scenarios selected for implementation will be particularly beneficial for the country, since the strategic value of all scenarios is over 70%, and national energy independence is ensured. The optimum choice naturally is the implementation of the whole portfolio, since this entails a strategic value of 100% and all strategic goals of this plan are reached. However, such an investment would be particularly costly. Therefore, the second scenario is recommended for implementation, since it requires a little less than 60% of the total investment capital, while, at the same time, it fully satisfies the vision and mission of the present strategic plan (i.e., national energy independence).

Through the strategic planning performed, it has been proven that the combination of GIS with Microsoft Project Server software can, therefore, be seen as a powerful tool for solving complex siting and decision-making issues of OWFs. Moreover, the GIS-Microsoft Project Server integration could be used as an excellent tool for displaying the results in useful maps and, also, in practical portfolio selection scenarios. The proposed methodological approach is essential, as it could be utilized efficiently and easily from academics, scientific researchers, renewable energy planners, developers, government agencies, consultants, and potential investors for the efficient and sustainable exploitation of the offshore wind potential and the relevant deployment of OWFs at different spatial planning scales. The need and the importance of such a strategic planning increase significantly over time, as the increasing population and improving living standards produce an increment on the energy demand. The present study could be further extended with a field investigation analysis of the proposed OWF sites, in order to verify their overall appropriateness. Thus, future work could include a thorough individual assessment of the proposed sites in conjunction with field investigation, in order to make the final selection/prioritization of sites for OWFs installation. Finally, it would be interesting to consider the utilization of energy storage systems in the proposed strategic plan for improving the penetration of RES into the electricity supply mix and maximizing the reliability of the OWF projects regardless weather conditions.

8. Conclusions

The present paper aims at the development of an integrated strategic planning methodology for OWFs in Greece, where portfolio analysis is used for the first time in relation to the siting of RES. More specifically, it focuses on a strategic plan that will ensure national energy independence for 25 years following its partial implementation, with significant economic, socio-political and environmental benefits for Greece.

A total of twenty criteria (exclusion and assessment) are employed for the siting of OWFs, which almost fully cover the economic, socio-political, technical and environmental issues that are related to such installations and applications. The majority of the criteria are in line with the international literature; however, there are also criteria proposed and applied for the first time in relation to siting issues, such as "seismic hazard zones" (exclusion criterion—on RES issues) and "electrical energy demand" (assessment criterion—on RES issues). In addition, "landscape protection/visual and acoustic disturbance" criterion is introduced, facilitating the combined use of two, relevant, previously suggested siting criteria ("distance from shore" and "visual and acoustic impacts").

Wind turbines of 5 MW capacity are taken into account, while the types of support structures selected are the TLB for depths of 50–200 m and the Hywind for depths of 200–500 m. For each OWF the precise mapping of the wind turbines is carried out by setting specific in-between distances in directions parallel and perpendicular to the prevailing wind direction. Then, the total investment cost for each project is calculated by estimating CAPEX, OPEX and DECEX.

In order to perform the portfolio analysis, the seven selected AC are evaluated by an EG. The highest weight is assigned to the "wind velocity" and "distance from high voltage electricity grid" criteria. Therefore, according to different expert participants the strategic policy orientation of the current plan focuses on the technical and economic dimension of the planning issue, based on the own high-experience. Next, the strategic impact of each OWF is evaluated on each AC and three different management scenarios are realized involving economic restrictions (investing 50%, 60%, and 75% of the total portfolio). After the application of the economic restrictions on the three scenarios, 11, 12 and 14 OWF projects are respectively selected for implementation. In all three scenarios, OWF3 presents the highest strategic value out of the selected projects and OWF14 the lowest, while the commonly selected projects in all 3 portfolios are OWF7, OWF8, OWF11, OWF12, OWF13, OWF14, OWF15, and OWF16. It is worth noting that, in all three cases, the energy independence of the country is achieved. The second scenario is recommended for implementation, since it requires less than 60% of the total investment capital and fully satisfies the vision and mission of the present strategic plan, which ensures national energy independence.

The proposed strategic planning for OWFs in Greece serves as a guide for the development of OWFs in Greece, according to criteria that are based on: (i) the special characteristics of the study area and the individual characteristics of each suitable site, (ii) the relevant provisions of the SFSPSD-RES, (iii) indisputable global experience, (iv) experts' opinion and (v) on extensive international literature review, which specializes on such topics. For this reason, the present paper does not simply recommend certain sites for the installation of OWFs, but areas that have been examined according to several parameters within a broader field of study, which is based on a guideline with clear strategic goals. Consequently, the current strategic planning framework can bridge important gaps between research, development and implementation on the complex RES siting and development issues. The relevant methodology includes distinct stages and can be applied in relation to various study areas and diverse spatial planning scales.

Author Contributions: Conceptualization, S.S. and D.G.V.; methodology, S.S., D.G.V. and E.L.; formal analysis, S.S. and E.L.; investigation, S.S.; data curation, S.S.; writing—original draft preparation, S.S., D.G.V.; writing—review and editing, S.S., D.G.V. and E.L.; visualization, S.S.; supervision, D.G.V. and E.L. All authors have read and agreed to the published version of the manuscript.

Funding: This research received no external funding.

Acknowledgments: The authors would like to thank Paraskevi Drakopoulou, GIS Specialist of the Lab of Geographical Information Systems at the Institute of Oceanography of the Hellenic Centre for Marine Research (HCMR) for providing wind velocity data, as well as all professors, researchers and spatial analysts who participated in the questionnaire survey.

Conflicts of Interest: The authors declare no conflict of interest.

References

1. Ng, C.; Ran, L. *Offshore Wind Farms: Technologies, Design and Operation*; Woodhead Publishing: Duxford, UK, 2016.

2. Wind Europe Business Intelligence; Remy, T.; Mbistrova, A. *Offshore Wind in Europe: Key Trends and Statistics 2017*; Wind Europe: Brussels, Belgium, 2018.
3. Wind Europe Business Intelligence; Selot, F.; Fraile, D.; Brindley, G. *Offshore Wind in Europe: Key Trends and Statistics 2018*; Wind Europe: Brussels, Belgium, 2019.
4. Ruccell, T. Floating Wind: Changing Gear. 2019. Available online: https://www.4coffshore.com/news/floating-wind--changing-gear-nid13899.html (accessed on 17 January 2020).
5. Castro-Santos, L.; Martins, E.; Soares, C.G. Economic comparison of technological alternatives to harness offshore wind and wave energies. *Energy* **2017**, *140*, 1121–1130. [CrossRef]
6. Castro-Santos, L. Decision variables for floating offshore wind farms based on life-cycle cost: The case study of Galicia (North-West of Spain). *Ocean Eng.* **2016**, *127*, 114–123. [CrossRef]
7. Costoya, X.; deCastro, M.; Santos, F.; Sousa, M.C.; Gomez-Gesteira, M. Projections of wind energy resources in the Caribbean for the 21st century. *Energy* **2019**, *178*, 356–367. [CrossRef]
8. Mahdy, M.; Bahaj, A.S. Multi criteria decision analysis for offshore wind energy potential in Egypt. *Renew. Energy* **2018**, *118*, 278–289. [CrossRef]
9. Mekonnen, A.D.; Gorsevski, P.V. A web-based participatory GIS (PGIS) for offshore wind farm suitability within Lake Erie, Ohio. *Renew. Sustain. Energy Rev.* **2015**, *41*, 162–177. [CrossRef]
10. Hong, L.; Möller, B. Offshore wind energy potential in China: Under technical, spatial and economic constraints. *Energy* **2011**, *36*, 4482–4491. [CrossRef]
11. Christoforaki, M.; Tsoutsos, T. Sustainable siting of an offshore wind park a case in Chania, Crete. *Renew. Energy* **2017**, *109*, 624–633. [CrossRef]
12. Stefanakou, A.-A.; Nikitakos, N. A decision support model for site selection of offshore wind farms. In Proceedings of the 9th International Scientific Conference on Energy and Climate Change, Athens, Greece, 12–14 October 2016.
13. Vagiona, D.G.; Kamilakis, M. Sustainable site selection for offshore wind farms in the South Aegean-Greece. *Sustainability* **2018**, *10*, 749. [CrossRef]
14. Vagiona, D.G.; Karanikolas, N.M. A multicriteria approach to evaluate offshore wind farms siting in Greece. *Glob. NEST J.* **2012**, *14*, 235–243.
15. Kim, C.-K.; Jang, S.; Kim, T.Y. Site selection for offshore wind farms in the southwest coast of South Korea. *Renew. Energy* **2018**, *120*, 151–162. [CrossRef]
16. Schallenberg-Rodríguez, J.; Montesdeoca, N.G. Spatial planning to estimate the offshore wind energy potential in coastal regions and islands. Practical case: The Canary Islands. *Energy* **2018**, *143*, 91–103. [CrossRef]
17. Möller, B. Continuous spatial modelling to analyse planning and economic consequences of offshore wind energy. *Energy Policy* **2011**, *39*, 511–517. [CrossRef]
18. Beacham, J.L.; Jensen, J.R.; Wang, Z. *A Feasibility Analysis of South Carolina Wind Resources for Electric Power Generation*; Institute for Public Service and Policy Research, University of South Carolina: Columbia, SC, USA, 2009.
19. Argin, M.; Yerci, V.; Erdogan, N.; Kucuksari, S.; Cali, U. Exploring the offshore wind energy potential of Turkey based on multicriteria site selection. *Energy Strateg. Rev.* **2019**, *23*, 33–46. [CrossRef]
20. Chaouachi, A.; Covrig, C.F.; Ardelean, M. Multi-criteria selection of offshore wind farms: Case study for the Baltic States. *Energy Policy* **2017**, *103*, 179–192. [CrossRef]
21. Kim, T.; Park, J.-I.; Maeng, J. Offshore wind farm site selection study around Jeju Island, South Korea. *Renew. Energy* **2016**, *94*, 619–628. [CrossRef]
22. Lee, K.-H.; Jun, S.-O.; Pak, K.-H.; Lee, D.-H.; Lee, K.-W.; Park, J.-P. Numerical optimization of site selection for offshore wind turbine installation using genetic algorithm. *Curr. Appl. Phys.* **2010**, *10*, S302–S306. [CrossRef]
23. Fetanat, A.; Khorasaninejad, E. A novel hybrid MCDM approach for offshore wind farm site selection: A case study of Iran. *Ocean Coast. Manage.* **2015**, *109*, 17–28. [CrossRef]
24. Wu, B.; Yip, T.L.; Xie, L.; Wang, Y. A fuzzy-MADM based approach for site selection of offshore wind farm in busy waterways in China. *Ocean Eng.* **2018**, *168*, 121–132. [CrossRef]
25. Christidis, T.; Law, J. Review: The use of Geographic Information Systems in wind turbine and wind energy research. *J. Renew. Sustain. Energy* **2012**, *4*, 012701. [CrossRef]
26. Waewsak, J.; Landry, M.; Gagnon, Y. Offshore wind power potential of the Gulf of Thailand. *Renew. Energy* **2015**, *81*, 609–626. [CrossRef]

27. Fred, D.; Forest, D. *Strategic Management: A Competitive Advantage Approach, Concepts and Cases*, 16th ed.; Pearson Education: Edinburgh, UK, 2017.
28. *Hellenic Wind Energy Association Wind energy statistics—First semester of 2018*; Hellenic Wind Energy Association (HWEA): Athens, Greece, 2018.
29. Georgopoulou, E.; Mirasgedis, S.; Sarafidis, Y.; Koutentaki, D.; Lalas, D.; Gakis, N. *National Energy Plan: Roadmap to 2050*; Institute for Environmental Research and Sustainable Development, National Observatory of Athens: Athens, Greece, 2014.
30. Ministry of Environment, Energy and Climate Change (MEECC). *Specific Framework for Spatial Planning and Sustainable Development for Renewable Energy Sources*; JMD 49828/2008, OGHE B′ 2464/3-12-08; MEECC: Athens, Greece, 2008.
31. 4C Offshore. Global offshore wind farms database. Available online: https://www.4coffshore.com/windfarms/ (accessed on 12 October 2019).
32. Lahtinen, T.; Hämäläinen, R.; Liesiö, J. Portfolio decision analysis methods in environmental decision making. *Environ. Modell. Softw.* **2017**, *94*, 73–86. [CrossRef]
33. Argin, M.; Yerci, V. The assessment of offshore wind power potential of Turkey. In Proceedings of the 9th International Conference on Electrical and Electronics Engineering (ELECO), Bursa, Turkey, 26–28 November 2015.
34. Ou, L.; Xu, W.; Yue, Q.; Ma, C.L.; Teng, X.; Dong, Y.E. Offshore wind zoning in China: Method and experience. *Ocean Coast. Manage.* **2018**, *151*, 99–108. [CrossRef]
35. Yue, C.-D.; Yang, M.-H. Exploring the potential of wind energy for a coastal state. *Energy Policy* **2009**, *37*, 3925–3940. [CrossRef]
36. Lynch, K.; Murphy, J.; Serri, L.; Airoldi, D. Site selection methodology for combined wind and ocean energy technologies in Europe. In Proceedings of the 4th International Conference on Ocean Energy, Dublin, Ireland, 17–19 October 2012.
37. Vasileiou, M.; Loukogeorgaki, E.; Vagiona, D.G. GIS-based multi-criteria decision analysis for site selection of hybrid offshore wind and wave energy systems in Greece. *Renew. Sustain. Energy Rev.* **2017**, *73*, 745–757. [CrossRef]
38. Kim, J.Y.; Kang, K.S.; Oh, K.Y.; Lee, J.S.; Ryu, M.S. A study on the site selection of offshore wind farm around Korean Peninsula. In Proceedings of the 3rd International Conference on Ocean Energy, Bilbao, Spain, 6–8 October 2010.
39. Myhr, A.; Bjerkseter, C.; Ågotnes, A.; Nygaard, T.A. Levelised cost of energy for offshore floating wind turbines in a life cycle perspective. *Renew. Energy* **2014**, *66*, 714–728. [CrossRef]
40. Independent Power Transmission Operator. Available online: http://www.admie.gr/ (accessed on 10 July 2018).
41. Hellenic Navy Hydrographic Service. Available online: https://www.hnhs.gr/el/?option=com_opencart&Itemid=269&route=common/home/ (accessed on 12 July 2018).
42. Hellenic Centre for Marine Research. Available online: http://www.hcmr.gr/el/ (accessed on 12 July 2018).
43. European statistical service: Eurostat. Available online: https://ec.europa.eu/eurostat/ (accessed on 23 August 2018).
44. GEODATA: Digital data. Available online: https://web.archive.org/web/20180315103323/http://geodata.gov.gr/ (accessed on 15 March 2018).
45. EMODnet: Your gateway to marine data in Europe. Available online: http://www.emodnet-humanactivities.eu/view-data.php/ (accessed on 15 July 2018).
46. Hellenic Ornithological Society. Available online: http://www.ornithologiki.gr/ (accessed on 10 July 2018).
47. MarineTraffic: Global Ship Tracking Intelligence | AIS Marine Traffic. Available online: https://www.marinetraffic.com/en/ais/home/centerx:25.7/centery:36.3/zoom:11 (accessed on 6 July 2018).
48. Technical Chamber of Greece: Official seismic hazard map. Available online: http://portal.tee.gr/portal/page/portal/SCIENTIFIC_WORK/ARTICLES/033/%D7%C1%D1%D4%C7%D3%20%D3%C5%C9%D3%CC%C9%CA%C7%D3%20%C5%D0%C9%CA%C9%CD%C4%D5%CD%CF%D4%C7%D4%C1%D3.htm/ (accessed on 18 July 2018).
49. Jonkman, J.; Butterfield, S.; Musial, W.; Scott, G. *Definition of a 5-MW Reference Wind Turbine for Offshore System Development*; National Renewable Energy Laboratory: Golden, CO, USA, 2009.

50. Fischer, T.; de Vries, W.; Schmidt, B. *Upwind WP4 Design Basis*; Institure of Aircraft Design Universität Stuttgart: Stuttgart, Germany, 2010.
51. Bjerkseter, C.; Agotnes, A. Levelised costs of energy for offshore floating wind turbine concepts. Master Thesis, Norwegian University of Life Sciences (NMBU), Oslo, Norway, 2013.
52. Myhr, A. Developing offshore floating wind turbines: The Tension-Leg-Buoy design. Philosophiae Doctor (PhD) Thesis, Norwegian University of Life Sciences (NMBU), Oslo, Norway, 2016. Ph.D. Thesis, Norwegian University of Life Sciences (NMBU), Oslo, Norway, 2016.
53. Nygaard, T.A.; Myhr, A.; Karl, J.M. A comparison of two conceptual designs for floating wind turbines. In Proceedings of the 3rd European Offshore Wind Conference & Exhibition, Stockholm, Sweden, 14–16 September 2009.
54. Hywind: The world's leading floating offshore wind solution. Available online: https://www.equinor.com/en/what-we-do/hywind-where-the-wind-takes-us.html/ (accessed on 28 August 2018).
55. Hywind Scotland Pilot Park—4C Offshore. Available online: https://www.4coffshore.com/windfarms/hywind-scotland-pilot-park-united-kingdom-uk76.html/ (accessed on 28 August 2018).
56. Pillai, A.C.; Chick, J.; Khorasanchi, M.; Barbouchi, S.; Johanning, L. Application of an offshore wind farm layout optimization methodology at Middelgrunden wind farm. *Ocean Eng.* **2017**, *139*, 287–297. [CrossRef]
57. Sun, H.; Yang, H.; Gao, X. Study on offshore wind farm layout optimization based on decommissioning strategy. *Energy Procedia* **2017**, *143*, 566–571. [CrossRef]
58. *Deutsche WindGuard GmbH Capacity Densities of European Offshore Wind Farms*; Federal Maritime and Hydrographic Agency: Hamburg, Germany, 2018.
59. Lynch, K.; Murphy, J. *Overview of Offshore Wind and Ocean Energy Technologies*; Hydraulics and Maritime Research Centre: Ireland, UK, 2012.
60. Stehly, T.; Heimiller, D.; Scott, G. *2016 Cost of Wind Energy Review*; National Renewable Energy Laboratory: Denver, CO, USA, 2017.
61. Gonzalez-Rodriquez, A.G. Review of offshore wind farm cost components. *Energy Sustain. Dev.* **2017**, *37*, 10–19. [CrossRef]
62. Ioannou, A.; Angus, A.; Brennan, F. A lifecycle techno-economic model of offshore wind energy for different entry and exit instances. *Appl. Energy* **2018**, *221*, 406–424. [CrossRef]
63. Brigham, E.F.; Houston, J.F. *Fundamentals of Financial Management*, Concise 8th ed.; Cengage Learning: Mason, OH, USA, 2015.
64. *BVG Associates Oil and Gas "Seize the Opportunity" Guides: Offshore Wind*; Scottish Enterprise: Glasgow, UK, 2016.
65. European Network of Transmission System Operators for Electricity: Net Electricity Generating Capacity. Available online: https://www.entsoe.eu/data/power-stats/net-gen-capacity/ (accessed on 29 October 2019).
66. Weston, D. Gode Wind construction starts. 2015. Available online: https://www.windpoweroffshore.com/article/1342947?utm_source=website&utm_medium=social (accessed on 22 June 2019).
67. Gode Wind 1 and 2—4C Offshore. Available online: https://www.4coffshore.com/windfarms/gode-wind-1-and-2-germany-de13.html (accessed on 22 June 2019).
68. Russell, T. OWFs supported by EIB's €17 billion; 2015. Available online: https://www.4coffshore.com/news/owfs-supported-by-eib27s-8017-billion-nid2431.html (accessed on 15 May 2019).

sustainability

MDPI

Article

Thermo-Poroelastic Analysis of Induced Seismicity at the Basel Enhanced Geothermal System

Sandro Andrés [1], David Santillán [1,*], Juan Carlos Mosquera [2] and Luis Cueto-Felgueroso [1]

1 Department of Civil Engineering: Hydraulics, Energy and Environment, Universidad Politécnica de Madrid, 28040 Madrid, Spain; sandro.andres@upm.es (S.A.); luis.cueto@upm.es (L.C.-F.)

2 Department of Continuum Mechanics and Theory of Structures, Universidad Politécnica de Madrid, 28040 Madrid, Spain; juancarlos.mosquera@upm.es

* Correspondence: david.santillan@upm.es

Received: 8 November 2019; Accepted: 29 November 2019; Published: 4 December 2019

Abstract: Geothermal energy has emerged as an alternative to ensure a green energy supply while tackling climate change. Geothermal systems extract the heat stored in the Earth's crust by warming up water, but the low rock permeability at exploitation depths may require the hydraulic stimulation of the rock fracture network. Enhanced Geothermal Systems (EGS) employ techniques such as hydro-shearing and hydro-fracturing for that purpose, but their use promotes anthropogenic earthquakes induced by the injection or extraction of fluids. This work addresses this problem through developing a computational 3D model to explore fault reactivation and evaluating the potential for earthquake triggering at preexisting geological faults. These are included in the model as frictional contacts that allow the relative displacement between both of its sides, governed by rate-and-state friction laws and fully coupled with thermo-hydro-mechanical equations. We apply our methodology to the Basel project, employing the on-site parameters and conditions. Our results demonstrate that earthquakes which occurred in December 2006 in Basel (Switzerland) are compatible with the geomechanical and frictional consequences of the hydraulic stimulation of the rock mass. The application of our model also shows that it can be useful for predicting fault reactivation and engineering injection protocols for managing the safe and sustainable operation of EGS.

Keywords: geothermal energy; induced seismicity; fault; Basel; poroelasticity

1. Introduction

The 2030 Agenda for Sustainable Development, adopted by world leaders in September 2015, was established by the United Nations and comprises 17 goals and 169 targets to be fulfilled by the ratifying countries. The Agenda aims to overcome all forms of poverty, while tackling climate change and environmental protection. Goal 7—"Affordable and Clean Energy"—aims to ensure access to affordable, reliable, sustainable, and modern energy for all [1]. Goal 7 goes hand-in-hand with Goal 13, "Climate Action", which aims to take urgent action to combat climate change and its impacts. Geothermal energy emerges as an alternative renewable energy to reach both goals [2,3], as it is affordable and clean. Geothermal systems extract the heat stored in the Earth's crust by warming up water or a mixture of water and gas. The fluid is circulated down through injection wells, heated by the contact with rocks, and returned to the surface through production wells to form a closed loop [4,5]. Hot water or steam is then transformed into a marketable product, such as electricity. Nevertheless, in most geothermal reservoirs, rock permeability at exploitation depths is very low, rendering geothermal projects economically infeasible. The challenge of permeability enhancement has been addressed by the so-called Enhanced Geothermal Systems (EGS) [2,6,7].

EGS enhance rock permeability through hydraulic stimulation of the rock mass fracture network. Two approaches are widely used: hydro-fracturing, which creates new fracture networks, and hydro-shearing, that reactivates preexisting joints [8,9]. A major environmental issue for these techniques is the risk of induced seismicity as a result of water injection and production [2]. Some major EGS experiences include the Soultz-sous-Forêts project (France) [10], the Cooper Basin project (Australia) [11], the Fenton Hill project (New Mexico, USA) [12], the Rosemanowes project (UK) [13,14], or the Ogachi project (Japan). Furthermore, a number of EGS demonstration projects have recently been launched in the US [15].

Hydro-fracturing, or hydraulic fracturing, enhances rock permeability by opening preexisting fractures and creating new ones. Water is injected within the rock mass under high pressure to open and create the new fractures by tensile failure [16–19]. This technique is broadly used for oil-well stimulation, although it is currently under debate due to environmental concerns [2]. Hydro-shearing increases rock permeability by initiating shear failure. Water is injected under high pressure, reducing normal stress across them and eventually triggering shear failure [9]. Since fault roughness guarantees slip is permanent after injection stops, permeability changes are also permanent [20]. The magnitude of seismic events triggered during hydro-shearing operations is typically small because fractures slip without increasing their aperture as the pore pressure remains bellow the minimum principal stress. Nonetheless, hydro-shearing is widely used on a global scale when conducting field experiments [21] and numerical simulations [8,22,23] to assess permeability evolution.

Most of the scientific community accepts that induced earthquakes may be triggered by water injection into the subsurface. Some examples are the disposal by injection into deep wells of waste water from oil and gas production [24,25], CO_2 sequestration in deep aquifers [26,27], or EGS facilities [28,29]. Earthquakes are the result of a fast slip event on a fault [30]. The onset of the slip depends on the shear stress on the fault and the frictional resistance, both of them affected by fluid injection [31]. On one hand, the increase in pore pressure puts effective stress on the fault decrease, whereas on the other hand, thermo-poro-mechanical effects increase shear stress on the fault [32,33]. The experience gained over the years on EGS projects has shown the usefulness of numerical models for simulating the nucleation and rupture of earthquakes under given injection protocols. Numerical simulations of induced earthquakes require coupling fluid flow, rock deformation, heat transfer, and a fault frictional response [34–36]. Highly nonlinear frictional laws and the disparity in time scales pose a major challenge for computational models. Over the past decade, there have been numerous contributions to thermo-hydro-mechanical modeling of induced seismicity [33,37]. Faults are typically simulated as 3D failure zones with slip-weakening rheology [38–40], or as frictional contact surfaces [41–46]. We adopt the latter approach, since it is consistent with the observed structure of faults [47]. We simulate faults as contact surfaces whose friction evolves according a rate-and-state friction law [48] that incorporates terms depending on the evolution of effective normal stress [49,50]. Our results suggest that injection protocols are relevant so they can be designed to minimize seismic risks [51].

A paradigmatic case is the Deep Heat Mining (DHM) project in Basilea (Switzerland) [52,53], a milestone EGS power station with an energy output capacity of 6 MW of electricity and 17 MW of heat. The station would provide electricity and heat for 10,000 dwellings [54]. The project required the drilling of two wells, Basel-1 at 5.000 m deep to inject cool water, and another well to return hot water back to the surface [55]. Once the former was built, hydraulic stimulation was performed in December 2006. However, it led to perceivable induced seismicity up to a local event magnitude of M_L 3.4, which exceeded the acceptable levels in the Basel urban area [56]. The seismic events led to the premature halt of the project, and finally, its withdrawal.

Here, we develop a fully implicit and monolithically coupled finite element model to simulate fault reactivation. Our model encompasses fully coupled heat transport, rock deformation, and fluid flow processes. Moreover, we describe fault as interfaces whose friction is governed by a laboratory derived rate-and-state friction law incorporating an effective normal stress-rate dependence. We analyze the effects of hydraulic stimulation on seismic risk through a Coulomb failure analysis. Measure of tendency to slip has proven very useful to understand some of the geomechanical challenges posed by subsurface energy technologies (e.g., [46,57–64]). We aim to analyze the hydro-shearing effects during the stimulation at the Basel-1 well using our three-dimensional (3D) model. We use, as input, the same stress-field and rock properties estimated for the main hydrogeological units at Basel-1, and simulate the stimulation using a transient well-head pressure and flow evolution similar to the values reported at the study site. The proposed model is a useful tool for engineers and practitioners to answer the essential question for the development of the geothermal energy as an affordable, clean, and safe renewable energy source: what is the optimal exploitation protocol that minimizes seismic risk and maximizes economic and energy performance?

2. Materials and Methods

In this section, we describe our thermo-hydro-geomechanical model, which fully couples heat transport, rock deformation, and fluid flow; faults are described as interfaces with friction governed by the rate-and-state law. We perform numerical simulations of the hydraulic stimulation operations conducted at the DHM project in December 2006 to show the ability of our model to characterize fault reactivation. We adopt as inputs the recorded well-head pressure and flow evolutions, as well as on-site material parameters.

2.1. Frictional Strength and Resistance of Faults

We employ the Amontons–Coulomb theory as a constitutive model of rock friction. The frictional strength, τ^*, that impedes sliding of a static, cohesionless contact interface is given by $\tau^* = \mu|\sigma|$ [65–67]. In general, the relation between these Coulomb magnitudes and the shear stress acting on the contact plane, τ, depends on the sliding regime. A static interface satisfies $\tau \leq \tau^*$ and for sliding under quasi-static conditions, the relationship $\tau \approx \tau^*$ holds.

In the fluid-saturated media, frictional strength is defined using the *effective* normal stress, $\sigma' = \sigma + p$, where σ is the *total* normal stress acting on the contact, and p is the pore pressure of the fluid. In the above and following expressions, tensile stresses are positive, and pore pressure is positive when above the atmospheric value. We assume that effective normal stresses remain compressive on contact surfaces.

Rate-and-State Models for Interfaces

Faults are often assumed to be well-oriented for failure but locked prior to injection, in such a way that the onset of slip on an inactive fault—reactivation—is essential to understand the geomechanics of induced seismicity.

Rate-and-state formulations gather the traditional concepts of static and dynamic friction by including the dependence of μ on the slip velocity and history of sliding [66,68–70]. These models were derived from laboratory experiments of unidirectional slip in the double-direct shear configuration, and account the response of μ after step changes in slip velocity or normal stress [71]. For a frictional interface that is sliding at velocity V, the definition of μ reads:

$$\mu = \mu^* + a \ln\left(\frac{V}{V^*}\right) + b \ln\left(\frac{\theta}{\theta^*}\right), \tag{1}$$

where μ^* is the steady-state coefficient at the reference slip velocity V^*, a is the direct-effect parameter, and b is the friction evolution parameter. θ is the state variable, and $\theta^*{=}D_c/V^*$ is its steady-state value, where D_c is the characteristic slip memory distance over which τ^* evolves once the system is perturbed [72].

Several definitions for θ have been proposed according to the rate-and-state friction models. Deep physical understandings and theoretical analyses [70,73,74], as well as comparisons with experiments of velocity steps [75,76], shear stress steps [77], and normal stress steps [78–81] have allowed us to elucidate the relative advantages and disadvantages of the proposed formulations. Our study focuses on those models that incorporate a dependence on the effective normal stress rate.

In such a sense, the Linker–Dieterich model [49] generalizes the basic rate-and-state laws by including a term in the state evolution equation, applicable to both the "slip" [69] and "aging" [68] laws [73]:

$$\begin{cases} \frac{d\theta}{dt} = -\frac{V\theta}{D_c}\ln\frac{V\theta}{D_c} - \frac{\alpha\theta}{b\sigma'}\frac{d\sigma'}{dt} \quad \text{(Slip law)} \\ \frac{d\theta}{dt} = 1 - \frac{V\theta}{D_c} - \frac{\alpha\theta}{b\sigma'}\frac{d\sigma'}{dt} \quad \text{(Aging law)} \end{cases} \tag{2}$$

In the above equations, the empirical parameter α controls the stressing-rate effect on the state variable, ranging from 0 to μ_0 [82]. Since we focus on the reactivation of faults which are initially at rest, we adopt a regularization of the rate-and-state models in the limit of zero slip speed, $V = 0$, proposed by Yang et al. [83]. μ is then defined as:

$$\mu(V,\theta) = \mu_0 + a\ln\left(\frac{V+V^*}{V^*}\right) + b\ln\left(\frac{\theta V^*}{D_c}\right), \tag{3}$$

with μ_0 being the initial friction coefficient. We implement in our model the equation for the aging law:

$$\frac{d\theta}{dt} = 1 - \frac{\theta\,(V+V^*)}{D_c} - \frac{\alpha\theta}{b\sigma'}\frac{d\sigma'}{dt}, \tag{4}$$

which is equivalent to the regularized model used by Tal et al. [84] with a threshold velocity $V^{th} = V^*$.

2.2. Thermo-Hydro-Mechanical 3D Model of Fault Reactivation

The rock is modeled as a thermo-poroelastic saturated material with single-phase flow. We adopt the classical theory of linear poroelasticity [85,86] and solve for the combination of fluid pressure, rock deformation, temperature, and frictional contact on the fault [31,32]. The solid and mass conservation, as well as mechanical equilibrium are coupled using the effective stress. The quasi-static Biot equations for a porous medium read:

$$\rho_f S\frac{\partial p}{\partial t} + \rho_f\alpha_B\frac{\partial\varepsilon_v}{\partial t} = \nabla\cdot\left(\rho_f\frac{k}{\eta_f}(\nabla p - \rho_f\mathbf{g})\right), \tag{5}$$

$$\nabla\cdot\sigma = \mathbf{0}, \tag{6}$$

where α_B is the Biot coefficient, $\varepsilon_v = \mathrm{tr}\,(\varepsilon)$ (with ε being the infinitesimal strain tensor) is the volumetric strain, k is the intrinsic permeability of the porous medium, η_f is the fluid dynamic viscosity, ρ_f is the fluid density, p is the pressure field, and σ is the total stress tensor. We consider a linear poroelastic material under small deformations, as well as plane strain conditions. Then, the effective stress tensor, $\sigma' = \sigma + \alpha_B p\mathbf{I}$, is a linear function of strains, $\sigma' = 2G\varepsilon_{el} + \lambda\mathrm{tr}(\varepsilon_{el})\mathbf{I}$, where λ and G are the Lamé constants, $\varepsilon_{el} = \varepsilon - \varepsilon_{th} = \frac{1}{2}\left(\nabla\mathbf{u} + \nabla\mathbf{u}^T\right) - \varepsilon_{th}$ is the elastic strain tensor, the result of subtracting the thermal strains to the total strain tensor ε, with $\mathbf{u}$ being the displacement field. The storage coefficient, $S = \phi\chi_f + (\alpha_B - \phi)\chi_s$, depends on the rock porosity, ϕ, and on the fluid and solid matrix compressibilities, χ_f and χ_s, where $\chi_s = (1-\alpha_B)/K$, and $K = \lambda + \frac{2}{3}G$ is the bulk modulus of the porous matrix.

Conservation of energy reduces to the heat equation [87]:

$$\begin{aligned} (\rho c)_{sat}\frac{\partial T}{\partial t} + \rho_f c_f\mathbf{v}\cdot\nabla T + \nabla\cdot\mathbf{q} &= Q, \\ \mathbf{q} &= -\kappa_{sat}\nabla T, \\ (\rho c)_{sat} &= \phi\rho c_s + (1-\phi)\rho_f c_f, \\ \kappa_{sat} &= \phi\kappa_s + (1-\phi)\kappa_f, \end{aligned} \tag{7}$$

where c_s is the heat capacity of the rock, c_f is the fluid heat capacity, κ_s is the thermal conductivity of the rock, and κ_f is the fluid thermal conductivity. $(\rho c)_{sat}$ y κ_{sat} are the saturated values of the product of density by heat capacity and thermal conductivity. Q includes the source or sink terms. Temperature changes propagate by diffusion and convection throughout the rock mass and the fluid. This changes produce thermal strains controlled by the thermal expansion coefficient α_T:

$$\varepsilon_{th} = \alpha_T(T - T_0), \tag{8}$$

where T is the temperature field, T_0 is the reference initial temperature, and α_T is the solid thermal expansion coefficient. We consider that the fluid keeps in a liquid state and its properties do not change with temperature or pressure, due to the high temperature and pressure conditions at the usual depths of the EGS projects. This couples the diffusive part of heat transport, while the convective part is modeled by introducing the Darcy velocity field **v** in Equation (7).

2.3. Case Study: The Deep Heat Mining Project in Basel, Switzerland

The DHM project in Basel was a milestone in geothermal energy. The know-how gained after the DHM project and the seismic events of 2006 boosted the development of EGS systems [52,53]. The first stage of the project was drilling the 5000 m depth Basel-1 well. The well reached a crystalline granitic rock basement at a temperature of 200 °C. The well was also employed for the hydraulic stimulation of the reservoir and the field characterization [56].

The number of fractures between 4629 m and 5000 m depth was between 0.2 and 0.3 fractures per meter [56]. The preferred fracture direction and orientation were NW–SE to NNW–SSE with dips greater than 60°, although the measures of the hypocenter locations during seismic events detected new families of fractures [88]. The events with greatest magnitude which rolled around in 2006 took place on a family with a deviation of 10° with respect of the maximum principal stress direction and a dip of 80°. The orientation of the principal stresses was deduced from acoustic geophysical studies within the Basel-1 well. The minimum principal stress, $\sigma_{h,min}$, had an orientation of 54 ± 14° and the maximum one, $\sigma_{h,max}$, 144 ± 14° [88]. These orientations are consistent with the in situ stress state in the upper Rhine Basin deduced from previous seismic events in the crystalline rocky massif [56].

The magnitude of the principal stresses were also quantified. The tectonic ratio of the minimum principal stress, $\sigma_{h,min}$, to the vertical one is 0.7, and the ratio of the maximum principal stress, $\sigma_{h,max}$, to the vertical one is 1.6, in such a way that the vertical stress is the intermediate principal stress [56]. Mechanic boundary conditions are defined by the expressions:

$$\begin{aligned} \sigma_v(d) &= (\rho(1-\phi) + \rho_f\phi)gd, \\ \sigma_{h,min}(d) &= Tect_{min} \cdot \sigma_v = 0.7\sigma_v, \\ \sigma_{h,max}(d) &= Tect_{max} \cdot \sigma_v = 1.6\sigma_v, \end{aligned} \tag{9}$$

where d is the depth. In these expressions, compressive stresses are assumed positive. The imposed stresses at the boundaries are assumed constant in time, which is a feasible hypothesis even if simulation time exceeds one decade [89]. The temperature at the bottom of the well is between 190 and 200 °C, with a thermal gradient of 40 °C/km [55].

2.4. The Basel 3D Model

The geometry of the 3D model domain is a 1.5 km^3 cube that is supposed to be homogeneous porous rock mass. In the center of the cube, we situated a sphere divided into two hemispheres with their intersection plane representing the main fault. The injection well is modeled as a cylinder with a diameter of 1 m and a height of 380 m whose geometry is subtracted from the solid cube. The well is vertical and located at a distance 50 m away from the center of the fault (Figure 1). The diameter of the simulated well is higher than the real one due to mathematical issues, where the diffusion of the injected fluid should not be simulated with elements which are too fine. The geometry of the model is described in Figure 1.

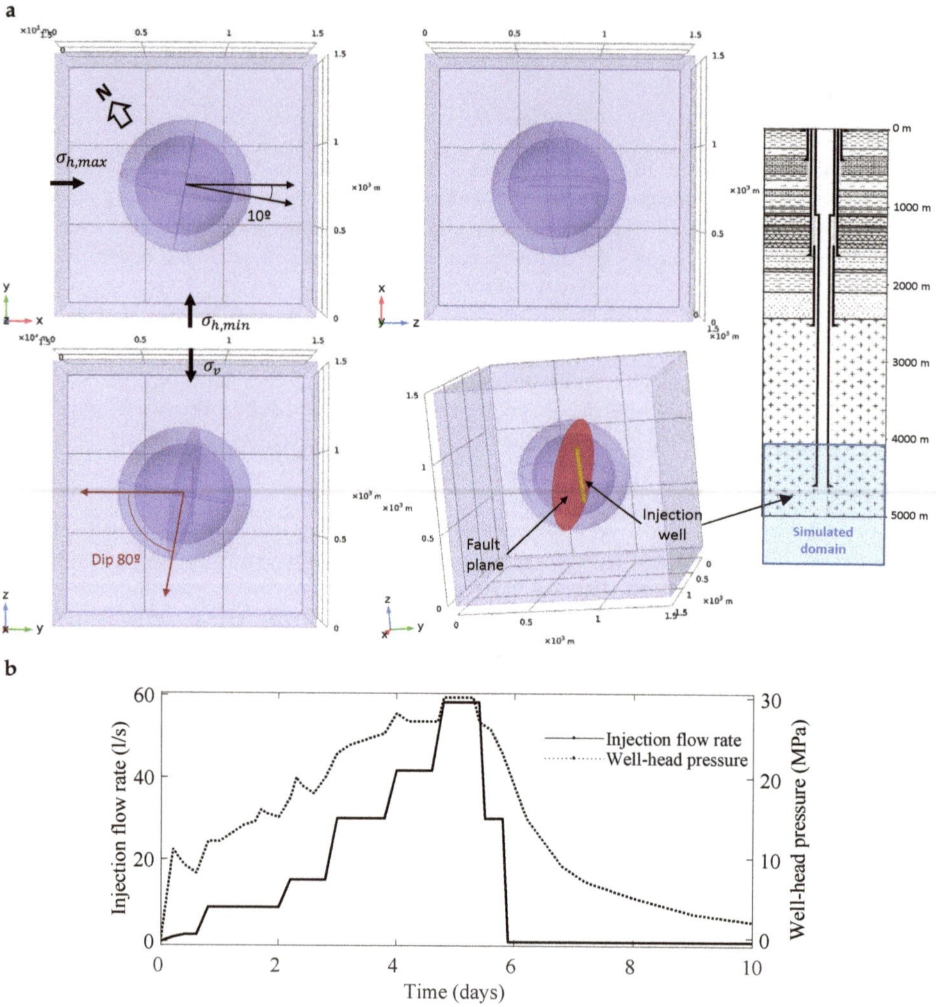

Figure 1. Scheme of the 3D Basel EGS model. In (**a**) we show the domain that is a 1.5 km^3 cube with the fault plane oriented 10° with respect to the x-axis and dip 80° to the SW. The domain is located between 4050 and 5550 m depth, while the injection section of the Basel-1 well extends from 4629 to 5000 m depth. In (**b**) we plot the injection protocol (left vertical axis) and injection pressure (right vertical axis) measured at the Basel-1 well, obtained from [56].

We focus on the depth range between 4050 and 5550 m, and the injection takes place through the well between 4620 to 5000 m depth. The x axis is parallel to the maximum principal stress ($\sigma_x = \sigma_{h,max}$), the y axis to the minimum principal stress ($\sigma_y = \sigma_{h,min}$), and the z axis to the intermediate principal stress. We apply vertical stress to simulate overburden strata according to $\sigma_v(z) = (\rho(1-\phi) + \rho_f\phi)g(5500 - z)$, where the z coordinate ranges between 0 and 1500 m in our model.

The complexity of the fracture network requires the adoption of some simplifications to define the orientation of the fault plane, based on the hypocenter locations estimated from the motion data records of the historical seismic activity [88]. We adopt the most unfavorable fracture family as the preferred sliding plane, which has a direction deviated approximately 10° with respect to the direction of the maximum principal stress, which dips 80° towards the SW. Most of the detected hypocenters do not separate more than 50 m with respect to the upper injection section of the well [55]. For that reason, the fault plane of the model is oriented 10° with respect to the x-axis and dipping 80° to the SW, with the injection well-located at a distance 50 m away from its center. We refine the mesh of the model in the area of the central sphere, as shown in Figure 2, and force mesh conformity throughout the fault plane. We use tetrahedral elements—quadratic for the solid mechanics equations and linear for fluid flow and heat transport.

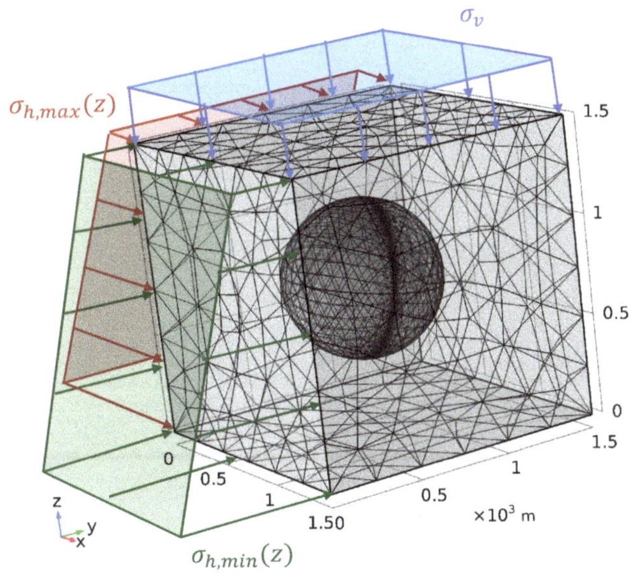

Figure 2. 3D finite element mesh and mechanical boundary conditions applied. At the exterior boundaries with no stresses applied, we impede the displacement in its normal direction.

Mechanic boundary conditions are equal to tectonic stresses and are imposed on three faces of the rectangular domain (Figure 2). We impose a reference hydrostatic pressure, $p = \rho_f g d = \rho_f g(5550 - z)$, along all external boundaries, where d is the depth, and the temperature is set at the geothermal gradient $T = 0.04\ [C^\circ/m] \cdot d = 0.04\ [C^\circ/m] \cdot (5550 - z)$.

We impede the displacements in the normal direction and impose no-flow (thermal and hydraulic) conditions on the other ones. At the fault plane, we impose the contact condition, allowing relative tangential displacements between its edges. We consider the fault is almost impermeable and has the same thermal properties as the rest of the domain, resulting in thermal continuity. We simplify the fluid flow around the injection well by imposing a volumetric flux along the boundary of an effective injection region (the lateral surface of the cylinder). During the simulations, temperature is also fixed in the injection well.

Based on the in-site measurements, we define feasible parameters for each physical process of the model.

The Young modulus, the Poisson ratio, and the density of the solid skeleton are usual values for crystalline rock formations of granitic type. The parameters related to the fluid, such as density or compressibility of water, are also usual values, unlike viscosity, whose value of 2.4×10^{-4} Pa·s, is lower than the viscosity of water at ambient temperature ($\eta_f = 10^{-3}$ Pa·s) due to the high temperatures at such depths [23]. Permeability and porosity have been chosen according to the characteristics of the rock mass. Lastly, thermal parameters have been taken from [90] that collects data of thermal conductivities and heat capacities from different materials. Coupling between the flow and mechanical problems have been included using the Biot coefficient $\alpha_B = 1$ and between the thermal and mechanical physics with the thermal expansion coefficient $\alpha_T = 8 \times 10^{-6}$.

The parameters of the "rate-and-state" model have been chosen within a feasible range to emulate the fault reactivation at a similar time scale. We consider the fault is arrested before reactivation, being the slip speed $V = 0$. Hence, Equations (1) and (2) can be simplified, so it is only necessary to define the parameters α, b, D_c, and V^*. We adopt the aging law and the rate-and-state parameters, $b = 0.03$, $D_c = 700$ μm, $V^* = 10^{-9}$ m/s, and $\mu_0 = 0.55$. Table 1 lists the parameters of the model.

Table 1. Parameters of Basel 3D model.

Parameter	Value	Unit	Description
E	20	GPa	Young Modulus of the rock
ν	0.25	–	Poisson ratio of the rock
ρ	2700	kg/m^3	Rock Density
$Tect_{max}$	1.6	–	High tectonic ratio
$Tect_{min}$	0.7	–	Low tectonic ratio
σ_h	$\sigma_v \cdot$Tect	MPa	Confinement stress
ρ_f	1000	kg/m^3	Fluid density
η_f	0.00024	Pa·s	Fluid viscosity
χ_f	4×10^{-10}	Pa^{-1}	Fluid compressibility
k	10^{-15}	m^2	Porous media permeability
ϕ	0.1	–	Porosity
κ_s	2.4	W/(m·K)	Solid thermal conductivity
κ_f	0.6	W/(m·K)	Fluid thermal conductivity
c_s	800	J/(kg·K)	Solid heat capacity
c_f	4200	J/(kg·K)	Fluid heat capacity
T_{amb}	293.15	K	Ambient temperature
α_B	1	–	Biot coefficient
α_T	8×10^{-6}	–	Thermal expansion coefficient
μ_0	0.55	–	Friction coefficient
c	0	MPa	Contact cohesion
a	0.005	–	Direct effect parameter
b	0.03	–	Friction evolution parameter
D_c	0.0007	m	Characteristic slip distance
V^*	10^{-9}	m/s	Reference velocity
α	0.2	–	Linker-Dieterich normal stress coefficient

The constitutive laws for fault strength are given by Equations (1) and (2), and the frictional contact on the fault is modeled using an Augmented Lagrangian formulation [91]. We solve, in a monolithically-coupled fashion, the field Equations (5)–(8) and the rate-and-state aging law (1) and (2) with the frictional contact variables [31,32].

3. Results and Discussion

3.1. Calibration

The hydraulic stimulation of the DHM geothermal project required the injection of 11,600 m^3 of water, following the protocol depicted in Figure 1b [56]. This volume was injected prior to the 8 December 2006 earthquake sequence. Since we simulated the on-site conditions during the injection operations, we imposed an inflow velocity at the injection well of the model that is the result of dividing the flow rate by the lateral surface of the cylinder $q_{iny}=Q_0/(2\pi r h)$, where r is the radius of the cylinder and h is the height where injection takes place. We assumed that the injection temperature remained constant and equal to ambient rock temperature.

Figure 3a,b shows a plot of the simulated evolutions of pressure and injected flow rate computed with our model, and the registered data at the Basel-1 injection well in 2006. We adopted as the flow boundary condition the measured injected flow rate at the injection well (see Figure 3a). Our computed injected pressure evolution initially differs from the measured values on-site, Figure 3b. The difference between pressure observed in real data and model results arises from the weakening and fracturing of the rock in the vicinity of the well, allowing water to flow through the rock matrix and its fractures. After 5 days of injection, when the largest earthquakes occurred, the difference between our simulated results and the measured data were drastically reduced, showing that our model is properly adjusted.

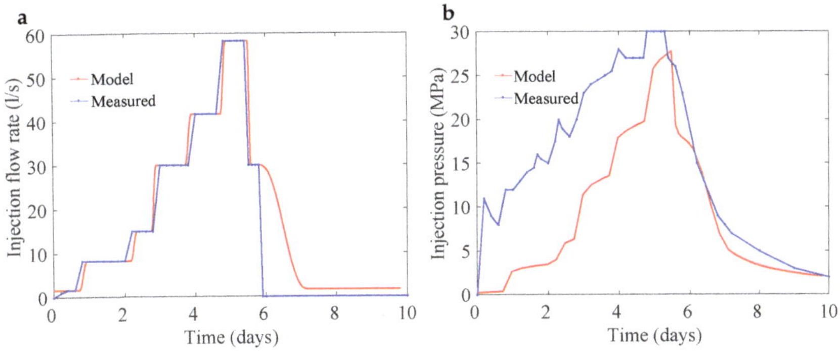

Figure 3. 3D model calibration results. In (**a**) we show the injection flow rate pattern used in the model (red line) that is similar to the real injection protocol from the Basel-1 data (blue line). In (**b**) we plot the computed injection pressure (red line) and the values measured in 2006 (blue line).

The results of pore pressure and temperature fields are included in Figure 4. We define a reference horizontal plane at 4800 m depth (z = 750 m), where the changes of pore pressure and temperature cause by the injection of cold water are displayed. We show in Figure 4a the locations of the reference plane, the fault plane, and the injection well. We also include the buildup of pore pressure around the injection well and near the fault plane at the time that reactivation occurs (t = 5.5 days). The injection cools down a small area around the well which does not reach the fault (Figure 4b). Since the time for heat diffusion is higher than the one for pressure diffusion, pore pressure changes around the well are much faster than temperature changes. As fault temperature remains constant during the stimulation phase (Figure 4b), pore pressure increases on the fault plane (Figure 4c). We explain in the next section how pore pressure on the fault plane controls the fault reactivation and the frictional properties of the contact.

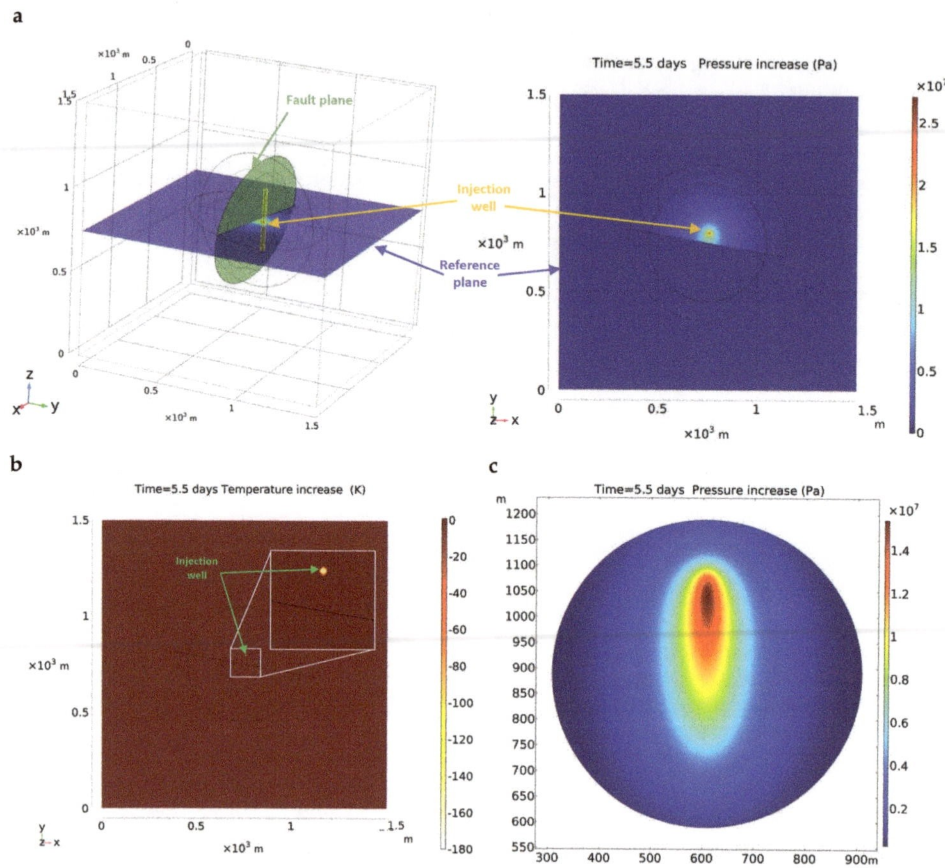

Figure 4. 3D model results at the horizontal reference plane. In (**a**) (left) we show the reference plane inside the model, the fault plane, and the injection well. The increase of pore pressure (right) in that plane shows the results of the pore pressure increase caused by the injection at the instant of fault reactivation (among 26 MPa). In (**b**) we display the increment of temperature due to the injection at the reference plane, with an inset that zooms the surroundings of the injection well and shows that the temperature diffusion is much slower than pressure propagation. In (**c**) we show the results of the pore pressure increase at the fault plane. The vertical axis of the image corresponds to the maximum slope line of the fault plane and the horizontal axis corresponds to a horizontal direction in the 3D model deviated 10° with respect to the x-axis.

3.2. Fault Reactivation

Fault reactivation is the onset of fault slip. It depends on the variables involved in fault stability, such as frictional strength $\mu|\sigma'|$ or shear stress τ. We quantify fault stability through the Coulomb Failure Function, defined as $CFF = \mu|\sigma'| - \tau$. The CFF equals to zero when the fault is at rest, and when CFF is less than zero the fault reactivates, given that the shear stress τ exceeds the frictional strength of the contact $\mu|\sigma'|$. Changes in Coulomb Failure Function can be used as a proxy for fault weakening ($\Delta CFF < 0$) or strengthening ($\Delta CFF > 0$). In that sense, we show in Figure 5 the increase in the failure function $\Delta CFF = CFF(t) - CFF(t = 0)$ on the fault plane, which indicates how the fault weakens due to the effect of fluid injection.

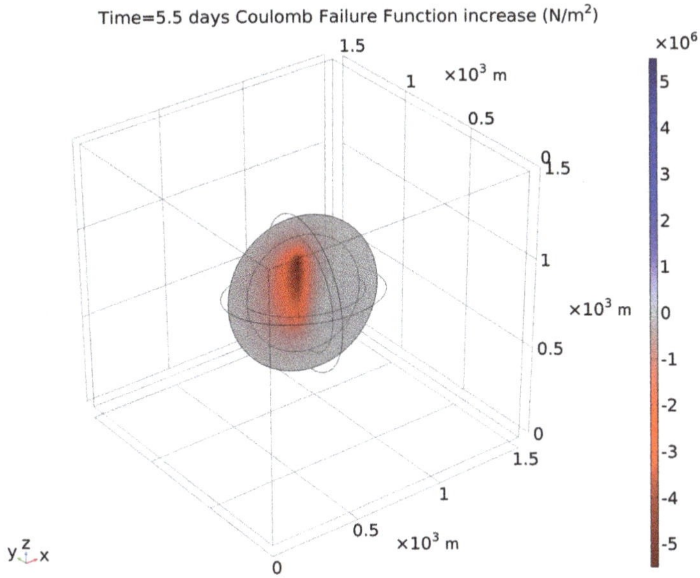

Figure 5. 3D model results on the fault plane 5.5 days after the onset of injection (the instant of fault reactivation). We display the increase in Coulomb Failure Function on the fault plane due to the injection $\Delta CFF = \Delta(\mu|\sigma'| - \tau) = CFF(t) - CFF(t = 0)$. The results show the fault weakening (red color, $\Delta CFF < 0$) due to the injection of cold water.

We include in Figure 6 our computed results with the 3D model for the variables involved in the frictional stability on the fault plane. The distribution of effective normal stress, $|\sigma'|$, acting on the fault plane (Figure 6a) at the onset of the slip shows that there is a decrease in fault effective compressions. Moreover, the spatial distribution is symmetric with respect to a vertical axis. The decrease in effective normal stress is almost the same as the increase in pore pressure (Figure 4c). Differences arise from poroelastic and thermal effects, and indicate that pore pressure changes dominate over thermal and poroelastic effects.

We plot the modulus of the shear stress τ on the fault at reactivation time in Figure 6b. Shear stress increases on the north side of the fault and decreases on the right side. This response is caused by the increase in pore pressure, as well as by the poroelastic effects accounted in our model [16–19,31,32]. The poroelastic effects are also coupled with the fault orientation and tectonic stresses, which contribute to the asymmetry of the results.

We show the value of the friction coefficient, μ, in Figure 6c, computed with the rate-and-state law (Equatons (3) and (4)). Since the fault is initially at rest ($V = 0$), the observed evolution of the friction coefficient is attributed to the decrease in the effective stress through the Linker-Dieterich term and the α-parameter in Equation (2) [79]. The Coulomb Failure Function CFF, Figure 6d, indicates that fault reactivation occurs after 5.5 days of injection. The asymmetry in the CFF distribution remarks the influence of shear stresses, in contrast with the symmetry of effective normal stress and friction coefficient that are directly pore-pressure dependent. This pattern of symmetry and asymmetry of stresses before and at reactivation also influence on the nucleation and rupture phases of the earthquake.

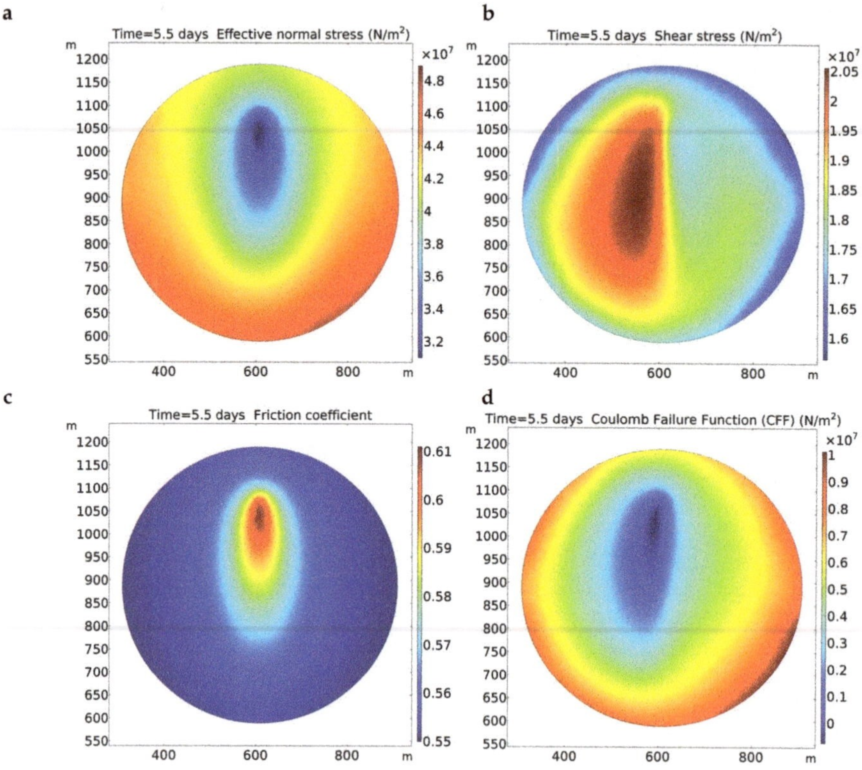

Figure 6. Display of 3D model results at the 2D fault plane 5.5 days after the injection starting (the instant of fault reactivation). In (**a**) we show the effective normal stress $|\sigma'|$ (positive values of effective normal stresses are compressive). In (**b**) we display the modulus of the shear stress τ, in (**c**) the friction coefficient μ obtained from the rate-and-state equations, and in (**d**) the Coulomb Failure Function CFF.

We illustrate the evolution of stresses and frictional variables up to the fault reactivation at a control point on the fault. The point is located at the left-half part of the fault, and it is the first point where the ratio of the acting shear stress to the effective normal stress equals the frictional strength—that is, it is the first fault point at which slip occurs (blue area where $CFF = 0$ in Figure 6d).

Figure 7 shows the evolution of the friction at the control point up to the reactivation. The mobilized friction (green line) is the ratio $\tau/|\sigma'|$ that normalizes the shear stress acting on the fault with the effective normal stress, and the friction coefficient μ (blue line) is a dimensionless frictional strength. When the mobilized friction equals frictional strength at 5.5 days ($\tau/|\sigma'| = \mu$), the fault reactivates.

We observe that due to the injection protocol that systematically increases the flow rate, and consequently, the pore pressure, the slopes of both the friction coefficient and the mobilized friction curves increase. Moreover, the friction coefficient changes from $\mu_0 = 0.55$ to 0.61 at the reactivation instant due to the effect of the variation of the effective normal stress on the friction coefficient. It delays the fault reactivation, which changes from 4.6 days if effective normal stress rate is disregarded, and to 5.5 days if the rate is accounted for. Therefore, the effect of normal stress rate on friction coefficient needs to be taken into account, as it partially controls the reactivation time. These results verify that our models properly reproduce the on-site reactivation of the reservoir's representative fault after 5.5 days of injection, elapsed from December 3 to 8.

Our methodology may be a useful tool for engineers, practitioners, and stakeholders to assess fault reactivation under real conditions of natural stresses, temperatures, and injection protocols. The application of our model to the DHM project in Basel has shown its ability to predict fault reactivation and demonstrated that the earthquake sequence occurred in December 2006 may have been caused by the hydraulic stimulation of the rock mass. Our model can also be useful for assessing new injection protocols, as well as stimulating and managing the operation of EGS system in the short and long term.

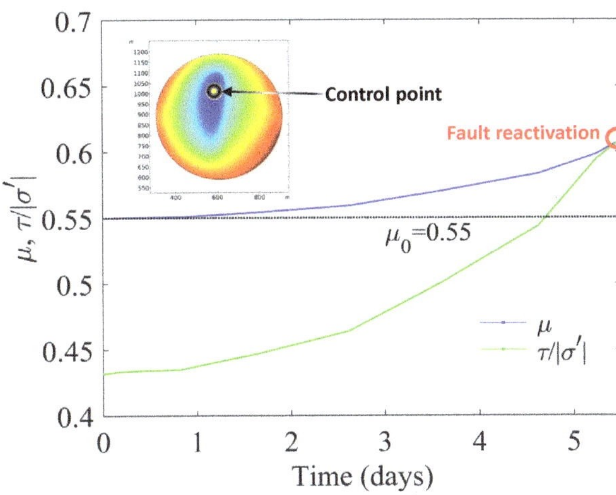

Figure 7. Evolution of the friction variables at the fault control point. The blue line represents the evolution of the friction coefficient μ as derived from the rate-and-state equations. The green line plots the evolution of the ratio $\tau/|\sigma'|$. The slopes of both curves are related to the different flow-rate levels of the injection protocol.

4. Conclusions

Geothermal energy emerges as an alternative renewable energy to ensure access to affordable, reliable, sustainable, and modern energy for all the world. In most of the geothermal deposits, rock permeability at exploitation depths is very low, rendering geothermal projects economically infeasible. This drawback has been solved by the so-called Enhanced Geothermal Systems (EGS), where rock permeability is enhanced through the so-called rock stimulation. One of the most widely used techniques is hydro-shearing, which reactivates preexisting joints by initiating shear failure. Water is injected under high pressure, reducing normal stress across them and eventually triggering shear failure. A major environmental issue for these techniques is induced seismicity as a result of water injection.

Here, we have presented a finite element model for the simulation of fault reactivation in poroelastic media with rate-and-state friction law. Our model monolithically couples fluid flow, rock mechanics, heat transport, and rate-and-state friction. We have conducted three-dimensional simulations of fault reactivation with frictional strength governed by a Linker–Dieterich law, embedded in a poroelastic homogenous media, and driven by fluid injection. The Linker–Dieterich law accounts for the effect of effective stress rate on the friction evolution, and is able to explain the observed delays in fault reactivation.

We applied our model to simulate the hydro-shearing effects during the stimulation at the Basel-1 well at the Deep Heat Mining geothermal project in Basilea (Switzerland). We adopted as input the stress field and rock properties estimated for the main hydrogeological units at Basel-1, and simulated the stimulation phase using a transient well-head pressure and flow evolution similar to the values reported at the study site. Our three-dimensional model satisfactorily reproduced the registered injection flow rate and pressure injection, as well as the time of the fault reactivation.

Our simulated results indicate that thermal effects are negligible during the stimulation phase. Temperature changes occur in a small area around the injection well and do not reach the fault plane. Nevertheless, pressure changes, together with poroelastic effects, weaken the fault, leading eventually to fault reactivation and the onset of slip. Our model was able to reproduce the instant at which fault reactivation occurred at the Basel-1 site, demonstrating that the earthquake sequence that occurred in December 2006 in Basel was caused by the hydraulic stimulation of the geothermal reservoir.

Our methodology emerges as a useful tool for engineers, practitioners, and stakeholders to assess fault reactivation under real conditions of natural stresses, temperatures, and injection protocols. The application of our model to the DHM project in Basel has shown its ability to predict fault reactivation in a real case. Our model can also be useful for assessing new injection protocols, as well as stimulating and managing the operation of an EGS system in the short and long term.

Author Contributions: Investigation, S.A., D.S., J.C.M., and L.C.-F.

Funding: This research was supported by the Universidad Politécnica de Madrid through the special programme for young scientists under grant VJIDOCUPM19DSS ("Programa Propio de I+D+I de la Universidad Politécnica de Madrid. Convocatoria de ayuda dirigida a jóvenes investigadores doctores para fortalecer sus planes de investigación"). DSS gratefully acknowledges funding from the Fundación BBVA though Becas Leonardo a Investigadores y Creadores Culturales 2019 (Proyecto realizado con la Beca Leonardo a Investigadores y Creadores Culturales 2019 de la Fundación BBVA).

Conflicts of Interest: The authors declare no conflict of interest. The founders had no role in the design of the study; in the collection, analyses, or interpretation of data; in the writing of the manuscript, or in the decision to publish the results.

References

1. Colglazier, W. Sustainable development agenda: 2030. *Int. J. Rock. Mech. Min. Sci.* **2015**, *349*, 1048–1050. [CrossRef] [PubMed]
2. MIT-Energy-Initiative. *The Future of Geothermal Energy: Impact of Enhanced Geothermal Systems (EGS) on the United States in the 21st Century*; Massachusetts Institute of Technology: Cambridge, MA, USA, 2006; pp. 1–372.
3. Mahbaz, S.; Dehghani-Sanij, A.; Dusseault, M.; Nathwani, J. Enhanced and integrated geothermal systems for sustainable development of Canada's northern communities. *Sustain. Energy Technol. Assess.* **2020**, *37*, 100565. [CrossRef]
4. Kazemi, A.; Mahbaz, S.; Dehghani-Sanij, A.; Dusseault, M.; Fraser, R. Performance Evaluation of an Enhanced Geothermal System in the Western Canada Sedimentary Basin. *Renew. Sustain. Energ. Rev.* **2019**, *113*, 109278. [CrossRef]
5. Kinney, C.; Dehghani-Sanij, A.; Mahbaz, S.; Dusseault, M.B.; Nathwani, J.S.; Fraser, R. Geothermal Energy for Sustainable Food Production in Canada's Remote Northern Communities. *Energies* **2019**, *12*, 4058. [CrossRef]
6. Soltani, M.; Kashkooli, F.M.; Dehghani-Sanij, A.R.; Kazemi, A.R.; Bordbar, N.; Farshchi, M.J.; Elmi, M.; Gharali, K.; Dusseault, M.B. A comprehensive study of geothermal heating and cooling systems. *Sustain. Cities Soc.* **2019**, *44*, 793–818. [CrossRef]
7. Soltani, M.; Moradi-Kashkooli, F.; Dehghani-Sanij, A.R.; Nokhosteen, A.; Ahmadi-Joughi, A.; Gharali, K.; Mahbaz, S.B.; Dusseault, M. A comprehensive review of geothermal energy evolution and development. *Int. J. Green Energy* **2019**, *16*, 971–1009. [CrossRef]
8. Rinaldi, A.P.; Rutqvist, J. Joint opening or hydroshearing? Analyzing a fracture zone stimulation at Fenton Hill. *Geothermics* **2019**, *77*, 83–98. [CrossRef]
9. Rinaldi, A.P.; Rutqvist, J.; Sonnenthal, E.L.; Cladouhos, T.T. Coupled THM Modeling of Hydroshearing Stimulation in Tight Fractured Volcanic Rock. *Transport Porous Media* **2014**, *108*, 131–150. [CrossRef]
10. Charléty, J.; Cuenot, N.; Dorbath, L.; Dorbath, C.; Haessler, H.; Frogneux, M. Large earthquakes during hydraulic stimulations at the geothermal site of Soultz-sous-Forêts. *Int. J. Rock. Mech. Min. Sci.* **2007**, *44*, 1091–1105, doi:10.1016/j.ijrmms.2007.06.003. [CrossRef]

11. Baisch, S.; Vörös, R.; Weidler, R.; Wyborn, D. Investigation of fault mechanisms during geothermal reservoir stimulation experiments in the Cooper Basin, Australia. *Bull. Seismol. Soc. Am.* **2009**, *99*, 148–158. [CrossRef]
12. Fehler, M.C. Stress control of seismicity patterns observed during hydraulic fracturing experiments at the Fenton Hill hot dry rock geothermal energy site, New Mexico. *Int. J. Rock. Mech. Min. Sci.* **1989**, *26*, 211–219. [CrossRef]
13. Parker, R. The Rosemanowes HDR project 1983–1991. *Geothermics* **1999**, *28*, 603–615. [CrossRef]
14. Pine, R.J.; Batchelor, A.S. Downward migration of shearing in jointed rock during hydraulic injections. *Int. J. Rock. Mech. Min. Sci.* **1984**, *21*, 249–263. [CrossRef]
15. Ziagos, J.; Phillips, B.R.; Boyd, L.; Jelacic, A.; Stillman, G.; Hass, E. A technology roadmap for strategic development of enhanced geothermal systems. In Proceedings of the 38th Workshop on Geothermal Reservoir Engineering, Stanford, CA, USA, 11–13 February 2013; pp. 11–13.
16. Santillán, D.; Juanes, R.; Cueto-Felgueroso, L. Phase field model of fluid–driven fracture in elastic media: Immersed fracture formulation and validation with analytical solutions. *J. Geophys. Res. Solid Earth* **2017**, *122*, 2565–2589. [CrossRef]
17. Santillan, D.; Mosquera, J.; Cueto-Felgueroso, L. Fluid-driven fracture propagation in heterogeneous media: Probability distributions of fracture trajectories. *Phys. Rev. E* **2017**, *96*, 053002. [CrossRef]
18. Santillan, D.; Mosquera, J.; Cueto-Felgueroso, L. Phase-field model for brittle fracture. Validation with experimental results and extension to dam engineering problems. *Eng. Fracture Mech.* **2017**, *178*, 109–125. [CrossRef]
19. Santillán, D.; Juanes, R.; Cueto-Felgueroso, L. Phase field model of hydraulic fracturing in poroelastic media: Fracture propagation, arrest, and branching under fluid injection and extraction. *J. Geophys. Res. Solid Earth* **2018**, *123*, 2127–2155. [CrossRef]
20. Riahi, A.; Damjanac, B. Numerical study of hydro-shearing in geothermal reservoirs with a preexisting discrete fracture network. In Proceedings of the Thirty–Eighth Workshop on Geothermal Reservoir Engineering, Stanford, CA, USA, 11–13 February 2013.
21. Jeanne, P.; Guglielmi, Y.; Rutqvist, J.; Nussbaum, C.; Birkholzer, J. Permeability Variations Associated With Fault Reactivation in a Claystone Formation Investigated by Field Experiments and Numerical Simulations. *J. Geophys. Res. Solid Earth* **2018**, *123*, 1694–1710. [CrossRef]
22. Finnila, A.; Dershowitz, W.; Doe, T.; McLaren, R. Hydro-Shearing and Hydraulic Fracturing for Enhanced Geothermal Systems in Archetypical Normal, Strike-Slip, and Thrust Faulting Terrains. *GRC Trans.* **2015**, *39*, 1–19.
23. Norbeck, J.H.; McClure, M.W.; Horne, R.N. Field observations at the Fenton Hill enhanced geothermal system test site support mixed-mechanism stimulation. *Geothermics* **2018**, *74*, 135–149. [CrossRef]
24. Weingarten, M.; Ge, S.; Godt, J.W.; Bekins, B.A.; Rubinstein, J.L. High-rate injection is associated with the increase in US mid-continent seismicity. *Science* **2015**, *348*, 1336–1340. [CrossRef] [PubMed]
25. Horton, S. Disposal of hydrofracking waste fluid by injection into subsurface aquifers triggers earthquake swarm in central Arkansas with potential for damaging earthquake. *Seismol. Res. Lett.* **2012**, *83*, 250–260. [CrossRef]
26. Juanes, R.; Hager, B.H.; Herzog, H.J. No geologic evidence that seismicity causes fault leakage that would render large-scale carbon capture and storage unsuccessful. *Proc. Natl. Acad. Sci. USA* **2012**, *109*, E3623. [CrossRef] [PubMed]
27. Vilarrasa, V.; Carrera, J. Geologic carbon storage is unlikely to trigger large earthquakes and reactivate faults through which CO2 could leak. *Proc. Natl. Acad. Sci. USA* **2015**, *112*, 5938–5943. [CrossRef] [PubMed]
28. Brodsky, E.E.; Lajoie, L.J. Anthropogenic seismicity rates and operational parameters at the Salton Sea Geothermal Field. *Science* **2013**, *341*, 543–546. [CrossRef] [PubMed]
29. Majer, E.L.; Baria, R.; Stark, M.; Oates, S.; Bommer, J.; Smith, B.; Asanuma, H. Induced seismicity associated with enhanced geothermal systems. *Geothermics* **2007**, *36*, 185–222. [CrossRef]
30. Scholz, C.H. Earthquakes and friction laws. *Nature* **1998**, *391*, 37. [CrossRef]
31. Cueto-Felgueroso, L.; Vila, C.; Santillan, D.; Mosquera, J.C. Numerical modeling of injection-induced earthquakes using laboratory-derived friction laws. *Water Resour. Res.* **2018**, *54*, 1–27. [CrossRef]
32. Cueto-Felgueroso, L.; Santillán, D.; Mosquera, J.C. Stick-slip dynamics of flow-induced seismicity on rate and state faults. *Geophys. Res. Lett.* **2017**, *44*, 4098–4106. [CrossRef]

33. De Simone, S.; Carrera, J.; Vilarrasa, V. Superposition approach to understand triggering mechanisms of post-injection induced seismicity. *Geothermics* **2017**, *70*, 85–97. [CrossRef]
34. McGarr, A. Maximum magnitude earthquakes induced by fluid injection. *J. Geophys. Res. Solid Earth* **2014**, *119*, 1008–1019. [CrossRef]
35. Mignan, A.; Karvounis, D.; Broccardo, M.; Wiemer, S.; Giardini, D. Including seismic risk mitigation measures into the Levelized Cost of Electricity in enhanced geothermal systems for optimal siting. *Appl. Energy* **2019**, *238*, 831–850. [CrossRef]
36. Scholz, C.H. *The Mechanics of Earthquakes and Faulting*; Cambridge University Press: Cambridge, UK, 2002.
37. Deichmann, N.; Giardini, D. Earthquakes Induced by the Stimulation of an Enhanced Geothermal System below Basel (Switzerland). *Seismol. Res. Lett.* **2009**, *80*, 784–798. [CrossRef]
38. Beck, M.; Seitz, G.; Class, H. Volume-based modelling of fault reactivation in porous media using a visco-elastic proxy model. *Transp. Porous Media* **2016**, *114*, 505–524. [CrossRef]
39. Borja, R.; Foster, C. Continuum mathematical modeling of slip weakening in geological systems. *J. Geophys. Res. Solid Earth* **2007**, *112*. [CrossRef]
40. Rutqvist, J.; Birkholzer, J.; Cappa, F.; Tsang, C.F. Estimating maximum sustainable injection pressure during geological sequestration of CO2 using coupled fluid flow and geomechanical fault-slip analysis. *Energy Conv. Manag.* **2007**, *48*, 1798–1807. [CrossRef]
41. Segura, J.M.; Carol, I. Coupled HM analysis using zero-thickness interface elements with double nodes. Part I: Theoretical model. *Int. J. Numer. Anal. Meth. Geomech.* **2008**, *32*, 2083–2101. [CrossRef]
42. Segura, J.M.; Carol, I. Coupled HM analysis using zero-thickness interface elements with double nodes. Part II: Verification and application. *Int. J. Numer. Anal. Meth. Geomech.* **2008**, *32*, 2103–2123. [CrossRef]
43. Ferronato, M.; Gambolati, G.; Janna, C.; Teatini, P. Numerical modelling of regional faults in land subsidence prediction above gas/oil reservoirs. *Int. J. Numer. Anal. Meth. Geomech.* **2008**, *32*, 633–657. [CrossRef]
44. Ghassemi, A.; Tao, Q. Thermo-poroelastic effects on reservoir seismicity and permeability change. *Geothermics* **2016**, *63*, 210–224. [CrossRef]
45. Jha, B.; Juanes, R. Coupled multiphase flow and poromechanics: A computational model of pore pressure effects on fault slip and earthquake triggering. *Water Resour. Res.* **2014**, *50*, 3776–3808. [CrossRef]
46. Morris, J.P.; Hao, Y.; Foxall, W.; McNab, W. A study of injection-induced mechanical deformation at the In Salah CO_2 storage project. *Int. J. Greenhouse Gas Control* **2011**, *5*, 270–280. [CrossRef]
47. Rice, J.R. Heating and weakening of faults during earthquake slip. *J. Geophys. Res. Solid Earth* **2006**, *111*. [CrossRef]
48. Dieterich, J.H. Earthquake nucleation on faults with rate-and state-dependent strength. *Tectonophysics* **1992**, *211*, 115–134. [CrossRef]
49. Dieterich, J.H.; Linker, F.M. Fault Stability Under Conditions of Variable Normal Stress. *Geophys. Res. Lett.* **1992**, *19*, 1691–1694. [CrossRef]
50. Kilgore, B.; Beeler, N.M.; Lozos, J.; Oglesby, D. Rock friction under variable normal stress. *J. Geophys. Res. Solid Earth* **2017**, *122*, 7042–7075. [CrossRef]
51. Andrés, S.; Santillán, D.; Mosquera, J.C.; Cueto-Felgueroso, L. Delayed weakening and reactivation of rate-and-state faults driven by pressure changes due to fluid injection. *J. Geophys. Res. Solid Earth* **2019**, doi:10.1029/2019JB018109. [CrossRef]
52. Wyss, R.; Link, K. Actual Developments in Deep Geothermal Energy in Switzerland. In Proceedings of the World Geothermal Congress 2015, Melbourne, Australia, 19–25 April 2015.
53. Meier, P.M.; Alcolea Rodríguez, A.; Bethmann, F. Lessons Learned from Basel: New EGS Projects in Switzerland Using Multistage Stimulation and a Probabilistic Traffic Light System for the Reduction of Seismic Risk. In Proceedings of the World Geothermal Congress 2015, Melbourne, Australia, 19–25 April 2015.
54. Swiss Seismological Service. DHM Project Description. 2019. Available online: http://www.seismo.ethz.ch/en/earthquakes/monitoring/geothermal-energy-basel/Project-Description/ (accessed on 3 December 2019).
55. Ladner, F.; Haring, M. Hydraulic Characteristics of the Basel 1 Enhanced Geothermal System. *GRC Trans.* **2009**, *33*, 199–203.
56. Haring, M.O.; Schanz, U.; Ladner, F.; Dyer, B.C. Characterisation of the Basel 1 enhanced geothermal system. *Geothermics* **2008**, *37*, 469–495. [CrossRef]

57. Jacquey, A.B.; Cacace, M.; Blöcher, G.; Watanabe, N.; Huenges, E.; Scheck-Wenderoth, M. Thermo-poroelastic numerical modelling for enhanced geothermal system performance: Case study of the Groß Schönebeck reservoir. *Tectonophysics* **2016**, *684*, 119–130. [CrossRef]
58. Kim, S.; Hosseini, S.A. Hydro-thermo-mechanical analysis during injection of cold fluid into a geologic formation. *Int. J. Rock. Mech. Min. Sci.* **2015**, *77*, 220–236. [CrossRef]
59. Li, C.; Laloui, L. Coupled multiphase thermo-hydro-mechanical analysis of supercritical CO_2 injection: Benchmark for the In Salah surface uplift problem. *Int. J. Greenhouse Gas Control* **2016**, *51*, 394–408. [CrossRef]
60. Rutqvist, J.; Birkholzer, J.; Tsang, C.F. Coupled reservoir-geomechanical analysis of the potential for tensile and shear failure associated with CO_2 injection in multilayered reservoir-caprock systems. *Int. J. Rock. Mech. Min. Sci.* **2008**, *45*, 132–143. [CrossRef]
61. De Simone, S.; Vilarrasa, V.; Carrera, J.; Alcolea, A.; Meier, P. Thermal coupling may control mechanical stability of geothermal reservoirs during cold water injection. *Phys. Chem. Earth* **2013**, *64*, 117–126. [CrossRef]
62. Vidal-Gilbert, S.; Tenthorey, E.; Dewhurst, D.; Ennis-King, J.; Van Ruth, P.; Hillis, R. Geomechanical analysis of the Naylor Field, Otway Basin, Australia: Implications for CO_2 injection and storage. *Int. J. Greenhouse Gas Control* **2010**, *4*, 827–839. [CrossRef]
63. Vilarrasa, V.; Bolster, D.; Olivella, S.; Carrera, J. Coupled hydromechanical modeling of CO_2 sequestration in deep saline aquifers. *Int. J. Greenhouse Gas Control* **2010**, *4*, 910–919. [CrossRef]
64. Vilarrasa, V.; Olivella, S.; Carrera, J.; Rutqvist, J. Long term impacts of cold CO_2 injection on the caprock integrity. *Int. J. Greenhouse Gas Control* **2014**, *24*, 1–13. [CrossRef]
65. Bowden, F.; Tabor, D. *The Friction and Lubrication of Solids I*; Clarendon Press: London, UK, 1950.
66. Baumberger, T.; Caroli, C. Solid friction from stick-slip down to pinning and aging. *Adv. Phys.* **2006**, *55*, 279–348. [CrossRef]
67. Barber, J.R. Multiscale Surfaces and Amontons' Law of Friction. *Tribol. Lett.* **2013**, *49*, 539–543. [CrossRef]
68. Dieterich, J.H. Modeling of rock friction: 1. Experimental results and constitutive equations. *J. Geophys. Res. Solid Earth* **1979**, *84*, 2161–2168. [CrossRef]
69. Ruina, A. Slip instability and state variable friction laws. *J. Geophys. Res. Solid Earth* **1983**, *88*, 10359–10370. [CrossRef]
70. Putelat, T.; Dawes, J.; Willis, J.R. On the microphysical foundations of rate-and-state friction. *J. Mech. Phys. Solids* **2011**, *59*, 1062–1075. [CrossRef]
71. Marone, C. Laboratory-derived friction laws and their application to seismic faulting. *Annu. Rev. Earth Planet. Sci.* **1998**, *26*, 643–696. [CrossRef]
72. Dieterich, J.H. Time-dependent friction and the mechanics of stick-slip. *Pure Appl. Geophys.* **1978**, *116*, 790–806. [CrossRef]
73. Rice, J.R.; Lapusta, N.; Ranjith, K. Rate and state dependent friction and the stability of sliding between elastically deformable solids. *J. Mech. Phys. Solids* **2001**, *49*, 1865–1898. [CrossRef]
74. Nakatani, M. Conceptual and physical clarification of rate and state friction: Frictional sliding as a thermally activated rheology. *J. Geophys. Res.* **2001**, *106*, 13347–13380. [CrossRef]
75. Rathbun, A.P.; Marone, C. Symmetry and the critical slip distance in rate and state friction laws. *J. Geophys. Res. Solid Earth* **2013**, *118*, 3728–3741. [CrossRef]
76. Bhattacharya, P.; Rubin, A.; Bayart, E.; Savage, H.; Marone, C. Critical evaluation of state evolution laws in rate and state friction: Fitting large velocity steps in simulated fault gouge with time-, slip-, and stress-dependent constitutive laws. *J. Geophys. Res. Solid Earth* **2015**, *120*, 6365–6385. [CrossRef]
77. Nagata, K.; Nakatani, M.; Yoshida, S. A revised rate- and state-dependent friction law obtained by constraining constitutive and evolution laws separately with laboratory data. *J. Geophys. Res.* **2012**, *117*, B02314. [CrossRef]
78. Linker, F.M.; Dieterich, J.H. Effects of Variable Normal Stress on Rock Friction: Observations and Constitutive Equations. *J. Geophys. Res.* **1992**, *97*, 4923–4940. [CrossRef]
79. Hong, T.; Marone, C. Effects of normal stress perturbations on the frictional properties of simulated faults. *Geochem. Geophys. Geosyst.* **2005**, *6*, 1–19. [CrossRef]
80. Kilgore, B.; Lozos, J.; Beeler, N.; Oglesby, D. Laboratory Observations of Fault Strength in Response to Changes in Normal Stress. *J. Appl. Mech.* **2012**, *79*, 03100701–03100710. [CrossRef]
81. Perfettini, H.; Molinari, A. A micromechanical model of rate and state friction: 2. Effect of shear and normal stress changes. *J. Geophys. Res. Solid Earth* **2017**, *122*, 2638–2652.

82. Perfettini, H.; Schmittbuhl, J.; Rice, R.J.; Cocco, M. Frictional response induced by time-dependent fluctuations of the normal loading. *J. Geophys. Res.* **2001**, *106*, 13455–13472. [CrossRef]
83. Yang, Z.; Zhang, H.P.; Marder, M. Dynamics of static friction between steel and silicon. *Proc. Natl. Acad. Sci. USA* **2008**, *105*, 13264–13268. [CrossRef] [PubMed]
84. Tal, Y.; Hager, B.H.; Ampuero, J.P. The effects of fault roughness on the earthquake nucleation process. *J. Geophys. Res. Solid Earth* **2018**, *123*, 437–456. [CrossRef]
85. Biot, M.A. General theory of three-dimensional consolidation. *J. Appl. Phys.* **1941**, *12*, 155–164. [CrossRef]
86. Rice, J.R.; Cleary, M.P. Some Basic Stress Diffusion Solutions for Fluid-Saturated Elastic Porous Media With Compressible Constituents. *Rev. Geophys. Space Phys.* **1976**, *14*, 227–241. [CrossRef]
87. Fourier, J.B.J. *Théorie Analytique de la Chaleur*; Chez Firmin Didot Pére et Fils: Paris, France, 1822.
88. Deichmann, N.; Krafta, T.; Evans, K. Identification of faults activated during the stimulation of the Basel geothermal project from cluster analysis and focal mechanisms of the larger magnitude events. *Geothermics* **2014**, *52*, 84–97. [CrossRef]
89. Gaucher, E.; Schoenball, M.; Heidbach, O.; Zang, A.; Fokker, P.A.; van Wees, J.D.; Kohl, T. Induced seismicity in geothermal reservoirs: A review of forecasting approaches. *Renew. Sustain. Energy Rev.* **2015**, *52*, 1473–1490. [CrossRef]
90. Cacace, M.; Jacquey, A. Flexible parallel implicit modelling of coupled thermal-hydraulic-mechanical processes in fractured rocks. *Solid Earth* **2017**, *8*, 921–941. [CrossRef]
91. COMSOL. *COMSOL Multiphysics Structural Mechanics Module User's Guide v5.2a*; Comsol: Stockholm, Sweden, 2016.

MDPI
St. Alban-Anlage 66
4052 Basel
Switzerland
Tel. +41 61 683 77 34
Fax +41 61 302 89 18
www.mdpi.com

Sustainability Editorial Office
E-mail: sustainability@mdpi.com
www.mdpi.com/journal/sustainability